MORNINGSIDE LECTURES IN MATHEMATICS

(Series Editors: Shing-Tung Yau, Lo Yang)

6. Lectures on the Analysis of Nonlinear Partial Differential Equations Vol. 6 (2023)
(Editors: Jean-Yves Chemin, Fanghua Lin, Ping Zhang)

5. Lectures on the Analysis of Nonlinear Partial Differential Equations Vol. 5 (2017)
(Editors: Jean-Yves Chemin, Fanghua Lin, Ping Zhang)

4. Lectures on the Analysis of Nonlinear Partial Differential Equations Vol. 4 (2016)
(Editors: Jean-Yves Chemin, Fanghua Lin, Ping Zhang)

3. Lectures on the Analysis of Nonlinear Partial Differential Equations Vol. 3 (2013)
(Editors: Fanghua Lin, Ping Zhang)

2. Lectures on the Analysis of Nonlinear Partial Differential Equations Vol. 2 (2011)
(Editors: Fanghua Lin, Ping Zhang)

1. Lectures on the Analysis of Nonlinear Partial Differential Equations Vol. 1 (2009)
(Editors: Fanghua Lin, Xueping Wang, Ping Zhang)

MLM 6

Morningside Lectures in Mathematics

Lectures on the Analysis of Nonlinear Partial Differential Equations Vol. 6

非线性偏微分方程 分析讲义 第六卷

○ Editors   Jean-Yves Chemin

Fanghua Lin

Ping Zhang

中国教育出版传媒集团

高等教育出版社·北京

TP International Press

图书在版编目（CIP）数据

非线性偏微分方程分析讲义 . 第六卷：英文 / （法）让-伊夫·舍曼，（美）林芳华，张平主编 . -- 北京：高等教育出版社，2023. 3

ISBN 978-7-04-059227-6

Ⅰ . ①非… Ⅱ . ①让… ②林… ③张… Ⅲ . ①非线性偏微分方程 – 英文 Ⅳ . ① O175.29

中国版本图书馆 CIP 数据核字（2022）第 146939 号

| 策划编辑 | 李　鹏 | 责任编辑 | 李　鹏 | 封面设计 | 杨伟露 | 版式设计 | 杜微言 |
| 责任校对 | 王　雨 | 责任印制 | 耿　轩 | | | | |

出版发行	高等教育出版社	网　址	http://www.hep.edu.cn
社　址	北京市西城区德外大街4号		http://www.hep.com.cn
邮政编码	100120	网上订购	http://www.hepmall.com.cn
印　刷	河北信瑞彩印刷有限公司		http://www.hepmall.com
开　本	787mm×1092mm 1/16		http://www.hepmall.cn
印　张	20.75		
字　数	500 千字	版　次	2023 年 3 月第 1 版
购书热线	010-58581118	印　次	2023 年 3 月第 1 次印刷
咨询电话	400-810-0598	定　价	168.00 元

本书如有缺页、倒页、脱页等质量问题，请到所购图书销售部门联系调换

版权所有　侵权必究

物 料 号　59227-00

Preface to the Sixth Volume

This book is a sequel to the previous volumes *Lectures on the Analysis of Nonlinear Partial Differential Equations*, Vol. I–V. In this sixth volume we have collected those lecture notes by Hajer Bahouri, Zeyu Lyu and François Vigneron, Raphaël Danchin, Claire David, Hongjie Dong, Reinhard Farwig, Fanghua Lin, Dongyi Wei and Zhifei Zhang. We appreciate very much for their time and efforts to provide us with these excellent notes. We believe these lecture notes will serve as valuable references on current developments in various research topics in nonlinear partial differential equations (PDE).

The notes collected here came initially from the seminars on Analysis in Partial Differential Equations. It was a part of the PDE program from Morningside Center of Mathematics and Harmonic Analysis and Its Application Center in the Academy of Mathematics and Systems Science, The Chinese Academy of Sciences (CAS) from the academic year 2019 to 2021. We would like to take this opportunity to thank the Morningside Center of Mathematics of CAS, Hua Loo-Keng Center for Mathematical Sciences of AMSS, as well as the Academy of Mathematics and Systems Science for the financial support.

Finally, we wish to thank Prof. Lo Yang and Prof. Nanhua Xi for their continuing support over all these years which was essential to the success of this PDE program. We would like also to thank Dr. Yao Xu from University of the Chinese Academy of Sciences for the organization and editorial works of these notes.

Jean-Yves Chemin in Paris
Fanghua Lin in New York
Ping Zhang in Beijing
June, 2022

Contents

Lectures on the Analysis of Nonlinear
Partial Differential Equations Vol. 6
MLM 6, pp. 1–29

Bubble and Profile Decompositions Methods: a Common Thread through Many Works in Geometry, Physics and Fluid Mechanics*

Hajer Bahouri †, Zeyu Lyu‡ and François Vigneron§

Abstract

Bubble and profile decompositions methods are powerful tools, which are deeply connected with compactness issues in function spaces. The spectrum of their applications is quite broad and encompass in particular questions in geometry, mathematical physics and fluid mechanics. In this text, we survey the fundamental principles of this theory and showcase its various applications by presenting a few landmark articles.

1 Introduction

1.1 Brief history of the bubble and profile decompositions

Bubble decompositions first appeared in the eighties, in the studies by Brézis-Coron in [21] and Struwe in [90], in the context of geometric problems. A decade later, they were also used in the framework of evolution equations (under the name of profile decompositions) in the works of Bahouri-Gérard [11] and Merle-Vega [76]. These later works are in the same vein as the result of P. Gérard [49]

---

*2000 Mathematics Subject Classification: 35R03, 35Q40.
  Key words and phrases: profile decompositions, partial differential equations.
  †CNRS & Sorbonne Université, Laboratoire Jacques-Louis Lions (LJLL), UMR 7598 CNRS, 4, Place Jussieu, 75005 Paris, France. E-mail: hajer.bahouri@ljll. math.upmc.fr
  ‡Sun Yat-Sen University, Xingang-Xi Road, School of Mathematics, Guangzhou, China. E-mail: lvzy3@mail2.sysu.edu.cn
  §Université de Reims Champagne-Ardenne, Laboratoire de Mathématiques de Reims, UMR 9008 CNRS, Moulin de la Housse, BP 1039, F-51687 Reims, France. E-mail: francois. vigneron@univ-reims.fr

that characterized (by means of profiles) the lack of compactness in the critical Sobolev embedding[1]:

$$\dot{H}^s(\mathbb{R}^d) \hookrightarrow L^p(\mathbb{R}^d) \quad \text{with} \quad 0 < s < d/2 \quad \text{and} \quad p = 2d/(d-2s). \tag{1.1}$$

The result of P. Gérard remains a cornerstone of the applications of profile decompositions to evolution equations. It states that a sequence $(u_n)_{n \geq 0}$ bounded in $\dot{H}^s(\mathbb{R}^d)$ can be decomposed, up to the extraction of a subsequence, in the following way:

$$u_n = \sum_{\ell=1}^{L} h_{\ell,n}^{-d/p} \phi^\ell \left(\frac{\cdot - x_{\ell,n}}{h_{\ell,n}} \right) + r_{n,L} \tag{1.2}$$

where $(\phi^\ell)_{\ell \geq 1}$ is a family of functions (called profiles) in $\dot{H}^s(\mathbb{R}^d)$, $(x^\ell)_{\ell \geq 1}$ is a family of cores[2] and $(\underline{h}^\ell)_{\ell \geq 1}$ is a family of scales. The remainders satisfy

$$\lim_{L \to +\infty} \left(\limsup_{n \to +\infty} \|r_{n,L}\|_{L^p(\mathbb{R}^d)} \right) = 0.$$

The profiles are "asymptotically orthogonal" in the sense that for $\ell \neq \ell'$

$$|\log(h_{\ell,n}/h_{\ell',n})| \to +\infty \text{ or } h_{\ell,n} = h_{\ell',n} \text{ and } |x_{\ell,n} - x_{\ell',n}|/h_{\ell,n} \to +\infty, \text{ as } n \to +\infty. \tag{1.3}$$

The following energy balance expresses this asymptotic orthogonality:

$$\|u_n\|_{\dot{H}^s(\mathbb{R}^d)}^2 = \sum_{\ell=1}^{L} \|\phi^\ell\|_{\dot{H}^s(\mathbb{R}^d)}^2 + \|r_{n,L}\|_{\dot{H}^s(\mathbb{R}^d)}^2 + o(1), \quad n \to \infty. \tag{1.4}$$

Since then many developments based on profile decompositions have been achieved by several authors. Some broaden the *functional framework of Sobolev embeddings*: see for instance the article of Jaffard [55] concerning Riesz potential spaces; the articles of Bahouri-Majdoub-Masmoudi [12, 13] and Bahouri-Perelman [14] about Orlicz spaces[3]; the works of Adimurthi-Tintarev [4], Bahouri-Cohen-Koch [9], Fieseler-Tintarev [39], Solimini [88] and Schindler-Tintarev [86] devoted to abstract functional frameworks including Sobolev, Besov, Triebel-Lizorkin, Lorentz, Hölder, BMO and BV spaces; the profile decompositions derived in anisotropic settings by Bahouri-Gallagher [10] and Bahouri-Chemin-Gallagher [8]; the result of Ben Ameur [17] describing the lack of compactness of the Sobolev embedding in Lebesgue spaces for the Heisenberg group and the articles of Christ-Shao [30], Frank-Lieb-Sabin [41] and Shao [87] concerning profile decompositions by means of Knapp examples, which are linked to the Fourier restriction inequalities initiated by Tomas-Stein [89, 94].

---

[1]This embedding can be established using several different methods; see *e.g.* [20, § IX.3] or [2]. Among others, one can mention the proof of Chemin-Xu [29].

[2]We follow here the terminology of P. Gérard [49] and call a *core* any real sequence $\underline{x}^\ell = (x_{\ell,n})_{n \in \mathbb{N}}$ of points in $\mathbb{R}^d$. Similarly, by *scale* $\underline{h}^\ell$, we denote any sequence $(h_{\ell,n})_{n \in \mathbb{N}}$ of positive real numbers.

[3]For an introduction to Orlicz spaces, we refer the reader to the monograph of Rao-Ren [83].

Other works investigate *nonlinear PDEs* that arise in geometry, physics and fluid mechanics. It is actually not possible to mention all the remarkable and interesting applications of profile decompositions in the study of nonlinear PDEs. Instead, we will limit ourselves to presenting a few of them in more details. We will pay a special attention to the founding article of Brézis-Coron [21], to the profile decomposition of P. Gérard [49], to the article of Gallagher [43] where the method of profile decompositions was applied in fluid mechanics for the first time and, finally, to the article of Hmidi-Keraani [53] that initiated the use of profiles in the study of the formation of singularities of solutions of L^2-critical nonlinear Schrödinger equations[4].

Even though we will not be able to review it thoroughly, we would also like to underline, in this introduction, the role of the outstanding article of Kenig-Merle [61] as it marked a great turning point in this theory and inspired many other results in the study of nonlinear evolution PDEs, either for global well-posedness problems or to solve various issues related to questions of blow-up, scattering or the soliton resolution conjecture,...

Overall, the profile decompositions are *a very versatile tool* that can serve many different purposes, ranging from the proof of the existence of extremals of some functionals or the computation of Sobolev constants to the construction of traveling waves for nonlinear evolution PDEs (such as for instance in the article of Gassot [47], where the author takes advantage of the profile decomposition of Ben Ameur [17] to construct families of traveling waves for the cubic Schrödinger equation on the Heisenberg group) or the determination of the quantized levels of energy where sequences of solutions of some elliptic PDEs concentrate at infinity, etc. Among others, one can mention the articles of Adimurthi-Druet [3], Druet-Hebey-Robert [37], Fieseler-Tintarev [39], Hutchings-Morgan-Ritoré-Ros [54], Struwe [90] and the references therein.

Let us point out, and we will discuss it further below, that the profile decompositions that are fit for *geometric* problems (actually rather called bubbles in this context) present some major differences with those involved in the study of evolution PDEs. In geometric questions, the decompositions of type (1.2) usually only include a finite number of bubbles, *i.e.* a finite number of profiles ϕ^ℓ. In most cases, these bubbles are explicitly known; they solve an elliptic PDE and, actually, which is the translated and rescaled version of some fundamental function. In such a case, all the bubbles have the same energy; thanks to the orthogonality in the balance of energy (1.4), this ensures for example that only a finite number may arise in the decomposition and that the energy will concentrate at well-defined quantized levels. Such a geometric rigidity is quite exceptional in the general framework of PDEs.

The study of the *lack of compactness* in critical Sobolev embedding in Lebesgue spaces as well as in Orlicz spaces was initiated by P.-L. Lions [72, 73] in the eighties

---

[4]Note that in the radial framework, Merle-Tsutsumi [75] obtained before the same result by another approach.

by means of defect measures[5]. In these pioneering works, the author highlighted that the Sobolev embeddings are not compact for two reasons. The first reason is the lack of compactness at infinity: a typical example is given by a traveling bump sequence $(u_n)_{n\in\mathbb{N}}$ defined by

$$u_n = \tau_{x_n}\varphi = \varphi(\cdot - x_n) \quad \text{with} \quad \varphi \in \mathcal{D}\setminus\{0\} \quad \text{and} \quad \|x_n\| \to \infty. \tag{1.5}$$

The second reason is related to concentration phenomena. In the case of homogeneous spaces like *e.g.* for the embedding $\dot{H}^s(\mathbb{R}^d) \hookrightarrow L^p(\mathbb{R}^d)$, it is illustrated by a sequence of rescaled functions of the following type

$$u_n = \delta_{h_n} u = \frac{1}{h_n^{d/p}}\varphi\left(\frac{\cdot}{h_n}\right), \tag{1.6}$$

where $\varphi \in \mathcal{D}\setminus\{0\}$ is a bump function and $(h_n)_{n\in\mathbb{N}}$ is a sequence of positive real numbers tending either to 0 or to infinity. In the case of inhomogeneous spaces, *e.g.* in the Sobolev embedding into the Orlicz space

$$H^1(\mathbb{R}^2) \hookrightarrow \mathcal{L}(\mathbb{R}^2) \tag{1.7}$$

where

$$\|u\|_{\mathcal{L}(\mathbb{R}^2)} = \inf\left\{\lambda > 0, \ \int_{\mathbb{R}^2}\left(e^{\frac{|u(x)|^2}{\lambda^2}} - 1\right)dx \le 1\right\},$$

the lack of compactness by concentration is an inhomogeneous phenomenon, which is illustrated by Moser's sequence $(f_{\alpha_n})_{n\in\mathbb{N}}$ defined by:

$$f_{\alpha_n}(x) = \begin{cases} 0, & \text{if} \quad |x| \ge 1, \\[2mm] -\dfrac{\log|x|}{\sqrt{2\alpha_n\pi}}, & \text{if} \quad e^{-\alpha_n} \le |x| \le 1, \\[2mm] \sqrt{\dfrac{\alpha_n}{2\pi}}, & \text{if} \quad |x| \le e^{-\alpha_n}, \end{cases} \tag{1.8}$$

where $(\alpha_n)_{n\in\mathbb{N}}$ denotes any sequence of positive real numbers tending to infinity. Observe that the above sequence $(f_{\alpha_n})_{n\in\mathbb{N}}$ also reads as follows:

$$f_{\alpha_n}(x) = \sqrt{\frac{\alpha_n}{2\pi}}\,\mathbf{L}\left(\frac{-\log|x|}{\alpha_n}\right) \quad \text{with} \quad \mathbf{L}(s) = \begin{cases} 0, & \text{if} \quad s \le 0, \\ s, & \text{if} \quad 0 \le s \le 1, \\ 1, & \text{if} \quad s \ge 1. \end{cases}$$

The Sobolev embedding (1.7) follows from the well-known Trudinger-Moser inequality due to Ruf [84] (see also [1]):

$$\sup_{\|u\|_{H^1}\le 1} \int_{\mathbb{R}^2}\left(e^{4\pi|u|^2} - 1\right)dx < \infty.$$

---

[5]The study of P.-L. Lions [72, 73] had a lot of impact and thereafter microlocal tools called H-measures or microlocal defect measures were introduced by Tartar [93] and P. Gérard [48] to investigate related problems. See also Murat-Tartar [79].

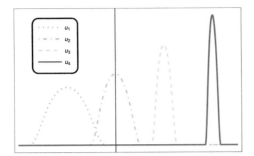 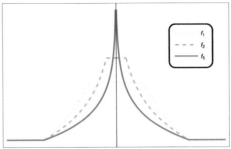

Figure 1 Left: a typical sequence (u_n) illustrating the default of compactness due to the simultaneous effect of translations (1.5) and concentration (1.6). Right: a few terms of Moser's (f_α) sequence (1.8).

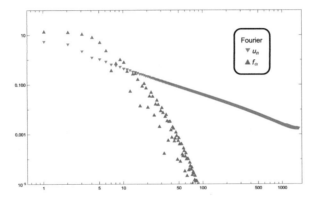

Figure 2 Typical frequency profile (modulus of the Fourier transform against the frequency) of a function u_n and f_α from Figure 1. An FFT was performed on a discretized version of each function. The units are arbitrary and in log-log scale.

It was characterized by means of profile decomposition in the spirit of (1.2) by Bahouri-Majdoub-Masmoudi [12, 13] and Bahouri-Perelman [14] for higher dimensions. The profiles involved in this setting are of the same type as the example by Moser and take the following form:

$$g_{\alpha_n}(x) \stackrel{\text{def}}{=} \sqrt{\frac{\alpha_n}{2\pi}}\, \psi\left(\frac{-\log|x - x_n|}{\alpha_n}\right),$$

where the fundamental profile ψ is defined on $\mathbb{R}$ and satisfies $\psi_{]-\infty,0]} = 0$ and $\psi' \in L^2(\mathbb{R})$, where $\alpha_n \to \infty$ and where, of course, the remainder term tends to 0 in $\mathcal{L}(\mathbb{R}^2)$. Observe that the sequences $(f_{\alpha_n})_{n\in\mathbb{N}}$ and $(g_{\alpha_n})_{n\in\mathbb{N}}$ are strikingly different from the scaling sequences (1.6) arising in (1.2) to describe the lack of compactness of the homogeneous embedding (1.1). Note however that in the two cases, the obstruction to compactness is expressed in the same manner in terms of defect measures, according to [72, 73]. Actually, contrary to the profiles $h_{\ell,n}^{-d/p} \phi^\ell\left(\frac{\cdot - x_{\ell,n}}{h_{\ell,n}}\right)$, which turn out to be asymptotically spectrally localized in rings, the sequence f_{α_n} remains spectrally spread out (see Figure 2). This example reveals the importance of *microlocal analysis* in the choice of a pertinent profile decomposition. In what follows, we will not pursue the analysis of this subtle phenomenon and refer the interested reader to [6].

In the rest of this text, we will focus on the *homogeneous* profile decomposition (1.2) because it can be generalized in many settings, including for instance the framework of critical embeddings Besov spaces[6], which arises naturally in the study of evolution equations. Note however that the proof of P. Gérard [49] that we will sketch below is mainly based on Fourier analysis. The description of the defect of compactness among more general function spaces requires other tools such as for instance wavelet decompositions (see for instance the book of Meyer [78] and the references therein for an introduction to wavelet theory) and ideas from the nonlinear approximation theory, which has been studied *e.g.* by DeVore [35]. For further details, one can consult [9, 55].

1.2 Layout

The sketch of proof of the profile decomposition of P. Gérard [49] is presented in Section 2. An introduction to the article of Brézis-Coron [21] is provided in Section 3 with a brief overview on the efficiency of bubble decompositions for geometric questions. The last section is dedicated to presenting targeted applications of profile decompositions in the analysis of nonlinear evolution PDEs. As mentioned above, despite the vast literature on the subject[7], we shall limit ourselves to a presentation of some ideas from the articles of Gallagher [43] and Hmidi-Keraani [53] with a few comments about related issues.

1.3 Notations

To avoid heaviness, C will denote a positive constant which may vary from line to line; we also use $A \lesssim B$ to denote an estimate of the form $A \leq CB$. For simplicity, we shall also still denote by (u_n) any subsequence of a sequence (u_n) and designate by $\circ(1)$ any sequence which tends to 0 as n goes to infinity.

2 Profile decompositions for critical Sobolev embeddings

In this section, we focus on the description of the lack of compactness of the critical embedding (1.1) of homogeneous Sobolev spaces into Lebesgue spaces. As mentioned above, such a description, initially due to P. Gérard [49], was generalized to many other pairs of function spaces including, for instance, Besov spaces $\dot{B}_{r,q}^s(\mathbb{R}^d)$, which come to play in several results related to fluid mechanics and mathematical physics. For instance, if we consider the critical Sobolev embedding $X \hookrightarrow Y$

---

[6]For an introduction to Besov spaces, one can for instance consult the monographs [7, 95] and the references therein.

[7]Among others, one can mention [11, 36, 38, 44, 45, 56, 58, 61, 62, 63, 66, 69, 76, 80, 82, 92] and the references therein.

where $X = \dot{B}^s_{r,q}(\mathbb{R}^d)$, $0 < s < d/r$, and where $Y = L^p(\mathbb{R}^d)$ with $p = rd/(d - rs)$ or $Y = \dot{B}^t_{a,b}(\mathbb{R}^d)$, $0 < t < d/a$, with $t < s$ and $d/r - s = d/a - t$, then any bounded sequence $(u_n)_{n\in\mathbb{N}}$ in X admits, up to the extraction of a subsequence, an asymptotically orthogonal decomposition of the type (1.2) where the remainder term $r_{n,L}$ satisfies

$$\lim_{L\to+\infty} \Big(\limsup_{n\to+\infty} \|r_{n,L}\|_Y \Big) = 0.$$

Even though all such profile decompositions start with a diagonal extraction procedure, the original strategy adopted by P. Gérard in [49], which uses Fourier analysis in a crucial way, cannot be generalized mutatis mutandis to Besov spaces $\dot{B}^s_{r,q}(\mathbb{R}^d)$ when $(r,q) \neq (2,2)$. Recall that $\dot{B}^s_{2,2}(\mathbb{R}^d)$ coincides with the Sobolev space $\dot{H}^s(\mathbb{R}^d)$. The characterization of critical Sobolev embeddings for general Besov spaces, as well as for Triebel-Lizorkin, Lorentz or Hölder spaces relies on wavelet decompositions and on nonlinear approximation theories (see Appendix A for a succinct introduction to this topic). In this section, we will not seek the maximal generality possible and we invite the interested reader to explore [8, 9, 35, 55, 64, 78] and the references therein. Neither will we discuss the results of inhomogeneous type that were mentioned briefly in the introduction and that revolve around Moser and Knapp examples; see for instance [12, 13, 30, 41].

2.1 Description of the default of compactness in critical homogeneous Sobolev embeddings

The result of P. Gérard [49] concerning the critical Sobolev embedding (1.1) can be stated precisely as follows:

Theorem 1 ([49]). *Let $0 < s < d/2$ and $p = 2d/(d - 2s)$ and consider $(u_n)_{n\in\mathbb{N}}$ a sequence of functions that is bounded in $\dot{H}^s(\mathbb{R}^d)$. Then up to extracting a subsequence (which we denote in the same way), there is a family of functions $(\phi^\ell)_{\ell\geq 1}$ in $\dot{H}^s(\mathbb{R}^d)$, and a family of cores $(x^\ell)_{\ell\geq 1}$, as well as a family of sequences of positive real numbers $((h_{\ell,n})_{n\in\mathbb{N}})_{\ell\geq 1}$, orthogonal in the sense of (1.3), such that for all integers $L \geq 1$*

$$u_n = \sum_{\ell=1}^{L} h_{\ell,n}^{-d/p} \phi^\ell \Big(\frac{\cdot - x_{\ell,n}}{h_{\ell,n}} \Big) + r_{n,L} \quad \text{with} \quad \lim_{L\to+\infty} \Big(\limsup_{n\to+\infty} \|r_{n,L}\|_{L^p(\mathbb{R}^d)} \Big) = 0. \tag{2.1}$$

In addition, one has the following orthogonality equality

$$\|u_n\|^2_{\dot{H}^s(\mathbb{R}^d)} = \sum_{\ell=1}^{L} \|\phi^\ell\|^2_{\dot{H}^s(\mathbb{R}^d)} + \|r_{n,L}\|^2_{\dot{H}^s(\mathbb{R}^d)} + o(1), \quad n \to \infty. \tag{2.2}$$

This result highlights the fact that translational and scaling invariance are the only features responsible for the lack of compactness in the homogeneous Sobolev embedding (1.1). Note that the decomposition (2.1) is not unique and can either be finite or infinite.

Remark 2.1. *Theorem 1 calls for some natural complementary statements.*

- *The scale invariance of the statement can be checked by straightforward computations of the norms:*

$$\left\| h_{\ell,n}^{-d/p} \phi^\ell \left(\frac{\cdot - x_{\ell,n}}{h_{\ell,n}} \right) \right\|_{\dot{H}^s} = \| \phi^\ell \|_{\dot{H}^s} \quad \text{and} \quad \left\| h_{\ell,n}^{-d/p} \phi^\ell \left(\frac{\cdot - x_{\ell,n}}{h_{\ell,n}} \right) \right\|_{L^p} = \| \phi^\ell \|_{L^p}.$$

- *The orthogonality condition* (1.3) *ensures both that the interaction between any pair of different elements in the profile decomposition* (2.1) *is asymptotically negligible in* $\dot{H}^s(\mathbb{R}^d)$ *as n goes to infinity and that the profiles* ϕ^ℓ *satisfy:*

$$\lim_{\ell \to \infty} \| \phi^\ell \|_{\dot{H}^s(\mathbb{R}^d)} = 0. \tag{2.3}$$

As we shall see in Section 4, such properties play a crucial role in the applications.

- *The profiles* ϕ^ℓ *can be obtained in the following way:*

$$h_{\ell,n}^{d/p} u_n (h_{\ell,n}(\cdot + x_{\ell,n})) \rightharpoonup \phi^\ell, \quad \text{as } n \to \infty, \tag{2.4}$$

i.e. as weak limits of translations and scaling transforms of the sequence $(u_n)_{n \in \mathbb{N}}$.

Theorem 1 allows, for example, to recover the fact that the best Sobolev constant

$$C = \inf \left\{ \| u \|_{\dot{H}^s(\mathbb{R}^d)} \, ; \, u \in \dot{H}^s(\mathbb{R}^d) \text{ and } \| u \|_{L^p(\mathbb{R}^d)} = 1 \right\}$$

is achieved as well as on the structure of the corresponding minimizing sequences. Indeed, taking advantage of the elementary inequality

$$\left| \left| \sum_{\ell=1}^{L} \alpha_\ell \right|^p - \sum_{\ell=1}^{L} |\alpha_\ell|^p \right| \leq C_L \sum_{\ell \neq k} |\alpha_\ell| |\alpha_k|^{p-1}, \tag{2.5}$$

one may deduce, under the assumptions of Theorem 1, that

$$\| u_n \|_{L^p(\mathbb{R}^d)}^p \xrightarrow{n \to \infty} \sum_{\ell=1}^{\infty} \| \phi^\ell \|_{L^p(\mathbb{R}^d)}^p .$$

The energy balance (2.2) obviously implies also that

$$\limsup_{n \to +\infty} \| u_n \|_{\dot{H}^s(\mathbb{R}^d)}^2 \geq \sum_{\ell=1}^{\infty} \| \phi^\ell \|_{\dot{H}^s(\mathbb{R}^d)}^2 .$$

This ensures easily the fact that C is achieved. For further details, one can consult the article of P.-L. Lions [73].

2.2   Sketch of proof of the profile decomposition by P. Gérard

The proof of Theorem 1 uses Fourier analysis in a crucial way. It comports three main steps:

(i) The first step is devoted to the extraction of the scales $(\underline{h}^\ell)_{\ell\geq 1}$ and consists (up to the extraction of subsequences) in splitting u_n as follows:

$$u_n = \sum_{\ell=1}^{L} u_n^\ell + r_n^L, \tag{2.6}$$

where, in the phrasing of P. Gérard in [49], each component $(|D|^s u_n^\ell)_{n\in\mathbb{N}}$ is $(h_{\ell,n})$-oscillating, *i.e.* is asymptotically spectrally localized in a ring of size $1/h_{\ell,n}$, with orthogonality between the scales in the sense that for $\ell \neq \ell'$

$$|\log(h_{\ell,n}/h_{\ell',n})| \overset{n\to\infty}{\longrightarrow} +\infty.$$

The purpose at this stage is to ensure that the remainder term r_n^L satisfies

$$\limsup_{n\to\infty} \|r_n^L\|_{\dot{B}_{2,\infty}^s(\mathbb{R}^d)} \to 0, \quad \text{as } L \to \infty.$$

This property allows one to get rid of the remainder r_n^L thanks to the refined Sobolev inequality due to Gérard-Meyer-Oru [50] (see also [7] for another proof):

$$\|f\|_{L^p(\mathbb{R}^d)} \lesssim \|f\|_{\dot{H}^s(\mathbb{R}^d)}^{2/p} \|f\|_{\dot{B}_{2,\infty}^s(\mathbb{R}^d)}^{1-2/p}.$$

To be more precise, let us start by introducing the concept of a sequence, which is oscillatory with respect to a scale $(h_n)_{n\in\mathbb{N}}$ and the concept of being unrelated to a scale.

Definition 2.2. *Let $(f_n)_{n\in\mathbb{N}}$ be a bounded sequence of functions of $L^2(\mathbb{R}^d)$. The sequence $(f_n)_{n\in\mathbb{N}}$ is said (h_n)-oscillating if*

$$\limsup_{n\to\infty} \left(\int_{h_n|\xi|\leq \frac{1}{R}} |\hat{f}_n(\xi)|^2 d\xi + \int_{h_n|\xi|\geq R} |\hat{f}_n(\xi)|^2 d\xi \right) \overset{R\to\infty}{\longrightarrow} 0.$$

The sequence $(f_n)_{n\in\mathbb{N}}$ is said unrelated to the scale (h_n) if, for all $0 < a < b$

$$\lim_{n\to\infty} \int_{a\leq h_n|\xi|\leq b} |\hat{f}_n(\xi)|^2 d\xi = 0.$$

Remark 2.3. *Observe that*

- *With the notations of Theorem 1, the sequences[8] $\left(h_{\ell,n}^{-(s+d/p)} (|D|^s \phi^\ell)\left(\frac{\cdot - x_{\ell,n}}{h_{\ell,n}} \right) \right)_{n\in\mathbb{N}}$ are $(h_{\ell,n})$-oscillating.*

---

[8]which, according to the fact that $0 < s < d/2$ and $p = 2d/(d-2s)$, are bounded sequence of functions of $L^2(\mathbb{R}^d)$.

- One can show that an L^2-bounded sequence $(f_n)_{n\in\mathbb{N}}$ is unrelated to any scale if and only if $\|f_n\|_{\dot{B}^0_{2,\infty}} \overset{n\to\infty}{\longrightarrow} 0$ (see [49] for a detailed proof).
- One can easily check that if $(f_n)_{n\in\mathbb{N}}$ is (h_n)-oscillating and $(g_n)_{n\in\mathbb{N}}$ is $(\tilde{h}_n)$-oscillating with

$$|\log(h_n/\tilde{h}_n)| \overset{n\to\infty}{\longrightarrow} +\infty,$$

then

$$\|f_n + g_n\|^2_{L^2(\mathbb{R}^d)} = \|f_n\|^2_{L^2(\mathbb{R}^d)} + \|g_n\|^2_{L^2(\mathbb{R}^d)} + o(1), \quad n \to \infty.$$

The key argument in the proof of Decomposition (2.6) consists to apply (recursively) the following proposition to the L^2-bounded sequence $(|D|^s u_n)_{n\in\mathbb{N}}$.

Proposition 2.4. *Let $(f_n)_{n\in\mathbb{N}}$ be a bounded sequence of functions of $L^2(\mathbb{R}^d)$ and (h_n) be a scale. Then up to extracting a subsequence (which we denote in the same way), there exists an L^2-bounded sequence $(g_n)_{n\in\mathbb{N}}$ such that*

- *The sequence $(g_n)_{n\in\mathbb{N}}$ is (h_n)-oscillating.*
- *The sequence $(f_n - g_n)_{n\in\mathbb{N}}$ is unrelated to (h_n).*

Proof. Consider the sequence of functions $(L_n)_{n\in\mathbb{N}}$ defined by

$$L_n \begin{cases}]1, \infty[\longrightarrow [0, \infty[, \\ R \longmapsto \displaystyle\int_{\frac{1}{R} \leq h_n|\xi| \leq R} |\hat{f}_n(\xi)|^2 d\xi. \end{cases}$$

For any integer n, it is straightforward that the function L_n is a non-decreasing bounded function. Helly's lemma (see for instance [40] or Appendix B in this paper) implies that there exists an increasing sequence of positive integers $\phi(n)$ such that

$$\forall R > 1, \quad L_{\phi(n)}(R) \overset{n\to\infty}{\longrightarrow} L(R),$$

and

$$\forall n \in \mathbb{N}^*, \quad L_{\phi(n)}(n) - \frac{1}{n} \leq L_{\phi(n)}(n) \leq L_{\phi(n)}(n) + \frac{1}{n}.$$

Since obviously the limit L is also a non-decreasing bounded function, we infer that there exists a positive real number ℓ such that $L(R) \overset{R\to\infty}{\longrightarrow} \ell$, and thus $L(n) \overset{n\to\infty}{\longrightarrow} \ell$. Now for any integer n, define g_n so that

$$\hat{g}_n(\xi) = \hat{f}_{\phi(n)}(\xi) \mathbf{1}_{\frac{1}{n} \leq \mathbf{h}_{\phi(n)}|\xi| \leq \mathbf{n}}.$$

Obviously $(g_n)_{n\in\mathbb{N}}$ is a bounded sequence of functions of $L^2(\mathbb{R}^d)$ which satisfies for $n \geq R \geq 1$

$$\int_{h_{\phi(n)}|\xi|\leq\frac{1}{R}} |\hat{g}_n(\xi)|^2 d\xi + \int_{h_{\phi(n)}|\xi|\geq R} |\hat{g}_n(\xi)|^2 d\xi$$

$$= L_{\phi(n)}(n) - L_{\phi(n)}(R) \overset{n\to\infty}{\longrightarrow} \ell - L_{\phi(n)}(R),$$

which according to Definition 2.2 ensures that the sequence $(g_n)_{n\in\mathbb{N}}$ is (h_n)-oscillating. Finally, since for all $b > a > 0$ and n sufficiently large, we have

$$\widehat{g}_n(\xi) = \widehat{f}_{\phi(n)}(\xi)\,, \quad \text{for } a \le h_{\phi(n)}|\xi| \le b\,,$$

we deduce that (up to a subsequence) the sequence $(f_n - g_n)_{n\in\mathbb{N}}$ is unrelated to (h_n). This completes the proof of the proposition.   □

By iterating the above extraction and taking advantage of the exhaustion method of Metivier-Schochet [77] one can complete the first step.

(ii) The second step is devoted to the extraction of the profiles and the cores in each $(h_{\ell,n})$-oscillating component. First, scale invariance allows for a reduction of the study to the case of a 1-oscillating sequence $(w_n)_{n\in\mathbb{N}}$, i.e. a sequence for which $h_{\ell,n} \equiv 1$. Next, one extracts a non-trivial weak limit in $\dot{H}^s(\mathbb{R}^d)$, among all possible translations:

$$w_n(\cdot + x_n^1) \overset{n\to\infty}{\rightharpoonup} \varphi^1\,.$$

Then, one considers the sequence $(r_n^1)_{n\in\mathbb{N}}$ defined by

$$r_n^1 = w_n - \varphi^1(\cdot - x_n^1)\,,$$

which is obviously 1-oscillating and satisfies $r_n^1(\cdot + x_n^1) \overset{n\to\infty}{\rightharpoonup} 0$. This readily implies that

$$\|w_n\|_{\dot{H}^s(\mathbb{R}^d)}^2 = \|\varphi^1\|_{\dot{H}^s(\mathbb{R}^d)}^2 + \|r_n^1\|_{\dot{H}^s(\mathbb{R}^d)}^2 + \mathrm{o}(1), \quad n\to\infty\,.$$

Arguing similarly, one extracts a weak limit of $(r_n^1)_{n\in\mathbb{N}}$ among its translations, namely

$$r_n^1(\cdot + x_n^2) \overset{n\to\infty}{\rightharpoonup} \varphi^2 \quad \text{with} \quad |x_n^1 - x_n^2| \overset{n\to\infty}{\to} \infty\,,$$

which now ensures that

$$w_n = \varphi^1(\cdot - x_n^1) + \varphi^2(\cdot - x_n^2) + r_n^2$$

and with

$$\|w_n\|_{\dot{H}^s(\mathbb{R}^d)}^2 = \|\varphi^1\|_{\dot{H}^s(\mathbb{R}^d)}^2 + \|\varphi^2\|_{\dot{H}^s(\mathbb{R}^d)}^2 + \|r_n^2\|_{\dot{H}^s(\mathbb{R}^d)}^2 + \mathrm{o}(1), \quad n\to\infty\,.$$

As long as all the $\dot{H}^s(\mathbb{R}^d)$ energy of the sequence is not spent, one can reiterate this construction and define the next cores and profiles.

(iii) The last step of the proof consists in showing that the process is indeed convergent, which we will not detail here.

3   Some insight on profiles in geometric issues

This section is primarily devoted to the introduction of the result of Brézis-Coron [21] where a profile (bubble) decomposition of sequences of solutions to an elliptic system was derived for the first time. This result was the starting point of a very active field of research tackling both geometric and variational problems. Another line of research consists in constructing solutions to nonlinear elliptic PDEs with a prescribed number of bubbles. However, we will not go further into all those interesting aspects that are at the crossroad of PDEs and Geometry. For recent advances, see *e.g.* [54], [74].

The founding result of Brézis-Coron [21] concerns the problem of finding surfaces $\Sigma \subset \mathbb{R}^3$ of constant mean curvatures spanned by a Jordan curve Γ. For that purpose, they consider surfaces parametrized by the unit disc D of $\mathbb{R}^2$, *i.e.* $\Sigma = u(\overline{D})$ where $u : \overline{D} \to \mathbb{R}^3$ solves the following system:

$$\begin{cases} \Delta u = 2u_x \wedge u_y & \text{on } D, \\ |u_x|^2 - |u_y|^2 = u_x \cdot u_y = 0 & \text{on } D, \\ u(\partial D) = \Gamma. \end{cases} \tag{3.1}$$

The vectors (u_x, u_y) constitute an orthogonal moving frame along Σ (see Figure 3). The first equation expresses the constancy of the mean curvature of Σ. The last one is the geometrical anchor that constraints the restriction of u to the circle ∂D to be a parametrization of Γ. This problem has a deep connection with physical models of soap bubbles leaning on a wire Γ, hence the name "bubbles" for the profiles exhibited in [21].

In [21], the authors investigate an asymptotic regime of (3.1) in the case when $\Gamma \to 0$. Namely, given a sequence of Jordan curves $(\Gamma_n)_{n \in \mathbb{N}}$ where $\Gamma_n \subset B(0, R_n)$ with $R_n \to 0$, they show that if the total area of the associated surfaces Σ_n remain bounded, then, up to a subsequence, the surfaces Σ_n converge either to the origin (collapse) or to a finite connected union of spheres of radius one, such that at least one of them contains the origin.

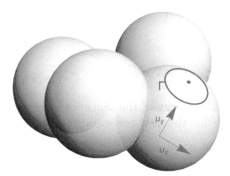

Figure 3   Solutions of (3.1) are parametrizations of surfaces of constant curvatures spanned by Γ. According to [21], if Γ shrinks to a point, then the solutions are, asymptotically, a finite union of spheric bubbles that contain this point.

Their approach relies on the following "bubble" decomposition:

Theorem 2 ([21]). *Let $(u^n)_{n \in \mathbb{N}}$ be a bounded sequence in $H^1(D; \mathbb{R}^3)$ of solutions of the following system*

$$\begin{cases} \Delta u^n = 2u_x^n \wedge u_y^n & \text{on } D, \\ u^n = \gamma^n & \text{on } \partial D, \end{cases} \tag{3.2}$$

where $\gamma^n \to 0$ in $H^{1/2}(\partial D)$. Then, there exists a finite family $(\omega^{(i)})_{i=1,\dots,p}$ in $H^1(\mathbb{R}^2)$, a family of cores $(\underline{a}^{(i)})_{i=1,\dots,p}$ in D, and a family of scales $((\epsilon_n^{(i)})_{n \in \mathbb{N}})_{i=1,\dots,p}$ in $]0, \infty[$ that converge to 0 and are orthogonal in the sense of (1.3), and such that, for some subsequence, still denoted by u_n for simplicity:

$$u^n(x) = \sum_{i=1}^{p} \omega^{(i)}\left(\frac{x - a_n^{(i)}}{\epsilon_n^{(i)}}\right) + r_n(x), \qquad \|r_n\|_{H^1(D)} \xrightarrow{n \to \infty} 0. \tag{3.3}$$

Moreover, one has the following orthogonal balance of energy:

$$\|\nabla u^n\|_{L^2(D)}^2 = \sum_{i=1}^{p} \|\nabla \omega^{(i)}\|_{L^2(\mathbb{R}^2)}^2 + o(1), \quad n \to \infty. \tag{3.4}$$

In contrast to the result of P. Gérard concerning any bounded sequence in $\dot{H}^s(\mathbb{R}^d)$, here the "bubble" decomposition is applied to a bounded sequence of solutions to an elliptic nonlinear system, which provides more rigidity.

More precisely, let us underline that the functions ω^i solve an analogous PDE on the whole plane, namely

$$\Delta \omega = 2\omega_x \wedge \omega_y \quad \text{on } \mathbb{R}^2. \tag{3.5}$$

As this equation is invariant by translations and scaling transforms, each bubble

$$\omega^{(i)}\left(\frac{x - a_n^{(i)}}{\epsilon_n^{(i)}}\right)$$

is therefore also a solution of (3.5). All the solutions of (3.5) are known explicitly and their energy $\int_{\mathbb{R}^2} |\nabla \omega|^2$ is quantized as an integer multiple of 8π (see [21, Lemma 0.1]). Consequently, thanks to the orthogonality of the bubbles (3.4), the decomposition (3.3) only includes a finite number p of bubbles, contrary to the decomposition (2.1), which can be infinite. As n goes to infinity, the energy (3.4) of the sequence $(u_n)_{n \in \mathbb{N}}$ therefore concentrates on quantized levels, which are multiple of 8π.

Remark 3.1. *Up to the extraction of a subsequence, the functions u^n concentrate around a finite number of points, which, with the previous notations, are given by*

$$a^{(i)} = \lim_{n \to \infty} a_n^{(i)}.$$

See Figure 3.

We refer the interested reader to [21] for a detailed proof of Theorem 2. The result is achieved by a diagonal extraction process, but in a different way from the proof of profile decomposition of P. Gérard. As we have already seen in Section 2.2, the proof of Theorem 1 is based on Fourier analysis, while the proof of Brézis-Coron [21] is direct and relies on a refined analysis of the sequence $(u_n)_{n \in \mathbb{N}}$ depending on the size of its energy. Roughly speaking, the proof of Theorem 2 comports two steps (assuming in addition that the sequence $(u_n)_{n \in \mathbb{N}}$ is also bounded in L^∞): (i) The first step consists in extracting the first bubble, whenever it is possible to do so. Assuming that the energy of the sequence $(u^n)_{n \in \mathbb{N}}$ is bounded from below, namely $\inf_n \|\nabla u^n\|_{L^2}^2 = \alpha > 0$, a preliminary family of interesting scales $\epsilon_n > 0$ and cores $a_n \in \bar{D}$ are constructed by applying the intermediate value theorem to the following functions that quantify the concentration:

$$Q_n(t) = \sup_{z \in \mathbb{R}^2} \int_{z+tD} |\nabla u^n|^2, \text{ for } t \geq 0,$$

where u^n are extended by zero outside of D. Obviously, each function Q_n is continuous and non-decreasing in t, and satisfies

$$Q_n(0) = 0 \quad \text{and} \quad Q_n(1) = Q_n(\infty) = \int_D |\nabla u^n|^2 \geq \alpha > 0.$$

So, if one chooses an intermediary value $v \in]0, \alpha[$, there exists $0 < \epsilon_n < 1$ and $a_n \in \bar{D}$ such that

$$Q_n(\epsilon_n) = \int_{a_n + \epsilon_n D} |\nabla u^n|^2 = v.$$

Alaoglu's theorem then provides a subsequence of scales and cores, denoted respectively by $\epsilon_n^{(1)}$ and $a_n^{(1)}$, such that

$$u^n(\epsilon_n^{(1)} z + a_n^{(1)}) \stackrel{n \to \infty}{\longrightarrow} \omega^{(1)}(z) \quad a.e. \ z \in \mathbb{R}^2,$$
$$\nabla_z u^n(\epsilon_n^{(1)} z + a_n^{(1)}) \stackrel{n \to \infty}{\rightharpoonup} \nabla_z \omega^{(1)}(z) \quad \text{in } L^2(\mathbb{R}^2).$$

One of the key points at this stage is the general identity [21, Lemma 2]

$$\left| \int_\Omega u \cdot v_x \wedge v_y \right| \leq c_0 \|\nabla u\|_{L^2} \|\nabla v\|_{L^2},$$

which allows one to use (locally) $u^n(\epsilon_n^{(1)} z + a_n^{(1)})$ itself as a test function of the equation (see [21, eq. 31]). In turn, this allows to convert the weak H^1 convergence of the sequence into a locally strong convergence (see [21, Lemma 3]). One can thus reason by contradiction and use the non-vanishing energy of (u^n) to ensure that $d(a_n^{(1)}, \partial D)/\epsilon_n^{(1)} \stackrel{n \to \infty}{\longrightarrow} \infty$. This means that $\omega^{(1)}$ is a solution of (3.5) on $\mathbb{R}^2$ and has therefore a quantized $\dot{H}^1$ energy.

If $\|u^n\|_{H^1(D)} < 8\pi$ then, necessarily, $\omega^{(1)} = 0$ and $\|u^n\|_{H^1(D)} \stackrel{n \to \infty}{\longrightarrow} 0$. In this case, the final statement holds directly, with no bubble i.e. $p = 0$. On the contrary, if $\|u^n\|_{H^1} \geq 8\pi$, then the profile $\omega^{(1)}$ is non-trivial and the first bubble takes the form:

$$\omega^{(1)}\left(\frac{\cdot - a_n^{(1)}}{\epsilon_n^{(1)}}\right).$$

(ii) The second step consists in successive iterations of the above construction, each extraction providing a new non-trivial profile, until the remaining energy passes below the threshold of 8π.

4 Profile decompositions and nonlinear evolution equations

The method of profile decompositions is now a prerequisite in the analysis of nonlinear evolution PDEs. It can be used to establish global well-posedness and study blow up issues. It is also helpful for the investigation of the qualitative behavior of the solutions and for the description of their asymptotic dynamics, as the time variable goes to infinity (in the case of global well-posedness) or as the time variable goes to the lifespan $T^* < \infty$ (in the case of blow-up at finite time). In the study of long-time behavior of global solutions, it is essential to decide whether they scatter or not, namely if these global solutions behave asymptotically like a solution of the corresponding linear solution or not. When the scattering fails, the next challenge[9] is to show that, asymptotically, the global solutions are a sum of decoupled solitons[10] and a scattering solution. There is an extensive literature on the subject which gained momentum from the remarkable work of Kenig-Merle [61]. Among others, one can mention [8, 10, 11, 13, 15, 36, 38, 43, 44, 45, 53, 56, 58, 60, 61, 62, 63, 66, 76, 80, 82, 85, 92] and the references therein.

In this review, we shall limit ourselves to a presentation of the article of Gallagher [43] and the article of Hmidi-Keraani [53]. Both articles were historical first as they each opened a new realm of applications to the method of profile decompositions: in [43], this strategy was used to investigate the global well-posedness issue in fluid mechanics, namely for the Navier-Stokes system while, in [53], the authors employed this technique to the problem of formation of singularities for the L^2-critical nonlinear Schrödinger equation.

4.1 Profile decompositions and fluid mechanics

Before stating the result of Gallagher [43], let us start by recalling briefly some basic facts about the three dimensional[11] incompressible Navier-Stokes equations

$$\begin{cases} \partial_t v + v \cdot \nabla v - \Delta v = -\nabla p & \text{in} \quad \mathbb{R}^+ \times \mathbb{R}^3, \\ \text{div } v = 0, \\ v_{|t=0} = v_0, \end{cases} \tag{NS}$$

---

[9]known as the soliton resolution conjecture.

[10]That is to say, well-localized solutions traveling at a fixed speed c, which can vanish or not.

[11]All the results presented here hold in the more general case of $\mathbb{R}^d$, with obvious adaptations in the functional spaces considered.

where v_0 is a divergence free vector field; $v(t,x)$ and $p(t,x)$ are respectively the velocity and the pressure of the fluid at time $t \geq 0$ and position $x \in \mathbb{R}^3$. The most fundamental result regarding the Cauchy problem was obtained by Leray in [70], who proved that for divergence free Cauchy data $v_0 \in L^2(\mathbb{R}^3)$, there is a global weak solution of (NS) such that

$$v \in L^\infty(\mathbb{R}_+; L^2(\mathbb{R}^3)) \cap L^2(\mathbb{R}_+; \dot{H}^1(\mathbb{R}^3))$$

and that satisfies the energy inequality

$$\|v(t)\|^2_{L^2(\mathbb{R}^3)} + 2 \int_0^t \|\nabla v(t')\|^2_{L^2(\mathbb{R}^3)} dt' \leq \|v_0\|^2_{L^2(\mathbb{R}^3)}.$$

The solutions constructed by Leray are not known to be unique, except in the case of a bidimensional ambient space (see [71]). Many studies exist on that problem of uniqueness: one may consult for instance the monographs of Bahouri-Chemin-Danchin [7], Chemin [25], Lemarié-Rieusset [68] and the references therein for an overview on this topic and in particular the founding article of Fujita-Kato [42], where the authors established that for $v_0 \in \dot{H}^{\frac{1}{2}}(\mathbb{R}^3)$, there exists a maximal time T^* and a unique solution v of (NS) stemming from v_0 such that

$$\forall T < T^*, \quad v \in E_T \overset{\text{def}}{=} C^0([0,T]; \dot{H}^{\frac{1}{2}}(\mathbb{R}^3)) \cap L^2([0,T]; \dot{H}^{\frac{3}{2}}(\mathbb{R}^3)) \qquad (4.1)$$

and that satisfies the energy inequality

$$\forall 0 \leq t < T^*, \quad \|v(t)\|^2_{\dot{H}^{\frac{1}{2}}(\mathbb{R}^3)} + 2 \int_0^t \|v(t')\|^2_{\dot{H}^{\frac{3}{2}}(\mathbb{R}^3)} dt' \leq \|v_0\|^2_{\dot{H}^{\frac{1}{2}}(\mathbb{R}^3)}.$$

Below, one will denote this unique solution by $v = \mathrm{NS}(v_0)$. Furthermore, there exists a universal positive constant c such that if

$$\|v_0\|_{\dot{H}^{\frac{1}{2}}(\mathbb{R}^3)} \leq c, \qquad (4.2)$$

then the associated solution v is global in time[12].

Many results of this type are known to hold. For instance it is possible to replace $\dot{H}^{\frac{1}{2}}(\mathbb{R}^3)$ by the larger Lebesgue space $L^3(\mathbb{R}^3)$ (see [24, 26, 51, 59, 97]) or by the Besov spaces $B_{p,\infty}^{-1+\frac{3}{p}}$ for finite p (see [22, 81]). To this day, the best known result on the uniqueness of solutions to (NS) is due to H. Koch and D. Tataru [65]. It is proved, as most results of this type, by a fixed point theorem in an appropriate Banach space. The smallness condition is the following:

$$\|u_0\|_{\mathrm{BMO}^{-1}(\mathbb{R}^3)} \overset{\text{def}}{=} \sup_{t>0} \, t^{\frac{1}{2}} \|e^{t\Delta} u_0\|_{L^\infty(\mathbb{R}^3)}$$

$$+ \sup_{\substack{x \in \mathbb{R}^3 \\ R>0}} \frac{1}{R^{\frac{3}{2}}} \left(\int_{[0,R^2] \times B(x,R)} |(e^{t\Delta} u_0)(t,y)|^2 \, dy dt \right)^{\frac{1}{2}} \leq c.$$

---

[12]Note for instance that in Chemin-Gallagher [27] and Chemin-Gallagher-Zhang [28], global well-posedness for the (NS) system was established for several examples of large data.

At this stage, one should emphasize that the Navier-Stokes system (NS) enjoys a scaling invariance property: defining the scaling operators, for any positive real number λ and any point x_0 of $\mathbb{R}^3$,

$$\Lambda_{\lambda,x_0} v(t,x) \stackrel{\text{def}}{=} \lambda v\big(\lambda^2 t, \lambda(x - x_0)\big), \tag{4.3}$$

if v solves (NS) with data v_0, then $\Lambda_{\lambda,x_0} v$ solves (NS) with data $\Lambda_{\lambda,x_0} v_0$.

The main result in [43] concerns the family of solutions of the Navier-Stokes system associated with a bounded sequence $(\varphi_n)_{n \in \mathbb{N}}$ of data in $\dot{H}^{\frac{1}{2}}(\mathbb{R}^3)$. According to the result of P. Gérard [49] (see Theorem 1 in this article), the sequence $(\varphi_n)_{n \in \mathbb{N}}$ admits, up to the extraction of a subsequence, a decomposition of the form (2.1), where of course $p = 3$. The goal of the following theorem, established by Gallagher in [43], is to investigate how the profile decomposition for (φ_n) is propagated by the Navier-Stokes equation.

Theorem 3 ([43]). *Let $(\varphi_n)_{n \in \mathbb{N}}$ be a family of divergence free vector fields, bounded in $\dot{H}^{\frac{1}{2}}(\mathbb{R}^3)$. One applies Theorem 1 to (φ_n) and denote by $(\phi^\ell)_{\ell \in \mathbb{N}}$ the corresponding sequence of profiles (after extracting a subsequence, which we denote in the same way). One adopts the other notations of this theorem. Let us also introduce*

$$v_n = \mathrm{NS}(\varphi_n) \quad \text{and} \quad V^\ell = \mathrm{NS}(\phi^\ell).$$

The following results hold[13].

1. *There exists a family $(T^\ell)_{\ell \in \mathbb{N}}$ of elements of $\mathbb{R}_+^* \cup \{+\infty\}$ and a finite subset $J \in \mathbb{N}$ such that*

$$\forall \ell \in \mathbb{N}, \ V^\ell \in E_{T^\ell} \quad \text{and} \quad \forall \ell \in \mathbb{N} \setminus J, \ T^\ell = +\infty. \tag{4.4}$$

Let us define $\tau_n \stackrel{\text{def}}{=} \min_{\ell \in J} h_{\ell,n}^2 T^\ell$.

2. *Then $\|v_n\|_{E_{\tau_n}}$ is bounded, and we have for all $0 \le t \le \tau_n$ and all integers $L \ge 1$:*

$$v_n(t,x) = V^0(t,x) + \sum_{\ell=1}^{L} \frac{1}{h_{\ell,n}} V^\ell \left(\frac{t}{h_{\ell,n}^2}, \frac{x - x_{\ell,n}}{h_{\ell,n}} \right) + w_n^L(t,x) + R_n^L(t,x), \tag{4.5}$$

where w_n^L solves the heat equation (with data $r_{n,L}$ given by Theorem 1)

$$\begin{cases} \partial_t w_n^L - \Delta w_n^L = 0 \quad \text{in} \quad \mathbb{R}_+ \times \mathbb{R}^3, \\ (w_n^L)_{|t=0} = r_{n,L}, \end{cases} \tag{4.6}$$

and where

$$\begin{cases} \lim_{L \to +\infty} \left(\limsup_{n \to +\infty} \|w_n^L\|_{L^\infty(\mathbb{R}_+, L^3(\mathbb{R}^3))} \right) = 0, \\ \lim_{L \to +\infty} \left(\limsup_{n \to +\infty} \|R_n^L\|_{E_{\tau_n}} \right) = 0. \end{cases}$$

---

[13] with E_T defined by (4.1).

This theorem was at the origin of many studies on the Navier-Stokes system and more generally on several equations arising in fluid mechanics; among others, one can mention [5, 8, 10, 32, 45, 46, 60, 82, 85] and the references therein.

Remark 4.1. *As a by-product of the decomposition (4.5), one can easily deduce that the life span T_n of v_n is bounded from below by the smallest of the life spans of each involved profile (up to scaling), namely*

$$T_n \geq \min_{\ell \in J} h_{\ell,n}^2 T^\ell = \tau_n \,.$$

Proof. Let $(\varphi_n)_{n \in \mathbb{N}}$ be a bounded sequence of divergence free vector fields in $\dot{H}^{\frac{1}{2}}(\mathbb{R}^3)$, and let $\phi^0 \in \dot{H}^{\frac{1}{2}}(\mathbb{R}^3)$ be any weak limit of (φ_n), then it follows from Theorem 1 that (up to the extraction of a subsequence) the following asymptotic orthogonal decomposition holds

$$\varphi_n = \phi^0 + \sum_{\ell=1}^{L} h_{\ell,n}^{-d/p} \phi^\ell \left(\frac{\cdot - x_{\ell,n}}{h_{\ell,n}} \right) + r_{n,L} \quad \text{with} \quad \lim_{L \to +\infty} \left(\limsup_{n \to +\infty} \|r_{n,L}\|_{L^3(\mathbb{R}^3)} \right) = 0 \,.$$

(4.7)

First observe that, in light of (2.3), one has $\|\phi^\ell\|_{\dot{H}^{\frac{1}{2}}(\mathbb{R}^3)} \leq c$ for ℓ sufficiently large, where c is the universal constant arising in (4.2). This readily ensures the first statement of Theorem 3.

Granted with the above decomposition for $(\varphi_n)_{n \in \mathbb{N}}$, the second statement then results from the scaling properties of the Navier-Stokes system, the orthogonality condition (1.3) and the common properties of the heat equation. More precisely, with the notations of Theorem 3, applying a perturbative argument, one can show that a good approximation for v_n is given by

$$v_n^{\mathrm{ap}}(t, \cdot) \overset{\text{def}}{=} V^0(t, \cdot) + \sum_{\ell=1}^{L} h_{\ell,n} V^\ell \left(\frac{t}{h_{\ell,n}^2}, \frac{\cdot - x_{\ell,n}}{h_{\ell,n}} \right) + e^{t\Delta} r_{n,L},$$

which is well-defined for all $0 \leq t \leq \tau_n$.   □

Let us end this section by presenting a brief proof (using profile decompositions) of a result of Rusin-Šverák [85] whose original proof did not use profile decompositions. This result can be stated as follows.

Theorem 4 ([85]). *Let ρ_{max} be the supremum of the set of $\rho > 0$ such that the Navier-Stokes system is globally well-posed for all v_0 in $\dot{H}^{\frac{1}{2}}(\mathbb{R}^3)$ satisfying $\|v_0\|_{\dot{H}^{\frac{1}{2}}(\mathbb{R}^3)} < \rho$. Then, if $\rho_{max} < \infty$, there exists a particular data $u_0 \in \dot{H}^{\frac{1}{2}}(\mathbb{R}^3)$, which generates a solution to (NS) that blows up at a finite time. Moreover $\|u_0\|_{\dot{H}^{\frac{1}{2}}(\mathbb{R}^3)} = \rho_{max}$.*

Proof. Assume that $\rho_{max} < \infty$. Then by definition, there exists a sequence $(v_0^n) \in \dot{H}^{\frac{1}{2}}(\mathbb{R}^3)$ such that

$$T^*(v_0^n) < \infty \quad \text{and} \quad \|v_0^n\|_{\dot{H}^{\frac{1}{2}}(\mathbb{R}^3)} \overset{n \to \infty}{\longrightarrow} \rho_{max} \,.$$

Invoking the profile decomposition for the sequence (v_0^n), one deduces from the orthogonal energy estimate (1.4) that the involved profiles (ϕ^ℓ) satisfy

$$\sum_{\ell \in \mathbb{N}} \|\phi^\ell\|_{\dot{H}^{\frac{1}{2}}(\mathbb{R}^3)}^2 \leq \rho_{max}^2.$$

The control of the life span given by the theorem of Gallagher stated above implies that, since one assumes that $T^*(v_0^n) < \infty$, there must be at least one profile ϕ^{ℓ_0} in (4.7) with a finite life span and such that

$$\|\varphi^{\ell_0}\|_{\dot{H}^{\frac{1}{2}}(\mathbb{R}^3)}^2 \geq \rho_{max}^2.$$

The two previous estimates are only compatible if and only if the profile decomposition of (v_0^n) includes exactly only one such profile φ^{ℓ_0} and if the remainder term tends to 0 strongly in $\dot{H}^{\frac{1}{2}}(\mathbb{R}^3)$, namely

$$\|\varphi^{\ell_0}\|_{\dot{H}^{\frac{1}{2}}(\mathbb{R}^3)} = \rho_{max},$$

which completes the proof of the theorem.   □

4.2   The blow-up result of Hmidi-Keraani

At first, the profile decompositions of P. Gérard in [49] found the way into the study of nonlinear evolution equations through the question of whether a given profile decomposition for a sequence of bounded data is propagated by the equation at hand, as was the case in the previous section (see for example Bahouri-Gérard [11], Gallagher [43], Gallagher-Gérard [44], Keraani [62],...). To the best of our knowledge, the first instance where these profile techniques were used for another purpose (at least in the framework of evolution equations) happened in the work of Hmidi-Keraani [53].

In [53], the authors were interested by the L^2-critical nonlinear Schrödinger equation:

$$\begin{cases} i\partial_t u + \Delta u + |u|^{\frac{4}{d}} u = 0 \text{ on } \mathbb{R}^d, \\ u_{|t=0} = u_0. \end{cases} \tag{4.8}$$

Many results have been devoted to the study of this equation (see for instance Bégout-Vargas [16], Bourgain [19], Cazenave [23], Colliander-Raynor-Sulem-Wright [33], Ginibre-Velo [52], Merle-Vega [76], and Sulem-Sulem [91]). In particular, it was established by Ginibre-Velo [52] that the Cauchy problem (4.8) is locally well-posed in[14] $H^1(\mathbb{R}^d)$ (see also the monograph of Cazenave [23] for further details), that is to say for all $u_0 \in H^1(\mathbb{R}^d)$, there exists $T > 0$ and a unique solution $u \in C([0,T]; H^1(\mathbb{R}^d))$ of the Cauchy problem (4.8) associated to u_0. Moreover defining the lifespan T^* of the solution u of (4.8) as the supremum of positive times T such that the Cauchy problem (4.8) has a solution $u \in C([0,T]; H^1(\mathbb{R}^d))$, we have the following blow-up criterion

$$T^* < \infty \Rightarrow \|\nabla u(t,\cdot)\|_{L^2(\mathbb{R}^d)} \xrightarrow{t \to T^*} \infty. \tag{4.9}$$

---

[14]Here $H^1(\mathbb{R}^d)$ stands for the inhomogeneous Sobolev space.

Finally, this unique solution enjoys the following conservation laws regarding the total mass:

$$\|u(t,\cdot)\|^2_{L^2(\mathbb{R}^d)} = \|u_0\|^2_{L^2(\mathbb{R}^d)} \tag{4.10}$$

and the balance of energy:

$$E(t) = \frac{1}{2}\int_{\mathbb{R}^d} |\nabla u(t,x)|^2 dx - \frac{d}{4+2d}\int_{\mathbb{R}^d} |u(t,x)|^{\frac{4}{d}+2} dx = E(0). \tag{4.11}$$

Nowadays, it is well-established that the global well-posedness for the Cauchy problem (4.8) is connected to the so-called ground-state, namely the unique positive radial solution of

$$\Delta Q - Q + |Q|^{\frac{4}{d}} Q = 0.$$

In particular, it turns out that the quantity $\|Q\|^2_{L^2(\mathbb{R}^d)}$ is the critical mass for the formation of singularities. Indeed on the one hand, it is well-known that there exists a solution u of (4.8) whose mass is equal to $\|Q\|^2_{L^2}$ and that blows up at finite time (see [23, 91]). On the other hand, if one combines the energy estimate (4.11) with the following inequality of Gagliardo-Nirenberg (due to Weinstein [96])

$$\|f\|^{\frac{4}{d}+2}_{L^{\frac{4}{d}+2}(\mathbb{R}^d)} \le C_d \|f\|^{\frac{4}{d}}_{L^2(\mathbb{R}^d)} \|\nabla f\|^2_{L^2(\mathbb{R}^d)} \quad \text{with} \quad C_d = \frac{d+2}{d}\|Q\|^{\frac{4}{d}}_{L^2(\mathbb{R}^d)}, \tag{4.12}$$

one can easily prove that for any data $u_0 \in H^1(\mathbb{R}^d)$ such that

$$\|u_0\|_{L^2(\mathbb{R}^d)} < \|Q\|_{L^2(\mathbb{R}^d)}, \tag{4.13}$$

the associated Cauchy problem (4.8) is globally well-posed. Indeed, in that case taking advantage of the conservation laws (4.10)—(4.11) and the Gagliardo-Nirenberg inequality (4.12), we readily gather that the solution at hand satisfies

$$\frac{1}{2}\|\nabla u(t,\cdot)\|_{L^2(\mathbb{R}^d)} \le E(0) + \frac{1}{2}\frac{\|u_0\|^{\frac{4}{d}}_{L^2(\mathbb{R}^d)}}{\|Q\|^{\frac{4}{d}}_{L^2(\mathbb{R}^d)}}\|\nabla u(t,\cdot)\|_{L^2(\mathbb{R}^d)},$$

which shows that under the assumption (4.13), $\|\nabla u(t,\cdot)\|_{L^2(\mathbb{R}^d)}$ remains uniformly bounded and thus, by virtue of the blow-up criterion (4.9), the solution is global in time.

In the radial framework, it was shown by Merle-Tsutsumi ([75]) that in case of blow-up, there is a minimal amount of concentration of the L^2-norm at the origin, namely that if u denotes a radial solution of the Cauchy problem (4.8), which blows-up at a finite time T^*, then for all $R > 0$,

$$\liminf_{t \to T^*} \int_{|x| \le R} |u(t,x)|^2 dx \ge \int_{\mathbb{R}^d} Q^2 dx.$$

In [53], Hmidi and Keraani established the following generalization:

Theorem 5 ([53]). *Let u be a solution to the Cauchy problem (4.8) and assume that it blows-up at a finite time T^*. Then given $\lambda(t)$ such that $\frac{\sqrt{T^*-t}}{\lambda(t)} \to 0$, as $t \nearrow T^*$, there exists $x(t)$ such that*

$$\int_{|x-x(t)|\leq\lambda(t)} |u(t,x)|^2 dx \geq \int_{\mathbb{R}^d} Q^2 dx.$$

Proof. The proof comports three steps that we will briefly sketch here. According to the blow-up criterion (4.9), if we assume that u is a solution of (4.8) that blows-up at a finite time T^*, then $\|\nabla u(t,\cdot)\|_{L^2(\mathbb{R}^d)} \overset{t\to T^*}{\longrightarrow} \infty$. In particular, there exists a sequence of times $(t_n)_{n\in\mathbb{N}}$ converging to T^* such that $\|\nabla u(t_n,\cdot)\|_{L^2(\mathbb{R}^d)} \overset{t\to T^*}{\longrightarrow} \infty$.
(i) In the first step, one rescales the sequence $(u(t_n,\cdot))_{n\in\mathbb{N}}$ as follows[15]

$$v_n = \rho_n^{\frac{d}{2}} u(t_n, \rho_n x) \quad \text{with} \quad \rho_n = \frac{\|\nabla Q\|_{L^2(\mathbb{R}^d)}}{\|\nabla u(t_n,\cdot)\|_{L^2(\mathbb{R}^d)}}. \tag{4.14}$$

This transform ensures that $(v_n)_{n\in\mathbb{N}}$ is a bounded sequence in $H^1(\mathbb{R}^d)$. Obviously the sequence $(\rho_n)_{n\in\mathbb{N}}$ tends to 0 as n goes to infinity. Then, according to the energy conservation law (4.11), one has

$$E(v_n) = \rho_n^2 E(0) \overset{n\to\infty}{\longrightarrow} 0,$$

which implies that

$$\frac{1}{2}\|\nabla v_n\|_{L^2(\mathbb{R}^d)} \overset{n\to\infty}{\longrightarrow} \frac{d}{4+2d}\|v_n\|_{L^{\frac{4}{d}+2}(\mathbb{R}^d)}^{\frac{4}{d}+2}.$$

Invoking (4.14), we infer that

$$\|v_n\|_{L^{\frac{4}{d}+2}(\mathbb{R}^d)}^{\frac{4}{d}+2} \overset{n\to\infty}{\longrightarrow} \frac{d+2}{d}\|\nabla Q\|_{L^2(\mathbb{R}^d)}^2.$$

(ii) In the second step, taking advantage of the profile decomposition method, one shows the existence of a sequence $(x_n)_{n\in\mathbb{N}}$ of points in $\mathbb{R}^d$ such that

$$v_n(\cdot + x_n) \rightharpoonup V \quad \text{with} \quad \|V\|_{L^2(\mathbb{R}^d)} \geq \|Q\|_{L^2(\mathbb{R}^d)}. \tag{4.15}$$

To achieve that goal, one starts by applying the profile decomposition of P. Gérard [49], which (up to the extraction of a subsequence) provides the following orthogonal decomposition:

$$v_n = \sum_{j=0}^{L} V^j(\cdot - x_n^j) + \mathcal{R}_n^L,$$

with, for all $2 < p < 2^*$, $\lim_{L\to+\infty}(\limsup_{n\to\infty}\|\mathcal{R}_n^L\|_{L^p(\mathbb{R}^d)}) = 0$ and such that

$$\|v_n\|_{L^2(\mathbb{R}^d)}^2 = \sum_{j=0}^{L}\|V^j\|_{L^2(\mathbb{R}^d)}^2 + \|\mathcal{R}_n^L\|_{L^2(\mathbb{R}^d)}^2 + o(1), \quad n\to\infty, \tag{4.16}$$

---

[15] As it will be seen further, the choice of the scaling ρ_n is crucial in the proof of the result.

as well as

$$\|\nabla Q\|_{L^2(\mathbb{R}^d)}^2 = \|\nabla v_n\|_{L^2(\mathbb{R}^d)}^2 = \sum_{j=0}^{L} \|\nabla V^j\|_{L^2(\mathbb{R}^d)}^2 + \|\nabla \mathcal{R}_n^L\|_{L^2(\mathbb{R}^d)}^2 + o(1), \quad n \to \infty.$$

(4.17)

Moreover, thanks to the orthogonality condition between the cores, one has

$$\sum_{j=0}^{\infty} \|V^j\|_{L^{\frac{4}{d}+2}(\mathbb{R}^d)}^{\frac{4}{d}+2} = \lim_{n \to \infty} \|v_n\|_{L^{\frac{4}{d}+2}(\mathbb{R}^d)}^{\frac{4}{d}+2} = \frac{d+2}{d} \|\nabla Q\|_{L^2(\mathbb{R}^d)}^2.$$

(4.18)

Then, Gagliardo-Nirenberg inequality (4.12) allows one to infer that

$$\sum_{j=0}^{\infty} \|V^j\|_{L^{\frac{4}{d}+2}(\mathbb{R}^d)}^{\frac{4}{d}+2} \leq C_d \left(\sup \|V^j\|_{L^2(\mathbb{R}^d)}^{\frac{4}{d}} \right) \sum_{j=0}^{\infty} \|\nabla V^j\|_{L^2(\mathbb{R}^d)}^2,$$

with $C_d = \frac{d+2}{d} \|Q\|_{L^2(\mathbb{R}^d)}^{\frac{4}{d}}$, which in view of (4.17) implies that

$$\sum_{j=0}^{\infty} \|V^j\|_{L^{\frac{4}{d}+2}(\mathbb{R}^d)}^{\frac{4}{d}+2} \leq C_d \left(\sup \|V^j\|_{L^2(\mathbb{R}^d)}^{\frac{4}{d}} \right) \|\nabla Q\|_{L^2(\mathbb{R}^d)}^2.$$

Since the series $\sum \|V^j\|_{L^2}^2$ is convergent as a consequence of the orthogonality equality (4.16), the supremum on the right-hand side is achieved for some index j_0. The latter estimate thus becomes

$$\|V^{j_0}\|_{L^2(\mathbb{R}^d)}^{\frac{4}{d}} \geq \frac{\sum_{j=0}^{\infty} \|V^j\|_{L^{\frac{4}{d}+2}(\mathbb{R}^d)}^{\frac{4}{d}+2}}{C_d \|\nabla Q\|_{L^2(\mathbb{R}^d)}^2}.$$

Combining this result with (4.18) and with the fact that $C_d = \frac{d+2}{d} \|Q\|_{L^2(\mathbb{R}^d)}^{\frac{4}{d}}$, one readily gathers that

$$\|V^{j_0}\|_{L^2(\mathbb{R}^2)} \geq \|Q\|_{L^2(\mathbb{R}^2)},$$

which completes the proof of the claim (4.15).

(iii) In the last step, one obtains the result by invoking the scale invariance of the equation and expressing v_n by means of u_n. □

A   Wavelet decompositions and nonlinear approximation theory

In this appendix, we introduce briefly some basic facts about wavelet decompositions and nonlinear approximation theory; for further details, one can consult [31, 34, 35, 67, 78] and the references therein.

Recall that wavelet decompositions (for instance Haar's system) have the form

$$f = \sum_{\lambda \in \Lambda} d_\lambda \psi_\lambda. \tag{A.1}$$

The set Λ comports both a scale index $j = j(\lambda)$ and a space index $k = k(\lambda)$. The wavelets can be normalized (for example) in the following way:

$$\psi_\lambda = \psi_{j,k} = 2^{rj} \psi(2^j \cdot - k),$$

where r is related to the scale invariance of the space at hand (see examples below). In the wavelet decomposition (A.1), the spatial index k provides an additional level of discretization compared to the Littlewood-Paley decomposition (for an introduction to Littlewood-Paley theory, see for instance [7, 18, 25, 68] and the references therein).

It is possible to characterize whether a function belongs to almost any classical functional space in terms of conditions pertaining only to $|d_\lambda|$. For instance, one has

$$\|f\|_{\dot{H}^s(\mathbb{R}^d)} \sim \|(d_\lambda)_{\lambda \in \Lambda}\|_{l^2} \quad \text{and} \quad \|f\|_{\dot{B}^t_{q,q}(\mathbb{R}^d)} \sim \|(d_\lambda)_{\lambda \in \Lambda}\|_{l^q}$$

if the previous decomposition is respectively normalized with $r = \frac{d}{2} - s$ for $\dot{H}^s(\mathbb{R}^d)$ or with $r = \frac{d}{q} - t$ for $\dot{B}^t_{q,q}(\mathbb{R}^d)$.

Nonlinear approximation theory, which is crucial in signal and image processing theories, consists in approximating a given function f by its N most significant components in the wavelet decomposition (A.1), namely

$$Q_N(f) = \sum_{\lambda \in E_N} d_\lambda \psi_\lambda, \tag{A.2}$$

where $E_N = E_N(f)$ is a subset of Λ with cardinal N. In many cases, E_N corresponds to the N largest wavelet coefficients. For instance, for the critical Sobolev embedding

$$\dot{B}^s_{p,p}(\mathbb{R}^d) \hookrightarrow \dot{B}^t_{q,q}(\mathbb{R}^d) \quad \text{with} \quad 0 < \frac{1}{p} - \frac{1}{q} = \frac{s-t}{d}, \tag{A.3}$$

it is straightforward that

$$\sup_{\|f\|_{\dot{B}^s_{p,p}(\mathbb{R}^d)} \leq 1} \|f - Q_N f\|_{\dot{B}^t_{q,q}(\mathbb{R}^d)} \leq C N^{-\frac{s-t}{d}}. \tag{A.4}$$

Indeed, if one denotes by $(d_m)_{m \geq 0}$ the decreasing rearrangement of $|d_\lambda|$, using the previous equivalence of norms and the fact that $p < q$, one can easily check that

$$\|f - Q_N f\|_{\dot{B}^t_{q,q}(\mathbb{R}^d)} \sim \left(\sum_{\lambda \notin E_N} |d_\lambda|^q \right)^{1/q} = \left(\sum_{m>N} |d_m|^q \right)^{1/q}$$

$$\leq \left(N^{-1} \sum_{m=1}^{N} |d_m|^p \right)^{1/p - 1/q} \left(\sum_{m>N} |d_m|^p \right)^{1/q}$$

$$\lesssim N^{-\frac{s-t}{d}} \|f\|_{\dot{B}^s_{p,p}(\mathbb{R}^d)}.$$

This completes the proof of (A.4).

B   Helly's lemma

For the convenience of the reader, let us briefly recall Helly's extraction lemma, which plays a role in the proof of the profile decomposition by P. Gérard (see §2.2). Further details can be found, for example in [40].

Proposition B.1. *For any sequence of increasing functions $f_n : \mathbb{R} \to [-1,1]$, there exists a subsequence which converges pointwise on $\mathbb{R}$.*

Proof. First, one considers the restrictions $f_n|_{\mathbb{Q}}$. The compactness of $[-1,1]$ ensures, by Bolzano-Weierstrass's theorem the existence of a first subsequence that converges at one given point in $\mathbb{Q}$. Iterating this idea, one can successively extract subsequences that converge simultaneously at a finite number of points in $\mathbb{Q}$. As $\mathbb{Q}$ is countable, Cantor's diagonal argument then provides one extraction $\psi : \mathbb{N} \to \mathbb{N}$ such that $(f_{\psi(n)})$ converges at every point of $\mathbb{Q}$. The map

$$f : \left| \begin{array}{ll} \mathbb{Q} \to & [-1,1] \\ q \mapsto & \lim_{n \to \infty} f_{\psi(n)}(q) \end{array} \right.$$

is obviously increasing and thus the sided limits

$$f(x-0) = \sup\{f(q)\,;\, q \in \mathbb{Q},\ q < x\} \quad \text{and} \quad f(x+0) = \inf\{f(q)\,;\, q \in \mathbb{Q},\ q > x\}$$

are well defined for any $x \in \mathbb{R}$ and satisfy $f(x-0) \le f(x+0)$. At all the points $x \in \mathbb{R}$ such that $f(x-0) = f(x+0)$ one defines $f(x)$ as this common value. For any $\varepsilon > 0$, the function $f|_{\mathbb{Q}}$ does not varies by more than ε on a small rational neighborhood whose adherence contains x. Convergence at two such rational points flanking x then ensures that for n large enough,

$$f(x) - 2\varepsilon \le f_{\psi(n)}(x) \le f(x) + 2\varepsilon$$

and thus $(f_{\psi(n)}(x))$ converges towards $f(x)$. The remaining set of "bad" points

$$B = \{x \in \mathbb{R}\,;\, f(x-0) < f(x+0)\}$$

is at most countable. Indeed, monotonicity implies that any choice of a rational ϑ_x such that $f(x-0) \le \vartheta_x \le f(x+0)$ induces an injective map $\vartheta : B \to \mathbb{Q}$. Then, using the compactness of $[f(x-0), f(x+0)]$ and Cantor's diagonal extraction once more, it is possible to extract a subsequence $(f_{\psi \circ \varphi(n)})$ that converges pointwise at any point of B and therefore, by construction, at any point of $\mathbb{R}$.   □

References

[1] S. Adachi, K. Tanaka, *Trudinger type inequalities in $\mathbb{R}^N$ and their best exponents*, Proc. Amer. Math. Soc., **128** (2000), 2051–2057.

[2] R. A. Adams, J. J. Fournier, *Sobolev Spaces*, Academic Press, 2003.

[3] Adimurthi, O. Druet, *Blow-up analysis in dimension 2 and a sharp form of Trudinger-Moser inequality*, Comm. Partial Differential Equations, **29** (2004), 295–322.

[4] Adimurthi, C. Tintarev, *Defect of compactness in spaces of bounded variation*, J. Funct. Anal., **271** (2016), 37–48.

[5] D. Albritton, *Blow-up criteria for the Navier-Stokes equations in non-endpoint critical Besov spaces*, Anal. PDE, **11** (2018), 1415–1456.

[6] H. Bahouri, *On the elements involved in the lack of compactness in critical Sobolev embedding*, Concentration analysis and applications to PDE, Trends Math., Birkhäuser, **260** (2013), 1–15.

[7] H. Bahouri, J.-Y. Chemin and R. Danchin, *Fourier Analysis and Nonlinear Partial Differential Equations*, Grundlehren der Mathematischen Wissenschaften, **343**, Springer-Verlag, 2011.

[8] H. Bahouri, J.-Y. Chemin and I. Gallagher, *On the stability of global solutions to the three dimensional Navier-Stokes equations*, Journal de l'École Polytechnique-Mathématique, **154** (2018), 843–911.

[9] H. Bahouri, A. Cohen and G. Koch, *A general wavelet-based profile decomposition in the critical embedding of function spaces*, Confluentes Mathematici, **3** (2011), 1–25.

[10] H. Bahouri, I. Gallagher, *On the stability in weak topology of the set of global solutions to the Navier-Stokes equations*, Archive for Rational Mechanics and Analysis, **209** (2013), 569–629.

[11] H. Bahouri, P. Gérard, *High frequency approximation of solutions to critical nonlinear wave equations*, Amer. J. Math., **121** (1999), 131–175.

[12] H. Bahouri, M. Majdoub and N. Masmoudi, *Lack of compactness in the 2D critical Sobolev embedding, the general case*, Journal de Mathématiques Pures et Appliquées, **101** (2014), 415–457.

[13] H. Bahouri, M. Majdoub and N. Masmoudi, *On the lack of compactness in the 2D critical Sobolev embedding*, Journal of Functional Analysis, **260** (2011), 208–252.

[14] H. Bahouri, G. Perelman, *A Fourier approach to the profile decomposition in Orlicz spaces*, Mathematical Research Letters, **21** (2014), 33–54.

[15] H. Bahouri, G. Perelman, *Global well-posedness for the derivative nonlinear Schrödinger equation*, Inventiones Mathematicae, **229** (2022), 639–688.

[16] P. Bégout, A. Vargas, *Mass concentration phenomena for the L^2-critical nonlinear Schrödinger equation*, Trans. Amer. Math. Soc., to appear.

[17] J. Ben Ameur, *Description du défaut de compacité de l'injection de Sobolev sur le groupe de Heisenberg*, Bulletin de la Société Mathématique de Belgique, **15**(4) (2008), 599–624.

[18] J.-M. Bony, *Calcul symbolique et propagation des singularités pour les équations aux dérivées partielles non linéaires*, Annales de l'École Normale Supérieure, **14** (1981), 209–246.

[19] J. Bourgain, *Refinements of Strichartz inequality and applications to 2D-NLS with critical nonlinearity*, Int. Math. Res. Not., **8** (1998), 253–283.

[20] H. Brézis, *Analyse fonctionnelle. Théorie et applications*, Dunod, 1999.

[21] H. Brézis, J.-M. Coron, *Convergence of solutions of H-Systems or how to blow bubbles*, Archive for Rational Mechanics and Analysis, **89** (1985), 21–86.
[22] M. Cannone, Y. Meyer and F. Planchon, *Solutions autosimilaires des équations de Navier-Stokes*, Séminaire "Équations aux Dérivées Partielles de l'École polytechnique", Exposé VIII, 1993–1994.
[23] T. Cazenave, *Semilinear Schrödinger Equations*, Courant Lecture Notes in Mathematics, **10**, AMS, 2003.
[24] J.-Y. Chemin, *Remarques sur l'existence globale pour le système de Navier-Stokes incompressible*, SIAM J. Math. Anal., **23** (1992), 20–28.
[25] J.-Y. Chemin, *Fluides parfaits incompressibles*, Astérisque, **230**, 1995.
[26] J.-Y. Chemin, *Théorèmes d'unicité pour le système de Navier-Stokes tridimensionnel*, Journal d'Analyse Mathématique, **77** (1999), 27–50.
[27] J.-Y. Chemin, I. Gallagher, *Large, global solutions to the Navier-Stokes equations, slowly varying in one direction*, Transactions of the American Mathematical Society, **362** (2010), 2859–2873.
[28] J.-Y. Chemin, I. Gallagher and P. Zhang, *Sums of large global solutions to the incompressible Navier-Stokes equations*, to appear in *Journal für die reine und angewandte Mathematik*.
[29] J.-Y. Chemin, C.-J. Xu, *Inclusions de Sobolev en calcul de Weyl-Hörmander et systèmes sous-elliptiques*, Annales Scientifiques de l'École Normale Supérieure, **30** (1997), 719–751.
[30] M. Christ, S. Shao, *Existence of extremals for a Fourier restriction inequality*, Analysis and PDE, **2** (2012), 261–312.
[31] A. Cohen, *Numerical analysis of wavelet methods*, Elsevier, 2003.
[32] M. Coles, S. Gustafson, *Solitary waves and dynamics for subcritical perturbations of energy critical NLS*, J. Publ. Res. Inst. Math. Sci., **58** (2020), 647–699.
[33] J. Colliander, S. Raynor, C. Sulem and J. D. Wright, *Ground state mass concentration in the L^2-critical nonlinear Schrödinger equation below H^1*, Math. Res. Lett., **12** (2005), 357–375.
[34] I. Daubechies, *Ten lectures on wavelets.* CBMS-NSF Reg. Conf. Series in Applied Maths., **61** (1992).
[35] R. DeVore, *Nonlinear approximation*, Acta Numerica, **7** (1998), 51–150.
[36] B. Dodson, *Global well-posedness and scattering for the defocusing, L^2-critical, nonlinear Schrödinger equation when $d = 2$*, Duke Math. J., **165** (2016), 3435–3516.
[37] O. Druet, E. Hebey and F. Robert, *Blow-up Theory for Elliptic PDEs in Riemannian Geometry*, Mathematical Notes, **45**, Princeton University Press, 2004.
[38] T. Duyckaerts, C. Kenig and F. Merle, *Universality of blow-up profile for small radial type II blow-up solutions of the energy-critical wave equation*, J. Eur. Math. Soc., **13** (2011), 533–599.
[39] K.-H. Fieseler, K. Tintarev, *Concentration Compactness. Functional-analytic Grounds and Applications*, Imperial College Press, 2007.
[40] S. Francinou, H. Gianella and S. Nicolas, *Oraux X - ENS, Analyse 2*, Cassini, 2009.

[41] R. Frank, E.-H. Lieb and J. Sabin, *Maximizers for the Stein-Tomas inequality*, Geometric and Functional Analysis, **26** (2016), 1095–1134.

[42] H. Fujita, T. Kato, *On the Navier-Stokes initial value problem I*, Archive for Rational Mechanics and Analysis, **16** (1964), 269–315.

[43] I. Gallagher, *Profile decomposition for solutions of the Navier-Stokes equations*, Bulletin de la Société Mathématique de France, **129** (2001), 285–316.

[44] I. Gallagher, P. Gérard, *Profile decomposition for the wave equation outside convex obstacles*, Journal de Mathématiques Pures et Appliquées, **80** (2001), 1–49.

[45] I. Gallagher, G. Koch and F. Planchon, *A profile decomposition approach to the $L_t^\infty(L_x^3)$ Navier-Stokes regularity criterion*, Mathematische Annalen, **355** (2013), 1527–1559.

[46] I. Gallagher, G. Koch and F. Planchon, *Blow-up of critical Besov norms at a potential Navier-Stokes singularity*, Comm. Math. Phys., **343** (2016), 39–82.

[47] L. Gassot, *On the radially symmetric traveling waves for the Schrödinger equation on the Heisenberg group*, Pure and Applied Analysis, **2** (2020), 739–794.

[48] P. Gérard, *Microlocal defect measures*, Communications in Partial Differential Equations, **16** (1991), 1761–1794.

[49] P. Gérard, *Description du défaut de compacité de l'injection de Sobolev*, ESAIM Control Optim. Calc. Var., **3** (1998), 213–233.

[50] P. Gérard, Y. Meyer and F. Oru, *Inégalités de Sobolev précisées*, Séminaire X-EDP, École Polytechnique (1996).

[51] Y. Giga, T. Miyakawa, *Solutions in L^r of the Navier-Stokes initial value problem*, Archive for Rational Mechanics and Analysis, **89** (1985), 267–281.

[52] J. Ginibre, G. Velo, *Scattering theory in the energy space for a class of nonlinear Schrödinger equations*, J. Math. Pure Appl., **64** (1984), 363–401.

[53] T. Hmidi, S. Keraani, *Remarks on the blowup for the L^2-critical nonlinear Schrödinger equations*, SIAM J. Math. Anal., **38** (2006), 1035–1047.

[54] M. Hutchings, F. Morgan, M. Ritoré and A. Ros, *Proof of the double bubble conjecture*, Annals of Mathematics, **155** (2002), 459–489.

[55] S. Jaffard, *Analysis of the lack of compactness in the critical Sobolev embeddings*, J. Funct. Anal., **161** (1999), 384–396.

[56] J. Jendrej, A. Lawrie, *Two-bubble dynamics for threshold solutions to the wave maps equation*, Invent. Math., **213** (2018), 1249–1325.

[57] H. Jia, C. Kenig, *Asymptotic decomposition for semilinear wave and equivariant wave map equations*, Amer. J. Math., **139** (2017), 1521–1603.

[58] H. Jia, B. Liu, W. Schlag and G. Xu, *Global center stable manifold for the defocusing energy critical wave equation with potential*, Amer. J. Math., **142** (2020), 1497–1557.

[59] T. Kato, *Strong L^p-solutions of the Navier-Stokes equation in $\mathbb{R}^m$ with applications to weak solutions*, Mathematische Zeitschrift, **187** (1984), 471–480.

[60] C. E. Kenig, G. Koch, *An alternative approach to the Navier-Stokes equations in critical spaces*, Annales de l'Institut Henri Poincaré (C) Non Linear Analysis, **28** (2011), 159–187.

[61] C. E. Kenig, F. Merle, *Global well-posedness, scattering and blow-up for the*

energy critical focusing non-linear wave equation, Acta Math., **201** (2008), 147–212.

[62] S. Keraani, *On the defect of compactness for the Strichartz estimates of the Schrödinger equations*, Journal of Differential equations, **175** (2001), 353–392.

[63] R. Killip, J. Murphy and M. Visan, *The final-state problem for the cubic-quintic NLS with nonvanishing boundary conditions*, Anal. PDE, **9** (2016), 1523–1574.

[64] G. Koch, *Profile decompositions for critical Lebesgue and Besov space embeddings*, Indiana University Mathematics Journal, **59** (2010), 1801–1830.

[65] H. Koch, D. Tataru, *Well-posedness for the Navier-Stokes equations*, Advances in Mathematics, **157** (2001), 22–35.

[66] J. Krieger, *On stability of type II blow up for the critical nonlinear wave equation on* $\mathbb{R}^{3+1}$, Mem. Amer. Math. Soc., **267** (2020).

[67] G. Kyriasis, *Nonlinear approximation and interpolation spaces*, J. Approx. Theory, **113** (2001), 110–126.

[68] P.-G. Lemarié-Rieusset, *Recent Developments in the Navier-Stokes Problem*, Chapman and Hall/CRC, 2002.

[69] E. Lenzmann, M. Lewin, *On singularity formation for the* L^2-*critical Boson star equation*, Nonlinearity, **24** (2011), 3515–3540.

[70] J. Leray, *Essai sur le mouvement d'un liquide visqueux emplissant l'espace*, Acta Mathematica, **63** (1933), 193–248.

[71] J. Leray, *Étude de diverses équations intégrales non linéaires et de quelques problèmes que pose l'hydrodynamique*, Journal de Mathématiques Pures et Appliquées, **12** (1933), 1–82.

[72] P.-L. Lions, *The concentration-compactness principle in the calculus of variations. The limit case. I*, Rev. Mat. Iberoamericana, **1** (1985), 145–201.

[73] P.-L. Lions, *The concentration-compactness principle in the calculus of variations. The limit case. II*, Rev. Mat. Iberoamericana, **1** (1985), 45–121.

[74] W. H. Meeks III, J. Pérez and G. Tinaglia, *Constant mean curvature surfaces*, Surveys in Differential Geometry, **21** (2016), 179–287.

[75] F. Merle, Y. Tsutsumi, L^2 *concentration of blowup solutions for the nonlinear Schrödinger equation with critical power nonlinearity*, J. Differential Equations, **84** (1990), 205–214.

[76] F. Merle, L. Vega, *Compactness at blow-up time for* L^2 *solutions of the critical nonlinear Schrödinger equation in 2D*, International Mathematical Research Notices, (1998), 399–425.

[77] G. Metivier, S. Schochet, *Interactions trilinéaires résonantes*, Séminaire "Équations aux Dérivées Partielles de l'École polytechnique", 1995—1996, 1–14.

[78] Y. Meyer, *Ondelettes et opérateurs*, Hermann, 1990.

[79] F. Murat, L. Tartar, *H-convergence*, Topics in the Mathematical Modelling of Composite Materials, Progr. Nonlinear Differential Equations Appl., **31**, Birkhäuser, 1997, 21–43.

[80] A. Nachman, I. Regev and D. Tataru, *A nonlinear Plancherel theorem with applications to global well-posedness for the defocusing Davey-Stewartson equation and to the inverse boundary value problem of Calderon*, Invent. Math.,

220 (2020), 395–451.

[81] F. Planchon, *Asymptotic behavior of global solutions to the Navier-Stokes equations in* $\mathbf{R}^3$, Rev. Mat. Iberoamericana, **14** (1998), 71–93.

[82] E. Poulon, *About the behavior of regular Navier-Stokes solutions near the blow up*, Bull. Soc. Math. France, **146** (2018), 355–390.

[83] M.-M. Rao, Z.-D. Ren, *Applications of Orlicz Spaces*, Monographs and Textbooks in Pure and Applied Mathematics, **250**, Marcel Dekker Inc., 2002.

[84] B. Ruf, *A sharp Trudinger-Moser type inequality for unbounded domains in* $\mathbb{R}^2$, J. Funct. Anal. **219** (2005), 340–367.

[85] W. Rusin, V. Šverák, *Minimal initial data for potential Navier-Stokes singularities*, Journal of Functional Analysis, **260** (2011), 879–891.

[86] I. Schindler, K. Tintarev, *An abstract version of the concentration compactness principle*, Rev. Mat. Complutense, **15** (2002), 417–436.

[87] S. Shao, *On existence of extremizers for the Tomas-Stein inequality for* S^1, Journal of Functional Analysis, **270** (2016), 3996–4038.

[88] S. Solimini, *A note on compactness-type properties with respect to Lorentz norms of bounded subset of a Sobolev space*, Annales de l'IHP Analyse non linéaire, **12** (1995), 319–337.

[89] E. M. Stein, *Harmonic Analysis: Real-Variable Methods, Orthogonality, Oscillatory Integrals*, Princeton University Press, 1993.

[90] M. Struwe, *A global compactness result for boundary value problems involving limiting nonlinearities*, Mathematische Zeitschrift, **187** (1984), 511–517.

[91] C. Sulem, P.-L. Sulem, *The Nonlinear Schrödinger Equation. Self-focusing and Wave Collapse*, Applied Mathematical Sciences, **139**, Springer-Verlag, 1999.

[92] T. Tao, *Concentration compactness for critical wave maps*, Ed. Joachim Krieger and Wilhelm Schlag, Bull. Amer. Math. Soc., **50** (2013), 655–662.

[93] L. Tartar, *H-measures, a new approach for studying homogenisation, oscillations and concentration effects in partial differential equations*, Proceedings of the Royal Society of Edinburgh, **115** (1990), 193–230.

[94] P. A. Tomas, *A restriction theorem for the Fourier transform*, Bull. Amer. Math. Soc., **81** (1975), 477–478.

[95] H. Triebel, *Theory of Function Spaces*, Birkhäuser, 1983.

[96] M. I. Weinstein, *Nonlinear Schrödinger equations and sharp interpolation estimates*, Comm. Math. Phys., **87** (1983), 567–576.

[97] F. Weissler, *The Navier-Stokes initial value problem in* L^p, Archive for Rational Mechanics and Analysis, **74** (1980), 219–230.

Lectures on the Analysis of Nonlinear
Partial Differential Equations Vol. 6
MLM 6, pp. 31–90

© Higher Education Press
and International Press
Beijing-Boston

The Inhomogeneous Incompressible Navier-Stokes Equations with Discontinuous Density: Three Different Approaches

Raphaël Danchin*

Abstract

The inhomogeneous incompressible Navier-Stokes equations (INS) that govern the evolution of viscous incompressible flows with non-constant density have received a lot of attention in the recent years. Here we propose three methods for proving existence and uniqueness of solutions with, possibly, discontinuous densities. As an application, we establish the persistence of interface regularity for patches of constant density, thus justifying the use of (INS) for modeling mixtures of incompressible viscous fluids.

Introduction

A number of recent works have been dedicated to the mathematical analysis of the following *inhomogeneous incompressible Navier-Stokes equations*[1]:

$$\begin{cases} \partial_t \varrho + \operatorname{div}(\varrho u) = 0 & \text{in } (0,T) \times \Omega, \\ \partial_t(\varrho u) + \operatorname{div}(\varrho u \otimes u) - \mu \Delta u + \nabla P = 0 & \text{in } (0,T) \times \Omega, \\ \operatorname{div} u = 0 & \text{in } (0,T) \times \Omega, \end{cases} \tag{1}$$

that correspond to the Eulerian description of the evolution of the velocity field $u = u(t,x) \in \mathbb{R}^d$, density $\varrho = \varrho(t,x) \in \mathbb{R}_+$ and pressure $P = P(t,x) \in \mathbb{R}$ of an incompressible flow with constant viscosity $\mu > 0$ in the domain Ω of $\mathbb{R}^d$.

---

*Université Paris-est, LAMA UMR 8050, 61, avenue du Général de Gaulle, 94010 Créteil Cedex, France. E-mail: danchin@univ-paris12.fr

[1] Throughout the text, we adopt the notation $(\operatorname{div}(a \otimes b))^j \overset{\text{def}}{=} \sum_i \partial_i(a^i b^j)$.

In the case where $\varrho \equiv 1$, the above system reduces to the well-known (*homogeneous*) *incompressible Navier-Stokes equations*:

$$\begin{cases} \partial_t u + \operatorname{div}(u \otimes u) - \mu \Delta u + \nabla P = 0 \text{ in } (0, T) \times \Omega, \\ \operatorname{div} u = 0 \qquad\qquad\qquad\qquad\qquad \text{ in } (0, T) \times \Omega. \end{cases} \tag{2}$$

System (1) is derived from the laws of classical physics: the first equation corresponds to the mass balance, the second equation is the momentum balance in the absence of external forces, and the third one, the incompressibility condition. The reader may refer to [3] or to the introduction of [32] for more details.

We are concerned with the existence and uniqueness issue of solutions to System (1) supplemented with homogeneous Dirichlet boundary conditions on $\partial\Omega$ for the velocity and initial data at time 0:

$$\rho|_{t=0} = \rho_0 \quad \text{and} \quad u|_{t=0} = u_0. \tag{3}$$

As we shall focus on nonsmooth solutions, System (1), (3) has to be understood in the distributional meaning: we require that for all $t \in [0, T)$ and for all function ϕ in $\mathcal{C}_c^\infty([0, T) \times \Omega; \mathbb{R})$,

$$\langle \rho(t), \phi(t) \rangle - \langle \rho_0, \phi_0 \rangle - \int_0^t \langle \rho, \partial_t \phi \rangle \, d\tau - \int_0^t \langle \rho u, \nabla \phi \rangle \, d\tau = 0, \tag{4}$$

$$\int_0^t \langle u, \nabla \phi \rangle \, d\tau = 0, \tag{5}$$

and that

$$\langle \rho(t) u(t), \varphi(t) \rangle - \langle \rho_0 u_0, \varphi_0 \rangle - \int_0^t \langle \rho u, \partial_t \varphi \rangle \, d\tau - \int_0^t \langle \rho u \otimes u, \nabla \varphi \rangle \, d\tau$$
$$+ \mu \int_0^t \langle \nabla u, \nabla \varphi \rangle \, d\tau = 0 \tag{6}$$

for all divergence-free function $\varphi \in \mathcal{C}_c^\infty([0, T) \times \Omega; \mathbb{R}^d)$, where the notation $\langle \cdot, \cdot \rangle$ designates the distribution bracket in Ω.

Smooth enough solutions of (1) with velocity vanishing at the boundary satisfy, for $t \in [0, T)$, the following balance involving the kinetic energy and the rate of dissipation:

$$\frac{1}{2} \int_\Omega \rho(t, x) |u(t, x)|^2 \, dx + \mu \int_0^t \int_\Omega |\nabla u(\tau, x)|^2 \, dx \, d\tau = \frac{1}{2} \int_\Omega \rho_0(x) |u_0(x)|^2 \, dx. \tag{7}$$

The mass equation and the incompressibility condition ensure that the density at time t may be deduced from ρ_0 by a measure preserving change of variables. Consequently, $\sup_x \rho(t, x)$ and $\inf_x \rho(t, x)$ are time independent and, for any smooth enough function $f : \mathbb{R}_+ \to \mathbb{R}$, we have

$$\int_\Omega f(\rho(t, x)) \, dx = \int_\Omega f(\rho_0(x)) \, dx \quad \text{for } t \in [0, T). \tag{8}$$

By taking advantage of those identities and of smart compactness arguments, one can establish the existence of global weak solutions with finite energy whenever the initial density ρ_0 is bounded and the initial kinetic energy $\frac{1}{2}\|\sqrt{\rho_0}u_0\|_{L^2}^2$ is finite. The case of constant density (that is (2)) goes back to the work by J. Leray [27] in 1934. The case with non-constant density $\rho_0 \in L^\infty$ bounded away from 0 has been elucidated by A. Kazhikhov [23], and J. Simon in [35] proved the existence of global weak solutions satisfying the energy inequality (that is, the left-hand side of (7) is less than or equal to the right-hand side) for general bounded and nonnegative densities. Then, P.-L. Lions [32] extended the result to the more general case of density-dependent viscosity (namely $\mu\Delta u$ is replaced with $\text{div}(\mu(\rho)(\nabla u + Du))$ for some given positive function μ).

In the case where the fluid domain is $\mathbb{R}^d$, System (1) is invariant for all $\lambda > 0$ by the rescaling

$$\rho \rightsquigarrow \rho_\lambda, \quad u \rightsquigarrow \lambda u_\lambda, \quad P \rightsquigarrow \lambda^2 P_\lambda \quad \text{where we agree that} \quad z_\lambda(t,x) := z(\lambda^2 t, \lambda x), \quad (9)$$

that is to say (ρ, u, P) fulfills (1) if and only if $(\rho_\lambda, \lambda u_\lambda, \lambda^2 P_\lambda)$ fulfills (1), for all $\lambda > 0$. This property will push us to solve the system in functional spaces endowed with norms having that scaling invariance. More generally, quantities invariant by (9) will play a decisive role in the existence theory of strong solutions to (1).

The main question we want to address here is whether System (1) allows to handle mixtures of incompressible viscous fluids with *different* densities. In other words, is it possible to solve the initial value problem for (1) supplemented with suitable velocity field and piecewise constant density ? The model situation we have in mind is that of an initial patch of density, that is

$$\rho_0 = \eta_1 1_{D_0} + \eta_2 1_{{}^c D_0} \quad (10)$$

for some bounded simply connected subdomain D_0 of Ω, and $\eta_1, \eta_2 \geq 0$.

If we assume in addition that $\sqrt{\rho_0}u_0$ is in L^2, then combining the weak solution theory recalled above and the Di Perna-Lions results in [19] for transport equations, one can construct a global weak solution (ρ, u) to (1) satisfying the energy inequality and

$$\rho(t) = \eta_1 1_{D_t} + \eta_2 1_{{}^c D_t} \quad \text{for all } t \geq 0,$$

where D_t is the image of D_0 by the generalized flow of u (see [32] for more details).

However, as in this context the flow is not regular, very few informations on D_t are available. In particular, it is not known whether the regularity of the boundary is preserved during the time evolution. Furthermore, no uniqueness, neither stability with respect to initial data is known in the weak solutions framework.

Our aim is to present three different approaches that have been used recently to investigate System (1) supplemented with discontinuous initial density, and that allow to have some insight on the evolution of density patches in a functional framework ensuring uniqueness of solutions. For expository purposes and because we are interested in global-in-time results, we shall often assume that the initial velocity is small and that the initial density is close to some positive constant. In

the last section, however, we shall give an example of construction of global strong
solutions in the case of large data.

The first two approaches that we here propose strongly rely on the notion
of *critical regularity* (that is, the norms we shall use to solve the system will
have the scaling invariance (9)) and of maximal regularity results for the following
evolutionary Stokes equations:

$$\begin{cases} \partial_t u - \Delta u + \nabla P = f \text{ in } (0,T) \times \Omega, \\ \operatorname{div} u = 0 \qquad\qquad \text{in } (0,T) \times \Omega, \\ u|_{t=0} = u_0 \qquad\quad \text{in } \Omega. \end{cases} \qquad (11)$$

By maximal regularity we mean that we shall consider functional spaces $E_0 \subset
\mathcal{S}'(\mathbb{R}^d)$ for u_0 and $F \subset \mathbb{R}_+ \times \mathbb{R}^d$ for f such that solutions to (11) fulfill

$$\|u\|_{L^\infty(\mathbb{R}_+;E_0)} + \|\partial_t u, \nabla^2 u, \nabla P\|_F \le C(\|u_0\|_{E_0} + \|f\|_F). \qquad (12)$$

The momentum equation of (1) will be seen as a perturbation of (11), which will
force us to assume the variations of density to be small.

In contrast, the last approach allows for large variations of density (even vacuum), and relies on completely different arguments: essentially energy estimates
and some modification of Gagliardo-Nirenberg inequalities. The price to pay is
that we need much more regularity for the velocity field than what is required
according to the scaling considerations. The gain is that one can consider general
densities that are just bounded and nonnegative.

In appendix, we recall definitions and results regarding Besov spaces and
Littlewood-Paley decomposition. We also give some details on Lagrangian coordinates, as they will play a central role for the three approaches, when proving
uniqueness.

As the different methods that we shall present will use the fact that the
viscosity is a constant, we shall always assume that $\mu = 1$ for notational simplicity.
This is of course not restrictive since one can change the unknown solution (ϱ, u, P)
to

$$(\widetilde{\varrho}, \widetilde{u}, \widetilde{P})(t,x) := (\varrho, \mu^{-1}u, \mu^{-2}P)(\mu^{-1}t, x).$$

Notation. Throughout the text, we shall often designate by C harmless positive
"constants" that may change from line to line. The notation $A \lesssim B$ means that
$A \le CB$ and $A \approx B$, that we have both $A \lesssim B$ and $B \lesssim A$.

For $z : \Omega \to \mathbb{R}^m$, we shall denote by $(Dz)_{1\le i,j\le m}$ the matrix of partial
(distributional) derivatives, and by ∇z, its transposed matrix.

For $p \in [1, +\infty]$ and $t \in (0, +\infty]$, we denote by $L^p_T(E)$ the space $L^p([0,T]; E)$
and by $\|\cdot\|_{L^p_T(E)}$, the corresponding norm. The index T is omitted if equal to
$+\infty$.

Finally, $\mathcal{C}_b([0,T]; E)$ designates the set of continuous functions on $[0,T]$ with
values in E, and, in the case where E is the dual of some Banach space, $\mathcal{C}_w([0,T]; E)$
designates the set of weakly $\star$ continuous functions from $[0,T]$ to E.

Acknowledgements. These notes have been written from a series of lectures given at Academy of Mathematics and Systems Science, CAS in Beijing, in October 2017. The author is grateful to Ping Zhang for supporting his invitation and suggesting him to write out this text.

1 First approach: critical regularity and endpoint maximal regularity

In this section and in the following one, the fluid domain Ω is $\mathbb{R}^d$, and we focus on small perturbations of the constant density state $\bar{\varrho} = 1$. Hence, setting $\varrho = 1 + a$ and restricting our attention to smooth enough solutions the system we have to solve recasts in

$$\begin{cases} \partial_t a + u \cdot \nabla a = 0, \\ (1+a)(\partial_t u + u \cdot \nabla u) - \mu \Delta u + \nabla P = 0, \\ \operatorname{div} u = 0, \end{cases} \tag{13}$$

supplemented with initial data a_0 and u_0 (with $\operatorname{div} u_0 = 0$) at time $t = 0$. As regards the boundary conditions, the functional framework we shall use will force the velocity to tend (weakly) to 0 at infinity.

The scaling invariance of System (13) is the same as for System (1), namely

$$a(t,x) \rightsquigarrow a(\lambda^2 t, \lambda x), \quad u(t,x) \rightsquigarrow \lambda u(\lambda^2 t, \lambda x), \quad P(t,x) \rightsquigarrow \lambda^2 P(\lambda^2 t, \lambda x), \quad \lambda > 0. \tag{14}$$

1.1 An abstract functional analysis approach

Let us start with general considerations that will help us to find appropriate functional frameworks for proving global existence (and, possibly, uniqueness) for small data.

Denote by A_0 and E_0, respectively, the functional spaces for a_0 and u_0. On the one hand, as a fulfills a transport equation, one can at most expect the initial regularity A_0 to be propagated (that is $a \in L^\infty_{loc}(\mathbb{R}_+; A_0)$). Furthermore, if the space A_0 contains some "regularity information", then the transport field u has to be at least with gradient in $L^1_{loc}(\mathbb{R}_+; L^\infty)$, and the generic estimate one can expect reads:

$$\|a(t)\|_{A_0} \leq \|a_0\|_{A_0} e^{C \int_0^t \|\nabla u\|_{L^\infty} \, d\tau}. \tag{15}$$

On the other hand, if we look at $(u, \nabla P)$ as the solution to the evolutionary Stokes system (11) with source term

$$f = -a \partial_t u - (1+a) u \cdot \nabla u \tag{16}$$

in some functional space F, and take u_0 in E_0, in such a way that the maximal regularity inequality (12) is fulfilled, then one can expect $(u, \nabla P)$ to be in the "solution space"

$$E := \left\{ (u, \nabla P), \, u \in L^\infty(\mathbb{R}_+; E_0) \quad \text{and} \quad \partial_t u, \nabla^2 u, \nabla P \in F \right\},$$

and that

$$\|(u, \nabla P)\|_E := \|u\|_{L^\infty(\mathbb{R}_+; E_0)} + \|\partial_t u, \nabla^2 u, \nabla P\|_F$$
$$\leq C(\|u_0\|_{E_0} + \|a\partial_t u + (1+a)u \cdot \nabla u\|_F).$$

In order to "close the estimate" in the case of a small solution, we need the continuity of the maps $(u, \nabla P) \mapsto u \cdot \nabla u$ and $(a, z) \mapsto az$ from E to F and from $L^\infty(\mathbb{R}_+; A_0) \times F$ to F, respectively, that is to say that for some $C > 0$, we have

$$\|u \cdot \nabla u\|_F \leq C\|(u, \nabla P)\|_E^2 \quad \text{and} \quad \|az\|_F \leq C\|a\|_{L^\infty(\mathbb{R}_+; A_0)}\|z\|_F. \quad (17)$$

The second inequality exactly means that $L^\infty(\mathbb{R}_+; A_0)$ is continuously embedded in the multiplier space[2] of F.

Now, assuming that we found spaces A_0, E_0, F and E fulfilling the above conditions, then we get the a priori estimate:

$$\|(u, \nabla P)\|_E \leq C(\|u_0\|_{E_0} + \|a\|_{L^\infty(\mathbb{R}_+; A_0)}\|\partial_t u\|_F + (1 + \|a\|_{L^\infty(\mathbb{R}_+; A_0)})\|(u, \nabla P)\|_E^2).$$

Therefore, if in addition we have[3]

$$\|a\|_{L^\infty(\mathbb{R}_+; A_0)} \leq C\|a_0\|_{A_0}, \quad (18)$$

then we conclude that

$$\|(u, \nabla P)\|_E \leq C(\|u_0\|_{E_0} + \|a_0\|_{A_0}\|\partial_t u\|_F + (1 + \|a_0\|_{A_0})\|(u, \nabla P)\|_E^2). \quad (19)$$

Given the definition of the norm in E, it is clear that if $\|u_0\|_{E_0}$ and $\|a_0\|_{A_0}$ are sufficiently small, then one can close the estimate for $(u, \nabla P)$.

Even though the above computations are formal, they give us a clear road map to finding spaces in which one can get global existence for (13). One can eventually expect the following generic statement:

Theorem 1.1. *Assume that the spaces A_0, E_0, F and E fulfill (12), (17) and (18). Then there exists a constant $c > 0$ such that if the data $a_0 \in A_0$ and $u_0 \in E_0$ with $\operatorname{div} u_0 = 0$ fulfill*

$$\|u_0\|_{E_0} + \|a_0\|_{A_0} \leq c \quad (20)$$

then (13) admits a global solution $(a, u, \nabla P)$ satisfying

$$\|a\|_{L^\infty(\mathbb{R}_+; A_0)} \leq C\|a\|_{A_0} \quad \text{and} \quad \|(u, \nabla P)\|_E \leq C\|u_0\|_{E_0}$$

for some absolute constant C.

---

[2]By definition, the multiplier space $\mathcal{M}(F)$ of F is the set of all distributions w satisfying $wz \in F$ for all $z \in F$. That set is endowed with the norm $\|w\|_{\mathcal{M}(F)} := \sup \|wz\|_F$, where the supremum is taken on all z in F with norm 1.

[3]For instance, that latter inequality is true if A_0 does not contain any regularity information (e.g. A_0 is a Lebesgue or a Lorentz space), or if (18) holds and $\|\nabla u\|_{L^1(\mathbb{R}_+; L^\infty)} \leq C\|(u, \nabla P)\|_E$ with a small enough right-hand side.

Inequality (19) suggests that, if restricting ourselves to spaces that have some scaling invariance, then that of $L^\infty(\mathbb{R}_+; A_0) \times E$ should be the one given by (14), and the spaces A_0 and E_0 are to be such that for all $\lambda > 0$, we have

$$\|a_0(\lambda \cdot)\|_{A_0} = \|a_0\|_{A_0} \quad \text{and} \quad \lambda \|u_0(\lambda \cdot)\|_{E_0} = \|u_0\|_{E_0}.$$

In what follows, such spaces will be named *critical for System* (13).

In the rest of the present section and in the next one, we shall give examples of functional spaces for which the above statement holds true and allows our considering density patches structures with sufficiently small discontinuities. We shall also exhibit supplementary conditions ensuring uniqueness and conservation of the regularity of density patches (the two questions being somehow connected).

1.2 Global well-posedness for slightly inhomogeneous incompressible viscous flows

Here we concentrate on the case where the space E_0 for the initial velocity u_0 is some homogeneous Besov space $\dot{B}^s_{p,r}(\mathbb{R}^d)$ (see the definition in Appendix). In this context, according to Proposition A.2, criticality means that $s = \frac{d}{p} - 1$. Furthermore, for reasons that will appear more clearly in a moment, we shall take $r = 1$. Then the corresponding spaces F and $E = E^p$ for the Stokes system turn out to be

$$F := L^1(\mathbb{R}_+; \dot{B}^{\frac{d}{p}-1}_{p,1}) \quad \text{and} \quad E^p := \left\{ (u, \nabla P) , \, u \in C_b(\mathbb{R}_+; \dot{B}^{\frac{d}{p}-1}_{p,1}) \right.$$

$$\left. \text{and} \, (\partial_t u, \nabla^2 u, \nabla P) \in F \right\}.$$

Indeed, as a consequence of Proposition 1.1 below, we will have if $(u, \nabla P)$ fulfills (11),

$$\|(u, \nabla P)\|_{E^p} \leq C \Big(\|u_0\|_{\dot{B}^{\frac{d}{p}-1}_{p,1}} + \|f\|_{L^1(\mathbb{R}_+; \dot{B}^{\frac{d}{p}-1}_{p,1})} \Big). \tag{21}$$

This will entail that ∇u is in $L^1(\mathbb{R}_+; \dot{B}^{\frac{d}{p}}_{p,1})$ and thus, because $\dot{B}^{\frac{d}{p}}_{p,1}$ is continuously embedded in the set C_b of continuous bounded functions, it will be possible to transport the regularity for a (in fact (18) will be satisfied). As regards A_0, we shall take the largest possible space, namely $\mathcal{M}(\dot{B}^{\frac{d}{p}-1}_{p,1}) \cap L^\infty$. Finally, as the space $\dot{B}^{\frac{d}{p}}_{p,1}$ is stable by product if $p < \infty$ (see e.g. [2], Chap. 2) and as $\operatorname{div} u = 0$, we have

$$\|u \cdot \nabla u\|_{\dot{B}^{\frac{d}{p}-1}_{p,1}} = \|\operatorname{div}(u \otimes u)\|_{\dot{B}^{\frac{d}{p}-1}_{p,1}} \leq C \|u \otimes u\|_{\dot{B}^{\frac{d}{p}}_{p,1}} \leq C \|u\|^2_{\dot{B}^{\frac{d}{p}}_{p,1}}, \tag{22}$$

and (17) is thus fulfilled. This motivates the following statement the proof of which will be carried out in the rest of the section.

Theorem 1.2. *Let* $1 \leq p < \infty$ *and* $d \geq 2$. *Let* a_0 *be in the multiplier space* $\mathcal{M}(\dot{B}_{p,1}^{\frac{d}{p}-1}(\mathbb{R}^d))$, *and let* u_0 *be in* $\dot{B}_{p,1}^{\frac{d}{p}-1}(\mathbb{R}^d)$ *with* $\operatorname{div} u_0 = 0$. *There exists a constant* c *depending only on* p *and* d *such that if*

$$\|a_0\|_{\mathcal{M}(\dot{B}_{p,1}^{\frac{d}{p}-1}) \cap L^\infty} + \|u_0\|_{\dot{B}_{p,1}^{\frac{d}{p}-1}} \leq c \qquad (23)$$

then (13) *has a global solution* $(a, u, \nabla P)$ *such that* $a \in L^\infty(\mathbb{R}_+; \mathcal{M}(\dot{B}_{p,1}^{\frac{d}{p}}) \cap L^\infty)$, $u \in L^\infty(\mathbb{R}_+; \dot{B}_{p,1}^{\frac{d}{p}-1})$ *and* $\nabla u \in L^1(\mathbb{R}_+; \dot{B}_{p,1}^{\frac{d}{p}})$.

If $1 \leq p < 2d$, *then* $(u, \nabla P)$ *is in* E^p, *the solution is unique, and there exists a constant* C *depending only on* p *and* d *such that*

$$\|a\|_{L^\infty(\mathcal{M}(\dot{B}_{p,1}^{\frac{d}{p}-1}) \cap L^\infty)} \leq C \|a_0\|_{\mathcal{M}(\dot{B}_{p,1}^{\frac{d}{p}-1}) \cap L^\infty} \quad and \quad \|(u, \nabla P)\|_{E^p} \leq C \|u_0\|_{\dot{B}_{p,1}^{\frac{d}{p}-1}}.$$

In order to prove that theorem, we shall first establish (21) and that the regularity $\mathcal{M}(\dot{B}_{p,1}^{\frac{d}{p}-1})$ is preserved by Lipschitz flows. Then, the existence part of the statement will be achieved by introducing a sequence of smooth approximate solutions and passing to the limit by means of compactness arguments. Proving uniqueness for $p \in [1, 2d)$ will require our rewriting System (13) in terms of Lagrangian coordinates.

1.3 Estimates in homogeneous Besov spaces for the heat flow

The final goal of this subsection is to prove (21). The result will be a consequence of a similar estimate for the following heat equation:

$$\begin{cases} \partial_t u - \Delta u = f \text{ in } \mathbb{R}_+ \times \mathbb{R}^d, \\ u|_{t=0} = u_0 \quad \text{on } \mathbb{R}^d. \end{cases} \qquad (24)$$

Let us focus on (24) for a while. Recall that whenever u_0 belongs to the Schwartz space $\mathcal{S}(\mathbb{R}^d)$ of tempered distributions and f is in $\mathcal{C}(\mathbb{R}_+; \mathcal{S}(\mathbb{R}^d))$, the (unique) solution to (24) in $\mathcal{C}^1(\mathbb{R}_+; \mathcal{S}(\mathbb{R}^d))$ is given by

$$u(t) = e^{t\Delta} u_0 + \int_0^t e^{(t-\tau)\Delta} f(\tau) \, d\tau, \qquad (25)$$

where the semi-group $(e^{t\Delta})_{t \geq 0}$ is defined on $\mathcal{S}$ by the relation

$$\mathcal{F}(e^{t\Delta} a)(\xi) \overset{\text{def}}{=} e^{-t|\xi|^2} \widehat{a}(\xi).$$

Being continuous on $\mathcal{S}$, operator $e^{t\Delta}$ has a transposed operator on $\mathcal{S}'$. As that latter operator coincides with $e^{t\Delta}$ on $\mathcal{S}$, we agree (slightly abusively) to denote it by $e^{t\Delta}$, too.

With that convention, using routine duality arguments, one may deduce that for general u_0 in $\mathcal{S}'$ and $f \in L^1_{loc}(\mathbb{R}_+; \mathcal{S}')$, relation (25) still provides the unique solution of (24) in $\mathcal{C}(\mathbb{R}_+; \mathcal{S}')$ *in the sense of distributions*, that is to say we have for any test function $\phi \in \mathcal{C}^1(\mathbb{R}_+; \mathcal{S})$ and $T \geq 0$:

$$\langle u(T), \phi(T) \rangle_{\mathcal{S}' \times \mathcal{S}} - \int_0^T \langle u, \partial_t \phi + \Delta\phi \rangle_{\mathcal{S}' \times \mathcal{S}} \, dt = \langle u_0, \phi(0) \rangle_{\mathcal{S}' \times \mathcal{S}} + \int_0^T \langle f, \phi \rangle_{\mathcal{S}' \times \mathcal{S}} \, dt.$$

We now want to address the maximal regularity issue for the heat equation in the homogeneous Besov spaces framework. The answer will come up as a straightforward consequence of the following lemma describing the action of the heat flow on spectrally localized functions.

Lemma 1.1. *There exist two positive constants c_0 and C such that for any $j \in \mathbb{Z}$, $p \in [1, \infty]$ and $\lambda \in \mathbb{R}_+$, we have for all $u \in \mathcal{S}'$ with $\dot{\Delta}_j u$ in L^p :*

$$\|e^{\lambda\Delta}\dot{\Delta}_j u\|_{L^p} \leq C e^{-c_0 \lambda 2^{2j}} \|\dot{\Delta}_j u\|_{L^p}.$$

Proof. Performing the rescaling $v(x) = u(2^{-j}x)$, we see that $\dot{\Delta}_0 v(2^j x) = \dot{\Delta}_j u(x)$ and $e^{2^{2j}\lambda\Delta}\dot{\Delta}_0 v(2^j x) = e^{\lambda\Delta}\dot{\Delta}_j u(x)$. Hence it suffices to consider the case $j = 0$.

Fix a function ϕ in $\mathcal{D}(\mathbb{R}^d \setminus \{0\})$ with value 1 on a neighborhood of the support of φ. We have

$$e^{\lambda\Delta}\dot{\Delta}_0 u = \mathcal{F}^{-1}\left(\phi e^{-\lambda|\cdot|^2}\mathcal{F}(\dot{\Delta}_0 u)\right)$$

$$= g_\lambda \star \dot{\Delta}_0 u \quad \text{with} \quad g_\lambda(x) \stackrel{\text{def}}{=} (2\pi)^{-d}\int e^{ix\cdot\xi}\phi(\xi)e^{-\lambda|\xi|^2}d\xi.$$

After integrating by parts, we get

$$g_\lambda(x) = (1 + |x|^2)^{-d}\int_{\mathbb{R}^d} e^{ix\cdot\xi}(\text{Id} - \Delta_\xi)^d\left(\phi(\xi)e^{-\lambda|\xi|^2}\right)d\xi.$$

Therefore, combining Leibniz and Faá-di-Bruno's formulae, we discover that

$$|g_\lambda(x))| \leq C(1 + |x|^2)^{-d}e^{-c_0\lambda},$$

which implies that

$$\|g_\lambda\|_{L^1} \leq C e^{-c_0\lambda}. \tag{26}$$

Then, the conclusion follows from the fact that the convolution product maps $L^1 \times L^p$ to L^p. □

As a consequence, we have the following result.

Proposition 1.1. *Let u_0 be in $\dot{B}_{p,1}^s$ and f be in $L_{loc}^1(\mathbb{R}_+; \dot{B}_{p,1}^s)$ for some $p \in [1, \infty]$ and $s \in \mathbb{R}$. Then the unique solution $u \in \mathcal{C}(\mathbb{R}_+; \mathcal{S}')$ of (24) belongs to $\mathcal{C}(\mathbb{R}_+; \dot{B}_{p,1}^s)$ (only to $\mathcal{C}_w(\mathbb{R}_+; \dot{B}_{p,1}^s)$ if $p = +\infty$) and satisfies for all $t > 0$:*

$$\|u(t)\|_{\dot{B}_{p,1}^s} + \int_0^t \|\nabla^2 u\|_{\dot{B}_{p,1}^s} \, d\tau \leq C\left(\|u_0\|_{\dot{B}_{p,1}^s} + \int_0^t \|f\|_{\dot{B}_{p,1}^s} \, d\tau\right). \tag{27}$$

Proof. If u satisfies (24) then for any $j \in \mathbb{Z}$,

$$\partial_t \dot{\Delta}_j u - \Delta \dot{\Delta}_j u = \dot{\Delta}_j f.$$

Hence, according to (25),

$$\dot{\Delta}_j u(t) = e^{t\Delta} \dot{\Delta}_j u_0 + \int_0^t e^{(t-\tau)\Delta} \dot{\Delta}_j f(\tau) \, d\tau.$$

Taking advantage of Lemma 1.1, we thus have

$$\|\dot{\Delta}_j u(t)\|_{L^p} \lesssim e^{-c_0 2^{2j} t} \|\dot{\Delta}_j u_0\|_{L^p} + \int_0^t e^{-c_0 2^{2j}(t-\tau)} \|\dot{\Delta}_j f(\tau)\|_{L^p} \, d\tau. \qquad (28)$$

Multiplying by 2^{js} and summing up over $j \in \mathbb{Z}$ yields

$$\sum_j 2^{js} \|\dot{\Delta}_j u(t)\|_{L^p} \lesssim \sum_j e^{-c_0 2^{2j} t} 2^{js} \|\dot{\Delta}_j u_0\|_{L^p} + \int_0^t e^{-c_0 2^{2j}(t-\tau)} 2^{js} \sum_j \|\dot{\Delta}_j f(\tau)\|_{L^p} \, d\tau$$

whence

$$\|u\|_{L_t^\infty(\dot{B}_{p,1}^s)} \lesssim \|u_0\|_{\dot{B}_{p,1}^s} + \|f\|_{L_t^1(\dot{B}_{p,1}^s)}.$$

Note that integrating (28) with respect to time yields

$$2^{2j} \|\dot{\Delta}_j u\|_{L_t^1(L^p)} \lesssim \left(1 - e^{-c_0 2^{2j} t}\right)\left(\|\dot{\Delta}_j u_0\|_{L^p} + \|\dot{\Delta}_j f\|_{L_t^1(L^p)}\right).$$

Therefore, multiplying by 2^{js}, using Bernstein inequality and summing up over j, we get

$$\|\nabla^2 u\|_{L_t^1(\dot{B}_{p,1}^s)} \lesssim \sum_j \left(1 - e^{-c_0 2^{2j} t}\right) 2^{js} \left(\|\dot{\Delta}_j u_0\|_{L^p} + \|\dot{\Delta}_j f\|_{L_t^1(L^p)}\right), \qquad (29)$$

which is even slightly better than what we wanted to prove, as $(1 - e^{-c 2^{2j} t})$ is bounded by 1 and tends to 0 when t goes to 0^+. $\qquad \square$

It is now easy to establish a similar result for the following Stokes system:

$$\begin{cases} \partial_t u - \Delta u + \nabla P = f & \text{in } (0,T) \times \mathbb{R}^d, \\ \operatorname{div} u = g & \text{in } (0,T) \times \mathbb{R}^d, \\ u|_{t=t_0} = u_0 & \text{on} \qquad \mathbb{R}^d. \end{cases} \qquad (30)$$

Proposition 1.2. *Let $u_0 \in \dot{B}_{p,1}^s$ and $f \in L^1(0,T; \dot{B}_{p,1}^s)$ with $p \in [1, \infty]$ and $s \in \mathbb{R}$. Let $g \in \mathcal{C}([0,T]; \dot{B}_{p,1}^{s-1})$ be such*

$$\nabla g \in L^1(0,T; \dot{B}_{p,1}^s(\mathbb{R}^d)) \quad \text{and} \quad \partial_t g = \operatorname{div} R_t \quad \text{with} \quad R_t \in L^1(0,T; \dot{B}_{p,1}^s)$$

and assume that the compatibility condition $g|_{t=0} = \operatorname{div} u_0$ on $\mathbb{R}^d$ is satisfied.

Then System (30) *has a unique solution* $(u, \nabla P)$ *with*

$$u \in \mathcal{C}([0, T); \dot{B}^s_{p,1}) \quad and \quad \partial_t u, \nabla^2 u, \nabla P \in L^1(0, T; \dot{B}^s_{p,1})$$

and the following estimate is valid:

$$\|u\|_{L^\infty(0,T;\dot{B}^s_{p,1})} + \|\partial_t u, \nabla^2 u, \nabla P\|_{L^1(0,T;\dot{B}^s_{p,1})} \le C\big(\|f, \nabla g, R_t\|_{L^1(0,T;\dot{B}^s_{p,1})} + \|u_0\|_{\dot{B}^s_{p,1}}\big).$$

Proof. It is only a matter of computing the pressure term, then solving the heat equation for u in terms of the data and of ∇P. To determine the pressure, we set

$$\nabla P := \mathcal{Q}f + \nabla g - \mathcal{Q}R_t \quad with \quad \mathcal{Q} := -(-\Delta)^{-1}\nabla\mathrm{div}.$$

Then, as $\mathcal{Q}$ maps $\dot{B}^s_{p,1}$ to itself (being a homogeneous Fourier multiplier of degree 0, see (160)), we get

$$\|\nabla P\|_{L^1(0,T;\dot{B}^s_{p,1})} \le C\|f, \nabla g, R_t\|_{L^1(0,T;\dot{B}^s_{p,1})}.$$

Next, we define u according to Proposition 1.1 to be the unique solution in $\mathcal{C}(\mathbb{R}_+; \mathcal{S}')$ of

$$\partial_t u - \Delta u = f - \nabla P, \qquad u|_{t=0} = u_0.$$

Then Proposition 1.1 guarantees that u fulfills the desired inequality.

Finally, we observe that, thanks to the definition of ∇P, we have

$$\partial_t \mathrm{div}\, u - \Delta \mathrm{div}\, u = \mathrm{div}\, f - \Delta P = \partial_t g - \Delta \mathrm{div}\, u.$$

Hence $\partial_t(\mathrm{div}\, u - g) \equiv 0$ and as $\mathrm{div}\, u_0 = g(0)$, we get $\mathrm{div}\, u - g \equiv 0$. □

1.4   Estimates for the transport equation

Our next task is to prove that if $u \in \mathcal{C}(\mathbb{R}_+; \dot{B}^{\frac{d}{p}-1}_{p,1})$ satisfies $\nabla u \in L^1(\mathbb{R}_+; \dot{B}^{\frac{d}{p}}_{p,1})$ and $\mathrm{div}\, u = 0$ then the regularity $\mathcal{M}(\dot{B}^{\frac{d}{p}-1}_{p,1})$ of the density is conserved during the evolution. So we consider the transport equation

$$\partial_t a + v \cdot \nabla a = 0, \qquad a|_{t=0} = a_0 \tag{31}$$

where v is a given time-dependent divergence free vector-field, and a_0 belongs to the multiplier space $\mathcal{M}(\dot{B}^s_{p,r})$.

It is well-known that under suitable regularity assumptions, (31) has a unique solution that may be expressed in terms of a_0 and of the flow of v. The existence of that flow is justified by the following version of the Cauchy-Lipschitz theorem.

Theorem 1.3. *Let Ω be an open subset of a Banach space E, and I be an open interval of $\mathbb{R}$. Assume that $v : I \times \Omega \to E$ is measurable and satisfies for all x and y in Ω:*

- *The map $t \mapsto \|v(t, x)\|_E$ is in $L^1_{loc}(I)$;*

- *There exists $L \in L^1_{loc}(I)$ so that $\|v(t,x) - v(t,y)\|_E \leq L(t)\|x - y\|_E$.*

Then, for any $t_0 \in I$ and $y_0 \in \Omega$ there exists a unique function $X_{y_0} \in \mathcal{C}(I;\Omega)$ satisfying

$$X_{y_0}(t) = y_0 + \int_{t_0}^{t} v(\tau, X_{y_0}(\tau))\, d\tau \quad \text{for all } t \in I. \tag{32}$$

Moreover, for all $t \in I$, the application $y \mapsto X_y(t)$ is a bi-Lipschitz diffeomorphism of Ω.

Now, assuming that the vector-field v in (31) satisfies the conditions of the above theorem with $\Omega = E = \mathbb{R}^d$ and denoting $X(t,y) := X_y(t)$, we have

$$\frac{d}{dt}(a \circ X) = (\partial_t a + v \cdot \nabla a) \circ X = 0, \qquad a \circ X|_{t=0} = a_0.$$

Hence the explicit solution of (31) is given by

$$a(t,x) = a_0(X^{-1}(t,x)). \tag{33}$$

That div $v = 0$ ensures that $X(t,\cdot)$ is measure preserving. Therefore estimating the multiplier norm of $a(t,\cdot)$ reduces to the study of the action of a Lipschitz measure preserving change of variables on homogeneous Besov norms. This is the aim of the following lemma.

Lemma 1.2. *For all $s \in (-1,1)$ and $1 \leq p, r \leq +\infty$, for all bi-Lipschitz measure preserving diffeomorphism ψ and function b in $\dot{B}^s_{p,r}$, we have $b \circ \psi$ in $\dot{B}^s_{p,r}$ and[4], for some constant C depending only on the regularity parameters,*

$$\|b \circ \psi\|_{\dot{B}^s_{p,r}} \leq C \|\nabla \psi^{\operatorname{sgn} s}\|_{L^\infty}^{|s|} \|b\|_{\dot{B}^s_{p,r}}, \quad \text{if } s \neq 0,$$

$$\|b \circ \psi\|_{\dot{B}^0_{p,r}} \leq C \sqrt{\|\nabla \psi\|_{L^\infty} \|\nabla \psi^{-1}\|_{L^\infty}} \, \|b\|_{\dot{B}^0_{p,r}}.$$

Proof. It is based on the characterization of Besov spaces by interpolation, which is recalled in Proposition A.1, and on similar inequalities for the norms in $\dot{W}^{-1,p}$, L^p and $\dot{W}^{1,p}$. Indeed, as ψ is measure preserving, the map $b \mapsto b \circ \psi$ is continuous on L^p and satisfies

$$\|b \circ \psi\|_{L^p} = \|b\|_{L^p}. \tag{34}$$

Next, as the chain rule implies $D(b \circ \psi) = (Db \circ \psi) \cdot D\psi$, we have

$$\|b \circ \psi\|_{\dot{W}^{1,p}} \leq \|D\psi\|_{L^\infty} \|b\|_{\dot{W}^{1,p}}. \tag{35}$$

Finally, we have

$$\|b \circ \psi\|_{\dot{W}^{-1,p}} = \sup_{\|\nabla w\|_{L^{p'}}=1} \int w(x)\, b(\psi(x))\, dx$$

$$= \sup_{\|\nabla w\|_{L^{p'}}=1} \int w(\psi^{-1}(y))\, b(y)\, dy$$

$$\leq \|b\|_{\dot{W}^{-1,p}} \sup_{\|\nabla w\|_{L^{p'}}=1} \|\nabla(w \circ \psi^{-1})\|_{L^{p'}}.$$

---

[4]The square root may be replaced with a log, see [37].

Therefore,

$$\|b \circ \psi\|_{\dot{W}^{-1,p}} \le \|b\|_{\dot{W}^{-1,p}} \|\nabla \psi^{-1}\|_{L^\infty}. \tag{36}$$

Then, proving the lemma for $s \in (0,1)$ stems from (34) and (35), and the definition of the norm in $\dot{B}^s_{p,r}$ by interpolation: let us introduce the notation

$$K(t,b) := \inf_{a+c=b} \left(\|a\|_{L^p} + t\|\nabla c\|_{L^p} \right).$$

We have

$$\|b \circ \psi\|_{\dot{B}^s_{p,r}} \approx \left(\int_{\mathbb{R}_+} \left(t^{-s} K(t, b \circ \psi) \right)^r \frac{dt}{t} \right)^{\frac{1}{r}}.$$

As $a + c = b$ if and only if $a \circ \psi + c \circ \psi = b \circ \psi$, Inequalities (34) and (35) imply that

$$K(t, b \circ \psi) \le K(t\|\nabla\psi\|_{L^\infty}^{-1}, b),$$

whence

$$\|b \circ \psi\|_{\dot{B}^s_{p,r}} \le C\|\nabla\psi\|_{L^\infty}^{s} \|b\|_{\dot{B}^s_{p,r}}. \tag{37}$$

Proving the result for $s \in (-1,0)$ relies on similar arguments (just replace (35) with (36)). We end up with

$$\|b \circ \psi\|_{\dot{B}^s_{p,r}} \le C\|\nabla\psi^{-1}\|_{L^\infty}^{-s} \|b\|_{\dot{B}^s_{p,r}}. \tag{38}$$

Finally, for the case $s = 0$, one can interpolate (with parameters $1/2$ and r) between (35) and (36). This is just a slight adaptation of Proposition A.1. $\square$

Granted with the above lemma, it is now easy to prove a priori estimates in multiplier spaces for the transport equation (31). We have:

Proposition 1.3. *Let $s \in (-1,1)$ and $1 \le p, r \le +\infty$, and let a be a solution to (31). Then, for all $t \ge 0$, we have*

$$\|a(t)\|_{\mathcal{M}(\dot{B}^s_{p,r})} \le Ce^{2 \int_0^t \|\nabla v\|_{L^\infty} d\tau} \|a_0\|_{\mathcal{M}(\dot{B}^s_{p,r})}. \tag{39}$$

Proof. It suffices to estimate the norm of $a(t)b$ in $\dot{B}^s_{p,r}$ for any b in $\dot{B}^s_{p,r}$. Now, taking advantage of (33), we see that

$$\|a(t)\,b\|_{\dot{B}^s_{p,r}} = \|(a_0\, b \circ X_t) \circ X_t^{-1}\|_{\dot{B}^s_{p,r}}.$$

It is well-known that

$$\|\nabla X_t^{\pm 1}\|_{L^\infty} \le \exp\left(\int_0^t \|\nabla v\|_{L^\infty}\, dt \right).$$

Therefore we have, thanks to Lemma 1.2 and to the definition of multiplier spaces,

$$\|(a_0 \, b \circ X_t) \circ X_t^{-1}\|_{\dot{B}^s_{p,r}} \le C\|(a_0 \, b \circ X_t)\|_{\dot{B}^s_{p,r}} \exp\left(\int_0^t \|\nabla v\|_{L^\infty} \, dt\right)$$

$$\le C\|a_0\|_{\mathcal{M}(\dot{B}^s_{p,r})} \|b \circ X_t\|_{\dot{B}^s_{p,r}} \exp\left(\int_0^t \|\nabla v\|_{L^\infty} \, dt\right)$$

$$\le C\|a_0\|_{\mathcal{M}(\dot{B}^s_{p,r})} \|b\|_{\dot{B}^s_{p,r}} \exp\left(2\int_0^t \|\nabla v\|_{L^\infty} \, dt\right),$$

which completes the proof. □

Remark 1.1. *Proposition 1.3 may be extended to any index s in $(-d/p', d/p)$ (and even $s = d/p$ if $r = 1$) if we assume that v is in $L^1_{loc}(\mathbb{R}_+; \dot{B}^{\frac{d}{p}}_{p,1})$ and replace $\|\nabla v\|_{L^\infty}$ by $\|\nabla v\|_{\dot{B}^{\frac{d}{p}}_{p,1}}$ in (39). The reader will find more details in* [12].

1.5 The existence scheme

We smooth out the initial velocity: $u_0^n \overset{\text{def}}{=} \dot{S}_n u_0$ (see the Appendix for the definition of the low frequency cut-off $\dot{S}_n$), then solve the following system:

$$\begin{cases} \partial_t a^n = -\text{div}\,(u^n a^n), \\ \partial_t u^n - \Delta u^n + \nabla P^n = -\dot{S}_n\big(\partial_t(a^n u^n) + \text{div}\,((1+a^n)u^n \otimes u^n)\big), \\ \text{div}\,u^n = 0 \end{cases} \quad (40)$$

supplemented with initial data $a^n|_{t=0} = a_0$ and $u^n|_{t=0} = u_0^n$.

For fixed $n \in \mathbb{N}$, solving the above system may be done iteratively from the linear Stokes system: consider the sequence $(a_q^n, u_q^n, P_q^n)_{q \in \mathbb{N}}$ given by $(a_0^n, u_0^n, P_0^n) := (a_0, \dot{S}_n u_0, 0)$ and

$$\begin{cases} \partial_t u_q^n - \Delta u_q^n + \nabla P_q^n = -\dot{S}_n\big(\partial_t(a_{q-1}^n u_{q-1}^n) + \text{div}\,((1+a_{q-1}^n)u_{q-1}^n \otimes u_{q-1}^n)\big), \\ \text{div}\,u_q^n = 0, \\ u_q^n|_{t=0} = \dot{S}_n u_0, \end{cases}$$

and $a_q^n := a_0 \circ (X_q^n(t, \cdot))^{-1}$, with X_q^n being the flow of u_q^n (that is globally defined provided we have $\nabla u_q^n \in L^1_{loc}(\mathbb{R}_+; L^\infty)$). Note that the cut-off operator $\dot{S}_n$ ensures that $(u_q^n, \nabla P_q^n)$ is smooth. Of course, all approximate solutions $(a_q^n, u_q^n, P_q^n)_{q \in \mathbb{N}}$ are global in time (since the above system is linear), and using the same type of estimates as those that will be described below for $(a^n, u^n, \nabla P^n)$, one can get bounds independent of q and n in the space $L^\infty(\mathbb{R}_+; L^\infty \cap \mathcal{M}(\dot{B}^{\frac{d}{p}-1}_{p,1}))$ for a^n, and in E^p for $(u^n, \nabla P^n)$.

As regards the convergence of the sequence $(a_q^n, u_q^n, \nabla P_q^n)_{q \in \mathbb{N}}$, the sole difficulty is for proving contraction because a^n is only bounded. However one may resort to estimates in the space $L_{loc}^\infty(\mathbb{R}_+; \dot{W}^{-1,\infty})$ for the transport equation. Indeed, if $b(0) = 0$ then for all $t \geq 0$, we have

$$\|b(t)\|_{W^{-1,\infty}} \leq \int_0^t e^{\int_\tau^t \|\nabla u\|_{L^\infty} d\tau'} \|(\partial_t b + \mathrm{div}\,(bv))(\tau)\|_{W^{-1,\infty}}\, d\tau.$$

That latter estimate may be achieved by duality from the standard $W^{1,1}$ estimate for the transport equation.

Finally, one may conclude that for all $n \in \mathbb{N}$, System (40) has a global solution with $(u^n, \nabla P^n)$ smooth and $a^n = a_0 \circ (X_t^n)^{-1}$, where X^n stands for the flow of u^n.

Let us next turn to the proof of uniform estimates for $(a^n, u^n, \nabla P^n)$. As regards a^n, it is obvious that

$$\|a^n(t)\|_{L^\infty} = \|a_0\|_{L^\infty} \quad \text{for all } t \geq 0, \tag{41}$$

and Proposition 1.3 immediately implies that, in the case $p > d/2$,

$$\|a^n(t)\|_{L^\infty(\mathcal{M}(\dot{B}_{p,1}^{\frac{d}{p}-1}))} \leq C e^{\int_0^t \|\nabla u^n\|_{L^\infty} d\tau} \|a_0\|_{\mathcal{M}(\dot{B}_{p,1}^{\frac{d}{p}-1})} \quad \text{for all } t \geq 0$$

which, according to the embedding $\dot{B}_{p,1}^{\frac{d}{p}} \hookrightarrow L^\infty$ implies that[5]

$$\|a^n\|_{L^\infty(\mathcal{M}(\dot{B}_{p,1}^{\frac{d}{p}-1}))} \leq C e^{C \int_0^{+\infty} \|\nabla u^n\|_{\dot{B}_{p,1}^{\frac{d}{p}}} d\tau} \|a_0\|_{\mathcal{M}(\dot{B}_{p,1}^{\frac{d}{p}-1})}. \tag{42}$$

Of course, owing to the definition of $\dot{S}_n$, we have

$$\forall n \in \mathbb{N}, \ \|u_0^n\|_{\dot{B}_{p,1}^{\frac{d}{p}-1}} \leq \|\mathcal{F}^{-1}\chi\|_{L^1} \|u_0\|_{\dot{B}_{p,1}^{\frac{d}{p}-1}}. \tag{43}$$

Hence, using Proposition 1.2 with $s = d/p - 1$, we discover that

$$\|(u^n, \nabla P^n)\|_{E^p} \leq C\left(\|u_0\|_{\dot{B}_{p,1}^{\frac{d}{p}-1}} + \|\partial_t(a^n u^n) + \mathrm{div}\,((1+a^n)u^n \otimes u^n)\|_{L^1(\dot{B}_{p,1}^{\frac{d}{p}-1})}\right). \tag{44}$$

As the function u^n is smooth and divergence free, one can use the fact that

$$\partial_t(a^n u^n) + \mathrm{div}\,((1+a^n)u^n \otimes u^n) = a^n \partial_t u^n + (1+a^n)\mathrm{div}\,(u^n \otimes u^n).$$

Therefore, using the definition of $\mathcal{M}(\dot{B}_{p,1}^{\frac{d}{p}-1})$ and the fact that the space $\dot{B}_{p,1}^{\frac{d}{p}}$ is stable by product of functions if $p < \infty$, we get that

$$\|(u^n, \nabla P^n)\|_{E^p} \leq C\Big(\|u_0\|_{\dot{B}_{p,1}^{\frac{d}{p}-1}} + \|a^n\|_{L^\infty(\mathcal{M}(\dot{B}_{p,1}^{\frac{d}{p}-1}))} \|\partial_t u^n\|_{L^1(\dot{B}_{p,1}^{\frac{d}{p}-1})}$$
$$+ \big(1 + \|a^n\|_{L^\infty(\mathcal{M}(\dot{B}_{p,1}^{\frac{d}{p}-1}))}\big) \|u^n\|_{L^2(\dot{B}_{p,1}^{\frac{d}{p}})}^2 \Big).$$

---

[5]That latter inequality is valid for any $1 \leq p < \infty$, see Remark 1.1.

Note that, arguing by interpolation, we have

$$\|u^n\|^2_{L^2(\dot{B}_{p,1}^{\frac{d}{p}})} \leq C\|u^n\|_{L^\infty(\dot{B}_{p,1}^{\frac{d}{p}-1})} \|\nabla^2 u^n\|_{L^1(\dot{B}_{p,1}^{\frac{d}{p}-1})}.$$

Hence, if we assume that both $\|u_0\|_{\dot{B}_{p,1}^{\frac{d}{p}-1}}$ and $\|a_0\|_{\mathcal{M}(\dot{B}_{p,1}^{\frac{d}{p}-1})}$ are small enough then we have for all $n \in \mathbb{N}$, taking C larger as the case may be,

$$\|a^n\|_{L^\infty(\mathcal{M}(\dot{B}_{p,1}^{\frac{d}{p}-1}))} \leq C\|a_0\|_{\mathcal{M}(\dot{B}_{p,1}^{\frac{d}{p}-1})} \quad \text{and} \quad \|(u^n, \nabla P^n)\|_{E^p} \leq C\|u_0\|_{\dot{B}_{p,1}^{\frac{d}{p}-1}}. \quad (45)$$

First step: Compactness and convergence of a subsequence. The bounds that have been proved so far, combined with classical functional analysis arguments (the compactness of the unit ball of Besov spaces for the weak $\star$ topology) already ensure that, up to extraction,

$$a^n \rightharpoonup a \ \text{ in } L^\infty(\mathbb{R}_+ \times \mathbb{R}^d) \quad \text{weak} \star \quad \text{and} \quad u^n \rightharpoonup u \ \text{ in } L^\infty(\mathbb{R}_+; \dot{B}_{p,1}^{\frac{d}{p}-1}) \quad \text{weak} \star, \quad (46)$$

together with other similar properties related to the definition of E^p. For example, we also have for all $r > 1$,

$$u \in L^r(\mathbb{R}_+; \dot{B}_{p,1}^{\frac{d}{p}-1+\frac{2}{r}}) \quad \text{and} \quad u^n \rightharpoonup u \ \text{ in } L^r(\mathbb{R}_+; \dot{B}_{p,1}^{\frac{d}{p}-1+\frac{2}{r}}) \quad \text{weak} \star. \quad (47)$$

This already enables us to pass to the limit in the left-hand side of the equations in (40). To pass to the limit in the other terms, we need to exhibit strong convergence properties. This will be achieved by compactness arguments.

We know at the same time that $(u^n)_{n\in\mathbb{N}}$ and $(\partial_t u^n)_{n\in\mathbb{N}}$ are bounded in $L^\infty(\mathbb{R}_+; \dot{B}_{p,1}^{\frac{d}{p}-1})$ and $L^1(\mathbb{R}_+; \dot{B}_{p,1}^{\frac{d}{p}-1})$, respectively. Therefore, for all $T > 0$, the sequence $(u^n)_{n\in\mathbb{N}}$ is bounded in $W^{1,1}(0,T; \dot{B}_{p,1}^{\frac{d}{p}-1})$.

Taking advantage of compact embedding, one can thus conclude (after using Cantor diagonal extraction process), that, up to extraction, for all $T > 0$ and ϕ in $\mathcal{C}_c^\infty(\mathbb{R}^d)$,

$$\phi u^n \longrightarrow \phi u \ \text{ in } L^2(0,T; B_{p,1}^{\frac{d}{p}-2}).$$

Interpolating with the uniform bounds allows to upgrade the convergence to $L^2(0,T; B_{p,1}^{\frac{d}{p}-\varepsilon})$ for all $\varepsilon > 0$, which allows to justify that, in the distributional meaning, we have

$$a^n u^n \to au, \quad u^n \otimes u^n \to u \otimes u \quad \text{and} \quad a^n u^n \otimes u^n \to au \otimes u,$$

and one can thus conclude that $\rho := 1 + a$ and u fulfill System (1) in the sense (4), (5), (6). Note that here the only limitation to pass to the limit is to have $\frac{d}{p} - \varepsilon > 0$ for some small enough $\varepsilon > 0$, that is to say, p is finite.

Second step: Upgrading the regularity of the solution. In order to complete the proof of the existence part in the general case $1 \leq p < \infty$, we still have to justify that a is in $L^\infty(\mathbb{R}_+; \mathcal{M}(\dot{B}_{p,1}^{\frac{d}{p}-1}))$ and that ∇u is in $L^1(\mathbb{R}_+; \dot{B}_{p,1}^{\frac{d}{p}})$. Indeed, whether the uniform bound in (45) ensures that, for all $t \in \mathbb{R}_+$, $a(t)$ is in the multiplier space $\mathcal{M}(\dot{B}_{p,1}^{\frac{d}{p}-1})$ is unclear, and, at this stage, one can only assert that $\nabla^2 u$, ∇P and $\partial_t u$ are bounded measures on $\mathbb{R}^+$, with values in $\dot{B}_{p,1}^{\frac{d}{p}-1}$.

In order to show that ∇u is in $L^1(\mathbb{R}_+; \dot{B}_{p,1}^{\frac{d}{p}})$, one can use the characterization of Besov norms by duality: remember (see [2, Prop. 2.29] and [36, 4.8.2]) that for all z in $\mathcal{S}_h'(\mathbb{R}^d)$, we have:

$$\|z\|_{\dot{B}_{p,1}^{\frac{d}{p}}} \approx \sup_{\theta} \langle z, \theta \rangle$$

where the supremum is taken on all θ in $\mathcal{S}(\mathbb{R}^d)$ with $\|\theta\|_{\dot{B}_{p',\infty}^{-\frac{d}{p'}}} \leq 1$. Consequently, we may write for all $j \in \{1, \ldots, d\}$,

$$\int_{\mathbb{R}_+} \|\partial_j u\|_{\dot{B}_{p,1}^{\frac{d}{p}}} \approx \sup_{\phi} \int_{\mathbb{R}_+} \int_{\mathbb{R}^d} \partial_j u \, \phi \, dx \, dt, \tag{48}$$

where the supremum is taken on all ϕ in $\mathcal{C}_c(\mathbb{R}_+; \mathcal{S}(\mathbb{R}^d))$ with $\|\phi\|_{L^\infty(\mathbb{R}_+; \dot{B}_{p',\infty}^{-\frac{d}{p'}})} \leq 1$.

So, let us fix such a function ϕ and observe that:

$$\int_{\mathbb{R}_+} \int_{\mathbb{R}^d} \phi \, \partial_j u \, dx = \int_{\mathbb{R}_+} \int_{\mathbb{R}^d} \phi \, \partial_j u^n \, dx + \int_{\mathbb{R}_+} \int_{\mathbb{R}^d} \phi \, \partial_j (u - u^n) \, dx. \tag{49}$$

On the one hand, according to (45) and (48), we have

$$\left| \int_{\mathbb{R}_+} \int_{\mathbb{R}^d} \phi \, \partial_j u^n \, dx \right| \leq C \|u_0\|_{\dot{B}_{p,1}^{\frac{d}{p}-1}}.$$

On the other hand, Property (47) ensures that the last term of (49) converges to 0 when n goes to infinity. Hence, we have for all ϕ in $\mathcal{C}_c(\mathbb{R}_+; \mathcal{S}(\mathbb{R}^d))$ with $\|\phi\|_{L^\infty(\mathbb{R}_+; \dot{B}_{p',\infty}^{-\frac{d}{p'}})} \leq 1$, the inequality:

$$\left| \int_{\mathbb{R}_+} \int_{\mathbb{R}^d} \phi \, \partial_j u \, dx \right| \leq C \|u_0\|_{\dot{B}_{p,1}^{\frac{d}{p}-1}},$$

and (48) allows to conclude that ∇u is in $L^1(\mathbb{R}_+; \dot{B}_{p,1}^{\frac{d}{p}})$, with norm bounded by $C\|u_0\|_{\dot{B}_{p,1}^{\frac{d}{p}-1}}$.

It is now easy to recover the regularity of the density. Indeed, as a is just transported by u, Proposition 1.3 (or Remark 1.1), the above bound for ∇u in $L^1(\mathbb{R}_+; \dot{B}_{p,1}^{\frac{d}{p}})$ and the fact that u_0 is small allows to conclude that a is in $L^\infty(\mathbb{R}_+; \mathcal{M}(\dot{B}_{p,1}^{\frac{d}{p}-1}))$ and fulfills

$$\|a\|_{L^\infty(\mathcal{M}(\dot{B}_{p,1}^{\frac{d}{p}-1}))} \leq C \|a_0\|_{\mathcal{M}(\dot{B}_{p,1}^{\frac{d}{p}-1})}.$$

We know how to prove that ∇P and $\partial_t u$ are in $L^1(\mathbb{R}_+; \dot{B}_{p,1}^{\frac{d}{p}-1})$ only in the case $1 \le p < 2d$. Then, the properties that we proved hitherto allow to justify that

$$\partial_t(au) + \mathrm{div}\,(au \otimes u) = a(\partial_t u + u \cdot \nabla u),$$

which implies that ∇P is a solution to

$$\mathrm{div}\left(\frac{\nabla P}{1+a}\right) = -\mathrm{div}\,(u \cdot \nabla u).$$

As the right-hand side is in $L^1(\mathbb{R}_+; \dot{B}_{p,1}^{\frac{d}{p}-1})$ and a is in $L^\infty(\mathbb{R}_+; \mathcal{M}(\dot{B}_{p,1}^{\frac{d}{p}-1}))$, the only bounded measure with values in $\dot{B}_{p,1}^{\frac{d}{p}-1}$ solution of the above equation is in $L^1(\mathbb{R}_+; \dot{B}_{p,1}^{\frac{d}{p}-1})$ (indeed the only solution going to 0 at infinity of the corresponding homogeneous elliptic equation is zero since $1+a$ is bounded and bounded away from 0). Finally, rewriting the momentum equation as

$$\partial_t u - \Delta u + \nabla P = \frac{a}{1+a}\left(\nabla P - \Delta u\right) - u \cdot \nabla u,$$

observing that the right-hand side is in $L^1(\mathbb{R}_+; \dot{B}_{p,1}^{\frac{d}{p}-1})$ and using Proposition 1.2 guarantees that u is in $\mathcal{C}_b(\mathbb{R}_+; \dot{B}_{p,1}^{\frac{d}{p}-1})$ and $\partial_t u$ is in $L^1(\mathbb{R}_+; \dot{B}_{p,1}^{\frac{d}{p}-1})$. This completes the proof of the existence part of Theorem 1.2.

1.6 Lagrangian coordinates and uniqueness

Throughout this part, we assume that $1 \le p < 2d$. Consider two solutions $(a^1, u^1, \nabla P^1)$ and $(a^2, u^2, \nabla P^2)$ of (13) with the regularity given by Theorem 1.2 and emanating from the same initial data. The usual way of proving uniqueness is to derive a priori estimates in the spirit of the proof of existence, for $\delta\!a \stackrel{\mathrm{def}}{=} a^2 - a^1$, $\delta\!u \stackrel{\mathrm{def}}{=} u^2 - u^1$ and $\delta\!P \stackrel{\mathrm{def}}{=} P^2 - P^1$, through the system satisfied by $(\delta\!a, \delta\!u, \nabla \delta\!P)$, namely

$$\begin{cases} \partial_t \delta\!a + u^1 \cdot \nabla \delta\!a = -\delta\!u \cdot \nabla a^2, \\ \partial_t \delta\!u - \Delta \delta\!u + \nabla \delta\!P = -a^1 \partial_t \delta\!u + \delta\!a\,(\partial_t u^2 + u^2 \cdot \nabla u^2) + (1+a^1)(u^1 \cdot \nabla \delta\!u + \delta\!u \cdot \nabla u^2), \\ \mathrm{div}\,\delta\!u = 0. \end{cases}$$

The unpleasant fact that occurs in PDE's of hyperbolic type (and also in the hyperbolic-parabolic case we deal with) is that the hyperbolic part of the system causes a loss of one derivative in the stability estimates: here $\delta\!u \cdot \nabla a^2$ cannot be more regular than ∇a^2. In our very low regularity setting, this loss is fatal since it implies that (at most) $\delta\!a$ can be estimated in the space of derivatives of bounded functions. Hence, one can hardly give a meaning to e.g. the term $\delta\!a\,\partial_t u^2$ seen as the product of a derivative of a bounded function and of a function in $\dot{B}_{p,1}^{\frac{d}{p}-1}$.

To overcome the difficulty, we shall recast System (1) *in Lagrangian coordinates*. To this end, we consider the flow X of u defined in (32). Remember that

∇u is in $L^1(\mathbb{R}_+; \dot{B}^{\frac{d}{p}}_{p,1})$ and that $\dot{B}^{\frac{d}{p}}_{p,1}$ is embedded in $\mathcal{C}_b$. Hence, the flow is indeed defined in a unique way for all time, and the map $X(t, \cdot)$ is a $\mathcal{C}^1$ diffeomorphism of $\mathbb{R}^d$. We are thus allowed to change the *Eulerian variable* x into the *Lagrangian variable* $y = X^{-1}(t, x)$ and the unknowns (a, u, P) of System (13) become

$$\bar{\rho}(t, y) := \rho(t, X(t, y)), \quad \bar{u}(t, y) := u(t, X(t, y)) \quad \text{and} \quad \bar{P}(t, y) := P(t, X(t, y)).$$

In Lagrangian coordinates, the convective derivatives $\partial_t + u \cdot \nabla_x$ is changed into a mere time derivative. Hence any function of the density becomes time-independent. This means that the only evolution equation that has to be considered is the one for $(\bar{u}, \bar{P})$. To find it, one can of course start from the chain rule which implies that for any function $F : [0, T] \times \mathbb{R}^d \to \mathbb{R}^d$ differentiable with respect to the space variable, we have, denoting $\bar{F}(t, y) := F(t, X(t, y))$,

$$D_y \bar{F}(t, y) = D_x F(t, x) \cdot D_y X(t, y) \quad \text{with} \quad y = X^{-1}(t, x).$$

Hence,

$$\nabla_x F(t, x) = {}^T A(t, y) \cdot \nabla_y \bar{F}(t, y) \quad \text{with} \quad A(t, y) := (DX(t, y))^{-1}.$$

Now, two important remarks are in order. First, one may rewrite X and A in terms of $\bar{u}$:

$$X(t, y) = y + \int_0^t \bar{u}(\tau, y) \, d\tau \quad \text{and thus} \quad A(t, y) = \left(\text{Id} + \int_0^t D_y \bar{u}(\tau, y) \, d\tau \right)^{-1}.$$
$$\tag{50}$$

Second, because the flow is measure preserving, one may rewrite the divergence operator as follows (see Lemma B.1):

$$\text{div}_x u = A : \nabla_y \bar{u} = \text{div}_y (A\bar{u}). \tag{51}$$

Consequently, in Lagrangian coordinates, System (13) becomes

$$\begin{cases} \rho_0 \partial_t \bar{u} - \text{div} \left(A^T A \nabla_y \bar{u} \right) + {}^T A \cdot \nabla_y \bar{P} = 0, \\ \text{div}_y (A\bar{u}) = 0, \end{cases} \tag{52}$$

with A defined in (50) and where ρ_0 stands for the initial density.

Conversely, one may prove that under our assumptions, if solving (52) in the space E^p then, provided

$$\int_0^T \|\nabla \bar{u}\|_{\dot{B}^{\frac{d}{p}}_{p,1}} \, d\tau \le c \tag{53}$$

where $c = c(d, p)$ stands for the constant in Lemma B.2, the map X defined in (50) is a $\mathcal{C}^1$ diffeomorphism for all $t \in [0, T]$, which allows to set $\rho(t, x) := \rho_0(X^{-1}(t, x))$, $u(t, x) := \bar{u}(t, X^{-1}(t, x))$ and $P(t, x) := \bar{P}(t, X^{-1}(t, x))$ and to check that (ρ, u, P) indeed satisfies (1) (see Proposition B.1 and the Appendix of [11]).

The advantage of Lagrangian coordinates for proving uniqueness or, more generally, stability estimates, is clear: we do not have to worry about the hyperbolic part of System (13), as it is time independent, and the equation for $\bar{u}$ is parabolic, at least for small enough time. This observation has been used in [11] in order to prove the whole Theorem 1.2 by means of the standard contracting mapping argument.

Here, we are going to use that reformulation in order to prove uniqueness: let $(\bar{u}_1, \bar{P}_1)$ and $(\bar{u}_2, \bar{P}_2)$ be the couples (u_1, P_1) and (u_2, P_2) in Lagrangian coordinates, and A_1 and A_2 be the corresponding matrices defined by (50). Fix some $T > 0$ so that (53) is fulfilled by $\bar{u}_1$ and $\bar{u}_2$.

The system satisfied by $\delta u := \bar{u}_2 - \bar{u}_1$ and $\nabla \delta P := \nabla \bar{P}_2 - \nabla \bar{P}_1$ reads

$$\begin{cases} \partial_t \delta u - \mu \Delta \delta u + \nabla \delta P = \delta f := \delta f_1 + \delta f_2 + \delta f_3 + \operatorname{div} \delta f_4 + \operatorname{div} \delta f_5, \\ \operatorname{div} \delta u = \delta g := \operatorname{div} \big((\operatorname{Id} - A_2) \delta u + (A_1 - A_2) \bar{u}_1 \big) \end{cases} \tag{54}$$

with

$$\delta f_1 := a_0 \partial_t \delta u, \quad \delta f_2 := (\operatorname{Id} - {}^T A_2) \nabla \delta P, \quad \delta f_3 := {}^T(A_1 - A_2) \nabla \bar{P}_1,$$
$$\delta f_4 := \big(A_2{}^T A_2 - A_1{}^T A_1 \big) \nabla \bar{u}_1 \quad \text{and} \quad \delta f_5 := \big(A_2{}^T A_2 - \operatorname{Id} \big) \nabla \delta u.$$

Bounding $(\delta u, \nabla \delta P)$ will stem from Proposition 1.2, which ensures that, if $\partial_t \delta g = \operatorname{div} \delta R_t$ then for all $T \geq 0$, we have

$$\|(\delta u, \nabla \delta P)\|_{E_T^p} \lesssim \|\delta f\|_{L_T^1(\dot{B}_{p,1}^{\frac{d}{p}-1})} + \|\delta g\|_{L_T^1(\dot{B}_{p,1}^{\frac{d}{p}})} + \|\delta R_t\|_{L_T^1(\dot{B}_{p,1}^{\frac{d}{p}-1})}. \tag{55}$$

Let us first bound $\delta f_1, \delta f_2, \delta f_3$ and $\delta f_4, \delta f_5$ in $L_T^1(\dot{B}_{p,1}^{\frac{d}{p}-1})$ and $L_T^1(\dot{B}_{p,1}^{\frac{d}{p}})$, respectively. From the definition of the multiplier space $\mathcal{M}(\dot{B}_{p,1}^{\frac{d}{p}-1})$, we readily have

$$\|\delta f_1\|_{L_T^1(\dot{B}_{p,1}^{\frac{d}{p}-1})} \leq \|a_0\|_{\mathcal{M}(\dot{B}_{p,1}^{\frac{d}{p}-1})} \|\partial_t \delta u\|_{L_T^1(\dot{B}_{p,1}^{\frac{d}{p}-1})}. \tag{56}$$

Next, using Inequality (165) and the fact that the product maps $\dot{B}_{p,1}^{\frac{d}{p}} \times \dot{B}_{p,1}^{\frac{d}{p}-1}$ to $\dot{B}_{p,1}^{\frac{d}{p}-1}$ if $p < 2d$ (see e.g. [2, Chap. 2]) yields

$$\|\delta f_2\|_{L_T^1(\dot{B}_{p,1}^{\frac{d}{p}-1})} \lesssim \|D\bar{u}_2\|_{L_T^1(\dot{B}_{p,1}^{\frac{d}{p}})} \|D\delta P\|_{L_T^1(\dot{B}_{p,1}^{\frac{d}{p}-1})}, \tag{57}$$

$$\|\delta f_5\|_{L_T^1(\dot{B}_{p,1}^{\frac{d}{p}})} \lesssim \|D\bar{u}_2\|_{L_T^1(\dot{B}_{p,1}^{\frac{d}{p}})} \|D\delta u\|_{L_T^1(\dot{B}_{p,1}^{\frac{d}{p}})}. \tag{58}$$

Inequality (169) ensures that

$$\|\delta f_3\|_{L_T^1(\dot{B}_{p,1}^{\frac{d}{p}})} \lesssim \|D\delta u\|_{L_T^1(\dot{B}_{p,1}^{\frac{d}{p}})} \|D\bar{P}_1\|_{L_T^1(\dot{B}_{p,1}^{\frac{d}{p}-1})} \tag{59}$$

whereas Inequalities (165) and (169) and the stability of $\dot{B}_{p,1}^{\frac{d}{p}}$ by product yield

$$\|\delta f_4\|_{L_T^1(\dot{B}_{p,1}^{\frac{d}{p}})} \lesssim \|D\delta u\|_{L_T^1(\dot{B}_{p,1}^{\frac{d}{p}})} \|D\bar{u}_1\|_{L_T^1(\dot{B}_{p,1}^{\frac{d}{p}})}. \tag{60}$$

In order to bound δg in $L_T^1(\dot{B}_{p,1}^{\frac{d}{p}})$, we shall use the fact that

$$\operatorname{div}\bar{u}_i = \operatorname{div}\left((\operatorname{Id} - A_i)\bar{u}_i\right) = D\bar{u}_i : (\operatorname{Id} - A_i).$$

Hence

$$\delta g = D\delta u : (\operatorname{Id} - A_2) - D\bar{u}_1 : (A_2 - A_1).$$

Now, easy computations based on (165) and (171) yield

$$\left\|D\delta u : (\operatorname{Id} - A_2)\right\|_{L_T^1(\dot{B}_{p,1}^{\frac{d}{p}})} \lesssim \|D\bar{u}_2\|_{L_T^1(\dot{B}_{p,1}^{\frac{d}{p}})}\|D\delta u\|_{L_T^1(\dot{B}_{p,1}^{\frac{d}{p}})}, \tag{61}$$

$$\left\|D\bar{u}_1 : (A_2 - A_1)\right\|_{L_T^1(\dot{B}_{p,1}^{\frac{d}{p}})} \lesssim \|D\bar{u}_1\|_{L_T^1(\dot{B}_{p,1}^{\frac{d}{p}})}\|D\delta u\|_{L_T^1(\dot{B}_{p,1}^{\frac{d}{p}})}. \tag{62}$$

Finally, to bound $\partial_t \delta g$, we decompose it into $\operatorname{div}(\delta R_1 + \delta R_2 + \delta R_3 + \delta R_4)$ with

$$\delta R_1 := -\partial_t A_2\,\delta u, \qquad \delta R_2 := (\operatorname{Id} - A_2)\partial_t\delta u,$$
$$\delta R_3 := \partial_t(A_1 - A_2)\,\bar{u}_1, \qquad \delta R_4 := (A_1 - A_2)\partial_t\bar{u}_1.$$

Using (165), (166) and product laws in Besov spaces, we see that

$$\|\delta R_1\|_{L_T^1(\dot{B}_{p,1}^{\frac{d}{p}-1})} \lesssim \|D\bar{u}_2\|_{L_T^1(\dot{B}_{p,1}^{\frac{d}{p}})}\|\delta u\|_{L_T^\infty(\dot{B}_{p,1}^{\frac{d}{p}-1})}, \tag{63}$$

$$\|\delta R_2\|_{L_T^1(\dot{B}_{p,1}^{\frac{d}{p}-1})} \lesssim \|D\bar{u}_2\|_{L_T^1(\dot{B}_{p,1}^{\frac{d}{p}})}\|\partial_t\delta u\|_{L_T^1(\dot{B}_{p,1}^{\frac{d}{p}-1})}. \tag{64}$$

In order to bound δR_3, it suffices to take advantage of (170). We get

$$\|\delta R_3\|_{L_T^1(\dot{B}_{p,1}^{\frac{d}{p}-1})} \lesssim \|D\delta u\|_{L_T^1(\dot{B}_{p,1}^{\frac{d}{p}})}\|\bar{u}_1\|_{L_T^\infty(\dot{B}_{p,1}^{\frac{d}{p}-1})}. \tag{65}$$

Finally, using again (169), we see that

$$\|\delta R_4\|_{L_T^1(\dot{B}_{p,1}^{\frac{d}{p}-1})} \lesssim \|D\delta u\|_{L_T^1(\dot{B}_{p,1}^{\frac{d}{p}})}\|\partial_t\bar{u}_1\|_{L_T^1(\dot{B}_{p,1}^{\frac{d}{p}-1})}. \tag{66}$$

One can now plug Inequalities (56) to (66) in (55). We end up with

$$\|(\delta u, \nabla\delta P)\|_{E_T^p} \leq C\left(\|a_0\|_{\mathcal{M}(\dot{B}_{p,1}^{\frac{d}{p}-1})} + \|D\bar{u}_2\|_{L_T^1(\dot{B}_{p,1}^{\frac{d}{p}})} + \|(\bar{u}_1, \nabla\bar{P}_1)\|_{E_T^p}\right)\|(\delta u, \nabla\delta P)\|_{E_T^p}.$$

As (23) and (53) are fulfilled and since $(\bar{u}_1, \nabla\bar{P}_1)$ is small, this implies that the left-hand side is null, whence uniqueness on $[0, T]$ for some small enough $T > 0$. From it and standard connectivity arguments, we obtain uniqueness on the whole $\mathbb{R}_+$. □

1.7 Application to density patches

Consider the special case where ρ_0 is given by (10). It is well-known that if D_0 is uniformly Lipschitz then 1_{D_0} is in $\mathcal{M}(B_{p,1}^s)$ whenever $-1 + \frac{1}{p} < s < \frac{1}{p}$ (see e.g.

[36]). As for compactly supported functions and for this range of indices, being in
$B^s_{p,1}$ is equivalent to be in $\dot{B}^s_{p,1}$ (see [12]), we have also $1_{D_0} \in \mathcal{M}(\dot{B}^s_{p,1})$.

Therefore provided u_0 is in $\dot{B}^{\frac{d}{p}-1}_{p,1}$ for some $d-1 < p < \infty$ and small enough,
and $|\eta_1 - \eta_2|$ is small enough, one can apply Theorem 1.2. From it and embedding,
we get that u is in $L^1_{loc}(\mathbb{R}_+; \mathcal{C}^1)$ hence has a $\mathcal{C}^1$ flow, and one can thus conclude
with no additional effort to the following statement.

Corollary 1.1. *Let D_0 be a uniformly Lipschitz (resp. $\mathcal{C}^1$) bounded domain of $\mathbb{R}^d$
and assume that ρ_0 is given by (10). There exists a constant $c = c(d,p)$ such that
for any divergence free vector field u_0 in $\dot{B}^{\frac{d}{p}-1}_{p,1}$ with $d-1 < p < 2d$, if*

$$|\eta_2 - \eta_1| + \|u_0\|_{\dot{B}^{\frac{d}{p}-1}_{p,1}} \le c$$

then (1) has a unique solution $(\rho, u, \nabla P)$ with $(u, \nabla P)$ in E^p and ρ given by

$$\rho(t, \cdot) = \eta_1 1_{D_t} + \eta_2 1_{{}^c D_t}$$

where $D_t := X(t, D_0)$ and $X(t, \cdot)$ stands for the flow of u at time t.

*Furthermore, the Lipschitz (resp. $\mathcal{C}^1$) regularity of the patch is preserved for
all time.*

1.8 Further remarks and open questions

For large u_0, we have local-in-time results in the same spirit as Theorem 1.2 and
Corollary 1.1. The proof consists in decomposing u into $u = e^{t\Delta}u_0 + \tilde{u}$, and
studying the system for $(a, \tilde{u}, \nabla P)$. In order to prove that the interval of existence
is not empty, one has to use the fact that the fluctuation goes to 0 when time
goes to 0. There, the improvement pointed out in (29) plays some role (see [11]
for more details).

As Proposition 1.2 may be adapted to more general domains of $\mathbb{R}^d$ like e.g.
bounded or exterior domains, or the half-space, one can extend Theorem 1.2 to
those settings (see [10] and [12] for more details).

It is easy to extend Theorem 1.2 to the case of density dependent viscosity,
provided we add the assumption that $\mu(\rho_0)$ is in the multiplier space $\mathcal{M}(\dot{B}^{\frac{d}{p}}_{p,1})$.
That latter assumption unfortunately precludes us from considering patches of
density if the viscosity is not a constant. In their recent work [34], M. Paicu and P.
Zhang succeeded in considering the case of a density dependent viscosity coefficient
provided the variations of viscosity and density are small enough. The proof
requires more regularity for the velocity field and relies on completely different
techniques (in particular propagation of tangential regularity along the patch).

Finally, our method is not sensitive to the space dimension. In particular,
we do not obtain anything better in dimension 2 even though global existence is
known for smoother data. Similarly, we do not know how to handle large jumps
of discontinuity in this critical regularity setting.

2 Second approach: subcritical data and standard maximal regularity

Having a regularity information for the density patches problem by means of the first approach strongly depends on the properties of multiplier spaces for Besov spaces. In fact, if instead of considering piecewise constant densities along C^1 interfaces one allows the density to be non-constant, then it is not clear that one can still get some information on the propagation of the structure. Here we propose another approach that shows that, somehow, having piecewise constant densities does not play any particular role as regards the well-posedness theory of (1): the relevant information is that the variations of the density in L^∞ norm are small.

The key to that second approach that has been initiated in [22] is another type of maximal regularity estimates for the Stokes system, that will be recalled in the next paragraph. Thanks to that, all the terms of the momentum equation of (13) (i.e. u, ∇u, $\partial_t u$ and $\nabla^2 u$) will be estimated in Lebesgue spaces of type $L^r(\mathbb{R}_+; L^p(\mathbb{R}^d))$.

In all that follows, those spaces will be just denoted by $L^r(L^p)$, and we shall sometimes use the notation $L^r_T(L^p)$ to designate $L^r(0, T; L^p(\mathbb{R}^d))$.

2.1 Classical maximal regularity for the Stokes system

The starting point is the following well-known caloric characterization of Besov spaces (see e.g. [2, Chap. 2]).

Lemma 2.1. *For all $s > 0$ and $1 \leq p, r \leq \infty$, we have the following equivalence:*

$$\|z\|_{\dot{B}^{-s}_{p,r}} \approx \left\| t^{\frac{s}{2}} \|e^{t\Delta} z\|_{L^p(\mathbb{R}^d)} \right\|_{L^r(\mathbb{R}_+, \frac{dt}{t})}.$$

Let us just give some heuristics of the proof in the case $r < \infty$ (the case $r = \infty$ being similar). We have

$$\int_{\mathbb{R}_+} \left(t^{\frac{s}{2}} \|e^{t\Delta} z\|_{L^p} \right)^r \frac{dt}{t} \approx \sum_{j \in \mathbb{Z}} 2^{-jsr} \int_{2^{-2(j+1)}}^{2^{-2j}} \|e^{t\Delta} z\|^r_{L^p} \frac{dt}{t}.$$

We observe that $e^{t\Delta}$ behaves, up to some small and rapidly decaying remainder, as a low frequency localization operator below frequency $1/\sqrt{t}$. Therefore, in the integral corresponding to j, it acts like the low frequency cut-off $\dot{S}_j$. Now, as $-s < 0$, we have (see e.g. [2, Chap. 2]),

$$\|z\|_{\dot{B}^{-s}_{p,r}} \approx \left\| 2^{-js} \|\dot{S}_j z\|_{L^p} \right\|_{\ell^r(\mathbb{Z})},$$

whence the result.

That relationship between Besov spaces and Lebesgue spaces through the heat kernel is one of the main ingredients for the proof of the following classical maximal regularity result for the Stokes system. As in the previous section, we allow for nonzero divergence field, since this will be useful for proving uniqueness.

Proposition 2.1. *Let* $u_0 \in \dot{B}_{p,r}^{2-\frac{2}{r}}$ *and* $f \in L^r(0,T;L^p)$ *with* $1 < p,r < \infty$. *Let* $g \in \mathcal{C}([0,T]; \dot{B}_{p,r}^{1-\frac{2}{r}})$ *be such that*

$$\nabla g \in L^r(0,T;L^p) \quad and \quad \partial_t g = \operatorname{div} R_t \quad with \quad R_t \in L^r(0,T;L^p),$$

and assume that the compatibility condition $g|_{t=0} = \operatorname{div} u_0$ *on* $\mathbb{R}^d$ *is satisfied.*

Then System (30) *has a unique solution* $(u, \nabla P)$ *with*

$$u \in \mathcal{C}([0,T]; \dot{B}_{p,r}^{2-\frac{2}{r}}) \quad and \quad \partial_t u, \nabla^2 u, \nabla P \in L^r(0,T;L^p)$$

and the following estimate is valid:

$$\|u\|_{L^\infty(0,T;\dot{B}_{p,r}^{2-\frac{2}{r}})} + \|\partial_t u, \nabla^2 u, \nabla P\|_{L^r(0,T;L^p)} \leq C \big(\|f, \nabla g, R_t\|_{L^r(0,T;L^p)} + \|u_0\|_{\dot{B}_{p,r}^{2-\frac{2}{r}}} \big).$$

Proof. As in the proof of Proposition 1.2, we define the gradient of the pressure to be

$$\nabla P = \mathcal{Q}f - \mathcal{Q}R_t + \nabla g.$$

As $\mathcal{Q}$ maps L^p to L^p for all $1 < p < \infty$, it is clear that ∇P satisfies the desired estimate. Then, we define u to be the solution of the following heat equation:

$$\partial_t u - \Delta u = f - \nabla P, \qquad u|_{t=0} = u_0.$$

As in the proof of Proposition 1.2, the compatibility condition at time $t = 0$ guarantees that $\operatorname{div} u = g$ for all $t \in [0,T]$. Furthermore, the classical maximal regularity result for the heat equation (see e.g. [26]) enables us to handle the right-hand side while the previous lemma takes care of the initial data. We end up with

$$\|u\|_{L^\infty(0,T;\dot{B}_{p,r}^{2-\frac{2}{r}})} + \|\partial_t u, \nabla^2 u\|_{L^r(0,T;L^p)}$$
$$\leq C \big(\|u_0\|_{\dot{B}_{p,r}^{2-\frac{2}{r}}} + \|\nabla P\|_{L^r(0,T;L^p)} + \|f\|_{L^r(0,T;L^p)} \big),$$

whence the desired estimate. $\qquad\square$

2.2 The basic existence theorem

Granted with Proposition 2.1, it is tempting to study whether the abstract global existence statement, Theorem 1.1, holds true with the spaces $A_0 = L^\infty$ and $E_0 = \dot{B}_{p,r}^{2-\frac{2}{r}}$ for the initial data. Here we take $F := L^r(\mathbb{R}_+; L^p)$ and $E := E^{p,r}$, with

$$E^{p,r} \stackrel{\text{def}}{=} \big\{ (u, \nabla P) \text{ with } u \in \mathcal{C}_b(\mathbb{R}_+; \dot{B}_{p,r}^{2-\frac{2}{r}}) \text{ and } \partial_t u, \nabla^2 u, \nabla P \in L^r(\mathbb{R}_+; L^p) \big\},$$

endowed with the norm

$$\|(u, \nabla P)\|_{E^{p,r}} \stackrel{\text{def}}{=} \|u\|_{L^\infty(\dot{B}_{p,r}^{2-\frac{2}{r}})} + \|(\partial_t u, \nabla^2 u, \nabla P)\|_{L^r(L^p)}. \tag{67}$$

We denote by $E_T^{p,r}$ the local version of $E^{p,r}$, pertaining to functions that are defined on the time interval $[0,T]$.

Under very general assumptions for the velocity field, the L^∞ norm of a is propagated by the transport equation (we do not even need ∇u to be in $L^1(\mathbb{R}_+; L^\infty)$) so that the only compatibility condition that has still to be checked is whether we do have

$$\|u \cdot \nabla u\|_{L^r(\mathbb{R}_+; L^p)} \leq C\|(u, \nabla P)\|^2_{E^{p,r}}.$$

For scaling reasons, that inequality can hold only if p and r are interrelated through

$$2 - \frac{2}{r} = \frac{d}{p} - 1. \tag{68}$$

Conversely, if (68) is fulfilled for some $1 < p, r < \infty$, then we have

$$\|\nabla u\|_{L^{\frac{dr}{2r-1}}} \leq C\|\nabla^2 u\|^{\frac{1}{2}}_{L^p}\|u\|^{\frac{1}{2}}_{\dot{B}^{2-\frac{2}{r}}_{p,r}}. \tag{69}$$

Indeed, decompose ∇u into low and high frequencies as follows:

$$\|\nabla u\|_{L^{\frac{dr}{2r-1}}} \leq \sum_{j \leq N} \|\dot{\Delta}_j \nabla u\|_{L^{\frac{dr}{2r-1}}} + \sum_{j > N} \|\dot{\Delta}_j \nabla u\|_{L^{\frac{dr}{2r-1}}}.$$

Then, using Bernstein inequality and the relation (68), we get

$$\sum_{j \leq N} \|\dot{\Delta}_j \nabla u\|_{L^{\frac{dr}{2r-1}}} \lesssim \sum_{j \leq N} 2^{j(1-\frac{1}{r})}\left(\|\dot{\Delta}_j \nabla u\|_{L^p} 2^{j(1-\frac{2}{r})}\right)2^{-j(1-\frac{2}{r})} \lesssim 2^{\frac{N}{r}}\|\nabla u\|_{\dot{B}^{1-\frac{2}{r}}_{p,\infty}}$$

and

$$\sum_{j > N} \|\dot{\Delta}_j \nabla u\|_{L^{\frac{dr}{2r-1}}} \lesssim \sum_{j > N} 2^{-\frac{j}{r}}\|\dot{\Delta}_j \nabla^2 u\|_{L^p} \lesssim 2^{-\frac{N}{r}}\|\nabla^2 u\|_{\dot{B}^0_{p,\infty}}.$$

Therefore, using obvious embedding,

$$\|\nabla u\|_{L^{\frac{dr}{2r-1}}} \lesssim 2^{\frac{N}{r}}\|\nabla u\|_{\dot{B}^{1-\frac{2}{r}}_{p,r}} + 2^{-\frac{N}{r}}\|\nabla^2 u\|_{L^p}$$

and taking the "best" N yields (69).

Now, using the embedding $\dot{W}^{1,\frac{dr}{2r-1}} \hookrightarrow L^{\frac{dr}{r-1}}$, we conclude by Hölder inequality that

$$\|u \cdot \nabla u\|_{L^r(\mathbb{R}_+; L^p)} \leq \|u\|_{L^{2r}(\mathbb{R}_+; L^{\frac{dr}{r-1}})}\|\nabla u\|_{L^{2r}(\mathbb{R}_+; L^{\frac{dr}{2r-1}})} \leq C\|(u, \nabla P)\|^2_{E^{p,r}}. \tag{70}$$

Let us emphasize that having $(u, \nabla P)$ in $E^{p,r}$ with (p, r) satisfying (68) does not quite ensure that ∇u is in $L^1(\mathbb{R}_+; L^\infty)$. As regards the existence issue for (13), we do not really care since preserving the L^∞ norm of a does not require that much regularity. The situation is different for the uniqueness issue owing to the very low regularity of a. Indeed, as in the previous section, the only way we found to prove uniqueness is to recast (13) in Lagrangian coordinates, which requires that ∇u is in $L^1_{loc}(\mathbb{R}_+; L^\infty)$. For $(u, \nabla P) \in E^{p,r}$, that property is achieved if and only if $r = 1$. However Proposition 2.1 fails in that endpoint case. Therefore, in order to get uniqueness, we will have to consider initial velocities with (slightly) subcritical regularity: more precisely, we will require in addition that $u_0 \in \dot{B}_{\tilde{p},r}^{2-\frac{2}{r}}$ for some $\tilde{p} > d$. Then Proposition 2.1 will yield $\nabla^2 u \in L^r(\mathbb{R}_+; L^{\tilde{p}})$. As we have $\nabla^2 u \in L^r(\mathbb{R}_+; L^p)$ from the beginning, using the following interpolation inequality (that may be proved in the same way as (69)):

$$\|\nabla u\|_{L^\infty} \le C \|\nabla^2 u\|_{L^p}^\theta \|\nabla^2 u\|_{L^{\tilde{p}}}^{1-\theta} \quad \text{with} \quad \theta := \frac{p}{d}\left(\frac{\tilde{p}-d}{\tilde{p}-p}\right) \tag{71}$$

will eventually ensure that $\nabla u \in L^r(\mathbb{R}_+; L^\infty)$.

The above considerations motivate the following statement that will be proved in the rest of this section.

Theorem 2.1. *Assume* $d \ge 2$. *Let* $a_0 \in L^\infty(\mathbb{R}^d)$ *and* $u_0 \in \dot{B}_{p,r}^{-1+\frac{d}{p}}(\mathbb{R}^d)$ *with* $\operatorname{div} u_0 = 0$, $p \overset{\text{def}}{=} \frac{dr}{3r-2}$ *and* $r \in (1,\infty)$. *There exists a positive constant* $c_0 = c_0(r,d)$ *so that if*

$$\|a_0\|_{L^\infty} + \|u_0\|_{\dot{B}_{p,r}^{-1+\frac{d}{p}}} \le c_0 \tag{72}$$

then (13) *has a global solution* $(a, u, \nabla P)$ *satisfying* $\|a(t)\|_{L^\infty} = \|a_0\|_{L^\infty}$ *for all* $t \ge 0$, *and* $(u, \nabla P) \in E^{p,r}$, *and there exists* C *so that*

$$\|(u, \nabla P)\|_{E^{p,r}} \le C \|u_0\|_{\dot{B}_{p,r}^{-1+\frac{d}{p}}}. \tag{73}$$

If in addition $u_0 \in \dot{B}_{\tilde{p},r}^{-1+\frac{d}{\tilde{p}}}$ *for some* $d < \tilde{p} \le \frac{dr}{r-1}$, *then* $(u, \nabla P)$ *also belongs to* $E^{\tilde{p},r}$, *satisfies*

$$\|(u, \nabla P)\|_{E^{\tilde{p},r}} \le C \|u_0\|_{\dot{B}_{\tilde{p},r}^{-1+\frac{d}{\tilde{p}}}}, \tag{74}$$

and the solution $(a, u, \nabla P)$ *is unique in the space* $L^\infty(\mathbb{R}_+ \times \mathbb{R}^d) \times \left(E^{p,r} \cap E^{\tilde{p},r}\right)$.

Proof. To start with, let us establish the existence part of Theorem 2.1 in the "regular case" where $u_0 \in \dot{B}_{p,r}^{-1+\frac{d}{p}} \cap \dot{B}_{\tilde{p},r}^{-1+\frac{d}{\tilde{p}}}$. To do this, we shall smooth out the data so as to construct a sequence $(a^n, u^n, \nabla P^n)_{n \in \mathbb{N}}$ of smooth maximal solutions on some time interval $[0, T^n)$ that may be finite. Then taking advantage of Proposition 1.2 first with $p = \frac{dr}{3r-2}$ and next with $d < \tilde{p} \le \frac{dr}{r-1}$ and of Inequality (70) (and the corresponding one with $\tilde{p}$ instead of p) allows to prove uniform a priori estimates in $E_{T^n}^{p,r} \cap E_{T^n}^{\tilde{p},r}$. Those estimates combined with (71) provide us with a control on ∇u^n in $L^1(0, T^n; L^\infty)$. Combining with a classical blow-up criterion, one can thus conclude that $T^n = +\infty$. Next, compactness arguments enable us to show that $(a^n, u^n, \nabla P^n)$ converges to some global solution $(a, u, \nabla P)$ to (13) in $E^{p,r} \cap E^{\tilde{p},r}$. As a last step, we show, resorting again to approximation and compactness arguments, that in the case where we only have $u_0 \in \dot{B}_{p,r}^{-1+\frac{d}{p}}$, then we still have a global solution.

First step: Construction of smooth solutions for approximate data. Assuming that $u_0 \in \dot{B}_{p,r}^{-1+\frac{d}{p}} \cap \dot{B}_{\tilde{p},r}^{-1+\frac{d}{\tilde{p}}}$ and that (72) is fulfilled, we smooth out the data a_0 and u_0 by means of nonnegative mollifiers: we get a sequence (a_0^n, u_0^n) of smooth data such that for all $n \in \mathbb{N}$,

$$\|a_0^n\|_{L^\infty} \le \|a_0\|_{L^\infty}, \quad \|u_0^n\|_{\dot{B}_{p,r}^{2-\frac{2}{r}}} \le C\|u_0\|_{\dot{B}_{p,r}^{2-\frac{2}{r}}} \quad \text{and} \quad \|u_0^n\|_{\dot{B}_{\tilde{p},r}^{2-\frac{2}{r}}} \le C\|u_0\|_{\dot{B}_{\tilde{p},r}^{2-\frac{2}{r}}}$$
(75)

with in addition

$$u_0^n \to u_0 \quad \text{in} \quad \dot{B}_{p,r}^{2-\frac{2}{r}} \cap \dot{B}_{\tilde{p},r}^{2-\frac{2}{r}} \quad \text{and} \quad a_0^n \rightharpoonup a_0 \quad \text{in} \quad L^\infty \text{ weak } \star.$$
(76)

Let $(a^n, u^n, \nabla P^n)$ be the corresponding smooth local maximal solution that is provided by e.g. [8], and denote by T^n its maximal time of existence.

Second step: Critical regularity estimates. It is obvious that

$$\|a^n(t)\|_{L^\infty} = \|a_0^n\|_{L^\infty} \le \|a_0\|_{L^\infty} \quad \text{for all } t \in [0, T^n),$$
(77)

and Proposition 2.1 ensures that

$$\|(u^n, \nabla P^n)\|_{E_{T^n}^{p,r}} \le C\Big(\|u_0^n\|_{\dot{B}_{p,r}^{2-\frac{2}{r}}} + \|a^n\|_{L^\infty(0,T^n \times \mathbb{R}^d)}\|\partial_t u^n\|_{L_{T^n}^r(L^p)}$$
$$+ (1 + \|a^n\|_{L^\infty(0,T^n \times \mathbb{R}^d)})\|u^n \cdot \nabla u^n\|_{L_{T^n}^r(L^p)}\Big).$$

Therefore, taking advantage of (70) and of (77), we discover that

$$\|(u^n, \nabla P^n)\|_{E_{T^n}^{p,r}} \le C\Big(\|u_0\|_{\dot{B}_{p,r}^{2-\frac{2}{r}}} + \|a_0\|_{L^\infty}\|\partial_t u^n\|_{L_{T^n}^r(L^p)}$$
$$+ (1 + \|a_0\|_{L^\infty})\|(u^n, \nabla P^n)\|_{E_{T^n}^{p,r}}\Big),$$

which implies, if c_0 is small enough in (72), that

$$\|(u^n, \nabla P^n)\|_{E_{T^n}^{p,r}} \le C\|u_0\|_{\dot{B}_{p,r}^{2-\frac{2}{r}}} \quad \text{on} \quad [0, T^n).$$
(78)

Third step: Subcritical regularity estimates. From Proposition 2.1, we also know that

$$\|(u^n, \nabla P^n)\|_{E_{T^n}^{\tilde{p},r}} \le C\Big(\|u_0^n\|_{\dot{B}_{\tilde{p},r}^{2-\frac{2}{r}}} + \|a^n\|_{L^\infty(0,T^n \times \mathbb{R}^d)}\|\partial_t u^n\|_{L_{T^n}^r(L^{\tilde{p}})}$$
$$+ (1 + \|a^n\|_{L^\infty(0,T^n \times \mathbb{R}^d)})\|u^n \cdot \nabla u^n\|_{L_{T^n}^r(L^{\tilde{p}})}\Big).$$

Now, from Hölder inequality with $\frac{1}{q} + \frac{1}{d} - \frac{1}{dr} = \frac{1}{p}$ (whence the upper bound for $\tilde{p}$), and obvious embedding, we discover that

$$\|u^n \cdot \nabla u^n\|_{L_{T^n}^r(L^{\tilde{p}})} \le \|u^n\|_{L_{T^n}^{2r}(L^{\frac{dr}{r-1}})}\|\nabla u^n\|_{L_{T^n}^{2r}(L^q)}.$$
(79)

Arguing exactly as for proving (69), we see that

$$\|\nabla u^n\|_{L^q} \le C\|\nabla^2 u^n\|_{L^{\tilde{p}}}^{\frac{1}{2}}\|u^n\|_{\dot{B}_{\tilde{p},r}^{2-\frac{2}{r}}}^{\frac{1}{2}}.$$
(80)

Keeping (78) in mind, one can thus conclude that

$$\|u^n \cdot \nabla u^n\|_{L^r_{T^n}(L^{\widetilde{p}})} \leq C\|(u^n, \nabla P^n)\|_{E^{p,r}_{T^n}}\|(u^n, \nabla P^n)\|_{E^{\widetilde{p},r}_{T^n}}$$

$$\leq C\|u_0\|_{\dot{B}^{2-\frac{2}{r}}_{p,r}}\|(u^n, \nabla P^n)\|_{E^{\widetilde{p},r}_{T^n}}.$$

Therefore, using (77), and assuming that (72) is fulfilled, we get the following additional uniform bound:

$$\|(u^n, \nabla P^n)\|_{E^{\widetilde{p},r}_{T^n}} \leq C\|u_0\|_{\dot{B}^{2-\frac{2}{r}}_{\widetilde{p},r}} \quad \text{for all } n \in \mathbb{N}. \tag{81}$$

Fourth step: Approximate solutions are global. Combining (78), (81) with the interpolation inequality (71), we discover that for all $n \in \mathbb{N}$, we have

$$\int_0^{T^n} \|\nabla u^n\|_{L^\infty} \, dt \leq C\|u_0\|_{\dot{B}^{2-\frac{2}{r}}_{p,r}}^\theta \|u_0\|_{\dot{B}^{2-\frac{2}{r}}_{\widetilde{p},r}}^{1-\theta} \quad \text{with} \quad \theta := \frac{p}{d}\left(\frac{\widetilde{p}-d}{\widetilde{p}-p}\right).$$

Hence, classical continuation criteria (see e.g. [8]) ensure that $T^n = +\infty$. In other words, the solution $(a^n, u^n, \nabla P^n)$ is global and we have for all $n \in \mathbb{N}$,

$$\|(u^n, \nabla P^n)\|_{E^{p,r}} \leq C\|u_0\|_{\dot{B}^{2-\frac{2}{r}}_{p,r}} \quad \text{and} \quad \|(u^n, \nabla P^n)\|_{E^{\widetilde{p},r}} \leq C\|u_0\|_{\dot{B}^{2-\frac{2}{r}}_{\widetilde{p},r}}. \tag{82}$$

Fifth step: Convergence. From (82), we know that $(\partial_t u^n)_{n \in \mathbb{N}}$ is bounded in $L^r(\mathbb{R}_+; L^p)$. Hence, using Ascoli-Arzelà Theorem and compact embeddings in Besov spaces ensures that there exists a subsequence, still denoted by $(a^n, u^n, \nabla P^n)_{n \in \mathbb{N}}$ and some $(a, u, \nabla P)$ with

$$a \in L^\infty(\mathbb{R}_+ \times \mathbb{R}^d), \quad u \in L^\infty(\mathbb{R}_+; \dot{B}^{2-\frac{2}{r}}_{p,r} \cap \dot{B}^{2-\frac{2}{r}}_{\widetilde{p},r}) \quad \text{and}$$

$$\partial_t u, \nabla^2 u, \nabla P \in L^r(\mathbb{R}_+; L^p \cap L^{\widetilde{p}})$$

such that

$$a^n \rightharpoonup a \quad \text{weak} * \text{ in } L^\infty(\mathbb{R}_+ \times \mathbb{R}^d),$$

$$(\partial_t u^n, \nabla^2 u^n) \rightharpoonup (\partial_t u, \nabla^2 u) \quad \text{and} \quad \nabla P^n \rightharpoonup \nabla P \quad \text{weakly in } L^r(\mathbb{R}_+; L^p \cap L^{\widetilde{p}}), \tag{83}$$

with in addition for all small enough $\eta > 0$,

$$u^n \to u \quad \text{strongly in } L^{2r}_{loc}(\mathbb{R}_+; L^{\frac{dr}{r-1}-\eta}_{loc}(\mathbb{R}^d)),$$

$$\nabla u^n \to \nabla u \quad \text{strongly in } L^{2r}_{loc}(\mathbb{R}_+; L^{\frac{dr}{2r-1}-\eta}_{loc}(\mathbb{R}^d)). \tag{84}$$

By construction, $(a^n, u^n, \nabla \Pi^n)$ satisfies

$$\int_0^\infty \int_{\mathbb{R}^d} a^n(\partial_t \phi + u^n \cdot \nabla \phi) \, dx \, dt + \int_{\mathbb{R}^d} \phi(0, x) a_0^n(x) \, dx = 0,$$

$$\int_0^\infty \int_{\mathbb{R}^d} \phi \operatorname{div} u^n \, dx \, dt = 0 \quad \text{and}$$

$$\int_0^\infty \int_{\mathbb{R}^d} \left\{ (1 + a^n)\left(u^n \cdot \partial_t \Phi + ((u^n \otimes u^n) : \nabla \Phi)\right) - \nabla u^n : \nabla \Phi \right\} dx \, dt \tag{85}$$

$$+ \int_{\mathbb{R}^d} u_0^n \cdot \Phi(0, x) \, dx = 0,$$

for all functions $\phi \in \mathcal{C}_c^{\infty}([0, +\infty) \times \mathbb{R}^d; \mathbb{R})$ and $\Phi \in \mathcal{C}_c^{\infty}([0, +\infty) \times \mathbb{R}^d; \mathbb{R}^d)$ with $\operatorname{div}\Phi \equiv 0$.

Putting (83) and (84) together, one can thus pass to the limit in all the terms of (85). Hence $(a, u, \nabla P)$ is a distributional solution of (1). Furthermore, the fact that ∇u is in $L^r([0, T]; L^{\infty})$ (use (71)) guarantees that the only solution to the transport equation of (1) is a, and that its L^{∞} norm is time independent. Finally, looking at $(u, \nabla P)$ as the (unique) solution of the Stokes system (30) with initial data $u_0 \in \dot{B}_{p,r}^{2-\frac{2}{r}} \cap \dot{B}_{\widetilde{p},r}^{2-\frac{2}{r}}$ and source term $f \in L^r(\mathbb{R}_+; L^p \cap L^{\widetilde{p}})$, one can conclude that u is in $\mathcal{C}(\mathbb{R}_+; \dot{B}_{p,r}^{2-\frac{2}{r}} \cap \dot{B}_{\widetilde{p},r}^{2-\frac{2}{r}})$, which completes the proof of the existence part of Theorem 2.1 in the subcritical case.

Sixth step: The critical regularity case. Here we only assume that u_0 is in $\dot{B}_{p,r}^{2-\frac{2}{r}}$ and that (72) is fulfilled. Then we set $u_0^n := \dot{S}_n u_0$ (where $\dot{S}_n$ is the low frequency cut-off defined in the appendix) so that we have

$$\|u_0^n\|_{\dot{B}_{p,r}^{2-\frac{2}{r}}} \leq C \|u_0\|_{\dot{B}_{p,r}^{2-\frac{2}{r}}} \quad \text{and} \quad u_0^n \in \dot{B}_{\widetilde{p},r}^{2-\frac{2}{r}}.$$

Thanks to the previous steps, the fact that (a_0, u_0^n) fulfills the smallness condition (72) and that u_0^n is in $\dot{B}_{\widetilde{p},r}^{2-\frac{2}{r}}$ ensure that (1) has a global solution $(a^n, u^n, \nabla P^n)$ in $L^{\infty}(\mathbb{R}_+ \times \mathbb{R}^d) \times (E^{p,r} \cap E^{\widetilde{p},r})$ satisfying the first inequality of (81). Then mimicking the fifth step, one can easily conclude that $(a^n, u^n, \nabla P^n)$ converges to some solution $(a, u, \nabla P)$ of (1) satisfying the properties of the first part of Theorem 2.1. □

2.3 Uniqueness

As in the previous section, we do not know how to prove uniqueness directly in our regularity framework where a is only bounded, and we have to use Lagrangian coordinates. Hence, we consider two solutions $(a^1, u^1, \nabla P^1)$ and $(a^2, u^2, \nabla P^2)$ in $L^{\infty}(0, T \times \mathbb{R}^d) \times (E_T^{p,r} \cap E_T^{\widetilde{p},r})$ with the same initial data satisfying (72), and denote by $(\bar{u}_i, \bar{P}_i)$ (with $i = 1, 2$) the functions (u_i, P_i) in Lagrangian coordinates. Let us emphasize once again that, as $(u_1, \nabla P_1)$ and $(u_2, \nabla P_2)$ belong to $E_T^{p,r} \cap E_T^{\widetilde{p},r}$, we know from (71) that ∇u^1 and ∇u^2 are in $L^1(0, T; L^{\infty})$, and thus have a unique bi-Lipschitz flow on $[0, T]$. We shall assume with no loss of generality that

$$\int_0^T \|D\bar{u}^i\|_{L^{\infty}} \, dt \leq \frac{1}{2} \quad \text{for } i = 1, 2.$$

Therefore, because

$$A_i(t) - \operatorname{Id} = -\left(\int_0^t D\bar{u}_i \, d\tau\right) \sum_{k=0}^{+\infty} \left(-\int_0^t D\bar{u}_i \, d\tau\right)^k \quad \text{for } i = 1, 2, \tag{86}$$

we get

$$\|A_i(t) - \operatorname{Id}\|_{L^{\infty}} \leq 2 \int_0^t \|D\bar{u}_i\|_{L^{\infty}} \, d\tau \quad \text{for all } t \in [0, T]. \tag{87}$$

Hence System (52) is fulfilled on $[0,T] \times \mathbb{R}^d$, and using the chain rule, one can prove that $(\bar{u}_1, \nabla \bar{P}_1)$ and $(\bar{u}_2, \nabla \bar{P}_2)$ belong to $E_T^{p,r} \cap E_T^{\bar{p},r}$.

Now, denoting $\delta u := \bar{u}_2 - \bar{u}_1$ and $\delta P := \bar{P}_2 - \bar{P}_1$, and applying Proposition 2.1 to System (54) (keeping the same definition for δf, δg and δR_t), we discover that for all $t \in [0,T]$,

$$\|(\delta u, \nabla \delta P)\|_{E_t^{p,r}} \le C\big(\|\delta f\|_{L_t^r(L^p)} + \|D\delta g\|_{L_t^r(L^p)} + \|\delta R_t\|_{L_t^r(L^p)}\big). \qquad (88)$$

Clearly the term corresponding to δf_1 may be absorbed by the left-hand side if $\|a_0\|_{L^\infty}$ is small enough. As regards, the terms δf_i for $i = 2,3,4,5$, using (86) and (87), we get for all $t \in [0,T]$,

$$\|\delta f_2\|_{L_t^r(L^p)} \le 2\|D\bar{u}_2\|_{L_t^1(L^\infty)}\|D\delta P\|_{L_t^r(L^p)},$$

$$\|\delta f_3\|_{L_t^r(L^p)} \lesssim \|D\delta u\|_{L_t^1(L^\infty)}\|D\bar{P}_1\|_{L_t^r(L^p)},$$

$$\|\text{div}\,\delta f_4\|_{L_t^r(L^p)} \lesssim \|D^2\delta u\|_{L_t^1(L^p)}\|D\bar{u}_1\|_{L_t^r(L^\infty)} + \|D\delta u\|_{L_t^1(L^\infty)}\|D^2\bar{u}_1\|_{L_t^r(L^p)},$$

$$\|\text{div}\,\delta f_5\|_{L_t^r(L^p)} \lesssim \|D^2\bar{u}_2\|_{L_t^1(L^p)}\|D\delta u\|_{L_t^r(L^\infty)} + \|D\bar{u}_2\|_{L_t^1(L^\infty)}\|D^2\delta u\|_{L_t^r(L^p)}.$$

In order to bound δg, we use the fact that

$$\delta g = D\delta u : (\text{Id} - A_2) - D\bar{u}_1 : (A_2 - A_1).$$

Therefore

$$\|D\delta g\|_{L_t^r(L^p)} \lesssim \|D\bar{u}_2\|_{L_t^1(L^\infty)}\|D^2\delta u\|_{L_t^r(L^p)} + \|D^2\bar{u}_2\|_{L_t^1(L^p)}\|D\delta u\|_{L_t^r(L^\infty)}$$
$$+ \|D^2\bar{u}_1\|_{L_t^r(L^p)}\|D\delta u\|_{L_t^1(L^\infty)} + \|D\bar{u}_1\|_{L_t^r(L^\infty)}\|D^2\delta u\|_{L_t^1(L^p)}.$$

Finally, to bound δR_t, we decompose it into

$$\delta R_t = -\partial_t A_2\,\delta u + (\text{Id} - A_2)\partial_t \delta u + \partial_t(A_1 - A_2)\,\bar{u}_1 + (A_1 - A_2)\partial_t \bar{u}_1.$$

Using (168), we easily get

$$\|\delta R_t\|_{L_t^r(L^p)} \lesssim \|D\bar{u}_2\|_{L_t^{2r}(L^{\frac{dr}{2r-1}})}\|\delta u\|_{L_t^{2r}(L^{\frac{dr}{r-1}})} + \|D\bar{u}_2\|_{L_t^1(L^\infty)}\|\partial_t \delta u\|_{L_t^r(L^p)}$$
$$+ \|D\delta u\|_{L_t^{2r}(L^{\frac{dr}{2r-1}})}\|\bar{u}_1\|_{L_t^{2r}(L^{\frac{dr}{r-1}})} + \|D\delta u\|_{L_t^1(L^\infty)}\|\partial_t \bar{u}_1\|_{L_t^r(L^p)},$$

whence, for some C_t that tends to 0 when t goes to 0,

$$\|(\delta u, \nabla \delta P)\|_{E_t^{p,r}} \le C_t\big(\|(\delta u, \nabla \delta P)\|_{E_t^{p,r}} + \|D\delta u\|_{L_t^1(L^\infty)}\big).$$

Similar computations lead to

$$\|(\delta u, \nabla \delta P)\|_{E_t^{\bar{p},r}} \le C_t\big(\|(\delta u, \nabla \delta P)\|_{E_t^{\bar{p},r}} + \|D\delta u\|_{L_t^1(L^\infty)}\big).$$

The last term of the above two inequalities may be bounded thanks to (71), and we thus get for all $t \in [0,T]$,

$$\|(\delta u, \nabla \delta P)\|_{E_t^{p,r} \cap E_t^{\bar{p},r}} \le C_t\|(\delta u, \nabla \delta P)\|_{E_t^{p,r} \cap E_t^{\bar{p},r}} \quad \text{with} \quad C_t \to 0 \quad \text{when} \quad t \to 0.$$

This yields uniqueness on a small time interval. Then standard connectivity arguments yield uniqueness on the whole $\mathbb{R}_+$. $\qquad\qquad \square$

In the subcritical regularity case, the unique solution $(a, u, \nabla P)$ provided by Theorem 2.1 satisfies $\nabla^2 u \in L^r(\mathbb{R}_+; L^{\tilde{p}})$. As $\dot{W}^{1,\tilde{p}}$ is continuously embedded in the homogeneous Hölder space $\dot{C}^{0,1-d/\tilde{p}}$ and as we know from (71) that ∇u is in $L^1(\mathbb{R}_+; L^\infty)$, one can deduce that the flow of u is in $C^{1,1-d/\tilde{p}}$. Therefore we readily get the following application to density patches.

Corollary 2.1. *Let u_0 satisfy the assumptions of Theorem 2.1 and assume that ρ_0 is given by (10) for some bounded domain D_0 of $\mathbb{R}^d$ with $C^{1,\beta}$ regularity, $0 \leq \beta \leq 1 - d/\tilde{p}$.*

Then, there exists a constant c (independent of the domain) such that if

$$|\eta_2 - \eta_1| + \|u_0\|_{\dot{B}^{2-\frac{2}{r}}_{p,r}} \leq c$$

then (1) has a unique global solution $(\rho, u, \nabla P)$ with $(u, \nabla P)$ in $E^{p,r}$ and ρ given by

$$\rho(t, \cdot) = \eta_1 1_{D_t} + \eta_2 1_{{}^c D_t}$$

where $D_t := X(t, D_0)$ and $X(t, \cdot)$ stands for the flow of u at time t. Furthermore, the domain D_t remains $C^{1,\beta}$ for all time.

Remark 2.1. *In the critical regularity case, even though it need not be in C^1, the flow of u is still well defined. In fact, one can show (see e.g. [2, Chap.3]) that it is in $C^{0,\alpha}$ for all $\alpha < 1$. Therefore, if we start with a Lipschitz boundary, then D_t remains $C^{0,\alpha}$ for all $\alpha < 1$ and $t > 0$.*

2.4 Global existence for data with one large velocity component

The aim of this paragraph is to improve the smallness condition of Theorem 2.1. In fact, we are going to show that one can afford to have one large component of the initial velocity (the vertical one u_0^d, say), provided the horizontal components $u_0^h = (u_0^1, \ldots, u_0^{d-1})$ of u_0, and a_0 are small accordingly. This nice feature of System (1) has been first noticed in [22]. The main idea is that, owing to $\operatorname{div} u = 0$, the horizontal and vertical components of the vector-field u fulfill

$$\begin{cases} \partial_t u^h - \Delta u^h + \nabla_h P = a\partial_t u^h - (1+a)(u^h \cdot \nabla_h u^h + u^d \partial_d u^h), \\ \partial_t u^d - \Delta u^d + \partial_d P = a\partial_t u^d - (1+a)(u^h \cdot \nabla_h u^d - u^d \operatorname{div}_h u^h), \end{cases} \tag{89}$$

where $\nabla_h := (\partial_1, \ldots, \partial_{d-1})$ and $\operatorname{div}_h u^h := \sum_{j=1}^{d-1} \partial_j u^j$.

As, in the above system, there is no quadratic term involving u^d only, it looks reasonable that if u^d is large but u^h is very small, then one can keep u^d under control and u^h very small for all time. This is exactly the spirit of the statement below.

Theorem 2.2. *Let $a_0 \in L^\infty$ and $u_0 \in \dot{B}^{-1+\frac{d}{p}}_{p,r}$ with $p \stackrel{\text{def}}{=} \frac{dr}{3r-2}$ and $r \in (1,\infty)$. There exist two positive constants $c_0 = c_0(r, d)$ and $C_1 = C_1(r, d)$ so that if*

$$\eta_0 \stackrel{\text{def}}{=} \left(\|a_0\|_{L^\infty} + \|u_0^h\|_{\dot{B}^{-1+\frac{d}{p}}_{p,r}}\right) \exp\left(C\|u_0^d\|^{2r}_{\dot{B}^{-1+\frac{d}{p}}_{p,r}}\right) \leq c_0 \tag{90}$$

then (13) *has a global solution* $(a, u, \nabla P)$ *with* $\|a(t)\|_{L^\infty} = \|a_0\|_{L^\infty}$ *for all* $t \geq 0$, *and* $(u, \nabla P) \in E^{p,r}$ *satisfying*

$$\|(u, \nabla P)\|_{E^{p,r}} \leq C\|u_0\|_{\dot{B}_{p,r}^{-1+\frac{d}{p}}} \quad and \quad \|u^h\|_{L^\infty(\dot{B}_{p,r}^{-1+\frac{d}{p}})} + \|\partial_t u^h, \nabla^2 u^h\|_{L^r(L^p)} \leq C\eta_0.$$

If in addition u_0 *is in* $\dot{B}_{\tilde{p},r}^{-1+\frac{d}{p}}(\mathbb{R}^d)$ *for some* $d < \tilde{p} \leq \frac{dr}{r-1}$ *then* $(u, \nabla P)$ *belongs to and is unique in* $E^{p,r} \cap E^{\tilde{p},r}$, *and*

$$\|(u, \nabla P)\|_{E^{\tilde{p},r}} \leq C\|u_0\|_{\dot{B}_{\tilde{p},r}^{2-\frac{2}{r}}} \exp\left(C\|u_0\|_{\dot{B}_{p,r}^{-1+\frac{d}{p}}}^{2r}\right).$$

Proof. We only write out the a priori estimates leading to the global existence statement. We shall use repeatedly the fact that

$$\|a\|_{L^\infty(\mathbb{R}_+ \times \mathbb{R}^d)} = \|a_0\|_{L^\infty} \quad \text{is small.} \tag{91}$$

First step: Estimates for $(u, \nabla P)$ *in terms of* ∇u^h. Taking advantage of Proposition 1.2 and of the rewriting of the convection term that has been pointed out in (89), we get, under assumption (91), for all $t \geq 0$,

$$\|u(t)\|_{\dot{B}_{p,r}^{-1+\frac{d}{p}}}^r + \int_0^t \|\partial_t u, \nabla^2 u, \nabla P\|_{L^p}^r \, d\tau$$

$$\lesssim \|u_0\|_{\dot{B}_{p,r}^{-1+\frac{d}{p}}}^r + \int_0^t \left(\|u^h \cdot \nabla_h u^h\|_{L^p}^r + \|u^d \partial_d u^h\|_{L^p}^r + \|u^h \cdot \nabla_h u^d\|_{L^p}^r \right.$$

$$\left. + \|u^d \mathrm{div}_h u^h\|_{L^p}^r \right) d\tau.$$

Therefore, using (70), we get

$$\|u(t)\|_{\dot{B}_{p,r}^{-1+\frac{d}{p}}}^r + \int_0^t \|\partial_t u, \nabla^2 u, \nabla P\|_{L^p}^r \, d\tau$$

$$\lesssim \|u_0\|_{\dot{B}_{p,r}^{-1+\frac{d}{p}}}^r + \int_0^t \|\nabla u^h\|_{L^{\frac{dr}{2r-1}}}^r \|\nabla u\|_{L^{\frac{dr}{2r-1}}}^r \, d\tau,$$

which combining with the interpolation inequality (69) and Young inequality eventually leads to

$$\|u(t)\|_{\dot{B}_{p,r}^{-1+\frac{d}{p}}}^r + \int_0^t \|\partial_t u, \nabla^2 u, \nabla P\|_{L^p}^r \, d\tau$$

$$\leq C\left(\|u_0\|_{\dot{B}_{p,r}^{-1+\frac{d}{p}}}^r + \int_0^t \|\nabla u^h\|_{L^{\frac{dr}{2r-1}}}^{2r} \|u\|_{\dot{B}_{p,r}^{-1+\frac{d}{p}}}^r \, d\tau \right). \tag{92}$$

Second step: Estimate of the pressure. In order to close the estimate (92) for all time, we need ∇u^h to be small enough in $L^{2r}(\mathbb{R}_+; L^{\frac{dr}{2r-1}})$. This later information will be achieved by applying maximal regularity estimates for the first equation of (89), seen as a heat equation. We first need to show that $\nabla_h P$ is small enough, though.

The goal of this step is to show that the whole gradient of P is small. As a start, let us rewrite the pressure as the solution of the following elliptic equation:

$$\operatorname{div}\left(\frac{\nabla P}{1+a}\right) = -\operatorname{div}\left(\frac{a}{1+a}\,\Delta u\right) - \operatorname{div}\left(u\cdot\nabla u\right).$$

Therefore, rewriting the convection term as in (89) and using the fat that $\nabla P(t,\cdot)$, $\Delta u(t,\cdot)$ and $(u\cdot\nabla u)(t,\cdot)$ are in L^p for almost all $t\in\mathbb{R}_+$, we may write

$$\nabla P = -\nabla\operatorname{div}\left(-\Delta\right)^{-1}\left(\frac{a}{1+a}\,\nabla P\right) + \nabla\operatorname{div}\left(-\Delta\right)^{-1}\left(\frac{a}{1+a}\,\Delta u\right)$$
$$+\nabla\operatorname{div}_h(-\Delta)^{-1}(u^h\cdot\nabla_h u^h + u^d\partial_d u^h)$$
$$+\nabla\partial_d(-\Delta)^{-1}(u^h\cdot\nabla_h u^d) - \nabla\partial_d(-\Delta)^{-1}(u^d\operatorname{div}_h u^h).$$

Hence, using the continuity of the Riesz operators from L^p to L^p,

$$\|\nabla P\|_{L^p} \lesssim \|a_0\|_{L^\infty}\left(\|\nabla P\|_{L^p} + \|\Delta u\|_{L^p}\right)$$
$$+\|u^h\cdot\nabla_h u^h + u^d\partial_d u^h\|_{L^p} + \|u^h\cdot\nabla_h u^d\|_{L^p} + \|u^d\operatorname{div}_h u^h\|_{L^p}.$$

Then, thanks to (70) and (91), we end up with

$$\|\nabla P\|_{L^p} \lesssim \|a_0\|_{L^\infty}\|\Delta u\|_{L^p} + \|\nabla u^h\|^2_{L^{\frac{dr}{2r-1}}} + \|\nabla u^h\|_{L^{\frac{dr}{2r-1}}}\|\nabla u^d\|_{L^{\frac{dr}{2r-1}}}. \tag{93}$$

Third step: Estimates of the horizontal components. We rewrite u^h as the solution to the following heat equation:

$$\partial_t u^h - \Delta u^h = -a\partial_t u^h - \nabla_h P - (1+a)(u^h\cdot\nabla_h u^h + u^d\partial_d u^h).$$

Then using the maximal regularity properties of the heat flow and (91), we get for all $t\geq 0$,

$$\|u^h(t)\|^r_{\dot{B}^{-1+\frac{d}{p}}_{p,r}} + \int_0^t \|\partial_t u^h, \nabla^2 u^h\|^r_{L^p}\,d\tau$$
$$\lesssim \|u_0^h\|^r_{\dot{B}^{-1+\frac{d}{p}}_{p,r}} + \int_0^t \left(\|\nabla_h P\|^r_{L^p} + \|u^h\cdot\nabla_h u^h + u^d\partial_d u^h\|^r_{L^p}\right)d\tau.$$

Combining with (70) and (93), the above inequality becomes:

$$\|u^h(t)\|^r_{\dot{B}^{-1+\frac{d}{p}}_{p,r}} + \int_0^t \|\partial_t u^h, \nabla^2 u^h\|^r_{L^p}\,d\tau$$
$$\lesssim \|u_0^h\|^r_{\dot{B}^{2-\frac{2}{r}}_{p,r}} + \|a_0\|^r_{L^\infty}\int_0^t \|\Delta u\|^r_{L^p}\,d\tau$$
$$+\int_0^t \|\nabla u^h\|^{2r}_{L^{\frac{dr}{2r-1}}}\,d\tau + \int_0^t \|\nabla u^h\|^r_{L^{\frac{dr}{2r-1}}}\|\nabla u^d\|^r_{L^{\frac{dr}{2r-1}}}\,d\tau,$$

and using (69) eventually leads to

$$\|u^h(t)\|^r_{\dot B^{-1+\frac{d}{p}}_{p,r}} + \int_0^t \|\partial_t u^h, \nabla^2 u^h\|^r_{L^p}\,d\tau$$

$$\lesssim \|u_0^h\|^r_{\dot B^{-1+\frac{d}{p}}_{p,r}} + \|a_0\|^r_{L^\infty}\int_0^t \|\Delta u\|^r_{L^p}\,d\tau$$

$$+ \int_0^t \|\nabla u^h\|^{2r}_{L^{\frac{dr}{2r-1}}}\,d\tau + \int_0^t \|\nabla u^d\|^{2r}_{L^{\frac{dr}{2r-1}}} \|u^h\|^r_{\dot B^{-1+\frac{d}{p}}_{p,r}}\,d\tau. \tag{94}$$

Fourth step: Bootstrap. At this stage, we assume that for some small enough constant $c > 0$, we have

$$\int_0^T \|\nabla u^h\|^{2r}_{L^{\frac{dr}{2r-1}}}\,dt \le c. \tag{95}$$

Then, Inequality (92) becomes (after using Gronwall lemma)

$$\|(u, \nabla P)\|_{E^{p,r}_T} \le C\|u_0\|_{\dot B^{2-\frac{2}{r}}_{p,r}}. \tag{96}$$

Therefore, reverting to Inequality (94) and using (69), we get

$$\|u^h\|^r_{L^\infty_T(\dot B^{-1+\frac{d}{p}}_{p,r})} + \int_0^T \|\partial_t u^h, \nabla^2 u^h\|^r_{L^p}\,d\tau$$

$$\lesssim \left(\|u_0^h\|_{\dot B^{-1+\frac{d}{p}}_{p,r}} + \|a_0\|_{L^\infty}\|u_0\|_{\dot B^{-1+\frac{d}{p}}_{p,r}}\right)^r$$

$$+ \|u^h\|^r_{L^\infty_T(\dot B^{-1+\frac{d}{p}}_{p,r})}\int_0^T \|\nabla^2 u^h\|^r_{L^p}\,d\tau + \int_0^t \|\nabla u^d\|^{2r}_{L^{\frac{dr}{2r-1}}}\|u^h\|^r_{\dot B^{-1+\frac{d}{p}}_{p,r}}\,d\tau.$$

The last term may be handled thanks to Gronwall inequality, and the last-but-one term may be absorbed by the left-hand side if c_0 has been chosen small enough (use a bootstrap argument). In the end making using of (96), we end up with

$$\|u^h\|_{L^\infty_T(\dot B^{-1+\frac{d}{p}}_{p,r})} + \|\partial_t u^h, \nabla^2 u^h\|_{L^r_T(L^p)}$$

$$\le C\left(\|u_0^h\|_{\dot B^{-1+\frac{d}{p}}_{p,r}} + \|a_0\|_{L^\infty}\|u_0\|_{\dot B^{-1+\frac{d}{p}}_{p,r}}\right)\exp\left(C\|u_0\|^{2r}_{\dot B^{-1+\frac{d}{p}}_{p,r}}\right). \tag{97}$$

Now, assuming that the right-hand side is small enough and using a standard bootstrap argument, it is easy to conclude that (96) is fulfilled for $T = +\infty$, and thus also (97). This completes the proof of the first part of the theorem (as clearly one only has to consider the case where u_0^d is *larger* than u_0^h).

Fifth step: Additional regularity. If we assume that u_0 is also in the space $\dot B^{2-\frac{2}{r}}_{\tilde p,r}$ with $d < \tilde p \le \frac{dr}{r-1}$, then applying Proposition 1.2 with f defined according to (16) and using (79) yields for all $T \ge 0$,

$$\|(u, \nabla P)\|^r_{E^{\tilde p,r}_T} \lesssim \|u_0\|^r_{\dot B^{2-\frac{2}{r}}_{\tilde p,r}} + \int_0^t \|\nabla u\|^r_{L^{\frac{dr}{2r-1}}}\|\nabla u\|^r_{L^q}\,dt.$$

Hence, thanks to the interpolation inequality (80) and Young inequality,

$$\|(u, \nabla P)\|_{E_T^{\widetilde{p},r}}^r \lesssim \|u_0\|_{\dot{B}_{\widetilde{p},r}^{2-\frac{2}{r}}}^r + \int_0^t \|\nabla u\|_{L^{\frac{2r}{2r-1}}}^{2r} \frac{dr}{} \|u\|_{\dot{B}_{\widetilde{p},r}^{2-\frac{2}{r}}}^r \, dt.$$

Then, using Gronwall inequality and (96), one can conclude that

$$\|(u, \nabla P)\|_{E_T^{\widetilde{p},r}}^r \le C\|u_0\|_{\dot{B}_{\widetilde{p},r}^{2-\frac{2}{r}}} \exp\left(C\|u_0\|_{\dot{B}_{\widetilde{p},r}^{2-\frac{2}{r}}}^{2r}\right).$$

In order to get uniqueness in the case $\widetilde{p} > d$, one just has to follow the proof of uniqueness in Theorem 2.1. $\qquad\square$

2.5 Further generalizations and open questions

- Larger values of r in Theorem 2.1 may be achieved (in particular p and r need not be interrelated). The idea goes as follows: from a formal view point, the a priori estimates that we used so far may be repeated whenever

$$t^\alpha \nabla^2 e^{t\Delta} u_0 \in L^r(\mathbb{R}_+; L^p) \quad \text{for some } \alpha \ge 0. \tag{98}$$

According to Lemma 2.1, this is equivalent to u_0 in $\dot{B}_{p,r}^{2-2\alpha-\frac{2}{r}}$, and critical regularity thus now means that

$$\frac{d}{p} = 3 - 2\alpha - \frac{2}{r}.$$

The above relation allows our taking $r \ge \frac{2p}{3p-d}$. With this viewpoint, the only limitation to give a meaning to the approximation process and to the notion of solution is that Δu has to be locally integrable in time, with values in some suitable Banach space. This is equivalent to asking that property for $t^\alpha \nabla^2 e^{t\Delta} u_0$, and thus, according to (98), to having $\alpha < 1 - 1/r$. A global existence statement based on that idea has been proved by J. Huang, M. Paicu and P. Zhang in [22].
- The same type of maximal regularity results are known to be true in more general domains than $\mathbb{R}^d$ (in particular in the half-space, or smooth bounded domains). Then, Besov spaces have to be defined by real interpolation from the homogeneous domain of the Stokes operator in L^p,

$$\dot{D}(A_p) := \{v : \Omega \to \mathbb{R}^d, \ \text{div}\, v = 0, \ \nabla^2 v \in L^p \ \text{and} \ v|_{\partial\Omega} = 0\},$$

as follows:

$$[L^p, \dot{D}(A_p)]_{1-\frac{1}{r}, r}.$$

It turns out to be possible to generalize Gagliardo-Nirenberg type inequality and to get a global existence statement similar to Theorem 2.1. More precisely, if Ω is a $\mathcal{C}^2$ bounded domain then $\dot{D}(A_p)$ coincides with $W^{2,p} \cap W_0^{1,p}$ (with div $v = 0$) and one can repeat what we presented here. In the half-space, one can even have an isotropic smallness condition similar to that of Theorem 2.2, see [15].
- As pointed out in a work in progress [14], it is possible to achieve global well-posedness with uniqueness and C^1 flow in a critical framework, if using maximal regularity estimates with Lorentz space regularity in time.

3 Third approach: time weighted energy method and vacuum

In this section, we abort the idea of scaling invariance and allow the initial velocity to have more regularity. At the same time, the density will be just bounded and will not need to be close to some positive constant. It will be even allowed to vanish without requiring a compatibility condition as in [6].

For expository purposes, we assume that the fluid domain is the two-dimensional unit torus $\mathbb{T}^2$, the reader being refereed to [13] for similar results in $\mathbb{T}^3$ or in bounded domains of $\mathbb{R}^d$ with $d = 2, 3$.

Our main goal is to prove the following statement, which is the generalization to inhomogeneous fluids with rough density of a well-known result for the constant density case (see [24, 27, 31]).

Theorem 3.1. *Consider any data* (ρ_0, u_0) *in* $L^\infty(\mathbb{T}^2) \times H^1(\mathbb{T}^2)$ *satisfying* $\operatorname{div} u_0 = 0$, $\rho_0 \not\equiv 0$ *and, for some constant* $\rho^* > 0$,

$$0 \leq \rho_0 \leq \rho^*.$$

Then, System (1) *supplemented with data* (ρ_0, u_0) *admits a unique global solution* $(\rho, u, \nabla P)$ *in the sense of* (4), (5), (6), *that satisfies the energy equality* (7), *the conservation of total mass and momentum, that is*

$$\int_{\mathbb{T}^2} \rho(t, x)\, dx = \int_{\mathbb{T}^2} \rho_0(x)\, dx \quad and$$

$$\int_{\mathbb{T}^2} (\rho u)(t, x)\, dx = \int_{\mathbb{T}^2} (\rho_0 u_0)(x)\, dx \quad for\ all\ \ t \geq 0,$$

and the following properties of regularity:

$$\rho \in L^\infty(\mathbb{R}_+; L^\infty(\mathbb{T}^2)), \quad u \in L^\infty(\mathbb{R}_+; H^1(\mathbb{T}^2)),$$
$$\sqrt{\rho}\,\partial_t u, \nabla^2 u, \nabla P \in L^2(\mathbb{R}_+; L^2(\mathbb{T}^2))$$

and also, for all $1 \leq r < 2$ *and* $1 \leq m < \infty$,

$$\nabla(\sqrt{t}P), \nabla^2(\sqrt{t}u) \in L^\infty(0, T; L^r(\mathbb{T}^2)) \cap L^2(0, T; L^m(\mathbb{T}^2)) \quad for\ all\ T > 0.$$

Furthermore, we have $\sqrt{\rho}\,u \in \mathcal{C}(\mathbb{R}_+; L^2(\mathbb{T}^2))$ *and* $\rho \in \mathcal{C}(\mathbb{R}_+; L^p(\mathbb{T}^2))$ *for all finite* p.

For proving existence *and* uniqueness while ρ is just bounded and likely to vanish, the main difficulty is to establish that, under the above hypotheses, the constructed velocity field u satisfies

$$u \in L^1_{loc}(\mathbb{R}_+; W^{1,\infty}). \tag{99}$$

In fact, if (99) is fulfilled, then it is possible to recast (1) in Lagrangian coordinates and to prove uniqueness in the same way as in the preceding sections.

To achieve (99), we shall first prove estimates in $L^\infty(\mathbb{R}_+; L^2)$ for ∇u, and in $L^2(\mathbb{R}_+; L^2)$ for $\nabla^2 u$, ∇P and $\sqrt{\rho}\,\partial_t u$. Then, using time weighted energy estimates in the spirit of [28], classical Sobolev embedding and estimates for the stationary Stokes system (so as to shift integrability from time variable to space variable), will eventually lead to (99).

3.1 High order energy estimates

For notational simplicity, we assume throughout that

$$\int_{\mathbb{T}^2} \rho_0 \, dx = 1 \quad \text{and} \quad \int_{\mathbb{T}^2} \rho_0 u_0 \, dx = 0. \tag{100}$$

Neither of those assumptions is restrictive, as one may rescale the density accordingly, and use the Galilean invariance of the system.

The main goal of this paragraph is to prove the following a priori estimate for global solutions to (1):

$$\|\nabla u(t)\|_{L^2}^2 + \frac{1}{2}\int_0^t \left(\|\sqrt{\rho}\,\partial_t u\|_{L^2}^2 + \frac{1}{4\rho^*}\|\nabla^2 u, \nabla P\|_{L^2}^2 \right) d\tau \le \left(e + \|\nabla u_0\|_{L^2}^2 \right)^{K_0} - e, \tag{101}$$

with $K_0 := \exp\left(C(\rho^*)^2 \|\sqrt{\rho_0}\,u_0\|_{L^2}^4 \log\left(e + \|\rho_0 - 1\|_{L^2}^2 + \frac{\rho^*}{\|\sqrt{\rho_0}\,u_0\|_{L^2}^2} \right) \right).$

To achieve our goal, we take the L^2 scalar product of the momentum equation of (1) with $\partial_t u$, and get

$$\frac{1}{2}\frac{d}{dt}\int_{\mathbb{T}^2} |\nabla u|^2 dx + \int_{\mathbb{T}^2} \rho|\partial_t u|^2 \, dx = -\int_{\mathbb{T}^2} (\rho u \cdot \nabla u)\cdot \partial_t u \, dx \tag{102}$$

$$\le \frac{1}{2}\int_{\mathbb{T}^2} \rho|\partial_t u|^2 \, dx + \frac{1}{2}\int_{\mathbb{T}^2} \rho|u \cdot \nabla u|^2 \, dx. \tag{103}$$

To estimate $\nabla^2 u$ and ∇P, we use the fact that

$$-\Delta u + \nabla P = -(\rho \partial_t u + \rho u \cdot \nabla u) \quad \text{and} \quad \operatorname{div} \Delta u = 0.$$

Hence we have

$$\|\nabla^2 u\|_{L^2}^2 + \|\nabla P\|_{L^2}^2 = \|\rho(\partial_t u + u \cdot \nabla u)\|_{L^2}^2 \le 2\rho^* \left(\int_{\mathbb{T}^2} \rho|\partial_t u|^2 \, dx + \int_{\mathbb{T}^2} \rho|u\cdot\nabla u|^2 \, dx \right).$$

Putting together with (102) thus yields

$$\frac{d}{dt}\int_{\mathbb{T}^2} |\nabla u|^2 dx + \frac{1}{2}\int_{\mathbb{T}^2} \rho|\partial_t u|^2 \, dx + \frac{1}{4\rho^*}\left(\|\nabla^2 u\|_{L^2}^2 + \|\nabla P\|_{L^2}^2 \right) \le \frac{3}{2}\int_{\mathbb{T}^2} \rho|u\cdot\nabla u|^2 \, dx. \tag{104}$$

At this stage, the standard way to bound the right-hand side is to use Hölder then Gagliardo-Nirenberg inequality to eventually get

$$\int_{\mathbb{T}^2} \rho |u \cdot \nabla u|^2 \, dx \leq \rho^* \|u\|_{L^4}^2 \|\nabla u\|_{L^4}^2$$

$$\leq C\rho^* \|u\|_{L^2} \|\nabla u\|_{L^2}^2 \|\nabla^2 u\|_{L^2}$$

$$\leq \frac{1}{12\rho^*} \|\nabla^2 u\|_{L^2}^2 + C(\rho^*)^3 \|u\|_{L^2}^2 \|\nabla u\|_{L^2}^4.$$

If the density is bounded away from 0, then one may use

$$\rho_* \|u\|_{L^2}^2 \leq \|\sqrt{\rho} u\|_{L^2}^2,$$

so that inserting the resulting inequality in (104) and taking advantage of (7) allows to close the H^1 estimates globally in time. In the vacuum case, the substitute to that argument is the following Desjardins' interpolation inequality (first proved in [18]) for functions in $H^1(\mathbb{T}^2)$ with total mass equal to 1:

$$\left(\int_{\mathbb{T}^2} \rho z^4 \, dx \right)^{\frac{1}{2}} \leq C \|\sqrt{\rho} z\|_{L^2} \|\nabla z\|_{L^2} \log^{\frac{1}{2}} \left(e + \|\rho - 1\|_{L^2}^2 + \frac{\rho^* \|\nabla z\|_{L^2}^2}{\|\sqrt{\rho} z\|_{L^2}^2} \right). \quad (105)$$

Applying that inequality to $z = u$ then using the mass conservation, the energy balance (7) and the fact that the function $z \mapsto z \log(A + 1/z)$ is increasing, we get

$$\|\sqrt{\rho} |u|^2\|_{L^2}^2 \leq C \|\sqrt{\rho_0} u_0\|_{L^2}^2 \|\nabla u\|_{L^2}^2 \log \left(e + \|\rho_0 - 1\|_{L^2}^2 + \rho^* \frac{\|\nabla u\|_{L^2}^2}{\|\sqrt{\rho_0} u_0\|_{L^2}^2} \right). \quad (106)$$

Therefore,

$$\int_{\mathbb{T}^2} \rho |u \cdot \nabla u|^2 dx$$

$$\leq \frac{1}{12\rho^*} \|\nabla^2 u\|_{L^2}^2 + C(\rho^*)^2 \|\sqrt{\rho_0} u_0\|_{L^2}^2 \|\nabla u\|_{L^2}^4 \log \left(e + \|\rho_0 - 1\|_{L^2}^2 + \rho^* \frac{\|\nabla u\|_{L^2}^2}{\|\sqrt{\rho_0} u_0\|_{L^2}^2} \right).$$

Then combining with (104) yields

$$\frac{d}{dt} \int_{\mathbb{T}^2} |\nabla u|^2 dx + \frac{1}{2} \int_{\mathbb{T}^2} \left(\rho |\partial_t u|^2 + \frac{1}{4\rho^*} (|\nabla^2 u|^2 + |\nabla P|^2) \right) dx$$

$$\leq C_0 \|\nabla u\|_{L^2}^2 \log(e + \|\nabla u\|_{L^2}^2) \|\nabla u\|_{L^2}^2$$

where $C_0 := C(\rho^*)^2 \|\sqrt{\rho_0} u_0\|_{L^2}^2 \log \left(e + \|\rho_0 - 1\|_{L^2}^2 + \frac{\rho^*}{\|\sqrt{\rho_0} u_0\|_{L^2}^2} \right).$

Denoting $f(t) := C_0 \|\nabla u(t)\|_{L^2}^2$ and

$$X(t) := \int_{\mathbb{T}^2} |\nabla u(t)|^2 \, dx + \frac{1}{2} \int_{\mathbb{T}^2} \left(\rho |\partial_t u|^2 + \frac{1}{4\rho^*} (|\nabla^2 u|^2 + |\nabla P|^2) \right) dx,$$

the above inequality rewrites

$$\frac{d}{dt} X \leq f X \log(e + X),$$

from which we easily get (101), using once more (7).

Let us now prove (105). Set $\widetilde{z} := z - \bar{z}$, fix some integer n, then decompose $\widetilde{z}$ into low and high frequencies according to Fourier series, as follows:

$$\widetilde{z} = \widetilde{z}_n + \widetilde{z}^n \quad \text{with} \quad \widetilde{z}_n(x) := \sum_{1 \le |k| \le n} \widehat{z}_k\, e^{2i\pi k \cdot x}.$$

Then we write

$$\left(\int_{\mathbb{T}^2} \rho z^4\, dx\right)^{\frac{1}{2}} = \left(\int_{\mathbb{T}^2} \left(\bar{z} + \widetilde{z}_n + \widetilde{z}^n\right)^2 \rho z^2 dx\right)^{\frac{1}{2}}$$

$$\le |\bar{z}|\|\sqrt{\rho}z\|_{L^2} + \|\widetilde{z}_n\|_{L^\infty}\|\sqrt{\rho}z\|_{L^2} + (\rho^*)^{\frac{1}{4}}\|\widetilde{z}^n\|_{L^4}\left(\int_{\mathbb{T}^2} \rho z^4\, dx\right)^{\frac{1}{4}}.$$

Hence

$$\left(\int_{\mathbb{T}^2} \rho z^4\, dx\right)^{\frac{1}{2}} \le 2|\bar{z}|\|\sqrt{\rho}\,z\|_{L^2} + 2\|\widetilde{z}_n\|_{L^\infty}\|\sqrt{\rho}\,z\|_{L^2} + \sqrt{\rho^*}\|\widetilde{z}^n\|_{L^4}^2. \tag{107}$$

From the embedding $\dot{H}^{\frac{1}{2}} \hookrightarrow L^4$, we easily get

$$\|\widetilde{z}^n\|_{L^4} \le Cn^{-1/2}\|\nabla z\|_{L^2},$$

and it is also clear that for all x in $\mathbb{T}^2$,

$$|\widetilde{z}_n(x)| \le \sum_{1 \le |k| \le n} |2\pi k \widehat{z}_k| \frac{|e^{2i\pi k \cdot x}|}{2\pi|k|}$$

$$\le \|\nabla z\|_{L^2}\left(\sum_{1 \le |k| \le n} \frac{1}{4\pi^2|k|^2}\right)^{1/2}$$

$$\le C\sqrt{\log n}\,\|\nabla z\|_{L^2}.$$

Therefore, reverting to (107), we conclude that

$$\int_{\mathbb{T}^2} \rho z^4\, dx \lesssim \bar{z}^2\|\sqrt{\rho}z\|_{L^2}^2 + \left(\log n\|\sqrt{\rho}z\|_{L^2}^2 + n^{-2}\rho^*\|\nabla z\|_{L^2}^2\right)\|\nabla z\|_{L^2}^2.$$

Taking for n the closest positive integer to $\frac{\sqrt{\rho^*}\|\nabla z\|_{L^2}}{\|\sqrt{\rho}z\|_{L^2}}$, we end up with

$$\left(\int_{\mathbb{T}^2} \rho z^4\, dx\right)^{\frac{1}{2}} \le C\|\sqrt{\rho}z\|_{L^2}\left(|\bar{z}| + \|\nabla z\|_{L^2}\log^{\frac{1}{2}}\left(e + \frac{\rho^*\|\nabla z\|_{L^2}^2}{\|\sqrt{\rho}z\|_{L^2}^2}\right)\right). \tag{108}$$

Finally, to bound the average $\bar{z}$, it suffices to use the following lemma:

Lemma 3.1. *Let $a : \mathbb{T}^2 \to \mathbb{R}$ be a nonnegative measurable function with total mass $M > 0$. Then we have for all z in $H^1(\mathbb{T}^2)$,*

$$\|z\|_{L^2} \le \frac{1}{M}\left|\int_{\mathbb{T}^2} az\, dx\right| + C\log^{\frac{1}{2}}\left(e + \frac{\|a - M\|_{L^2}}{M}\right)\|\nabla z\|_{L^2}. \tag{109}$$

Proof. We start with the obvious inequality

$$\|z\|_2 \le |\bar{z}| + \|\nabla z\|_2 \quad \text{with} \quad \bar{z} := \int_{\mathbb{T}^2} z\, dx, \tag{110}$$

then we use the fact that

$$M\bar{z} = \int_{\mathbb{T}^2} az\, dx + \int_{\mathbb{T}^2} (M-a)\tilde{z}\, dx \quad \text{with} \quad \tilde{z} := z - \bar{z}. \tag{111}$$

Then, exactly as above, we decompose $\tilde{z}$ into $\tilde{z}_n + \tilde{z}^n$, then write that

$$\left| \int_{\mathbb{T}^2} (a-M)\tilde{z}\, dx \right| \le M\|\tilde{z}_n\|_{L^\infty} + \|a - M\|_{L^2}\|\tilde{z}^n\|_{L^2},$$

and take the "best" n. The details are left to the reader. $\qquad\square$

3.2   Time weighted estimates

That ∇u is in $L^2(\mathbb{R}_+; H^1)$ does not quite ensure that the flow is Lipschitz. Recall however that in the constant density case (or even if the density is positive and sufficiently smooth, see [8]), having u_0 in H^1 guarantees that ∇u is in $L^1_{loc}(\mathbb{R}_+; H^s)$ for all $s < 2$, and the flow is thus almost in $\mathcal{C}^2$.

In order to recover that information in our case, one cannot differentiate the momentum equation of (1) with respect to the space variable, since $\nabla\rho$ would appear. On the other side, differentiating along the flow (i.e. taking convective derivative) should be harmless as $(\partial_t + u \cdot \nabla)\rho = 0$. We shall see below that just taking one time derivative and using a sufficient number of integration by parts also works. Indeed, differentiating the momentum equation once in time yields

$$\rho\partial_{tt}^2 u + \rho u \cdot \nabla\partial_t u - \Delta\partial_t u + \nabla\partial_t P = f \quad \text{with} \quad f := -\rho_t\partial_t u - \rho_t u \cdot \nabla u - \rho\partial_t u \cdot \nabla u. \tag{112}$$

Hence, taking the scalar product with $\partial_t u$,

$$\frac{1}{2}\frac{d}{dt}\int_{\mathbb{T}^2} \rho|\partial_t u|^2\, dx + \int_{\mathbb{T}^2} |\nabla\partial_t u|^2\, dx = \int_{\mathbb{T}^2} f \cdot \partial_t u\, dx.$$

The above computation requires the term $\|\sqrt{\rho}\partial_t u\|_{L^2}$ to have some limit for t tending to 0. This is equivalent to the fact that there exists some functions P_0 and g in L^2 such that

$$-\Delta u_0 + \nabla P_0 = \sqrt{\rho_0}\, g.$$

In other words, u_0 has to be (at least) in H^2, and has to satisfy some compatibility condition that forces u_0 to be an harmonic function inside vacuum patches if any.

From a scaling point of view, it is not a surprise that having $\partial_t u$ in $L^\infty_{loc}(\mathbb{R}_+; L^2)$ is equivalent to $\nabla^2 u$ in L^2. For the same reason, the H^1 (spatial) regularity we assumed for u_0 corresponds to just half a time derivative. In fact, as regards the heat flow, this is almost equivalent to the fact that $\sqrt{t}\,\partial_t u$ is in L^2 since

$$\int_0^{+\infty} \sqrt{t}\|\partial_t(e^{t\Delta}u_0)\|_{L^2}\, dt = \int_0^{+\infty} \sqrt{t}\|e^{t\Delta}\Delta u_0\|_{L^2}\, dt \approx \|\Delta u_0\|_{\dot{B}_{2,\infty}^{-1}}.$$

As $L^2 \hookrightarrow \dot{B}^0_{2,\infty}$, in the case we are interested in where ∇u_0 is in L^2, arguing by density, we see that we have in addition

$$\lim_{t \to 0} \sqrt{t} \|\partial_t(e^{t\Delta} u_0)\|_{L^2} = 0.$$

Based on the above heuristics, we plan to bound $\sqrt{\rho t}\, \partial_t u$ and $\sqrt{t}\, \nabla \partial_t u$ in $L^\infty([0,T]; L^2)$ and $L^2([0,T]; L^2)$, respectively. Hence we take the L^2 scalar product of (112) with $t\, \partial_t u$, and get

$$\frac{1}{2} \frac{d}{dt} \int_{\mathbb{T}^2} \rho t |\partial_t u|^2 \, dx + \int_{\mathbb{T}^2} t |\nabla \partial_t u|^2 \, dx = \sum_{i=1}^{5} I_i, \tag{113}$$

with

$$I_1 = \frac{1}{2} \int_{\mathbb{T}^2} \rho |\partial_t u|^2 \, dx, \tag{114}$$

$$I_2 = -\int_{\mathbb{T}^2} t \rho_t |\partial_t u|^2 \, dx, \tag{115}$$

$$I_3 = -\int_{\mathbb{T}^2} \left(\sqrt{t}\, \rho_t u \cdot \nabla u \right) \cdot \left(\sqrt{t}\, \partial_t u \right) dx, \tag{116}$$

$$I_4 = -\int_{\mathbb{T}^2} \left(\sqrt{t}\, \rho \partial_t u \cdot \nabla u \right) \cdot \left(\sqrt{t}\, \partial_t u \right) dx, \tag{117}$$

$$I_5 = -\int_{\mathbb{T}^2} \left(\sqrt{t}\, \rho u \cdot \nabla \partial_t u \right) \cdot \left(\sqrt{t}\, \partial_t u \right) dx. \tag{118}$$

To bound I_2, we write that

$$I_2 = \int_{\mathbb{T}^2} t \, \mathrm{div}\,(\rho u) |\partial_t u|^2 \, dx \leq 2 \int_{\mathbb{T}^2} t \rho |u| \, |\nabla \partial_t u| \, |\partial_t u| \, dx$$

$$\leq \frac{1}{10} \|\sqrt{t}\, \nabla \partial_t u\|_{L^2}^2 + C\rho^* \|\sqrt{\rho t}\, \partial_t u\|_{L^2}^2 \|u\|_{L^\infty}^2.$$

To handle I_3, we use the continuity equation and perform an integration by parts:

$$I_3 = -\int_{\mathbb{T}^2} \left(\sqrt{t}\, \rho_t u \cdot \nabla u \right) \cdot \left(\sqrt{t}\, \partial_t u \right) dx = -\int_{\mathbb{T}^2} t \rho u \cdot \nabla[(u \cdot \nabla u) \cdot \partial_t u] \, dx.$$

Hence

$$I_3 \leq \int_{\mathbb{T}^2} t \rho |u| \left(|\nabla u|^2 |\partial_t u| + |u| \, |\nabla^2 u| \, |\partial_t u| + |u| \, |\nabla u| \, |\nabla \partial_t u| \right) dx =: I_{31} + I_{32} + I_{33}. \tag{119}$$

To bound I_{31}, we just write that for all $t \in [0,T]$,

$$I_{31} = \int_{\mathbb{T}^2} \sqrt{\rho t}\, |u| |\nabla u| |\nabla u| \, |\sqrt{\rho t}\, \partial_t u| \, dx \leq \|u\|_{L^\infty}^2 \|\sqrt{\rho t}\partial_t u\|_{L^2}^2 + CT\rho^* \|\nabla u\|_{L^4}^4. \tag{120}$$

For I_{32}, we have

$$I_{32} = \int_{\mathbb{T}^2} t\rho|u|^2|\nabla^2 u|\,|\partial_t u|\,dx \le \rho^* T\|\nabla^2 u\|_{L^2}^2 + \|u\|_{L^\infty}^4\|\sqrt{\rho t}\,\partial_t u\|_{L^2}^2.$$

For I_{33}, we have

$$I_{33} = \int_{\mathbb{T}^2} t\rho|u|^2|\nabla u|\,|\nabla\partial_t u|\,dx \le C\int_{\mathbb{T}^2} t\rho^2|u|^4|\nabla u|^2\,dx + \frac{1}{10}\int_{\mathbb{T}^2}|\nabla\sqrt{t}\,\partial_t u|^2\,dx.$$

To handle the term I_4, we write that for all $0 \le t \le T$,

$$I_4 \le \|\nabla u\|_{L^2}\|\sqrt{\rho t}\,\partial_t u\|_{L^4}^2$$

$$\le (\rho^*)^{3/4}\|\nabla u\|_{L^2}\|\sqrt{\rho t}\,\partial_t u\|_{L^2}^{\frac{1}{2}}\|\sqrt{t}\,\partial_t u\|_{L^6}^{\frac{3}{2}}$$

$$\le C(\rho^*)^{3/4}\|\nabla u\|_{L^2}\|\sqrt{\rho t}\,\partial_t u\|_{L^2}^{\frac{1}{2}}\|\sqrt{t}\,\nabla\partial_t u\|_{L^2}^{\frac{3}{2}}$$

$$\le \frac{1}{10}\|\sqrt{t}\,\nabla\partial_t u\|_{L^2}^2 + C(\rho^*)^3 T\|\sqrt{\rho}\partial_t u\|_{L^2}^2\|\nabla u\|_{L^2}^4.$$

Finally, for I_5, we observe that

$$I_5 = \int_{\mathbb{T}^2} |\rho u|\,|\sqrt{t}\,\nabla\partial_t u|\,|\sqrt{t}\,\partial_t u|\,dx \le \frac{1}{10}\|\nabla\sqrt{t}\,\partial_t u\|_{L^2}^2 + C\rho^*\|u\|_{L^\infty}^2\|\sqrt{\rho t}\,\partial_t u\|_{L^2}^2.$$

So altogether, we get for some constant C_{T,ρ^*} depending only on ρ^* and T,

$$\frac{d}{dt}\|\sqrt{\rho t}\,\partial_t u\|_{L^2}^2 + \|\nabla\sqrt{t}\,\partial_t u\|_{L^2}^2$$

$$\le C\big((1+\rho^*)\|u\|_{L^\infty}^2 + \|u\|_{L^\infty}^4\big)\|\sqrt{\rho t}\,\partial_t u\|_{L^2}^2$$

$$+ C_{T,\rho^*}\big(\|\nabla u\|_{L^4}^4 + \|\nabla^2 u\|_{L^2}^2 + \|u\|_{L^\infty}^4\|\nabla u\|_{L^2}^2 + \|\sqrt{\rho}\partial_t u\|_{L^2}^2(1+\|\nabla u\|_{L^2}^4)\big).$$

Therefore, using Gronwall Lemma,

$$\|\sqrt{\rho t}\,\partial_t u\|_{L^2}^2 + \int_0^t \|\nabla\sqrt{t}\,\partial_t u\|_{L^2}^2\,d\tau$$

$$\le C_{T,\rho*}\int_0^t e^{C(1+\rho^*)\int_\tau^t\left(\|u\|_{L^\infty}^2+\|u\|_{L^\infty}^4\right)d\tau'}\big(\|\nabla u\|_{L^4}^4$$

$$+\|\nabla^2 u\|_{L^2}^2 + \|u\|_{L^\infty}^4\|\nabla u\|_{L^2}^2 + \|\sqrt{\rho}\partial_t u\|_{L^2}^2(1+\|\nabla u\|_{L^2}^4)\big)\,d\tau.$$

By taking advantage of the energy equality (7), of the H^1 estimate (101) and of Sobolev embedding, it is clear that the right-hand side may be bounded just in terms of the data and of t: there exists a nondecreasing continuous function h depending only on the data, so that $h(0) = 0$ and for all $t \ge 0$,

$$\|\sqrt{\rho t}\,\partial_t u\|_{L^2} + \int_0^t \|\nabla\sqrt{\tau}\,\partial_t u\|_{L^2}^2\,d\tau \le h(t). \qquad (121)$$

From (121), one can deduce that $\sqrt{t}\partial_t u$ is in $L^2_{loc}(\mathbb{R}_+; L^q)$ for all $q < \infty$. Indeed, denoting by $\overline{(\partial_t u)}$ the average of $\partial_t u$, one can write that

$$\int_{\mathbb{T}^2} \rho\partial_t u\,dx = \overline{(\partial_t u)} + \int_{\mathbb{T}^2}\rho(\partial_t u - \overline{(\partial_t u)})\,dx.$$

Hence,

$$|\overline{(\partial_t u)}| \leq \|\rho\|_{L^2} \|\nabla \partial_t u\|_{L^2} + \|\sqrt{\rho}\, \partial_t u\|_{L^2}.$$

Consequently, by Sobolev embedding, and because $\|\rho\|_{L^2}$ is time independent, we have for all $T > 0$ and $q < \infty$,

$$\|\sqrt{t}\, \partial_t u\|_{L^2(0,T;L^q)} \leq (C_q + \|\rho_0\|_{L^2}) \|\sqrt{t}\, \nabla \partial_t u\|_{L^2(0,T;L^2)} + \|\sqrt{\rho t}\, \partial_t u\|_{L^2(0,T;L^2)}. \tag{122}$$

3.3 Shift of integrability

Let us rewrite the momentum equation of (1) as follows:

$$\begin{cases} -\Delta \sqrt{t}\, u + \nabla \sqrt{t}\, P = -\sqrt{t}\, \rho \partial_t u - \sqrt{t}\, \rho u \cdot \nabla u, \\ \operatorname{div} \sqrt{t}\, u = 0. \end{cases} \tag{123}$$

As ρ is bounded, (121) ensures that for all $T > 0$, $\sqrt{t}\, \rho \partial_t u$ is in $L^\infty(0,T;L^2)$, and that $\rho \sqrt{t}\, \partial_t u$ is in $L^2(0,T;L^q)$ for all finite q. Therefore, by Hölder inequality, we have

$$\|\rho \sqrt{t}\, \partial_t u\|_{L^p(0,T;L^r)} \leq C_{0,T} \quad \text{for all } p \in [2,\infty] \text{ and } 2 \leq r < \frac{2p}{p-2}. \tag{124}$$

Similarly, the bounds for ∇u in $L^\infty(0,T;L^2) \cap L^2(0,T;H^1)$ imply that

$$\|\nabla u\|_{L^p(0,T;L^r)} \leq C_{0,T} \quad \text{for all } p \geq 2 \text{ and } r < \frac{2p}{p-2}, \tag{125}$$

and as, obviously, u is bounded in all spaces $L^q(0,T;L^r)$ (except $q = r = \infty$), one can conclude that

$$\|\sqrt{t}\, \rho u \cdot \nabla u\|_{L^p(0,T;L^r)} \leq C_{0,T} \quad \text{for all } p \in [2,\infty] \text{ and } 2 \leq r < \frac{2p}{p-2}. \tag{126}$$

Then, the maximal regularity estimate for the Stokes system implies that

$$\|\nabla^2 \sqrt{t}\, u, \nabla \sqrt{t}\, P\|_{L^p(0,T;L^r)} \leq C_{0,T} \quad \text{for all } 2 \leq p \leq \infty \text{ and } 2 \leq p < \frac{2p}{p-2}. \tag{127}$$

3.4 The proof of existence

We shall take advantage of classical results for (1) supplemented with smooth data with positive density, and of compactness arguments. More precisely, consider a sequence (ρ_0^n, u_0^n) of smooth approximate data with

$$\operatorname{div} u_0^n = 0, \quad \text{and} \quad n^{-1} \leq \rho_0^n \leq \rho,$$

and such that

$$u_0^n \to u_0 \text{ in } H^1, \quad \rho_0^n \rightharpoonup \rho_0 \text{ in } L^\infty \text{ weak} \star \text{ and } \rho_0^n \to \rho_0 \text{ in } L^p, \text{ if } p < \infty. \tag{128}$$

Then, in light of the results of [8, 25], there exists a unique global smooth solution (ρ^n, u^n, P^n) corresponding to data (ρ_0^n, u_0^n), and satisfying $n^{-1} \leq \rho^n \leq \rho^*$.

Being smooth, the triplet (ρ^n, u^n, P^n) satisfies the estimates of the previous subsection, namely for some constant C_0 depending only on the initial data (ρ_0, u_0),

$$\|\sqrt{\rho^n}\partial_t u^n, \nabla P^n\|_{L^2(\mathbb{R}_+ \times \mathbb{T}^2)} + \|\nabla u^n\|_{L^\infty(\mathbb{R}_+;L^2) \cap L^2(\mathbb{R}_+;H^1)} \leq C_0 \tag{129}$$

and, for all $T > 0$ and constant $C_{0,T}$ depending only on the initial data and on T,

$$\sup_{t \in [0,T]} \left(t \|\sqrt{\rho^n(t)}\partial_t u^n(t)\|_2^2 \right) + \int_0^T t \|\nabla \partial_t u^n\|_2^2 \, dt \leq C_{0,T}, \tag{130}$$

$$\|\nabla^2 \sqrt{t}\, u^n, \nabla \sqrt{t}\, P^n\|_{L^p(0,T;L^r)} \leq C_{0,T} \quad \text{for all } 2 \leq p \leq \infty \text{ and } 2 \leq p < \frac{2p}{p-2}. \tag{131}$$

In order to glean some compactness for u^n, we are going to use the following interpolation inequality (see the proof in [13])

$$\|z\|_{H^{\frac{1}{2}-\alpha}(0,T;L^2)}^2 \leq \|z\|_{L^2(0,T;L^2)}^2 + C_{\alpha,T}\|\sqrt{t}\, \partial_t z\|_{L^2(0,T;L^2)}^2, \qquad \alpha \in (0,1/2) \tag{132}$$

with $C_{\alpha,T}$ depending only on α and on T.

Combining that inequality with Lemma 3.1, Inequality (122) and the uniform bounds (129), (130), (131), we deduce that $(u^n)_{n \in \mathbb{N}}$ is bounded in $H^{\frac{1}{2}-\alpha}(0,T;L^2)$ for all $T > 0$ and $\alpha \in (0,1/2)$. Then, interpolating with (129), one can conclude that $(u^n)_{n \in \mathbb{N}}$ is bounded in, say, $H^{\frac{1}{8}}(0,T \times \mathbb{T}^2)$ for all $T > 0$. Finally, taking advantage of the compact embedding $H^{\frac{1}{8}}(0,T \times \mathbb{T}^2) \hookrightarrow L^2(0,T \times \mathbb{T}^2)$, we see that, up to subsequence, $u^n \to u$ in $L^2_{loc}(\mathbb{R}_+;L^2)$ for some u that, in addition, satisfies (129) (as regards ∇P^n and u^n).

For the density, we have $\rho^n \rightharpoonup \rho$ in $L^\infty(\mathbb{R}_+ \times \mathbb{T}^2)$ and thus $0 \leq \rho \leq \rho^*$. All those informations are more than enough to justify that (ρ, u) is a weak solution to (1) in the meaning of (4), (5), (6), and that (131) is fulfilled by $(u, \nabla P)$. Now, arguing as in [15], page 2405, one can show that for all $t \in \mathbb{R}_+$, we have

$$(\rho^n(t))^2 \rightharpoonup (\rho(t))^2 \text{ in } L^\infty, \tag{133}$$

which eventually implies that

$$\rho^n \to \rho \text{ in } \mathcal{C}(\mathbb{R}_+;L^p) \text{ for all finite } p. \tag{134}$$

Indeed, it suffices to prove that $\rho^n(t) \to \rho(t)$ strongly in L^2, for all $t \geq 0$. Then we write that

$$\int_{\mathbb{T}^2} (\rho^n - \rho)^2 \, dx = \int_{\mathbb{T}^2} (\rho^n)^2 \, dx - 2 \int_{\mathbb{T}^2} \rho \rho^n \, dx + \int_{\mathbb{T}^2} \rho^2 \, dx.$$

For the first term in the right-hand side, we may use (133) and for the second one, we may use the weak convergence in L^2.

Now, we note that (122) implies that

$$\sqrt{t}\,\partial_t u^n \to \sqrt{t}\,\partial_t u \quad \text{in} \quad L^2_{loc}(\mathbb{R}_+; L^2).$$

Hence the strong convergence results for $(\rho^n)_{n\in\mathbb{N}}$ ensure that $(\sqrt{t\rho^n}\,\partial_t u^n)_{n\in\mathbb{N}}$ converges in the sense of distributions to $\sqrt{t\rho}\,\partial_t u$ in $L^2_{loc}(\mathbb{R}_+; L^2)$. Therefore (130) is satisfied by (ρ, u). Similarly, (ρ, u) satisfies (129).

Of course, being smooth, the solution (ρ^n, u^n, P^n) fulfills (7), the total mass and momentum conservation as well as $\|\rho^n\|_{L^\infty} \le \rho^*$, and thus

$$\|\sqrt{\rho^n(t)}\,u^n(t)\|_2^2 + 2\int_{t_0}^t \|\nabla u^n\|_2^2\,d\tau = \|\sqrt{\rho^n(t_0)}\,u^n(t_0)\|_2^2 \quad \text{for all } t, t_0 \ge 0. \quad (135)$$

The fact that $u^n \to u$ in $L_{2,loc}(\mathbb{R}_+; H^1)$ guarantees that, for all $t \ge 0$, the second term converges to $\int_{t_0}^t \|\nabla u\|_2^2\,d\tau$. Next, (130) guarantees that $\nabla\partial_t u^n$ is bounded in $L^2(t_0, T \times \mathbb{T}^2)$ for all $0 < t_0 < T$. Then combining with (122) gives boundedness of $\partial_t u^n$ in $L^2(t_0, T; H^1)$. Now, because u^n is bounded in $L^\infty(t_0, T; H^1)$ thanks to (101), one can conclude that $u^n \to u$ strongly in $\mathcal{C}([t_0, T]; L^p)$ for all $p < \infty$. Combining with (134) ensures that $\sqrt{\rho^n}u^n$ tends to $\sqrt{\rho}u$ in $\mathcal{C}(]0, +\infty[; L^2)$, and that one can pass to the limit in all the terms of (135). We eventually get

$$\|\sqrt{\rho(t)}\,u(t)\|_2^2 + 2\int_{t_0}^t \|\nabla u\|_2^2\,d\tau = \|\sqrt{\rho(t_0)}\,u(t_0)\|_2^2 \quad \text{for all } t \ge t_0 \ge 0. \quad (136)$$

Finally, to get the strong continuity of $\sqrt{\rho}u$ at $t = 0$, we notice that the uniform bounds ensure the weak continuity, and that (136) gives

$$\|\sqrt{\rho(t)}\,u(t)\|_2^2 \longrightarrow \|\sqrt{\rho_0}\,u_0\|_2^2 \quad \text{for } t \to 0.$$

This completes the proof of the existence. □

3.5 Uniqueness

Recall that, in Lagrangian coordinates, the density is time independent, and the other two equations of System (1) rewrite

$$\begin{cases} \rho_0\partial_t\bar{u} - \operatorname{div}\left(A^T A\nabla\bar{u}\right) + {}^T A\cdot\nabla\bar{P} = 0, \\ \operatorname{div}\left(A\bar{u}\right) = {}^T A : \nabla\bar{u} = 0, \end{cases} \quad (137)$$

with

$$A = \sum_{k=0}^{+\infty}(-1)^k\left(\int_0^t Du(\tau, \cdot)\,d\tau\right)^k. \quad (138)$$

In our regularity framework, that latter system is equivalent to (1) whenever (say)

$$\int_0^T \|\nabla u\|_{L^\infty}\,d\tau \le \frac{1}{2}. \quad (139)$$

Now, consider two solutions (ρ_1, u_1, P_1) and (ρ_2, u_2, P_2) of (1) on $[0, T] \times \mathbb{T}^2$, emanating from the same initial data, satisfying (139) and, for $i = 1, 2$,

$$\sqrt{t}\nabla u_i \in L^2(0, T; L^\infty), \quad \nabla u_i \in L^1(0, T; L^\infty) \cap L^4(0, T; L^3) \cap L^2(0, T; L^6),$$

$$u_i \in L^4(0, T; L^\infty), \quad \sqrt{t}\nabla P_i \in L^2(0, T; L^4) \quad \text{and} \quad \sqrt{t}\partial_t u_i \in L^{4/3}(0, T; L^6). \tag{140}$$

The couples $(\bar{u}_1, \bar{P}_1)$ and $(\bar{u}_2, \bar{P}_2)$ (that is the velocity and the pressure in Lagrangian coordinates) fulfill also (139) (taking T smaller as the case may be) and (140). Then, $\delta A := A_2 - A_1$, $\delta u := \bar{u}_2 - \bar{u}_1$ and $\delta P := \bar{P}_2 - \bar{P}_1$ satisfy

$$\begin{cases} \rho_0 \partial_t \delta u - \mathrm{div}\,(A_1{}^T A_1 \nabla \delta u) + {}^T A_1 \cdot \nabla \delta P = \mathrm{div}\,((A_2{}^T A_2 - A_1{}^T A_1)\nabla \bar{u}_2) + {}^T \delta A \cdot \nabla \bar{P}_2, \\ \mathrm{div}\,(A_1 \delta u) = \mathrm{div}\,(\delta A\, \bar{u}_2) = {}^T \delta A : \nabla \bar{u}_2, \\ \delta u|_{t=0} = 0. \end{cases}$$

In order to prove that $\delta u \equiv 0$, it looks natural to take the L^2 scalar product of the above equation with δu. However, we have to take into consideration the fact that the term corresponding to ${}^T A_1 \nabla \delta P$ does not vanish since $\mathrm{div}\,(A_1 \delta u) \neq 0$. To overcome that difficulty, we decompose δu into

$$\delta u = w + z, \tag{141}$$

where w stands for a suitable solution to:

$$\mathrm{div}\,(A_1 w) = \mathrm{div}\,(\delta A\, u_2) = {}^T \delta A : \nabla \bar{u}_2, \qquad w|_{t=0} = 0, \tag{142}$$

and z fulfills $z(0) = 0$ and

$$\begin{cases} \rho_0 \partial_t z - \mathrm{div}\,(A_1{}^T A_1 \nabla z) + {}^T A_1 \cdot \nabla \delta P \\ \qquad = \mathrm{div}\,((A_2{}^T A_2 - A_1{}^T A_1)\nabla \bar{u}_2) + {}^T \delta A \cdot \nabla \bar{P}_2 - \rho_0 \partial_t w + \mathrm{div}\,(A_1{}^T A_1 \nabla w), \\ \mathrm{div}\,(A_1 z) = 0. \end{cases} \tag{143}$$

The construction of w is based on the following lemma.

Lemma 3.2. *Let A be a matrix valued function on $[0, T] \times \mathbb{T}^2$ satisfying*

$$\det A \equiv 1. \tag{144}$$

There exists a constant c such that if

$$\|\mathrm{Id} - A\|_{L^\infty(0, T; L^\infty)} + \|\partial_t A\|_{L^2(0, T; L^6)} \leq c \tag{145}$$

then, for all function $g : [0, T] \times \mathbb{T}^2 \to \mathbb{R}$ satisfying $g \in L^2(0, T \times \mathbb{T}^2)$ and

$$g = \mathrm{div}\, R \quad \text{with} \quad R \in L^4(0, T; L^2) \quad \text{and} \quad \partial_t R \in L^{4/3}(0, T; L^{3/2}),$$

the equation

$$\mathrm{div}\,(Aw) = g \quad \text{in} \quad [0, T] \times \mathbb{T}^2 \tag{146}$$

admits a solution w in the space

$$X_T := \Big\{ v \in L^4(0,T; L^2(\mathbb{T}^2)) , \ \nabla v \in L^2(0,T; L^2(\mathbb{T}^2))$$

$$\text{and } \partial_t v \in L^{4/3}(0,T; L^{3/2}(\mathbb{T}^2)) \Big\}$$

satisfying the following inequalities:

$$\|w\|_{L^4(0,T;L^2)} \le C\|R\|_{L^4(0,T;L^2)}, \quad \|\nabla w\|_{L^2(0,T;L^2)} \le C\|g\|_{L^2(0,T;L^2)}$$

$$\text{and } \|\partial_t w\|_{L^{4/3}(0,T;L^{3/2})} \le C\|R\|_{L^4(0,T;L^2)} + C\|\partial_t R\|_{L^{4/3}(0,T;L^{3/2})}, \tag{147}$$

and also $w(t_0) = 0$ for any $t_0 \in [0,T]$ such that $g(t_0) = 0$.

Proof. For any $v \in X_T$, we set

$$\Phi(v) := \nabla \Delta^{-1} \mathrm{div}\big((\mathrm{Id} - A)v + R\big).$$

It is obvious that $\Phi(v)$ satisfies the linear equation

$$\mathrm{div}\, w = \mathrm{div}\,((\mathrm{Id} - A)v) + g \quad \text{in } (0,T) \times \mathbb{T}^2.$$

Furthermore the operator $\nabla \Delta^{-1}\mathrm{div}$ is a self-map on $L^2(\mathbb{T}^2)$ with norm at most 2 (and even 1 if we restrict to mean free functions). Hence we have

$$\|\Phi(v)\|_{L^4(0,T;L^2)} \lesssim \|\mathrm{Id} - A\|_{L^\infty(0,T;L^\infty)} \|v\|_{L^4(0,T;L^2)} + \|R\|_{L^4(0,T;L^2)}.$$

Then using the fact that (144) implies that

$$\mathrm{div}\,(Az) = A^T : \nabla z, \tag{148}$$

we get

$$\|\nabla \Phi(v)\|_{L^2(0,T;L^2)} \le \|\mathrm{Id} - A\|_{L^\infty(0,T;L^\infty)} \|\nabla v\|_{L^2(0,T;L^2)} + \|g\|_{L^2(0,T;L^2)}.$$

And finally, because $\partial_t((\mathrm{Id} - A)v) = (\mathrm{Id} - A)\partial_t v - \partial_t A\, v$, we have

$$\|\partial_t(\Phi(v))\|_{L^{\frac{4}{3}}(0,T;L^{\frac{3}{2}})} \lesssim \|(\mathrm{Id} - A)\partial_t v\|_{L^{\frac{4}{3}}(0,T;L^{\frac{3}{2}})} + \|\partial_t A\, v\|_{L^{\frac{4}{3}}(0,T;L^{\frac{3}{2}})}$$

$$+ \|\partial_t R\|_{L^{\frac{4}{3}}(0,T;L^{\frac{3}{2}})}$$

$$\lesssim \|\mathrm{Id} - A\|_{L^\infty(0,T;L^\infty)} \|\partial_t v\|_{L^{\frac{4}{3}}(0,T;L^{\frac{3}{2}})}$$

$$+ \|\partial_t A\|_{L^2(0,T;L^6)} \|v\|_{L^4(0,T;L^2)} + \|\partial_t R\|_{L^{\frac{4}{3}}(0,T;L^{\frac{3}{2}})}.$$

This proves that Φ maps X_T to X_T, if c in (145) is small enough. Then, obvious variations on the above computations give for any couple (v_1, v_2) in X_T^2,

$$\|\Phi(v_2) - \Phi(v_1)\|_{X_T} \le C\big(\|\mathrm{Id} - A\|_{L^\infty(0,T;L^\infty)} + \|\partial_t A\|_{L^2(0,T;L^6)}\big)\|v_2 - v_1\|_{X_T}$$

$$\le \frac{1}{2}\|v_2 - v_1\|_{X_T}.$$

Hence, applying the standard Banach fixed point theorem in X_T provides a solution to the equation $\Phi(v) = v$. Then looking back at the above computations in the case $\Phi(v) = v$ gives the desired inequalities. $\qquad\square$

Let us continue the proof of uniqueness. We take for w the solution provided by the above lemma with $A = A_1$, $g = {}^T\delta A : \nabla \bar{u}_2$ and $R = \delta A \, \bar{u}_2$ and get some solution w to (142) such that $w(0) = 0$ (because obviously $g(0) = 0$) and

$$\|w\|_{L^4(0,T;L^2)} \leq C\|\delta A \, \bar{u}_2\|_{L^4(0,T;L^2)}, \quad \|\nabla w\|_{L^2(0,T;L^2)} \leq C\|{}^T\delta A : \nabla\bar{u}_2\|_{L^2(0,T;L^2)}$$

$$\text{and} \quad \|\partial_t w\|_{L^{4/3}(0,T;L^{3/2})} \leq C\|\delta A \, \bar{u}_2\|_{L^4(0,T;L^2)} + C\|\partial_t(\delta A \, \bar{u}_2)\|_{L^{4/3}(0,T;L^{3/2})}.$$
$$(149)$$

Let us bound the right-hand side of (149). Regarding ${}^T\delta A : \nabla\bar{u}_2$, one can use the fact that if both u_1 and u_2 fulfill (139), then we have

$$\sup_{t\in[0,T]} \|t^{-1/2}\delta A\|_{L^2} \leq C \sup_{t\in[0,T]} \left\|t^{-1/2}\int_0^t \nabla\delta u \, d\tau\right\|_{L^2} \leq C\|\nabla\delta u\|_{L^2(0,T;L^2)}. \quad (150)$$

This stems from Hölder inequality and Identity (171) in the appendix. Therefore, thanks to (140) and (150), we have

$$\|{}^T\delta A : \nabla\bar{u}_2\|_{L^2(0,T\times\mathbb{T}^2)} \leq \sup_{t\in[0,T]} \|t^{-1/2}\delta A(t)\|_2 \|t^{1/2}\nabla\bar{u}^2\|_{L^2(0,T;L^\infty)}$$

$$\leq c(T)\|\nabla\delta u\|_{L^2(0,T;L^2)} \quad \text{with} \quad c(T) \to 0 \quad \text{for} \quad T \to 0.$$

Similarly,

$$\|\delta A \, \bar{u}_2\|_{L^4(0,T;L^2)} \leq \|t^{-1/2}\delta A\|_{L^\infty(0,T;L^2)} \|t^{1/2}\bar{u}_2\|_{L^4(0,T;L^\infty)},$$

whence, using (140), (149) and (150) gives

$$\|w\|_{L^4(0,T;L^2)} \leq c(T)\|\nabla\delta u\|_{L^2(0,T;L^2)}. \quad (151)$$

Finally, in order to bound $\partial_t(\delta A \, \bar{u}_2)$ in $L^{\frac{4}{3}}(0,T;L^{\frac{3}{2}})$, we use that

$$\partial_t(\delta A \, \bar{u}_2) = \delta A \, \partial_t \bar{u}_2 + \partial_t(\delta A) \, \bar{u}_2.$$

Thanks to (140) and (150),

$$\|\delta A \, \partial_t u_2\|_{L^{\frac{4}{3}}(0,T;L^{\frac{3}{2}})} \leq \|t^{-1/2}\delta A\|_{L^\infty(0,T;L^2)} \|t^{1/2}\partial_t\bar{u}_2\|_{L^{\frac{4}{3}}(0,T;L^6)}$$

$$\leq c(T)\|\nabla\delta u\|_{L^2(0,T;L^2)}.$$

One can bound the other term as follows:

$$\|\partial_t(\delta A) \, \bar{u}_2\|_{L^{\frac{4}{3}}(0,T;L^{\frac{3}{2}})} \leq \|\partial_t\delta A\|_{L^2(0,T\times\mathbb{T}^2)} \|\bar{u}_2\|_{L^4(0,T;L^6)}.$$

Differentiating (171) with respect to t and using (139) for u^1 and u^2, we see that

$$\|\partial_t\delta A\|_{L^2} \leq C\left(\|\nabla\delta u\|_{L^2} + \left\|t^{-1/2}\int_0^t \nabla\delta u \, d\tau\right\|_{L^2} \left(\|t^{1/2}\nabla\bar{u}_1\|_{L^\infty} + \|t^{1/2}\nabla\bar{u}_2\|_{L^\infty}\right)\right).$$

Therefore

$$\|\partial_t\delta A\|_{L^2(0,T\times\mathbb{T}^2)} \leq C\|\nabla\delta u\|_{L^2(0,T\times\mathbb{T}^2)},$$

and thus, owing to (140),

$$\|\partial_t \delta A\, \bar{u}_2\|_{L^{\frac{4}{3}}(0,T;L^{\frac{3}{2}})} \le c(T)\|\nabla \delta u\|_{L^2(0,T\times\mathbb{T}^2)}.$$

Altogether, this gives

$$\|w\|_{L^4(0,T;L^2)} + \|\nabla w\|_{L^2(0,T\times\mathbb{T}^2)} + \|\partial_t w\|_{L^{4/3}(0,T;L^{3/2})} \le c(T)\|\nabla \delta u\|_{L^2(0,T\times\mathbb{T}^2)} \tag{152}$$

with $c(T)$ going to 0 when T tends to 0.

Next, to estimate z, we test (143) by z, using the fact that

$$\int_{\mathbb{T}^2} ({}^T A_1 \cdot \nabla \delta P) \cdot z\, dx = -\int_{\mathbb{T}^2} \operatorname{div}(A_1 z)\, \delta P\, dx = 0.$$

So we have

$$\frac{1}{2}\frac{d}{dt}\int_{\mathbb{T}^2} \rho_0 |z|^2\, dx + \int_{\mathbb{T}^2} |{}^T A_1 \cdot \nabla z|^2 dx = \sum_{k=1}^4 I_k, \tag{153}$$

where

$$I_1 := \int_{\mathbb{T}^2} \left((A_1 {}^T A_1 - A_2 {}^T A_2) \cdot \nabla \bar{u}_2\right) \cdot \nabla z\, dx,$$

$$I_2 := \int_{\mathbb{T}^2} \left(({}^T A_1 - {}^T A_2) \cdot \nabla \bar{P}_2\right) \cdot z\, dx,$$

$$I_3 := -\int_{\mathbb{T}^2} \rho_0\, \partial_t w \cdot z\, dx,$$

$$I_4 := -\int_{\mathbb{T}^2} ({}^T A_1 \nabla w) \cdot ({}^T A_1 \nabla z)\, dx.$$

We have, by virtue of (138) and (139),

$$I_1 = \int_{\mathbb{T}^2} \operatorname{div}\left((\delta A\, {}^T A_2 + A_1\, {}^T \delta A)\nabla \bar{u}_2\right) \cdot z\, dx$$

$$\le \int_{\mathbb{T}^2} |\delta A\, {}^T A_2 + A_1\, {}^T \delta A|\, |\nabla \bar{u}_2|\, |\nabla z|\, dx$$

$$\le C\|t^{-1/2}\delta A\|_{L^2}\|t^{1/2}\nabla \bar{u}_2\|_{L^\infty}\|\nabla z\|_{L^2}.$$

Therefore, thanks to (140) and (150),

$$\int_0^T I_1(t)\, dt \le C\|t^{-1/2}\delta A\|_{L^\infty(0,T;L^2)}\|t^{1/2}\nabla \bar{u}_2\|_{L^2(0,T;L^\infty)}\|\nabla z\|_{L^2(0,T\times\mathbb{T}^2)}$$

$$\le c(T)\|\nabla \delta u\|_{L^2(0,T\times\mathbb{T}^2)}\|\nabla z\|_{L^2(0,T\times\mathbb{T}^2)}. \tag{154}$$

Next,

$$I_2 \le C\|t^{-1/2}\delta A\|_{L^2}\|t^{1/2}\nabla \bar{P}_2\|_4\|z\|_{L^4}, \tag{155}$$

whence, according to (150) and Sobolev embedding

$$\int_0^T I_2(t)\, dt \le \|t^{-1/2}\delta A\|_{L^\infty(0,T;L^2)}\|t^{1/2}\nabla Q^2\|_{L^2(0,T;L^4)}\|z\|_{L^2(0,T;L^4)}$$

$$\le c(T)\|\nabla \delta u\|_{L^2(0,T\times\mathbb{T}^2)}\|z\|_{L^2(0,T;H^1)}.$$

At this stage, one may use (109) and get for some constant C depending only on ρ_0,

$$\|z\|_{H^1} \leq C(\|\sqrt{\rho_0}z\|_{L^2} + \|\nabla z\|_{L^2}). \tag{156}$$

Therefore, one concludes that

$$\int_0^T I_2(t)\,dt \leq c(T)\left(\|\sqrt{\rho_0}\,z\|_{L^\infty(0,T;L^2)} + \|\nabla z\|_{L^2(0,T\times\mathbb{T}^2)}\right)\|\nabla\delta u\|_{L^2(0,T\times\mathbb{T}^2)}.$$

Next, using Hölder inequality, one can write that

$$\int_0^T I_3(t)\,dt \leq \|\rho_0\|_{L^\infty}^{3/4}\|\partial_t w\|_{L^{4/3}(0,T;L^{3/2})}\|\rho_0^{1/4}z\|_{L^4(0,T;L^3)}.$$

Note that from Hölder inequality and the Sobolev embedding $H^1(\mathbb{T}^2) \hookrightarrow L^6(\mathbb{T}^2)$, we have

$$\|\rho_0^{1/4}z\|_{L^4(0,T;L^3)} \leq \|\sqrt{\rho_0}\,z\|_{L^\infty(0,T;L^2)}^{1/2}\|z\|_{L^2(0,T;L^6)}^{1/2}$$
$$\leq C\|\sqrt{\rho_0}\,z\|_{L^\infty(0,T;L^2)}^{1/2}\|z\|_{L^2(0,T;H^1)}^{1/2}.$$

Then, taking advantage of (156) and (152), we conclude that

$$\int_0^T I_3(t)\,dt \leq c(T)\left(\|\sqrt{\rho_0}\,z\|_{L^\infty(0,T;L^2)} + \|\nabla z\|_{L^2(0,T\times\mathbb{T}^2)}\right)^{1/2}$$
$$\cdot\|\sqrt{\rho_0}\,z\|_{L^\infty(0,T;L^2)}^{1/2}\|\nabla\delta u\|_{L^2(0,T\times\mathbb{T}^2)}.$$

Finally, integrating by parts, and using (152) and (139),

$$\int_0^T I_4(t)\,dt \leq \frac{1}{2}\int_0^T \|^T A_1 \cdot \nabla z\|_{L^2}^2\,dt + \frac{1}{2}\int_0^T \|^T A_1 \cdot \nabla w\|_{L^2}^2\,dt$$
$$\leq \frac{1}{2}\int_0^T \|^T A_1 \cdot \nabla z\|_{L^2}^2\,dt + c(T)\int_0^T \|\nabla\delta u\|_{L^2}^2\,dt.$$

So altogether, this gives for all small enough $T > 0$,

$$\sup_{t\in[0,T]} \|\sqrt{\rho_0}z(t)\|_{L^2}^2 + \int_0^T \|\nabla z\|_{L^2}^2\,dt \leq c(T)\int_0^T \|\nabla\delta u\|_{L^2}^2\,dt. \tag{157}$$

Combining with (152), we conclude that

$$\int_0^T \|\nabla\delta u\|_{L^2}^2\,dt \leq c(T)\int_0^T \|\nabla\delta u\|_{L^2}^2\,dt.$$

Hence $\nabla\delta u \equiv 0$ on $[0,T]\times\mathbb{T}^2$ if T is small enough.

Then, plugging that information in (157) yields

$$\|\sqrt{\rho_0}\,z\|_{L^\infty(0,T;L^2)} + \|\nabla z\|_{L^2(0,T\times\mathbb{T}^2)} = 0.$$

Combining with Lemma 3.1 finally implies that $z \equiv 0$ on $[0, T] \times \mathbb{T}^2$, and (152) clearly yields $w \equiv 0$. Therefore we proved that for small enough $T > 0$,

$$u^1 \equiv u^2 \quad \text{on} \ [0, T] \times \mathbb{T}^2.$$

Reverting to Eulerian coordinates, we conclude that the two solutions of (INS) coincide on $[0, T] \times \mathbb{T}^2$. Then, standard connectivity arguments yield uniqueness on the whole $\mathbb{R}_+$.

3.6   Application to the density patches problem

As a consequence of Theorem 3.1, we obtain the following answer to Lions' question stated in [32], page 34:

Corollary 3.1. *Assume that $\rho_0 = 1_{D_0}$ for some bounded open subset D_0 of $\mathbb{T}^2$ with $\mathcal{C}^{1,\alpha}$ regularity for some $\alpha \in (0, 1)$. Then, for any divergence free initial velocity u_0 in $H^1(\mathbb{T}^2)$, the unique global solution $(\rho, u, \nabla P)$ provided by Theorem 3.1 is such that for all $t \geq 0$,*

$$\rho(t, \cdot) = 1_{D_t} \quad \text{with} \ \ D_t := X(t, D_0),$$

where $X(t, \cdot)$ stands for the flow of u, that is the unique solution of (32).

Furthermore, D_t has $\mathcal{C}^{1,\alpha}$ regularity with a control of the Hölder norm in terms of the initial data.

Proof. Assume that D_0 corresponds to the level set $\{f_0 = 0\}$ of some function $f_0 :$ $\mathbb{T}^2 \to \mathbb{R}$ with $\mathcal{C}^{1,\alpha}$ regularity. Then, we have $D_t := f_t^{-1}(\{0\})$ with $f_t := f \circ X(t, \cdot)$. Fix some $T > 0$. Theorem 3.1 and interpolation imply that we have

$$\sqrt{t}\nabla^2 u \in L_{2+\varepsilon}(0, T; L_{1/\varepsilon}(\mathbb{T}^2)) \quad \text{for all small enough } \varepsilon > 0.$$

By Sobolev embedding with respect to the space variable, and Hölder inequality with respect to the time variable, one can conclude that ∇u is in $L^1(0, T; \mathcal{C}^{0,\beta})$ for all $\beta \in (0, 1)$. Consequently the flow $X(t, \cdot)$ is in $\mathcal{C}^{1,\beta}$ for all $\beta \in (0, 1)$, which implies that f_t is in $\mathcal{C}^{1,\alpha}$ provided that $\alpha < 1$. $\square$

Remark 3.1. *Just writing that the triplet (ρ, u, P) constructed in the above corollary is a solution of (1) in the sense of (4), (5), (6), we see that System (1) is fulfilled into the (time dependent) fluid domain D_t, and that the following two conditions are satisfied at the interface:*

$$[u]|_{\partial D_t} = 0 \quad \text{and} \quad [\mathcal{T} \cdot n]|_{\partial D_t} = 0,$$

where $\mathcal{T} := -P\mathrm{Id} + Du + \nabla u$ is the viscous stress tensor.

In other words, Corollary 3.1 may also be seen as a solution to some free boundary problem, of water wave type.

3.7 Further remarks and open questions

For simplicity, we made a rather strong regularity assumption on the velocity. However, as in the dimension 2 the critical regularity index in the Sobolev setting is 0, it is natural to study whether Theorem 3.1 may be adapted to the case where u_0 is only in H^s for some $s \in (0,1)$ (or, even better, in the Besov space $B^0_{2,1}$). A possibility may be to use the same type of time weighted estimates as in [33]. Of course, a similar question is relevant in dimension 3, but with $s > 1/2$. In this regard, one can mention the recent paper by P. Zhang [38] where global existence is proved for small velocities in $\dot{B}^{\frac{1}{2}}_{2,1}(\mathbb{R}^3)$ and general densities in $L^\infty(\mathbb{R}^3)$, just bounded away from zero (the question of uniqueness and of Lipschitz regularity of the flow in this context is open, though).

Surprisingly, we do not know how to prove Theorem 3.1 in the whole plane $\mathbb{R}^2$ (or, more generally, in unbounded domains). Of course, having a local-in-time result in that setting is not a big deal, but we lack a control on low frequencies to extend it to all time (technically speaking, we need a substitute for Lemma 3.1).

From the very beginning, our approach relies on the fact that the viscosity coefficient is a constant: whether one may adapt Inequality (101) (even locally in time) to the case where $\mu = \mu(\rho)$ and ρ is only bounded (even if away from vacuum) is totally unclear.

In most evolutionary PDE's, proving stability estimates works essentially the same as uniqueness. In our case however, unless it is assumed that the initial density is bounded away from zero, we do not know if the stability of the solution by perturbation of the initial density holds true.

A   Littlewood-Paley decomposition and functional spaces

Here we give a short presentation of the Besov spaces that have been used in Sections 1 and 2. More details may be found in e.g. [2, 36].

A standard way to define the Besov spaces on $\mathbb{R}^d$ is to resort to the Littlewood-Paley decomposition, a dyadic localization procedure in the frequency space for tempered distributions over $\mathbb{R}^d$. To define it, one may start from some smooth radial nonincreasing function χ with $\operatorname{Supp} \chi \subset B(0, \frac{4}{3})$ and $\chi \equiv 1$ on $B(0, \frac{3}{4})$, then set $\varphi(\xi) = \chi(\xi/2) - \chi(\xi)$ so that

$$\chi + \sum_{j \in \mathbb{N}} \varphi(2^{-j}\cdot) = 1 \ \text{ in } \ \mathbb{R}^d \qquad \text{and} \qquad \sum_{j \in \mathbb{Z}} \varphi(2^{-j}\cdot) = 1 \ \text{ in } \ \mathbb{R}^d \setminus \{0\}.$$

The homogeneous *dyadic blocks* $\dot{\Delta}_j$ are defined by

$$\dot{\Delta}_j u \overset{\text{def}}{=} \varphi(2^{-j}D)u \overset{\text{def}}{=} \mathcal{F}^{-1}(\varphi(2^{-j}\cdot)\mathcal{F}u) \overset{\text{def}}{=} 2^{jd}h(2^j\cdot) \star u \quad \text{with} \quad h \overset{\text{def}}{=} \mathcal{F}^{-1}\varphi.$$

We also introduce the low frequency cut-off operator $\dot{S}_j$:

$$\dot{S}_j u \overset{\text{def}}{=} \chi(2^{-j}D)u \overset{\text{def}}{=} \mathcal{F}^{-1}(\chi(2^{-j}\cdot)\mathcal{F}u) \overset{\text{def}}{=} 2^{jd}\check{h}(2^j\cdot) \star u \quad \text{with} \quad \check{h} \overset{\text{def}}{=} \mathcal{F}^{-1}\chi.$$

Operators $\dot{\Delta}_j$ and $\dot{S}_j$ are continuous on all spaces L^p, with norm *independent* of j, a property that would fail if taking a rough cut-off function χ. Even though $\dot{\Delta}_j$ and $\dot{S}_j$ are not projectors, the following fundamental quasi-orthogonality property is fulfilled:

$$\dot{\Delta}_j \dot{\Delta}_k = 0 \quad \text{if} \quad |j - k| > 1. \tag{158}$$

The homogeneous Littlewood-Paley decomposition for u reads

$$u = \sum_j \dot{\Delta}_j u. \tag{159}$$

The above equality has to be understood in the sense of tempered distribution *modulo polynomials*. In order to rule out polynomials, one can restrict ourselves to the set $\mathcal{S}'_h$ of tempered distributions u such that

$$\lim_{j \to -\infty} \|\dot{S}_j u\|_{L^\infty} = 0 \quad \text{with} \quad \dot{S}_j \overset{\text{def}}{=} \chi(2^{-j}D).$$

That condition on the low frequencies forces u to tend to 0 at infinity, in the sense of distributions.

One may now define *homogeneous Besov spaces* as follows.

Definition A.1. *For $s \in \mathbb{R}$ and $1 \leq p, r \leq \infty$, we set*

$$\|u\|_{\dot{B}^s_{p,r}} \overset{\text{def}}{=} \left(\sum_{j \in \mathbb{Z}} 2^{rjs} \|\dot{\Delta}_j u\|^r_{L^p} \right)^{\frac{1}{r}} \quad \text{if} \ \ r < \infty \quad \text{and} \quad \|u\|_{\dot{B}^s_{p,\infty}} \overset{\text{def}}{=} \sup_{j \in \mathbb{Z}} 2^{js} \|\dot{\Delta}_j u\|_{L^p}.$$

We then define the homogeneous Besov space $\dot{B}^s_{p,r}$ to be the subset of distributions $u \in \mathcal{S}'_h$ such that $\|u\|_{\dot{B}^s_{p,r}} < \infty$.

It is obvious that the norm of $\dot{B}^s_{2,2}$ is equivalent to that of the homogeneous Sobolev space $\dot{H}^s$. One may also show that the norm in $\dot{B}^r_{\infty,\infty}$ coincides with that of the homogeneous Hölder space $\dot{C}^{0,r}$ if $r \in (0,1)$. On the other side, unless $p = 2$, Lebesgue spaces do not belong to the Besov spaces family. In fact, for general $p \in [1, +\infty]$, one just has the following (optimal) embeddings:

$$\dot{B}^0_{p,\min(p,2)} \hookrightarrow L^p \hookrightarrow \dot{B}^0_{p,\max(p,2)}.$$

Let us list classical properties of Besov spaces that have been used repeatedly in these notes (see the proof in e.g. [2]):

- The space $\dot{B}^s_{p,r}$ is complete if (and only if) $s < d/p$ or $s \leq d/p$ and $r = 1$.
- Fatou property: if $(u_n)_{n \in \mathbb{N}}$ is a bounded sequence of functions of $\dot{B}^s_{p,r}$ that converges in $\mathcal{S}'$ to some $u \in \mathcal{S}'_h$, then $u \in \dot{B}^s_{p,r}$ and $\|u\|_{\dot{B}^s_{p,r}} \leq C \liminf \|u_n\|_{\dot{B}^s_{p,r}}$.
- Duality: If u is in $\mathcal{S}'_h$ then we have

$$\|u\|_{\dot{B}^s_{p,r}} \leq C \sup_\phi \langle u, \phi \rangle$$

where the supremum is taken over those ϕ in $\mathcal{S} \cap \dot{B}^{-s}_{p',r'}$ such that $\|\phi\|_{\dot{B}^{-s}_{p',r'}} \leq 1$.

- The following inequality is satisfied for all $1 \leq p, r_1, r_2, r \leq \infty$, $s_1 \neq s_2$:

$$\|u\|_{\dot{B}_{p,r}^{\theta s_2 + (1-\theta)s_1}} \lesssim \|u\|_{\dot{B}_{p,r_1}^{s_1}}^{1-\theta} \|u\|_{\dot{B}_{p,r_2}^{s_2}}^{\theta}, \qquad \theta \in (0,1).$$

- The subspace $\mathcal{S}_0$ of Schwartz functions with Fourier transform supported away from 0 is dense in $\dot{B}_{p,r}^s$ whenever p and r are finite.
- For any smooth homogeneous of degree m function A on $\mathbb{R}^d \setminus \{0\}$ if the *Fourier multiplier* $A(D)$ defined by $A(D)u \overset{\text{def}}{=} \mathcal{F}^{-1}(A\mathcal{F}u)$ maps $\mathcal{S}_h'$ to itself then

$$A(D) : \dot{B}_{p,r}^s \to \dot{B}_{p,r}^{s-m}. \tag{160}$$

In particular, the derivative operator ∂_j maps $\dot{B}_{p,r}^s$ to $\dot{B}_{p,r}^{s-1}$.

Let us denote by $\dot{W}^{1,p}$ the completion of $\mathcal{S}_0$ for the norm $\|\nabla \cdot \|_{L^p}$. The following result of interpolation has been used several times.

Proposition A.1. *Let $p \in [1,+\infty]$. For any function u in $L^p + \dot{W}^{1,p}$ and $t > 0$, let us denote*

$$K(t,u) := \inf_{\substack{u=a+b \\ a \in L^p,\, b \in \dot{W}^{1,p}}} \left(\|a\|_{L^p} + t\|\nabla b\|_{L^p} \right).$$

For $s \in (0,1)$, let us set $\|u\|_{[L^p, \dot{W}^{1,p}]_{s,\infty}} := \sup_{t>0} \|t^{-s} K(t,u)\|_{L^p(\mathbb{R}^d)}$ and

$$\|u\|_{[L^p, \dot{W}^{1,p}]_{s,r}} := \left(\int_0^{+\infty} \|t^{-s} K(t,u)\|_{L^p(\mathbb{R}^d)}^r \frac{dt}{t} \right)^{\frac{1}{r}}, \qquad 1 \leq r < +\infty.$$

Then, we have for some constant $C \geq 1$ depending only on s,

$$C^{-1}\|u\|_{[L^p, \dot{W}^{1,p}]_{s,r}} \leq \|u\|_{\dot{B}_{p,r}^s} \leq C\|u\|_{[L^p, \dot{W}^{1,p}]_{s,r}}.$$

Proof. Let us first assume that u belongs to the space $\dot{B}_{p,r}^s$. Then, we fix some $t > 0$ and $j \in \mathbb{Z}$ so that $2^{-j} \leq t < 2^{-j+1}$ and observe that $u = (\text{Id} - \dot{S}_j)u + \dot{S}_j u$. Therefore, owing to the definition of $K(t,u)$, we have

$$K(t,u) \leq t\|\nabla \dot{S}_j u\|_{L^p} + \|(\text{Id} - \dot{S}_j)u\|_{L^p}.$$

Then, observing that

$$\dot{S}_j u = \sum_{k<j} \dot{\Delta}_k u \quad \text{and} \quad (\text{Id} - \dot{S}_j)u = \sum_{k \geq j} \dot{\Delta}_k u,$$

using Bernstein inequality and the fact that $t^{-s} \approx 2^{js}$ and $t \approx 2^{-j}$, we discover that

$$\begin{aligned}
&t^{-s} K(t,u) \\
&\lesssim \sum_{k<j} 2^{(k-j)(1-s)} \left(2^{k(s-1)} \|\nabla \dot{\Delta}_k u\|_{L^p} \right) + \sum_{k \geq j} 2^{(j-k)s} \left(2^{ks} \|\dot{\Delta}_k u\|_{L^p} \right) \\
&\lesssim \sum_{k \in \mathbb{Z}} \left(2^{s(j-k)} 1_{\mathbb{Z}^-}(j-k) + 2^{(s-1)(j-k)} 1_{\mathbb{N}^*}(j-k) \right) \left(2^{ks} \|\dot{\Delta}_k u\|_{L^p} \right). \tag{161}
\end{aligned}$$

Now, we have in the case $r < \infty$ (the case $r = \infty$ being left to the reader):

$$\left(\int_0^{+\infty} \| t^{-s} K(t,u) \|_{L^p(\mathbb{R}^d)}^r \frac{dt}{t} \right)^{\frac{1}{r}} \lesssim \left(\sum_{j \in \mathbb{Z}} 2^j \int_{2^{-j}}^{2^{-j+1}} \| t^{-s} K(t,u) \|_{L^p(\mathbb{R}^d)}^r dt \right)^{\frac{1}{r}}$$

$$\lesssim \left(\sum_{j \in \mathbb{Z}} \left(\sup_{2^{-j} \le t < 2^{-j+1}} \| t^{-s} K(t,u) \|_{L^p(\mathbb{R}^d)} \right)^r \right)^{\frac{1}{r}}.$$

Hence, plugging (161) in the right-hand side and using a convolution inequality for series yields

$$\| u \|_{[L^p, \dot{W}^{1,p}]_{s,r}} \le C \| u \|_{\dot{B}^s_{p,r}}.$$

Let us now prove the reverse inequality. For simplicity, we only consider the case $r = +\infty$. Let us fix some $j \in \mathbb{Z}$ and consider a in L^p and b in $\dot{W}^{1,p}$ so that

$$u = a + b \quad \text{and} \quad \| a \|_{L^p} + 2^{-j} \| \nabla b \|_{L^p} \le 2K(2^{-j}, u).$$

Then, we have

$$\| \dot{\Delta}_j u \|_{L^p} \le \| \dot{\Delta}_j a \|_{L^p} + \| \dot{\Delta}_j b \|_{L^p} \lesssim \| \dot{\Delta}_j a \|_{L^p} + 2^{-j} \| \nabla \dot{\Delta}_j b \|_{L^p} \lesssim K(2^{-j}, u).$$

Therefore,

$$2^{js} \| \dot{\Delta}_j u \|_{L^p} \lesssim (2^j)^s K(2^{-j}, u),$$

whence the result, after taking the supremum on j. $\qquad \square$

As an exercise to the reader, we leave the proof of the fact that homogeneous Besov norms have the following scaling invariance property:

Proposition A.2. *For any $s \in \mathbb{R}$ and $(p,r) \in [1, +\infty]^2$ there exists a constant C such that for all positive λ and $u \in \dot{B}^s_{p,r}$, we have*

$$C^{-1} \lambda^{s - \frac{d}{p}} \| u \|_{\dot{B}^s_{p,r}} \le \| u(\lambda \cdot) \|_{\dot{B}^s_{p,r}} \le C \lambda^{s - \frac{d}{p}} \| u \|_{\dot{B}^s_{p,r}}.$$

Let us finally come to classical embeddings:

Proposition A.3.   1. *If $s \in \mathbb{R}$, $1 \le p_1 \le p_2 \le \infty$ and $1 \le r_1 \le r_2 \le \infty$, then $\dot{B}^s_{p_1, r_1} \hookrightarrow \dot{B}^{s - d(\frac{1}{p_1} - \frac{1}{p_2})}_{p_2, r_2}$.*

   2. *For any $1 \le p, r_1, r_2 \le \infty$ and $-\frac{d}{p'} < s' < s$, the embedding of $\dot{B}^s_{p, r_1}$ in $\dot{B}^{s'}_{p, r_2}$ is locally compact. More precisely, for any $\varphi \in \mathcal{S}$, the map $u \mapsto \varphi u$ is compact from $\dot{B}^s_{p, r_1}$ to $\dot{B}^{s'}_{p, r_2}$.*

   3. *The space $\dot{B}^{\frac{d}{p}}_{p,1}$ is continuously embedded in the set $\mathcal{C}_b$ of bounded continuous functions. If $p < \infty$ then all functions of $\dot{B}^{\frac{d}{p}}_{p,1}$ tend to 0 at infinity.*

Proof. The embedding is a consequence of the convolution property $L^1 \star L^p \to L^p$ which implies that

$$\|\dot{\Delta}_j u\|_{L^p} \leq \|2^{jd} h(2^j \cdot)\|_{L^1} \|u\|_{L^p} = \|h\|_{L^1} \|u\|_{L^p}.$$

As for the second property, we just have to use that, owing to Bernstein inequality,

$$\|\dot{\Delta}_j u\|_{L^{p_2}} \leq C 2^{j(\frac{d}{p_1} - \frac{d}{p_2})} \|\dot{\Delta}_j u\|_{L^{p_1}}.$$

Finally, the above inequality with $p_1 = p$ and $p_2 = \infty$ implies that whenever u is in $\dot{B}_{p,1}^{\frac{d}{p}}$, the series $\sum_{j \in \mathbb{Z}} \dot{\Delta}_j u$ converges *uniformly* to u. As each term of the series is continuous and bounded, the same property holds for u. Finally, if $p < \infty$ then each term $\dot{\Delta}_j u$ goes to 0 at infinity (because it is in L^p), hence so does u.

The reader may refer to [2] (combined with Prop. 2.1.3 of [12]) for the proof of the compact embedding. $\qquad\square$

B Lagrangian coordinates

Here we establish estimates in Besov spaces for the flow of time-dependent vector-fields. As recalled in Section 1, if $v : [0,T) \times \mathbb{R}^d \to \mathbb{R}^d$ is measurable, such that $t \mapsto v(t,x)$ is in $L^1(0,T)$ for all $x \in \mathbb{R}^d$ with, in addition, $\nabla v \in L^1(0,T;L^\infty)$ then it has a unique flow X satisfying

$$X(t,y) = y + \int_0^t v(\tau, X(\tau,y)) \, d\tau \quad \text{for all } t \in [0,T) \text{ and } y \in \mathbb{R}^d.$$

Furthermore, for all $t \in [0,T)$, the map $X(t,\cdot)$ is a $\mathcal{C}^1$-diffeomorphism over $\mathbb{R}^d$ and if we set $\bar{v}(t,y) := v(t, X(t,y))$ then X fulfills

$$X(t,y) := y + \int_0^t \bar{v}(\tau,y) \, d\tau. \tag{162}$$

Of course, differentiating (162) with respect to the y variable yields

$$DX(t,y) = \text{Id} + \int_0^t D\bar{v}(\tau,y) \, d\tau. \tag{163}$$

Conversely, if X is defined by (162), then in order to go back to Eulerian coordinates, we need that for all $t \in [0,T]$, $X(t,\cdot)$ is a $\mathcal{C}^1$ diffeomorphism of $\mathbb{R}^d$. As throughout the paper, we are always in the situation where $\bar{v}$ is in $L^2(0,T;L^\infty)$ and satisfies

$$\int_0^T \|\nabla \bar{v}\|_{L^\infty} \, dt < 1,$$

that property of X is ensured by the following classical result that we here prove for the reader's convenience.

Proposition B.1. *Let $\psi : \mathbb{R}^d \to \mathbb{R}^d$ be in $\mathcal{C}^1$ and assume that $D\psi$ is invertible at every point of $\mathbb{R}^d$. If in addition ψ goes to infinity at infinity and $\|D\psi - \text{Id}\|_{L^\infty} < 1$, then ψ is a $\mathcal{C}^1$ diffeomorphism of $\mathbb{R}^d$.*

Proof. As ψ is locally invertible, it suffices to show that it is onto and one-to-one. To prove the first property, fix some $y \in \mathbb{R}^d$. Because ψ is continuous and goes to infinity at infinity, the minimization problem $\inf_{x \in \mathbb{R}^d} |\psi(x) - y|^2$ has a least one solution x_y. Then computing the differential of $x \mapsto |\psi(x) - y|^2$ at x_y, which is $2D\psi(x_y) \cdot (\psi(x_y) - y)$, and using that $D\psi(x_y)$ is invertible, we see that we must have $\psi(x_y) - y = 0$.

In order to show that ψ is one-to-one, consider $(x, x') \in \mathbb{R}^d \times \mathbb{R}^d$ so that $\psi(x) = \psi(x')$, and write that

$$x' - x = (x' - \psi(x')) - (x - \psi(x)).$$

Hence,

$$|x' - x| \le \|\text{Id} - D\psi\|_{L^\infty} |x' - x|$$

which, given that $\|\text{Id} - D\psi\|_{L^\infty} < 1$, implies that $x' = x$. ☐

A fundamental property that we used all the time is:

Lemma B.1. *Let H be a measurable vector-field over $\mathbb{R}^d$ and X be a C^1 measure preserving diffeomorphism. Let $\bar{H} := H \circ X$. Then, we have*

$$\text{div}_x H(x) = \text{div}_y (A\bar{H})(y) \quad \text{with} \quad x := X(y) \quad \text{and} \quad A(y) := (DX(y))^{-1}.$$

Proof. This stems from the following series of computations (based on integrations by parts, (138) and on the fact that X is measure preserving) which hold true for any scalar test function ϕ:

$$\int_{\mathbb{R}^d} \phi(x) \, \text{div}_x H(x) \, dx = - \int_{\mathbb{R}^d} D_x\phi(x) \cdot H(x) \, dx$$

$$= - \int_{\mathbb{R}^d} D_x\phi(X(y)) \cdot H(X(y)) \, dy$$

$$= - \int_{\mathbb{R}^d} D_y\bar{\phi}(y) \cdot A(y) \cdot \bar{H}(y) \, dy$$

$$= \int_{\mathbb{R}^d} \bar{\phi}(y) \, \text{div}_y (A\bar{H})(y) \, dy.$$

Performing the change of variable $x = X(y)$ in the last integral gives the desired result. ☐

Let us finally establish a few flow estimates that have been used in Section 1.

Lemma B.2. *Let $p \in [1, \infty)$. There exists a constant $c = c(d, p) > 0$ such that if*

$$\int_0^T \|D\bar{v}\|_{\dot{B}^{\frac{d}{p}}_{p,1}} \, d\tau \le c, \tag{164}$$

then we have for all $t \in [0, T]$,

$$\|\text{Id} - A(t)\|_{\dot{B}^{\frac{d}{p}}_{p,1}} \lesssim \|D\bar{v}\|_{L^1(0,t;\dot{B}^{\frac{d}{p}}_{p,1})}, \tag{165}$$

$$\|\partial_t A(t)\|_{\dot{B}^{\frac{d}{p}}_{p,1}} \lesssim \|D\bar{v}(t)\|_{\dot{B}^{\frac{d}{p}}_{p,1}}. \tag{166}$$

Proof. Proving (165) relies on the following identity:

$$A(t) = \left(\mathrm{Id} + \int_0^t D\bar{v}\,d\tau\right)^{-1} = \sum_{k\in\mathbb{N}}(-1)^k\left(\int_0^t D\bar{v}\,d\tau\right)^k \qquad (167)$$

and on the fact that the space $\dot{B}_{p,1}^{\frac{d}{p}}$ is stable by product, if $p < \infty$.

In order to prove the second inequality, we just have to use again the stability of $\dot{B}_{p,1}^{\frac{d}{p}}$ by product and the fact that differentiating (167) with respect to time gives

$$\partial_t A(t) = -D\bar{v} \cdot \sum_{k\in\mathbb{N}}(-1)^k\left(\int_0^t D\bar{v}\,d\tau\right)^k. \qquad (168)$$

This completes the proof of the lemma. $\qquad\qquad\qquad\qquad\qquad\qquad\qquad\Box$

Lemma B.3. *Let $\bar{v}_1$ and $\bar{v}_2$ be two vector-fields satisfying (164), and $\delta v := \bar{v}_2 - \bar{v}_1$. Denote $A_i(t,y) := (X_i(t,x))^{-1}$ for $x = X_i(t,y)$. Then, we have for all $p \in [1,\infty)$ and $t \in [0,T]$:*

$$\|A_2 - A_1\|_{L_t^\infty(\dot{B}_{p,1}^{\frac{d}{p}})} \lesssim \|D\delta v\|_{L_t^1(\dot{B}_{p,1}^{\frac{d}{p}})}, \qquad (169)$$

$$\|\partial_t(A_2 - A_1)\|_{L_t^1(\dot{B}_{p,1}^{\frac{d}{p}})} \lesssim \|D\delta v\|_{L_t^1(\dot{B}_{p,1}^{\frac{d}{p}})}. \qquad (170)$$

Proof. In order to prove the first inequality, we just have to use the fact that $\dot{B}_{p,1}^{\frac{d}{p}}$ is stable by product and that (167) implies that

$$A_2 - A_1 = \sum_{k\geq 1}\left(C_2^k - C_1^k\right) = \left(\int_0^t D\delta v\,d\tau\right)\sum_{k\geq 1}\sum_{j=0}^{k-1}C_1^j C_2^{k-1-j}$$

$$\text{with}\quad C_i(t) = -\int_0^t D\bar{v}_i\,d\tau. \qquad (171)$$

Proving the second inequality relies on similar arguments, once the above identity has been differentiated with respect to time. The details are left to the reader. $\quad\Box$

References

[1] S. Antontsev, A. Kazhikhov and V. Monakhov: *Boundary Value Problems in Mechanics of Nonhomogeneous Fluids*, Studies in Mathematics and Its Applications, **22**, North-Holland, 1990.

[2] H. Bahouri, J.-Y. Chemin and R. Danchin: *Fourier Analysis and Nonlinear Partial Differential Equations*, Grundlehren der mathematischen Wissenschaften, **343**, Springer, 2011.

[3] G. K. Batchelor: *An Introduction to Fluids Dynamics*, Cambridge University Press, 1967.

[4] J.-M. Bony: Calcul symbolique et propagation des singularités pour les équations aux dérivées partielles non linéaires, *Annales Scientifiques de l'École Normale Supérieure*, **14**(4), (1981), 209–246.

[5] J.-Y. Chemin: Théorèmes d'unicité pour le système de Navier-Stokes tridimensionnel, *Journal d'Analyse Mathématique*, **77**, (1999), 27–50.

[6] Y. Cho and H. Kim: Unique solvability for the density-dependent Navier-Stokes equations, *Nonlinear Anal.*, **59**(4), (2004), 465–489.

[7] W. Craig, X. Huang and Y. Wang: Global well-posedness for the 3D inhomogeneous incompressible Navier-Stokes equations, *J. Math. Fluid Mech.*, **15**(4), (2013), 747–758.

[8] R. Danchin: Local and global well-posedness results for flows of inhomogeneous viscous fluids, *Adv. Diff. Eq.*, **9** (2004), 353–386.

[9] R. Danchin: Density-dependent incompressible fluids in bounded domains, *J. Math. Fluid Mech.*, **8**(3), (2006), 333–381.

[10] R. Danchin and P. B. Mucha: A critical functional framework for the inhomogeneous Navier-Stokes equations in the half-space, *J. Funct. Anal.*, **256**(3), (2009), 881–927.

[11] R. Danchin and P. B. Mucha: A Lagrangian approach for the incompressible Navier-Stokes equations with variable density, *Communications on Pure and Applied Mathematics*, **65**(10), (2012), 1458–1480.

[12] R. Danchin and P. B. Mucha: Critical functional framework and maximal regularity in action on systems of incompressible flows, *Mémoires de la Société Mathématique de France*, **143** (2015).

[13] R. Danchin and P. B. Mucha: The incompressible Navier-Stokes equations in vacuum, *Communications on Pure and Applied Mathematics*, **72**, (2019), 1351–1385.

[14] R. Danchin and S. Wang: Global unique solutions for the inhomogeneous Navier-Stokes equation with only bounded density, in critical regularity spaces, arXiv: 2201. 11011.

[15] R. Danchin and P. Zhang: Inhomogeneous Navier-Stokes equations in the half-space, with only bounded density, *J. Funct. Anal.*, **267**, (2014), 2371–2436.

[16] R. Danchin and X. Zhang: On the persistence of Hölder regular patches of density for the inhomogeneous Navier-Stokes equations, *Journal de l'Ecole Polytechnique*, **4**, (2017), 781–811.

[17] B. Desjardins: Global existence results for the incompressible density-dependent Navier-Stokes equations in the whole space, *Differential Integral Equations*, **10**(3), (1997), 587–598.

[18] B. Desjardins: Regularity of weak solutions of the compressible isentropic Navier-Stokes equations, *Comm. Partial Differential Equations*, **22**(5-6), (1997), 977–1008.

[19] R. Di Perna and P.-L. Lions: Ordinary differential equations, transport theory and Sobolev spaces, *Inventiones Mathematicae*, **98**, (1989), 511–547.

[20] H. Fujita and T. Kato: On the Navier-Stokes initial value problem I, *Archive for Rational Mechanics and Analysis*, **16**, (1964), 269–315.

[21] F. Gancedo and E. Garcia-Juarez: Global regularity of 2D density patches for inhomogeneous Navier-Stokes, *Archive for Rational Mechanics and Analysis*, **229**, (2018), 339–360.

[22] J. Huang, M. Paicu and P. Zhang: Global well-posedness of incompressible

inhomogeneous fluid systems with bounded density or non-Lipschitz velocity, *Archive for Rational Mechanics and Analysis*, **209**(2), (2013), 631–682.

[23] A. Kazhikhov: Solvability of the initial-boundary value problem for the equations of the motion of an inhomogeneous viscous incompressible fluid, *Dokl. Akad. Nauk SSSR*, **216**, (1974), 1008–1010.

[24] O. Ladyzhenskaya: Solution "in the large" of the non-stationary boundary value problem for the Navier-Stokes system with two space variables, *Comm. Pure Appl. Math.*, **12**, (1959), 427–433.

[25] O. Ladyzhenskaya and V. Solonnikov: Unique solvability of an initial and boundary value problem for viscous incompressible inhomogeneous fluids, *J. Sov. Math.*, **9**(5), (1978), 697–749.

[26] P.-G. Lemarié-Rieusset: *Recent Developments in the Navier-Stokes Problem,* Chapman & Hall/CRC, 2002.

[27] J. Leray: Sur le mouvement d'un liquide visqueux remplissant l'espace, *Acta Mathematica,* **63**, (1934), 193–248.

[28] J. Li: Local existence and uniqueness of strong solutions to the Navier-Stokes equations with nonnegative density, *J. Differential Equations*, **263**, (2017), 6512–6536.

[29] X. Liao and P. Zhang: On the global regularity of 2D density patch for inhomogeneous incompressible viscous flow, *Archive for Rational Mechanics and Analysis*, **220**(2), (2016), 937–981.

[30] X. Liao and P. Zhang: Global regularities of two-dimensional density patch for inhomogeneous incompressible viscous flow with general density, *Communications on Pure and Applied Mathematics*, **72**(4), (2019), 835–884.

[31] J.-L. Lions and G. Prodi: Un théorème d'existence et unicité dans les équations de Navier-Stokes en dimension 2, *Comptes-Rendus Acad. Sci. Paris*, **248**, (1959), 3519–3521.

[32] P.-L. Lions: *Mathematical Topics in Fluid Dynamics, Vol. 1, Incompressible Models*, Oxford University Press, 1996.

[33] M. Paicu, P. Zhang and Z. Zhang: Global unique solvability of inhomogeneous Navier-Stokes equations with bounded density, *Comm. Partial Differential Equations*, **38**, (2013), 1208–1234.

[34] M. Paicu and P. Zhang: Striated regularity of 2-D inhomogeneous incompressible Navier-Stokes system with variable viscosity, *Comm. Math. Phys.*, **376**(1), (2020), 385–439.

[35] J. Simon: Nonhomogeneous viscous incompressible fluids: existence of velocity, density, and pressure, *SIAM J. Math. Anal.*, **21**(5), (1990), 1093–1117.

[36] H. Triebel: *Interpolation Theory, Function Spaces, Differential Operators*, North-Holland Mathematical Library, **18**, North-Holland Publishing Co., 1978.

[37] M. Vishik: Hydrodynamics in Besov spaces, *Archive for Rational Mechanics and Analysis*, **145**(3), (1998), 197–214.

[38] P. Zhang: Global Fujita-Kato solution of 3-D inhomogeneous incompressible Navier-Stokes system, *Adv. Math.*, **363**, (2020), 107007, 43 pp.

Lectures on the Analysis of Nonlinear
Partial Differential Equations Vol. 6
MLM 6, pp. 91–124

Wandering across the Weierstrass Function, While Revisiting Its Properties[*]

Claire David[†]

In honor of Jean-Yves Chemin

Abstract

The Weierstrass function is known as one of these so-called pathological mathematical objects, continuous everywhere, while nowhere differentiable. In the sequel, we have chosen, first, to concentrate on the unconventional history of this function, a function breaking with the mathematical canons of classical analysis of the XIXth century. We recall that it then took nearly a century for new mathematical properties of this function to be brought to light. It has since been the object of a renewed interest, mainly as regards the box-dimension of the related curve. We place ourselves in this vein, and, thanks to our result of 2018, which shows that this value can be obtained in a simple way, without calling for theoretical background in dynamic systems theory, we put forward the link between the non-differentiability and the value of the box-dimension of the curve.

Introduction

The Weierstrass function, introduced in the second part of the nineteenth century by Karl Weierstrass [KH16], [Wei75], is known as one of these so-called pathological mathematical objects, continuous everywhere, while nowhere differentiable; given $\lambda \in \,]0,1[$, and b such that $\lambda b > 1 + \frac{3\pi}{2}$, it is the sum of the uniformly convergent trigonometric series

$$x \in \mathbb{R} \mapsto \sum_{n=0}^{+\infty} \lambda^n \cos\left(\pi \, b^n \, x\right).$$

[*]AMS Subject Classification: 37F20, 28A80, 05C63.

[†]Sorbonne Université, CNRS, UMR 7598, Laboratoire Jacques-Louis Lions, 4, place Jussieu 75005, Paris, France.

The story of this function, and its introduction, by Karl Weierstrass, is of interest. It has to be placed in both a mathematical and a historical context. On the mathematical point of view, of course, much better than done by Bernhard Riemann in 1861 [Dar75], because the proof of the non-differentiability was given to the whole community, it challenged all the existing theories that went back to André-Marie Ampère at the beginning of the century, and led a new impulse that aroused, in the community, the emergence of new functions bearing the same type of properties.

In the historical point of view, it coincides with the global upgrade, material, moral and conceptual, initiated by Prussia in the XIXth century, within the framework of German unity, upgrade which is certainly behind the appointment of Karl Weierstrass, a former high-school teacher, as Professor at the Friedrich-Wilhelm University of Berlin.

Karl Weierstrass had distinguished himself by his results on Abelian functions [Wei54], [Wei56]: the German University could not miss such a talent. This choice proved more than just right. The introduction of the Weierstrass function has made history. Its impact lasts since, even if it took a while before new properties came to light.

Actually, in addition to its nowhere differentiability, an interesting feature of the function is its self similarity properties. After the works of A. S. Besicovitch and H. D. Ursell [BU37], it is Benoît Mandelbrot [Man77] who particularly highlighted the fractal properties of the Weierstrass Curve. He also conjectured that the Hausdorff dimension of the graph is $D_{\mathcal{W}} = 2 + \frac{\ln \lambda}{\ln b}$.

In the view of all that we have evoked, it seemed important to us to consider the Weierstrass function under the prism of an historical perspective, as we expose it in section 1, all the more as interesting discussions still occupy the mathematician community, and us in particular.

For instance, in [Dav18], we have showed that, in the case where $b = N_b$ is an integer, and contrarily to existing work on the subject, the box-counting dimension (or Minkowski dimension) of the Weierstrass Curve, which happens to be equal to its Hausdorff dimension [KMPY84], [BBR14], can be obtained in a simple way, without calling for theoretical background in dynamic systems theory, as it is usually the case. At stake are prefractals, by means of a sequence of graphs, that converge towards the Weierstrass Curve. This sequence of graphs enables one to show nice geometric properties, since, for any natural integer m, the consecutive vertices of the mth-order graph $\Gamma_{\mathcal{W}_m}$ are the vertices of simple not self-intersecting polygons with N_b sides, as it is exposed in section 2, polygons which play a part in the determination of the box-counting dimension of the curve.

Also, we improve or retrieve more classical results, and rather simply, as exposed in the sequel: in section 3, we put the light on the fact that our result concerning the box-dimension of the graph also gives an explicit lower bound, which is not given in existing works. Furthermore, we give a new proof of the non-differentiability of the Weierstrass function in the aforementioned case.

1 An historical overview: From Ampère and well-established beliefs, to the so-called pathological objects

In 1806, André-Marie Ampère [Amp06] gave what he considered as a "proof", according to which, for a given curve, it is always possible, except in a finite number of points, to calculate the slope. This "proof", that one can find in the Mathematics books of the time, served as a reference until the mid-nineteenth century.

The beginning of the memoir of André-Marie Ampère [Amp06].

This lasted a certain time, until the 1860's to be exact; let us quote the French mathematician Jean Gaston Darboux [Dar75]:

"Until the appearance of Riemann's memoir on trigonometric series, no doubt had been raised about the existence of the derivative of continuous functions. Excellent, illustrious geometers, among whom Ampère, had tried to give rigorous proofs of the existence of the derivative. These attempts were, no doubt, far from being satisfactory; but, I repeat, no doubt had been expressed about the very existence of a derivative for continuous functions."

Gaston Darboux of course refered to the mention, in 1861, by Bernhard Riemann, the Professor at the University of Göttingen, of the existence of a continuous function that would not be nowhere differentiable:

$$x \mapsto \mathcal{R}(x) = \sum_{n=1}^{+\infty} \frac{\sin n^2 x}{n^2}.$$

It is not clear whether Riemann gave a proof. If he did so, there is no mention of it in the literature of the time. And no one, at that time too, knew how to obtain it.

About two years later, during the winter 1863—1864, the former high school teacher (1842—1855) Karl Weierstrass, who had been appointed in 1856 Professor at what would then become the Friedrich-Wilhelm University of Berlin (the Königliches Gewerbeinstitut), gave a course on the theory of analytic functions. In this peculiar course took place the first evocation of a new function, continuous everywhere, and nowhere differentiable, which would then be called after him "Weierstrass function". How did this function come to Weierstrass's mind? Some, like J.-P. Kahane [Kah64], suggest that it could be attributed to the Riemann function, for which he did not know how to prove the non-differentiable feature. Without taking sides, it may simply come from the fact that these questions, that were in the air, aroused interest in the mathematical community of the time. To use terminology currently in vogue, it is what historians today call "circulation of ideas".

It is interesting to note that the appointment of Karl Weierstrass as Professor coincides with the global upgrade, material, moral and conceptual, initiated by Prussia. Prussia wanted the German science to dominate the world. So, when the Austrian Minister of Education, Leopold Graf von Thun und Hohenstein, proposed to Karl Weierstrass the creation of a chair, in the university of his choice, with an annual salary of 2000 gulden [KH16], Berlin immediately made a counter offer. This is the culmination of the regeneration Prussian process, launched in 1806, after the defeat of Iena against Napoleon.

In 1864, therefore, the Friedrich-Wilhelm University attributed a chair to Karl Weierstrass, at the exact moment when Bismarck began the German unification (War of Duchies). Everything was then connected: science, industry, prosperity, military and political power.

Beyond this configuration, what is of main interest to us is the specific story of the function, and, if one can say, its emergence in the mathematical community of the time. This of course leads one to consider the oldest known evidence, which can be found in a fac-similé of manuscript notes taken by Hermann Amandus Schwarz, 20 years old, who attended the course (ABBAW, Nachlass Schwarz, Nr. 29, Archivs der Berlin-Brandenburgischen Akademie der Wissenschaften, [KH16]):

"It is not proved that such functions have derivatives. Proofs are erroneous if I show that there are such functions which are continuous in the above sense, but do not possess a derivative in any point."

But one had to wait until 1872, July 18th, for the first official (oral) presentation of the aforementioned Weierstrass function, at the Berlin Academy of Sciences, by Karl Weierstrass himself.

As regards the first written reference, it occurred in a letter written by Karl Weierstrass to Paul-Gustav du Bois-Reymond, in 1873 [Wei73]:

The translation is the following:

"Dear Colleague,

In your last paper, published by Borchardt, you expose my proof showing that the function (...) was everywhere non-differentiable under the conditions I gave. I agree with everything."

BRIEFE VON K. WEIERSTRASS AN PAUL DU BOIS-REYMOND.

Berlin, 23. November 1873.
Potsdamer Str. No. 40.

Verehrter Herr Kollege!

In Ihrer neuesten, mir von Borchardt mitgeteilten Abhandlung[1] haben Sie meinen Beweis, daß die Funktion

$$\sum_{n=0}^{\infty} a^n \cos\left(b^n \pi x\right)$$

unter den angegebenen Bedingungen an keiner Stelle einen bestimmten Differential-Koeffizienten besitze, aufgenommen. Damit bin ich völlig einverstanden,

Beginning of the letter written by Karl Weierstrass to
P.-G. du Bois-Reymond [Wei73].

One may then wonder what was Weierstrass's point of view, on the Riemann function? He laid the emphasis upon, of course, the lack of proof, but, also, on the lack of precision: was the $\mathcal{R}$ function non-differentiable everywhere, or at certain points only:

Es wäre zunächst nach meiner Ansicht zweckmäßig ausdrücklich zu erwähnen, daß RIEMANN bereits im Jahre 1861 einigen seiner Zuhörer die durch die Reihe

$$\sum_{n=1}^{\infty} \frac{\sin\left(n^2 x\right)}{n^2}$$

dargestellte Funktion als eine solche, die keine Ableitung besitze, bezeichnet, seinen Beweis dafür aber niemandem mitgeteilt, sondern nur gelegentlich geäußert habe, derselbe sei aus der Theorie der elliptischen Funktionen zu holen. Auch sei nichts darüber bekannt, ob Riemann behauptet habe, seine Funktion besitze an keiner Stelle einen bestimmten Differentialquotienten, — im Kreise von Riemanns Schülern schien man wenigstens von der Existenz solcher Funktionen nichts gewußt zu haben, wie aus einer Äußerung HANKELS (Untersuchungen über

Second extract of the letter written by Karl Weierstrass to
P.-G. du Bois-Reymond [Wei73].

"It seems appropriate to recall that Riemann presented this function to his students in 1861. This function is not differentiable, yet, the proof has not been communicated to anyone, it has been said that this could be done with the theory of elliptic functions. It is also not known whether Riemann claimed that his function was non-differentiable everywhere, or at certain points only."

This remark is all the more interesting, since it was not until the 1970's that the differentiable character of the $\mathcal{R}$ function at specific rational multiples of π, of the form:

$$\frac{2\,p+1}{2\,q+1}\,\pi, \quad p,\ q \text{ integers}$$

was proved, by Joseph Gerver [Ger70].

As concerns the first publication, it took place in 1875, in the Crelle Journal, through an article written by P.-G. du Bois-Reymond [BR75]:

J o u r n a l

für die

reine und angewandte Mathematik.

In zwanglosen Heften.

Als Fortsetzung des von

A. L. C r e l l e

gegründeten Journals

herausgegeben

unter Mitwirkung der Herren

Schellbach, Kummer, Kronecker, Weierstrass

von

C. W. B o r c h a r d t.

Mit thätiger Beförderung hoher Königlich-Preussischer Behörden.

Neunundsiebzigster Band.

In vier Heften.

Mit einer Figurentafel.

Berlin, 1875.

Druck und Verlag von Georg Reimer.

Ganz etwas Anderes scheinen mir aber die Functionen zu bedeuten, die Herr *Weierstrass* seinen Bekannten mittheilt, die in *keinem* Punkte einen Differentialquotienten besitzen, was noch von keiner der vorher angeführten Functionen nachgewiesen worden ist, und welche bei ihrer grossen Einfachheit und scheinbaren Unverfänglichkeit ahnen lassen, eine wie verbreitete Eigenschaft die Nichtdifferentiirbarkeit der Functionen sein mag. Hier sind nicht besondere Zahlenarten, die doch schliesslich immer isolirt auftreten, mit gewissen Singularitäten behaftet, sondern diese sind durch das ganze Grössengebiet des Arguments gleichförmig und gleichsam stetig vertheilt**).

Um meine Zweifel zu zerstreuen, hatte Herr *Weierstrass* die Güte, mir ein Beispiel einer solchen Function mitzutheilen, und ich glaube mir die Fachgenossen zu Dank zu verpflichten, wenn ich es hier, wo es als Beispiel einer durchweg stetigen Function, die nicht zur folgenden Classe gehört, an seinem Platze ist, wörtlich nach der Aufzeichnung des Verfassers abdrucken lasse:

„Es sei x eine reelle Veränderliche, a eine ungerade ganze Zahl, b eine positive Constante, kleiner als Eins, und

$$f(x) = \sum_{0}^{\infty}(b^n\cos(a^n x)\pi);$$

so ist $f(x)$ eine stetige Function, von der sich zeigen lässt, dass sie, sobald der Werth des Products ab eine gewisse Grenze übersteigt, *an keiner Stelle einen bestimmten Differentialquotienten hat.*

Extract of the article of P.-G. du Bois-Reymond
in the Crelle Journal [BR75].

"The functions exposed by Mr. Weierstrass to his usual audience appear to me as being far different, since they possess nowhere a derivative; this has never before been proved; and despite an appearance of great simplicity, and as inconceivable as it may seem, they do not possess this expected property of differentiability. This does not concern isolated points, which could present singularities, but intervals evenly distributed throughout the field of study. To dissipate my doubts, Mr. Weierstrass was kind enough to give me an example of such a function, and I am very grateful to him; it is an example of a function, continuous everywhere, which does not belong to the usual classes of functions. Listen how the author exposes it: "Given a real number x, an odd integer a, and a positive constant b, smaller than one (...) then $f(x)$ is a function continuous everywhere which, as soon as the product $a\,b$ exceeds a known value, is nowhere differentiable.""

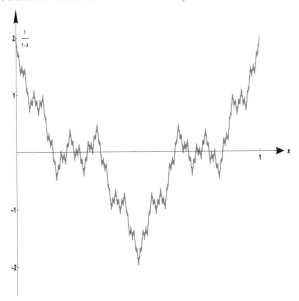

The Weierstrass Curve, in the case $N_b = 3$, $\lambda = \dfrac{1}{2}$.

The impulse given by Weierstrass has led, from the 1870's, to the emergence of other functions of that type. One may quote, for instance, the one proposed by Jean Gaston Darboux [Dar75], [Dar79]:

$$x \mapsto \mathcal{D}arboux(x) = \sum_{n=1}^{+\infty} \frac{\sin\left((n+1)!\,x\right)}{n!}.$$

Jean Gaston Darboux proved the non-differentiability of his function (see [Dar75], pages 107–108). The $(n+1)!$ instead of an $n!$ may intrigue. One has to look at the (non completely explicit) proof to understand that if an $n!$ had been substituted to the original $(n+1)!$, an $n+1$ factor crucial in the non-differentiable feature would have been reported missing.

More precisely: by introducing a strictly positive integer N, Darboux uses a

Claire David

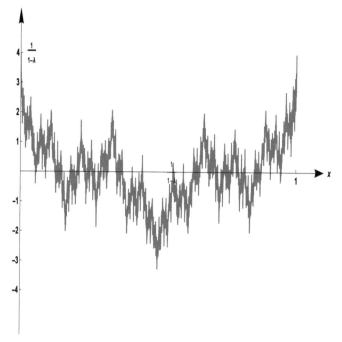

The Weierstrass Curve, in the case $N_b = 3$, $\lambda = \dfrac{3}{4}$.

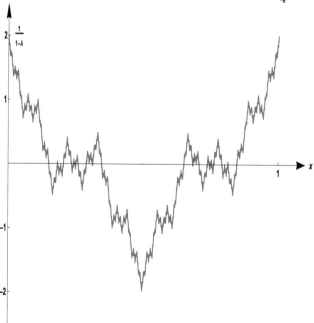

The Weierstrass Curve, in the case $N_b = 7$, $\lambda = \dfrac{1}{2}$.

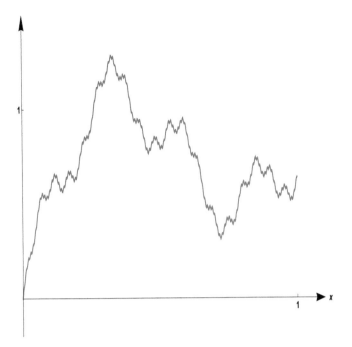

The Darboux Curve.

decomposition of his function of the form

$$Darboux = \phi_N + \psi_N$$

where, for any real number x,

$$\phi_N(x) = \sum_{n=1}^{N-1} \frac{\sin((n+1)!\,x)}{n!}, \quad \psi_N(x) = \sum_{n=N}^{+\infty} \frac{\sin((n+1)!\,x)}{n!}.$$

Given two strictly positive numbers h and ε such that

$$N \times N! \times h = 2\,\varepsilon$$

and due to the second order Taylor expansion that the reader will have of course applied:

$$\phi_N(x+h) - \phi_N(x)$$

$$= \sum_{n=1}^{N-1} \left\{ h\,(n+1)!\,\frac{\cos((n+1)!\,x)}{n!} - \frac{h^2}{2}\,((n+1)!)^2\,\frac{\sin((n+1)!\,x)}{n!} \right\} + o\left(h^2\right),$$

one "easily " (to use Darboux's terms) gets

$$\frac{\phi_N(x+h) - \phi_N(x)}{h}$$

$$= \frac{1}{h} \sum_{n=1}^{N-1} \left\{ h(n+1)! \frac{\cos((n+1)!x)}{n!} - \frac{h^2}{2}((n+1)!)^2 \frac{\sin((n+1)!x)}{n!} \right\} + o(h)$$

$$= \sum_{n=1}^{N-1} \left\{ (n+1)! \frac{\cos((n+1)!x)}{n!} - \frac{h}{2}((n+1)!)^2 \frac{\sin((n+1)!x)}{n!} \right\} + o(h)$$

$$= \sum_{n=1}^{N-1} \left\{ (n+1) \cos((n+1)!x) - \frac{h}{2}(n+1)(n+1)! \sin((n+1)!x) \right\} + o(h)$$

$$= \sum_{n=2}^{N} \left\{ n \cos(n!x) - \frac{h}{2} n \cdot n! \sin(n!x) \right\} + o(h).$$

Something is not clear in the original proof, because, instead of our previous expression, Darboux writes:

$$\frac{\phi_N(x+h) - \phi_N(x)}{h} = \sum_{n=???}^{N} n \cos(n!x) - \varepsilon \sin(N!x) + \omega(N, \varepsilon)$$

(we have written ??? for the lower bound in the sum, since the original text is not readable, one can hardly see if it is a "1", an "r" , a "x"), and where ω denotes a function such that, for a given ε,

$$\lim_{N \to +\infty} \omega(N, \varepsilon) = 0.$$

So, with our current terminology, ω corresponds to a sum of "$o(\cdot)$", and details are reported missing.

The main point of the proof given by Darboux is in fact to point out that, for the values of the real number x such that

$$\lim_{N \to +\infty} \sin(N!x) = 0,$$

the limit

$$\lim_{N \to +\infty} \sum_{n=???}^{N} n \cos(n!x)$$

does not exist.

Very elegantly, Darboux quoted Riemann, Schwarz and some others, but not Weierstrass ...

One finds, after, another example given in 1877 by Ulisse Dini [Din77], [Din78]:

$$x \mapsto Dini(x) = \sum_{n=1}^{+\infty} \frac{a^n \cos(1 \times 3 \times 5 \times \cdots \times (2n-1)x)}{1 \times 3 \times 5 \times \cdots \times (2n-1)}, \quad a > 1 + \frac{3\pi}{2}.$$

As a result, the existence of these functions casts a chill on the mathematical community. Let us recall what Charles Hermite wrote, in one of his numerous letters to Thomas Stieltjes, in 1893 ([Cor05], letter 374):

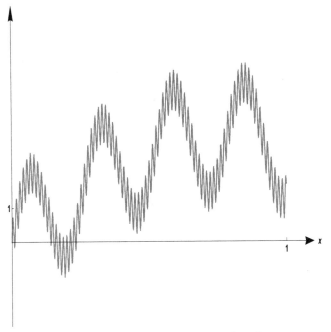

The Dini Curve, in the case $\alpha = \dfrac{3}{2} + \dfrac{3\pi}{2}$.

"I turn away with fright and horror from this lamentable plague of continuous functions that have no derivatives."

As for Poincaré [Poi90], he stated that:

"Logic sometimes creates monsters. For half a century, one has seen the birth of strange functions, functions that look as little as possible as the honest ones, the useful ones. No more continuity, or continuity, but no derivatives, etc ... Even more, from the logical point of view, those strange functions appear as the most general ones, while those one may fall on by chance are relegated as special cases. They only have a tiny corner left."

Yet, and it is very important, contrary to the erroneous interpretations found in the literature ([JP15], page 4), Poincaré never described Weierstrass's work as offensive to common sense [Poi98]:

"To begin with, I shall quote a note read at the Berlin Academy on July 18, 1872, and where Weierstrass gave examples of continuous functions of a real argument which, for any value of this argument, do not possess a finite derivative. A hundred years ago, such a function would have been regarded as an outrage to common sense. A continuous function, one would have said, is in essence susceptible of being represented by a curve, and a curve obviously always has a tangent."

What Poincaré said about these functions was, nevertheless, rather hard [Poi99]:

"Formerly, when new functions arose, it was because they were devoted to some practical purpose; today, they are invented expressly to put in default the reasoning of our fathers, and we will never get out of it."

Since then, the Weierstrass function has kept arousing interest. If this interest was initially due to its nowhere differentiability, its fractal properties, brought to light about ninety years later by B. Mandelbrot [Man82], pages 388–390, made the community consider it from a new angle. Mandelbrot was looking for an approximation of the Brownian motion, which accounts for its interest in the function introduced by Weierstrass.

By moving to a slightly more general frame, Mandelbrot thus chose to consider the related complex function defined, for any real number x, by

$$\mathcal{W}_c(x) = \frac{1}{\sqrt{1-w^2}} \sum_{n=0}^{+\infty} w^n \, e^{2 i \pi b^n x}$$

where

$$b > 1, \quad w = \frac{1}{b^H} = b^{D_W - 2}, \quad 1 < 2 - H = D_W < 2.$$

After an introductory comparison with the Brownian motion, B. Mandelbrot placed himself on the point of view of physics, and, especially, to study the function's spectra: for each frequency f of the form $f = b^n$, $n \in \mathbb{N}^\star$ the spectral line of energy, i.e. the one that results from emission or absorption of light in a narrow range of frequencies, given by

$$\frac{1}{1-w^2} \, w^{2n}$$

yields a cumulative energy in frequencies $f \geq b^n$ of

$$\sum_{k=n}^{+\infty} \frac{1}{1-w^2} \, w^{2k} = \frac{1}{(1-w^2)^2} \, w^{2n} = \frac{1}{(1-w^2)^2} \, \frac{1}{b^{2nH}} = \frac{1}{(1-w^2)^2} \, \frac{1}{f^{2H}}.$$

B. Mandelbrot recalled then that, since "a function's derivative is obtained by multiplying its kth Fourier coefficient by k", for physicists looking at the formal derivative of the complex Weierstrass function, the $b^{n^{th}}$ Fourier coefficient has an amplitude squared equal to

$$\frac{1}{1-w^2} \, w^{2n} \, b^{2n}.$$

Thus, the cumulative energies for frequencies greater or equal than b^n are infinite, which enable physicists to obtain the non-differentiability of the $\mathcal{W}$ function as an "intuitively obvious" feature.

B. Mandelbrot then explained that, if "the total high frequency energy is infinite", it is thus "catastrophic for the theory", echoing the 1900's theory of Rayleigh and Jeans of blackbody radiation. By resuming his comparison with Brownian motion, and for the purpose of future applications, B. Mandelbrot thus proposed to take into account a modified version of the function, one that would soon be called Weierstrass-Mandelbrot one, defined, for any real number x, by

$$\mathcal{W}_\mathcal{M}(x) = \frac{1}{\sqrt{1-w^2}} \sum_{n=-\infty}^{+\infty} w^n \left\{ e^{2 i \pi b^n x} - 1 \right\}.$$

Better than the classical Weierstrass function, the $\mathcal{W_M}$ function, still continuous everywhere, while nowhere differentiable, bears a scaling property and is self-affine: $\forall\, m \in \mathbb{Z}, \forall\, x \in \mathbb{R}$,

$$\mathcal{W_M}\left(b^m\, x\right) = \frac{1}{\sqrt{1-w^2}} \sum_{n=-\infty}^{+\infty} w^n \left\{ e^{2\, i\, \pi\, b^{m+n}\, x} - 1 \right\}$$

$$= \frac{1}{w^m}\, \mathcal{W_M}(x)$$

$$= b^{m\, H}\, \mathcal{W_M}(x).$$

To better stick real modelling, B. Mandelbrot then proposed to randomize the function, which enables one to approximate fractional Brown functions.

And as it has often been the case, B. Mandelbrot's intuition is proved to be right: the Weierstrass-Mandelbrot function has practical applications. It was for instance shown in the 1990's that the function could be used in the modelling of turbulence [HSR92].

As for the classical Weierstrass function, it still occupies mathematicians. At stake is particularly the determination of the dimension of the Weierstrass Curve, whether one considers the box (or Minkowski-Bouligand) one, or the Hausdorff one. The value of the box-dimension, and how to obtain it, was first found in the works of J. L. Kaplan et al. [KMPY84], or in the book of K. Falconer [Fal86] (Example 11.3). Both box and Hausdorff dimensions are discussed in the paper of F. Przytycki and M. Urbański [PU89]. An intermediate discussion, by means of a new dimension index, is proposed in the one by T.-Y. Hu and K.-S. Lau [HL93]. As for the Hausdorff dimension, a proof is given by B. Hunt [Hun98] in 1998 in the case where arbitrary phases are included in each cosinusoidal term of the summation. Recently, K. Barański, B. Bárány and J. Romanowska [BBR14] proved that, for any value of the real number b, there exists a threshold value λ_b belonging to the interval $\left]\frac{1}{b}, 1\right[$ such that the aforementioned dimension is equal to $D_\mathcal{W}$ for every b in $\left]\lambda_b, 1\right[$. In [Kel17], G. Keller proposes what appears as a much simpler and very original proof. Results by W. Shen [She18] go further than the ones of [BBR14].

One may note that Weierstrass's work is not self-evident. It hits hard a whole academic tradition, mindful of order and classicism, resisting the challenge of what was considered obvious and acquired. Nearly a century will be necessary for the mathematical community to take seriously and start exploiting the very rich potential offered by the nowhere differentiability of the Weierstrass function. It is not a coincidence that the discovery of our Berlin professor meets a real and renewed interest when it is associated to the work on Brownian motion, thanks to Mandelbrot. The random, the erratic, the breaking of sense and direction definitely make their entry into the so-called "serious" science. This goes hand in hand with the extension of the notion of dimension. One might go further, and extend this constant to the whole of thought and knowledge, in the twentieth century, all disciplines combined, including arts and letters. Now, this movement of deciphering the irrational goes on. In the same vein, our contribution will now try to put forward the link between the non-differentiability and the value of the box-dimension of the curve.

2 Basic properties of the Weierstrass function— Towards the graph

In the sequel, we aim at describing some geometric properties of the Weierstrass Curve, properties which will be useful especially as regards Theorem 3.1.

We place ourselves in the euclidian plane of dimension 2, referred to a direct orthonormal frame. The usual Cartesian coordinates are (x, y).

Notation. In the following, λ and b are two real numbers such that

$$0 < \lambda < 1 \quad , \quad b = N_b \in \mathbb{N} \quad \text{and} \quad \lambda N_b > 1.$$

We will consider the Weierstrass function $\mathcal{W}$, defined, for any real number x, by

$$\mathcal{W}(x) = \sum_{n=0}^{+\infty} \lambda^n \cos\left(2\pi N_b^n x\right).$$

Definition 2.1. Weierstrass Curve. We will call **Weierstrass Curve** the restriction to $[0, 1[\times\mathbb{R}$ of the graph of the Weierstrass function, and denote it by $\Gamma_{\mathcal{W}}$.

Property 2.1. Periodic properties of the Weierstrass function. *For any real number x:*

$$\mathcal{W}(x+1) = \sum_{n=0}^{+\infty} \lambda^n \cos\left(2\pi N_b^n x + 2\pi N_b^n\right) = \sum_{n=0}^{+\infty} \lambda^n \cos\left(2\pi N_b^n x\right) = \mathcal{W}(x).$$

The study of the Weierstrass function can be restricted to the interval $[0, 1[$.

The restriction $\Gamma_{\mathcal{W}}$ to $[0, 1[\times\mathbb{R}$, of the Weierstrass Curve, is approximated by prefractals (sequence of graphs, built through an iterative process).

To this purpose, we introduce the iterated function system of the family of C^∞ maps from $\mathbb{R}^2$ to $\mathbb{R}^2$:

$$\{T_0, \ldots, T_{N_b-1}\}$$

where, for any integer i belonging to $\{0, \ldots, N_b - 1\}$, and any (x, y) of $\mathbb{R}^2$,

$$T_i(x, y) = \left(\frac{x+i}{N_b}, \lambda y + \cos\left(2\pi\left(\frac{x+i}{N_b}\right)\right)\right).$$

Remark 2.1. For any i of $\{0, \ldots, N_b - 1\}$, the map T_i is not a contraction.

The **Gluing Lemma [BD85]** does not apply, but:

Lemma 2.2. *For any integer i belonging to $\{0, \ldots, N_b - 1\}$, the map T_i is a bijection of the graph of the Weierstrass function on $\mathbb{R}$.*

Proof. Let us consider $i \in \{0, \ldots, N_b - 1\}$. Consider a point $(y, \mathcal{W}(y))$ of $\Gamma_{\mathcal{W}}$, and let us look for a real number x of $[0,1]$ such that

$$T_i(x, \mathcal{W}(x)) = (y, \mathcal{W}(y)).$$

One has

$$y = \frac{x + i}{N_b}.$$

Then

$$x = N_b\, y - i.$$

This enables one to obtain

$$\mathcal{W}(x) = \mathcal{W}(N_b\, y - i) = \sum_{n=0}^{+\infty} \lambda^n \cos\left(2\,\pi\, N_b^{n+1}\, y - 2\,\pi\, N_b^n\, i\right)$$

$$= \sum_{n=0}^{+\infty} \lambda^n \cos\left(2\,\pi\, N_b^{n+1}\, y\right)$$

and

$$T_i(x, \mathcal{W}(x)) = \left(\frac{x+i}{N_b}, \lambda\,\mathcal{W}(x) + \cos\left(2\,\pi\,\left(\frac{x+i}{N_b}\right)\right)\right)$$

$$= \left(y, \sum_{n=0}^{+\infty} \lambda^{n+1} \cos\left(2\,\pi\, N_b^{n+1}\, y\right) + \cos\left(2\,\pi\, y\right)\right)$$

$$= \left(y, \sum_{n=0}^{+\infty} \lambda^n \cos\left(2\,\pi\, N_b^n\, y\right)\right)$$

$$= (y, \mathcal{W}(y)).$$

There exists thus a unique real number x such that

$$T_i(x, \mathcal{W}(x)) = (y, \mathcal{W}(y)).$$

$\square$

Property 2.3.

$$\Gamma_{\mathcal{W}} = \bigcup_{i=0}^{N_b - 1} T_i(\Gamma_{\mathcal{W}}).$$

Proof. This immediately comes from Lemma 2.2. $\square$

Definition 2.2. Word, on the graph $\Gamma_{\mathcal{W}}$. Let m be a strictly positive integer. We will call **number-letter** any integer $\mathcal{M}_i$ of $\{0, \ldots, N_b - 1\}$, and **word of length** $|\mathcal{M}| = m$, on the graph $\Gamma_{\mathcal{W}}$, any set of number-letters of the form

$$\mathcal{M} = (\mathcal{M}_1, \ldots, \mathcal{M}_m).$$

We will write

$$T_{\mathcal{M}} = T_{\mathcal{M}_1} \circ \cdots \circ T_{\mathcal{M}_m}.$$

Definition 2.3. For any integer i belonging to $\{0, ..., N_b - 1\}$, let us denote by

$$P_i = (x_i, y_i) = \left(\frac{i}{N_b - 1}, \frac{1}{1 - \lambda} \cos \left(\frac{2 \pi i}{N_b - 1} \right) \right)$$

the fixed point of the map T_i.

We will denote by V_0 the ordered set (according to increasing abscissa), of the points

$$\{P_0, ..., P_{N_b - 1}\}$$

since, for any i of $\{0, ..., N_b - 2\}$,

$$x_i \leq x_{i+1}.$$

The set of points V_0, where, for any i of $\{0, ..., N_b - 2\}$, the point P_i is linked to the point P_{i+1}, constitutes an oriented graph (according to increasing abscissa), that we will denote by $\Gamma_{\mathcal{W}_0}$. V_0 is called the set of vertices of the graph $\Gamma_{\mathcal{W}_0}$.

For any natural integer m, we set

$$V_m = \bigcup_{i=0}^{N_b - 1} T_i \left(V_{m-1} \right).$$

The set of points V_m, where two consecutive points are linked, is an oriented graph (according to increasing abscissa), which we will denote by $\Gamma_{\mathcal{W}_m}$. V_m is called the set of vertices of the graph $\Gamma_{\mathcal{W}_m}$. We will denote, in the following, by $\mathcal{N}_m^{\mathcal{S}}$ the number of vertices of the graph $\Gamma_{\mathcal{W}_m}$, and we will write

$$V_m = \left\{ \mathcal{S}_0^m, \mathcal{S}_1^m, \ldots, \mathcal{S}_{\mathcal{N}_m^{\mathcal{S}} - 1}^m \right\}.$$

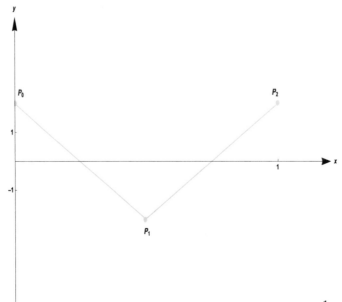

The fixed points P_0, P_1, P_2, and the graph $\Gamma_{\mathcal{W}_0}$, in the case where $\lambda = \dfrac{1}{2}$, and $N_b = 3$.

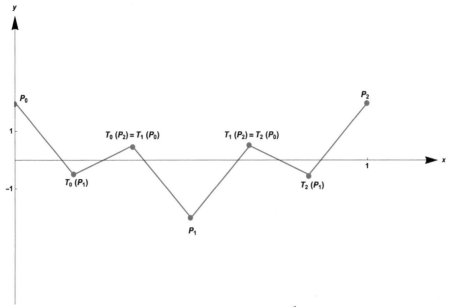

The graph $\Gamma_{\mathcal{W}_1}$, in the case where $\lambda = \dfrac{1}{2}$, and $N_b = 3$.
$T_0(P_2) = T_1(P_0)$ and $T_1(P_2) = T_2(P_0)$.

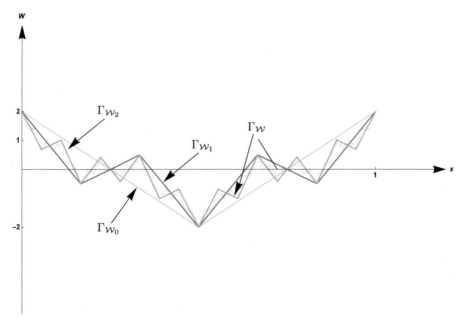

The graphs $\Gamma_{\mathcal{W}_0}$, $\Gamma_{\mathcal{W}_1}$, $\Gamma_{\mathcal{W}_2}$, $\Gamma_{\mathcal{W}}$, in the case where $\lambda = \dfrac{1}{2}$, and $N_b = 3$.

Property 2.4. *For any natural integer m,*

$$V_m \subset V_{m+1}.$$

Property 2.5. *For any integer i belonging to $\{0, ..., N_b - 2\}$,*

$$T_i\left(P_{N_b-1}\right) = T_{i+1}\left(P_0\right).$$

Proof. Since

$$P_0 = \left(0, \frac{1}{1-\lambda}\right), \quad P_{N_b-1} = \left(\frac{N_b-1}{N_b-1}, \frac{1}{1-\lambda}\cos\left(\frac{2\pi\left(N_b-1\right)}{N_b-1}\right)\right) = \left(1, \frac{1}{1-\lambda}\right),$$

one has

$$\begin{cases} T_i\left(P_{N_b-1}\right) = \left(\dfrac{1+i}{N_b}, \dfrac{\lambda}{1-\lambda} + \cos\left(2\pi\left(\dfrac{1+i}{N_b}\right)\right)\right), \\[3mm] T_{i+1}\left(P_0\right) = \left(\dfrac{i+1}{N_b}, \dfrac{\lambda}{1-\lambda} + \cos\left(2\pi\left(\dfrac{i+1}{N_b}\right)\right)\right). \end{cases}$$

$\square$

Property 2.6. *The sequence $\left(\mathcal{N}_m^{\mathcal{S}}\right)_{m\in\mathbb{N}}$ is an arithmetico-geometric one, with $\mathcal{N}_0^{\mathcal{S}} = N_b$ as first term:*

$$\forall\, m \in \mathbb{N} : \quad \mathcal{N}_{m+1}^{\mathcal{S}} = N_b\,\mathcal{N}_m^{\mathcal{S}} - \left(N_b - 2\right).$$

Proof. This result comes from the fact that each graph $\Gamma_{\mathcal{W}_m}$, $m \in \mathbb{N}^\star$, is built from its predecessor $\Gamma_{\mathcal{W}_{m-1}}$ by applying the N_b maps T_i, $0 \leq i \leq N_b - 1$, to the vertices of $\Gamma_{\mathcal{W}_{m-1}}$. Since, for any i of $\{0, ..., N_b - 2\}$,

$$T_i\left(P_{N_b-1}\right) = T_{i+1}\left(P_0\right),$$

the $N_b - 2$ points appear twice if one takes into account the images of the $\mathcal{N}_{m-1}$ vertices of $\Gamma_{\mathcal{W}_{m-1}}$ by the whole set of maps T_i, $0 \leq i \leq N_b - 1$. $\square$

Definition 2.4. Vertices of the graph $\Gamma_{\mathcal{W}}$. Two points X and Y of $\Gamma_{\mathcal{W}}$ will be called **vertices** of the graph $\Gamma_{\mathcal{W}}$ if there exists a natural integer m such that

$$(X, Y) \in V_m^2.$$

Definition 2.5. Consecutive vertices on the graph $\Gamma_{\mathcal{W}}$. Two points X and Y of $\Gamma_{\mathcal{W}}$ will be called **consecutive vertices** of the graph $\Gamma_{\mathcal{W}}$ if there exist a natural integer m, and an integer j of $\{0, ..., N_b - 2\}$, such that

$$X = \left(T_{i_1} \circ \cdots \circ T_{i_m}\right)(P_j) \quad \text{and} \quad Y = \left(T_{i_1} \circ \cdots \circ T_{i_m}\right)(P_{j+1}),$$
$$\{i_1, \ldots, i_m\} \in \{0, ..., N_b - 1\}^m$$

or

$$X = \left(T_{i_1} \circ T_{i_2} \circ \cdots \circ T_{i_m}\right)\left(P_{N_b-1}\right) \quad \text{and} \quad Y = \left(T_{i_1+1} \circ T_{i_2} \circ \cdots \circ T_{i_m}\right)(P_0).$$

Remark 2.2. It is important to note that X and Y cannot be in the same time the images of P_j and P_{j+1}, $0 \le j \le N_b - 2$, by $T_{i_1} \circ \cdots \circ T_{i_m}$, $(i_1, \ldots, i_m) \in \{0, \ldots, N_b - 2\}$, and of P_k and P_{k+1}, $0 \le k \le N_b - 2$, by $T_{p_1} \circ \cdots \circ T_{p_m}$, $(p_1, \ldots, p_m) \in \{0, \ldots, N_b - 2\}$. This result can be proved by induction, since, for any pair of integers (j, k) of $\{0, \ldots, N_b - 2\}^2$, for any i_m of $\{0, \ldots, N_b - 2\}$, and any p_m of $\{0, \ldots, N_b - 2\}$,

$$(i_m \neq p_m \quad \text{and} \quad j \neq k) \Longrightarrow (T_{i_m}(P_j) \neq T_{p_m}(P_k) \quad \text{and} \quad T_{p_m}(P_j) \neq T_{i_m}(P_k)).$$

Each map T_i, $0 \le i \le N_b - 1$ is indeed injective. Since the vertices of the initial graph Γ_{W_0} are distinct, one gets the expected result.

Property 2.7. *For any natural integer m, the N_m^S consecutive vertices of the graph Γ_{W_m} are, also, the vertices of N_b^m simple polygons $\mathcal{P}_{m,j}$, $0 \le j \le N_b^m - 1$, with N_b sides. For any integer j such that $0 \le j \le N_b^m - 1$, one obtains each polygon $\mathcal{P}_{m,j}$ by linking the point number j to the point number $j+1$ if $j = i \bmod N_b$, $0 \le i \le N_b - 2$, and the point number j to the point number $j - N_b + 1$ if $j = -1 \bmod N_b$.*

In the same way, the $N_m^S - 2$ consecutive vertices of the graph Γ_{W_m}, distinct of P_0 and $P_{N_b - 1}$, are the vertices of $N_b^m - 1$ simple polygons $\mathcal{Q}_{m,j}$, $1 \le j \le N_b^m - 2$, with N_b sides. For any integer j such that $1 \le j \le N_b^m - 2$, one obtains each polygon $\mathcal{Q}_{m,j}$ by linking the point number j to the point number $j+1$ if $j = i \bmod N_b$, $1 \le i \le N_b - 1$, and the point number j to the point number $j - N_b + 1$ if $j = 0 \bmod N_b$.

These polygons generate a Borel set of $\mathbb{R}^2$.

Property 2.8. *For any natural integer m, and any integer $j \in \{0, \ldots, N_b^m - 1\}$, there exists a word $\mathcal{M}_{m,j}^{\mathcal{P}}$ of length m such that the set of consecutive vertices of each N_b-gon $\mathcal{P}_{m,j}$ is of the form*

$$\left\{ T_{\mathcal{M}_{m,j}^{\mathcal{P}}}(P_k) \right\}_{0 \le k \le N_b - 1}.$$

In the same way, for any natural integer m, and any integer $j \in \{1, \ldots, N_b^m - 2\}$, there exists a word $\mathcal{M}_{m,j}^{\mathcal{Q}}$ of length m such that the set of consecutive vertices of each N_b-gon $\mathcal{Q}_{m,j}$ is of the form

$$\left\{ T_{\mathcal{M}_{m,j}^{\mathcal{Q}}}(P_{k+1}) \right\}_{0 \le k \le N_b - 1}.$$

Proof. The above result is obtained by induction.

It is obvious that, for $m = 1$, the consecutive vertices of the N_b-gons $\mathcal{P}_{1,0}$, $\mathcal{P}_{1,1}$, ..., $\mathcal{P}_{1,N_b - 1}$ are the respective images $T_0(P_0)$, $T_0(P_1)$, ..., $T_0(P_{N_b-1})$, ..., $T_{N_b-1}(P_0)$, $T_{N_b-1}(P_1)$, ..., $T_{N_b-1}(P_{N_b-1})$.

Now, given a natural integer m, let us assume that, for any integer $j \in \{1, \ldots, N_b^m - 2\}$, there exists a word $\mathcal{M}_{m,j}^{\mathcal{Q}}$ of length m such that the set of consecutive vertices of each N_b-gon $\mathcal{P}_{m,j}$ is of the form

$$\left\{ T_{\mathcal{M}_{m,j}^{\mathcal{P}}}(P_k) \right\}_{0 \le k \le N_b - 1}.$$

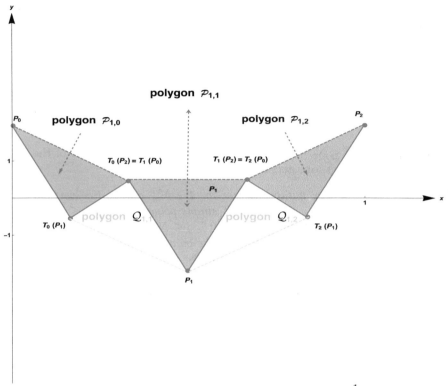

The polygons $\mathcal{P}_{1,0}$, $\mathcal{P}_{1,1}$, $\mathcal{P}_{1,2}$, $\mathcal{Q}_{1,1}$, $\mathcal{Q}_{1,2}$ in the case where $\lambda = \dfrac{1}{2}$, and $N_b = 3$.

Since

$$V_{m+1} = \bigcup_{i=0}^{N_b-1} T_i\left(V_m\right)$$

the set of consecutive vertices of each N_b-gon $\mathcal{P}_{m+1,j}$ is thus of the form

$$\left\{T_0 \circ T_{\mathcal{M}_{m,j}^{\mathcal{P}}}\left(P_k\right), \ldots, T_{N_b-1} \circ T_{\mathcal{M}_{m,j}^{\mathcal{P}}}\left(P_k\right)\right\}_{0 \leq k \leq N_b-1}$$

which naturally yields the searched result at the step $m+1$:

$$T_{\mathcal{M}_{m+1,0}^{\mathcal{P}}} \quad = \quad T_0 \circ T_{\mathcal{M}_{m,0}^{\mathcal{P}}},$$

$$\vdots$$

$$T_{\mathcal{M}_{m+1,N_b-1}^{\mathcal{P}}} \quad = \quad T_{N_b-1} \circ T_{\mathcal{M}_{m,0}^{\mathcal{P}}},$$

$$\vdots$$

$$T_{\mathcal{M}_{m+1,N_b^{m+1}-N_b}^{\mathcal{P}}} \quad = \quad T_0 \circ T_{\mathcal{M}_{m,N_b-1}^{\mathcal{P}}},$$

$$\vdots$$

$$T_{\mathcal{M}_{m+1,N_b^{m+1}-1}^{\mathcal{P}}} = T_{N_b-1} \circ T_{\mathcal{M}_{m,N_b-1}^{\mathcal{P}}}.$$

The second part of the property can be proved similarly.   □

Notation. For any natural integer m, we will respectively denote by

$$\{\mathcal{M}^{\mathcal{P}}_{m,j}\}_{0\le j\le N_b^m-1}, \quad \{\mathcal{M}^{\mathcal{Q}}_{m,j}\}_{0\le j\le N_b^m-1}$$

the ordered sets of the words of length m related to the sets of N_b-gons $\mathcal{P}_{m,j}$, $0\le j\le N_b^m-1$ and $\mathcal{Q}_{m,j}$, $1\le j\le N_b^m-2$ as given in Property 2.8.

Property 2.9. *The set* $\bigcup_{m\in\mathbb{N}} V_m$ *is dense in* $\Gamma_{\mathcal{W}}$.

Proof. Since the function $\mathcal{W}$ is continuous, it suffices to remark that the set of the abscissae of the vertices is dense in $[0,1]$. Given a natural integer i, let us denote by A_i the set of the abscissae of V_i. The set A_i is transformed into $\frac{A_i}{N_b}$ by the map T_0, then, this set is shifted by T_1, ..., T_{N_b-1}, and this produces a new set of points, the distance between two consecutive new points having been divided by N_b.

Formally, as exposed in the above, for any natural integer m, and any integer $j\in\{0,\dots,N_b^m-1\}$, there exists a word $\mathcal{M}^{\mathcal{P}}_{m,j}$ of length m such that the set of consecutive vertices of each N_b-gon $\mathcal{P}_{m,j}$ is of the form

$$\left\{T_{\mathcal{M}^{\mathcal{P}}_{m,j}}(P_k)\right\}_{0\le k\le N_b-1}.$$

Let us write $T_{\mathcal{M}^{\mathcal{P}}_{m,j}}$ under the form

$$T_{\mathcal{M}^{\mathcal{P}}_{m,j}} = T_{i_m}\circ T_{i_{m-1}}\circ\cdots\circ T_{i_1}$$

where $(i_1,\dots,i_m)\in\{0,\dots,N_b-1\}^m$.

One has then

$$x\left(T_{\mathcal{M}^{\mathcal{P}}_{m,j}}(P_k)\right) = \frac{x_k}{N_b^m} + \sum_{p=1}^{m}\frac{i_p}{N_b^p}, \quad x\left(T_{\mathcal{M}^{\mathcal{P}}_{m,j}}(P_{k+1})\right) = \frac{x_{k+1}}{N_b^m} + \sum_{p=1}^{m}\frac{i_p}{N_b^p}$$

and thus

$$x\left(T_{\mathcal{M}^{\mathcal{P}}_{m,j}}(P_{k+1})\right) - x\left(T_{\mathcal{M}}(P_k)\right) = \frac{1}{(N_b-1)N_b^m}.$$

One deduces then

$$[0,1] = \bigcup_{k=0}^{(N_b-1)(N_b^m-1)}\left[\frac{k}{(N_b-1)N_b^m}, \frac{k+1}{(N_b-1)N_b^m}\right]$$

$$= \bigcup_{0\le j\le N_b^m-1,\, 0\le k\le N_b-1}\left[x\left(T_{\mathcal{M}^{\mathcal{P}}_{m,j}}(P_k)\right), x\left(T_{\mathcal{M}^{\mathcal{P}}_{m,j}}(P_{k+1})\right)\right].$$

Let us now consider a point $X=(x,\mathcal{W}(x))$ of $\Gamma_{\mathcal{W}}$, and a strictly positive number ε. Due to the continuity of the Weierstrass function, there exists a natural integer m_0 such that, for any $m\ge m_0$,

$$\forall\,x'\in[0,1]:\quad |x-x'|\le\frac{1}{(N_b-1)N_b^m} \implies |\mathcal{W}(x)-\mathcal{W}(x')|\le\varepsilon.$$

By using our preliminary results, one deduces the existence of a natural integer $m_1 \geq m_0$ such that, for any $m \geq m_1$, the real number x belongs to an interval of the form

$$\left[\frac{k}{(N_b - 1)\, N_b^m}, \frac{k+1}{(N_b - 1)\, N_b^m} \right], \quad 0 \leq k \leq (N_b - 1)\,(N_b^m - 1)$$

or, equivalently

$$\left[x\left(T_{\mathcal{M}_{m,j}^{\mathcal{P}}}\, (P_k) \right), x\left(T_{\mathcal{M}_{m,j}^{\mathcal{P}}}\, (P_{k+1}) \right) \right], \quad 0 \leq j \leq N_b^m - 1,\, 0 \leq k \leq N_b - 1.$$

Thus

$$\left| \mathcal{W}(x) - \mathcal{W}\left(x\left(T_{\mathcal{M}_{m,j}^{\mathcal{P}}}\, (P_k) \right) \right) \right| \leq \varepsilon$$

which yields the expected density result. $\square$

Definition 2.6. Polygonal domain delimited by the graph $\Gamma_{\mathcal{W}_m}$, $m \in \mathbb{N}$.
For any natural integer m, we will call **polygonal domain delimited by the graph $\Gamma_{\mathcal{W}_m}$**, and denote by $\mathcal{D}\,(\Gamma_{\mathcal{W}_m})$, the reunion of the N_b^m polygons $\mathcal{P}_{m,j}$, $0 \leq j \leq N_b^m - 1$ and $\mathcal{Q}_{m,j}$, $1 \leq j \leq N_b^m - 2$.

Remark 2.3. The introduction of this polygonal domain arises naturally as one builds the Weierstrass Curve. In the literature, one can already find approximating polygons, for instance in the case of the Peano curve, as introduced by W. Wunderlich [Wun73]. Such a notion was then adopted by H. Sagan [Sag86], [Sag94]. As showed by H. Sagan, among other advantages, such polygons enable to obtain the exact coordinates of nodal points, which is of course also the case for the Weierstrass Curve. The term "approximating" is justified in so far as the polygons approximate the considered curve uniformly. In our case, we have chosen a slightly different, whatever equivalent, definition of convergence.

Definition 2.7. Convergence of the sequence of polygonal domains $(\mathcal{D}(\Gamma_{\mathcal{W}_m}))_{m \in \mathbb{N}}$. We will say that the sequence of polygonal domains $(\mathcal{D}\,(\Gamma_{\mathcal{W}_m}))_{m \in \mathbb{N}}$ converges towards the graph $\Gamma_{\mathcal{W}}$ if, when the integer m tends towards infinity, the Lebesgue measure of all polygons $\mathcal{P}_{m,j}$, $0 \leq j \leq N_b^m - 1$ and $\mathcal{Q}_{m,j}$, $1 \leq j \leq N_b^m - 2$, tends towards zero.

Property 2.10. *For any natural integer m, the vertices of the N_b-gons $\mathcal{P}_{m,j}$, $0 \leq j \leq N_b^m - 1$, are not self-intersecting.*

Proof. Let us prove, by induction, that the vertices of the N_b-gons $\mathcal{P}_{m,j}$, $0 \leq j \leq N_b^m - 1$ are not self-intersecting.
    For any integer i belonging to $\{0, ..., N_b - 1\}$,

$$P_i = (x_i, y_i) = \left(\frac{i}{N_b - 1}, \frac{1}{1 - \lambda} \cos\left(\frac{2\,\pi\,i}{N_b - 1} \right) \right).$$

Thus, for any integer i belonging to $\{0, ..., N_b - 2\}$,

$$y_{i+1} - y_i = \frac{1}{1 - \lambda} \left\{ \cos\left(\frac{2\pi(i+1)}{N_b - 1}\right) - \cos\left(\frac{2\pi i}{N_b - 1}\right) \right\}$$

$$= -\frac{2}{1 - \lambda} \sin\left(\frac{2\pi(i+1+i)}{2N_b - 1}\right) \sin\left(\frac{2\pi(i+1-i)}{2N_b - 1}\right)$$

$$= -\frac{2}{1 - \lambda} \sin\left(\frac{\pi(2i+1)}{N_b - 1}\right) \sin\left(\frac{\pi}{N_b - 1}\right).$$

For the values of the integer i such that

$$\frac{\pi(2i+1)}{N_b - 1} \leq \pi + 2p\pi, \qquad p \in \left\{0, 1, ..., \frac{N_b}{2(N_b - 1)}\right\}$$

i.e.

$$i \leq \frac{N_b - 2}{2} + p(N_b - 1), \qquad p \in \left\{0, 1, ..., \frac{N_b}{2(N_b - 1)}\right\},$$

one gets

$$y_{i+1} - y_i \leq 0.$$

To this point, one may note that the compatibility condition

$$\frac{N_b - 2}{2} + p(N_b - 1) \leq N_b - 1$$

leads to

$$p \leq \frac{N_b}{2(N_b - 1)}.$$

The sole entire admissible value for the integer p is thus 0.
    In the same way, one shows that, for the values of the integer i such that

$$\pi \leq \frac{\pi(2i+1)}{N_b - 1} \leq 2\pi$$

i.e.

$$\frac{N_b - 2}{2} \leq i \leq \frac{2N_b - 3}{2},$$

one gets

$$y_{i+1} - y_i \geq 0.$$

This proves that the set $\{P_0, P_1, ..., P_{N_b - 1}\}$ belongs to a non-self-intersecting continuous closed loop in the plane. One may also note that since the sequences $(y_i)_{0 \leq i \leq \frac{N_b - 2}{2}}$ and $(y_i)_{\frac{N_b - 2}{2} \leq i \leq \frac{2N_b - 3}{2}}$ are respectively non-increasing and non-decreasing, the polygon $P_0 P_1 ... P_{N_b - 1}$ is convex. One has then just to use the self-similarity of the graph, and reason by induction; for any strictly positive integer m,

$$V_m = \bigcup_{0 \leq i \leq N_b - 1} T_i(V_{m-1}).$$

By assuming that the points of V_{m-1} belong to a non-self-intersecting continuous closed loop in the plane, it is also the case of their images $T_i(V_{m-1})$ by each map T_i, $0 \leq i \leq N_b - 1$. For any integer i belonging to $\{1, \ldots, N_b - 2\}$, $T_i(V_{m-1})$ and $T_{i+1}(V_{m-1})$ have exactly one common vertex, which happens to be the last point of $T_i(V_{m-1})$, and the first one of $T_{i+1}(V_{m-1})$. Moreover, $T_i(V_{m-1})$ and $T_{i+1}(V_{m-1})$ are ordered sets, according to increasing abscissae.

The proof is done in a similar way for the vertices of the N_b-gons $\mathcal{Q}_{m,j}$, $1 \leq j \leq N_b^m - 2$.

$\square$

Definition 2.8. Edge relation, on the graph $\Gamma_{\mathcal{W}}$. Given a natural integer m, two points X and Y of $\Gamma_{\mathcal{W}_m}$ will be called **adjacent** if and only if X and Y are two consecutive vertices of $\Gamma_{\mathcal{W}_m}$. We will write

$$X \underset{m}{\sim} Y.$$

This edge relation ensures the existence of a word $\mathcal{M} = (\mathcal{M}_1, \ldots, \mathcal{M}_m)$ of length m, such that X and Y both belong to the iterate:

$$T_{\mathcal{M}} V_0 = (T_{\mathcal{M}_1} \circ \cdots \circ T_{\mathcal{M}_m}) V_0.$$

Given two points X and Y of the graph $\Gamma_{\mathcal{W}}$, we will say that X and Y are **adjacent** if and only if there exists a natural integer m such that

$$X \underset{m}{\sim} Y.$$

Proposition 2.11. Addresses, on the Weierstrass Curve. *Given a strictly positive integer m, and a word $\mathcal{M} = (\mathcal{M}_1, \ldots, \mathcal{M}_m)$ of length $m \in \mathbb{N}^*$, on the graph $\Gamma_{\mathcal{W}_m}$, for any integer j of $\{1, \ldots, N_b - 2\}$, any $X = T_{\mathcal{M}}(P_j)$ of $V_m \setminus V_0$, i.e. distinct from one of the N_b fixed point P_i, $0 \leq i \leq N_b - 1$, has exactly two adjacent vertices, given by*

$$T_{\mathcal{M}}(P_{j+1}) \quad and \quad T_{\mathcal{M}}(P_{j-1})$$

where

$$T_{\mathcal{M}} = T_{\mathcal{M}_1} \circ \cdots \circ T_{\mathcal{M}_m}.$$

By convention, the adjacent vertices of $T_{\mathcal{M}}(P_0)$ are $T_{\mathcal{M}}(P_1)$ and $T_{\mathcal{M}}(P_{N_h-1})$, those of $T_{\mathcal{M}}(P_{N_b-1})$, $T_{\mathcal{M}}(P_{N_b-2})$ and $T_{\mathcal{M}}(P_0)$.

3 From the box-counting dimension to the non-differentiability

Notation. For any integer j belonging to $\{0, \ldots, N_b - 1\}$, any natural integer m, and any word $\mathcal{M}$ of length m, we set

$$T_{\mathcal{M}}(P_j) = (x(T_{\mathcal{M}}(P_j)), y(T_{\mathcal{M}}(P_j))),$$

$$L_m = x(T_{\mathcal{M}}(P_{j+1})) - x(T_{\mathcal{M}}(P_j)) = \frac{1}{(N_b - 1) N_b^m},$$

$$h_{j,m} = y(T_{\mathcal{M}}(P_{j+1})) - y(T_{\mathcal{M}}(P_j)).$$

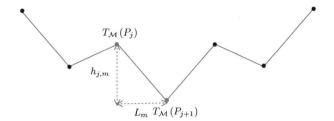

3.1 Box-counting dimension

Notation. We will denote by

$$D_W = 2 + \frac{\ln \lambda}{\ln N_b}$$

the Hausdorff dimension of Γ_W (see [BBR14], [Kel17]).

Definition 3.1. Box-counting dimension. By definition of the box-counting dimension D_W (we refer, for instance, to [Fal86]), one has

$$D_W = - \lim_{\delta \to 0+} \frac{\ln N_\delta (\Gamma_W)}{\ln \delta}$$

where $N_\delta (\Gamma_W)$ is any of the following:

(i) the smallest number of sets of diameter at most δ that cover Γ_W on $[0, 1[$;
(ii) the smallest number of closed balls of radius δ that cover Γ_W on $[0, 1[$;
(iii) the smallest number of cubes of side δ that cover Γ_W on $[0, 1[$;
(iv) the number of δ-mesh cubes that intersect Γ_W on $[0, 1[$;
(v) the largest number of disjoint balls of radius δ with centers in Γ_W on $[0, 1[$.

Theorem 3.1. An upper bound and a lower bound, for the box-dimension of the Weierstrass Curve [Dav18]. *For any integer j belonging to $\{0, 1, \ldots, N_b - 2\}$, each natural integer m, and each word M of length m, let us consider the rectangle, whose sides are parallel to the horizontal and vertical axes, of width*

$$L_m = x\left(T_M\left(P_{j+1}\right)\right) - x\left(T_M\left(P_j\right)\right) = \frac{1}{(N_b - 1)\, N_b^m}$$

and height $|h_{j,m}|$, such that the points $T_M\left(P_j\right)$ and $T_M\left(P_{j+1}\right)$ are two vertices of this rectangle.
    We set:

$$\eta_W = 2\,\pi^2 \left\{ \frac{(2\,N_b - 1)\,\lambda\,(N_b^2 - 1)}{(N_b - 1)^2\,(1 - \lambda)\,(\lambda\,N_b^2 - 1)} + \frac{2\,N_b}{(\lambda\,N_b^2 - 1)\,(\lambda\,N_b^3 - 1)} \right\}.$$

$$
C_1(N_b) = \begin{cases}
(N_b - 1)^{2-D_W} \left\{ \dfrac{2}{1-\lambda} \sin\left(\tfrac{\pi}{N_b-1}\right) \right. \\
\qquad \cdot \min_{0 \le j \le N_b - 1} \left| \sin\left(\tfrac{\pi(2j+1)}{N_b-1}\right) \right| \left. - \dfrac{2\pi}{N_b(N_b-1)} \dfrac{1}{\lambda N_b - 1} \right\} \qquad \text{if } N_b \text{ is odd,} \\[2ex]
(N_b - 1)^{2-D_W} \max \left\{ \dfrac{2}{1-\lambda} \sin\left(\tfrac{\pi}{N_b-1}\right) \right. \\
\qquad \cdot \min_{0 \le j \le N_b - 1} \left| \sin\left(\tfrac{\pi(2j+1)}{N_b-1}\right) \right| - \dfrac{2\pi}{N_b(N_b-1)} \dfrac{1}{\lambda N_b - 1}, \dfrac{4}{N_b^2} \dfrac{1 - N_b^{-2}}{N_b^2 - 1} \right\} \\
\qquad\qquad\qquad\qquad\qquad\qquad\qquad\qquad\qquad\qquad\qquad\qquad\qquad \text{if } N_b \text{ is even}
\end{cases}
$$

and

$$
C_2(N_b) = \eta_W (N_b - 1)^{2-D_W}.
$$

Then

$$
C_1(N_b)\, L_m^{2-D_W} \le |h_{j,m}| \le C_2(N_b)\, L_m^{2-D_W}.
$$

Proof. **Sketch of proof** (for the detailed proof, we refer to [Dav18]).

The proof is based on the fact that, given a strictly positive integer m, and two points X and Y of V_m such that

$$
X \underset{m}{\sim} Y
$$

there exists a word $\mathcal{M}$ of length $|\mathcal{M}| = m$, on the graph Γ_W, and an integer j of $\{0, \ldots, N_b - 2\}^2$, such that

$$
X = T_{\mathcal{M}}(P_j), \quad Y = T_{\mathcal{M}}(P_{j+1}).
$$

By writing $T_{\mathcal{M}}$ under the form

$$
T_{\mathcal{M}} = T_{i_m} \circ T_{i_{m-1}} \circ \cdots \circ T_{i_1}
$$

where $(i_1, \ldots, i_m) \in \{0, \ldots, N_b - 1\}^m$, one gets

$$
x(T_{\mathcal{M}}(P_j)) = \frac{x_j}{N_b^m} + \sum_{k=1}^{m} \frac{i_k}{N_b^k}, \quad x(T_{\mathcal{M}}(P_{j+1})) = \frac{x_{j+1}}{N_b^m} + \sum_{k=1}^{m} \frac{i_k}{N_b^k}
$$

and

$$
\begin{cases}
y(T_{\mathcal{M}}(P_j)) = \lambda^m y_j + \displaystyle\sum_{k=1}^{m} \lambda^{m-k} \cos\left(2\pi\left(\frac{x_j}{N_b^k} + \sum_{\ell=0}^{k} \frac{i_{m-\ell}}{N_b^{k-\ell}}\right)\right), \\[3ex]
y(T_{\mathcal{M}}(P_{j+1})) = \lambda^m y_{j+1} + \displaystyle\sum_{k=1}^{m} \lambda^{m-k} \cos\left(2\pi\left(\frac{x_{j+1}}{N_b^k} + \sum_{\ell=0}^{k} \frac{i_{m-\ell}}{N_b^{k-\ell}}\right)\right).
\end{cases}
$$

$\square$

Notation. Given a natural integer m, we set

$$
h_m = L_m^{2-D_W} = \frac{N_b^{(D_W-2)m}}{(N_b - 1)^{2-D_W}},
$$

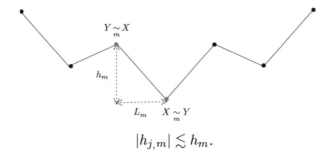

$$|h_{j,m}| \lesssim h_m.$$

Remark 3.1. Comparison with previous results: explicit lower and upper bounds.

It is worth noting that our result gives explicit lower and upper bounds for the quantity $|h_{j,m}|$, which enables one to obtain then the value of the box-counting dimension of the graph $\Gamma_{\mathcal{W}}$. Especially concerning the lower bound, such a result does not appear in the existing literature on the subject. Even if the result of G. Hardy [Har11], [Har16], at first destined to show the non-differentiability of the Weierstrass function, is not referenced among the ones related to the calculation of the box-dimension, one may note that he is the first to give a (non-explicit) upper bound and prove that, for any value of the real number x, and $\eta \to 0^+$,

$$\mathcal{W}(x+\eta) - \mathcal{W}(x) = \mathcal{O}\left(|\eta|^{2-D_{\mathcal{W}}}\right)$$

(see cite [Har11], Theorem 1.3.2, page 303).

Later, in [KMPY84], the authors rely on non-explicit lower-bound estimates. In [Hun98], as concerns the lower bound, the author calls for strictly positive constants K and K' which, again, are not given explicitly (see section 3, page 798). In [She18], also on the Hausdorff dimension of the graph, the estimates are so scattered that it is extremely difficult to reconstruct explicit ones. In [Kel17], again, there isn't any explicit lower bound, but general constants K_1 and $K(K_1)$.

Corollary 3.2. *The box-counting dimension of the graph $\Gamma_{\mathcal{W}}$ is exactly $D_{\mathcal{W}}$.*

Proof. (i) Given a strictly positive integer m, let us first consider the subdivision of the interval $[0, 1[$ into

$$N_m = \frac{1}{L_m} = (N_b - 1) N_b^m$$

sub-intervals of length L_m. One has to determine a natural integer $\tilde{N}_m$ such that the graph of $\Gamma_{\mathcal{W}}$ on $[0, 1[$ can be covered by $N_m \times \tilde{N}_m$ squares of side length L_m. The difficulty is indeed to cover not only the approached m^{th}-order graph $\Gamma_{\mathcal{W}m}$, but any $(m+p)^{th}$-order graph $\Gamma_{\mathcal{W}m+p}$, $p \in \mathbb{N}$, and, thus, $\Gamma_{\mathcal{W}}$.

This is achieved thanks to the Hölder condition satisfied by the Weierstrass function [Zyg02]:

$$\forall (x, y) \in [0, 1]^2 : \quad |\mathcal{W}(x) - \mathcal{W}(y)| \lesssim |x - y|^{2-D_{\mathcal{W}}}.$$

Thus, given two adjacent vertices X and Y of the mth-order graph $\Gamma_{\mathcal{W}m}$, all the points of the Weierstrass Curve that are between X and Y belong to a rectangle of height equal to $h_m = L_m^{2-D_{\mathcal{W}}}$, and of width L_m. A convenient cover of the Weierstrass Curve between X and Y requires at most

$$\frac{h_m}{L_m} \quad \text{squares of side length } L_m.$$

To cover the Weierstrass Curve on the semi-opened interval $[0,1[$ thus requires at most

$$N_m \frac{h_m}{L_m} = \frac{1}{L_m} \frac{h_m}{L_m} = \frac{h_m}{L_m^2} \leq \frac{C_2 L_m^{2-D_{\mathcal{W}}}}{L_m^2} = C_2 L_m^{-D_{\mathcal{W}}}.$$

Let us now consider a strictly positive real number δ such that

$$L_m < \delta \leq L_{m-1}$$

then, a δ-cover of $\Gamma_{\mathcal{W}}$ on $[0,1[$ is at most constituted of

$$C_2 L_m^{-D_{\mathcal{W}}}$$

squares of side L_m.

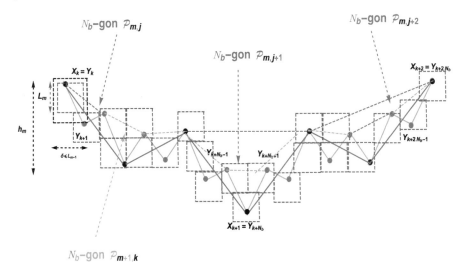

Given three consecutive vertices of $\Gamma_{\mathcal{W}_m}$, X_k, X_{k+1}, X_{k+2}, where k denotes a generic natural integer, $Y_{k+1}, \ldots, Y_{k+N_b-1}$ are the points of $V_{m+1} \setminus V_m$ such that: $Y_{k+1}, \ldots, Y_{k+N_b-1}$ are between X_k and X_{k+1}, and by $Y_{k+N_b+1}, \ldots, Y_{k+2N_b-1}$, the points of $V_{m+1} \setminus V_m$ such that: $Y_{k+N_b+1}, \ldots, Y_{k+2N_b-1}$ are between X_{k+1} and X_{k+2}. In magenta, one can see the δ-cover of squares of side L_m, the δ-cover of squares of side δ.

Hence

$$N_\delta\left(\Gamma_{\mathcal{W}}\right) \leq \frac{C_2(N_b) L_m^{-D_{\mathcal{W}}}}{N_b^{1-D_{\mathcal{W}}}}$$

which yields

$$-\limsup_{\delta \to 0^+} \frac{\ln N_\delta\left(\Gamma_{\mathcal{W}}\right)}{\ln \delta} \leq -\limsup_{m \to +\infty} \frac{\ln \dfrac{C_2(N_b) L_m^{-D_{\mathcal{W}}}}{N_b^{1-D_{\mathcal{W}}}}}{\ln L_m} = -\limsup_{m \to +\infty} \frac{\ln L_m^{-D_{\mathcal{W}}}}{\ln L_m} = D_{\mathcal{W}}.$$

(ii) <u>Conversely</u>, given a strictly positive real number δ such that

$$L_{m+1} \leq \delta < L_m, \quad m \in \mathbb{N}^\star$$

any square of side δ intersects at most N_b polygons $\mathcal{P}_{m+1,j}, 0 \leq j \leq N_b^{m+1} - 1$ that occur at step $m+1$ in the construction of $\Gamma_{\mathcal{W}}$ on $[0,1[$. Due, again, to the Hölder condition satisfied by the Weierstrass function [Zyg02], an N_b-gon $\mathcal{P}_{m+1,j}$, $0 \leq j \leq N_b^{m+1} - 1$, can be inscribed in a rectangle of height at most equal to h_{m+1}, and of width L_{m+1}, which contains all the points of the curve that are between the extreme vertices of $\mathcal{P}_{m+1,j}$.

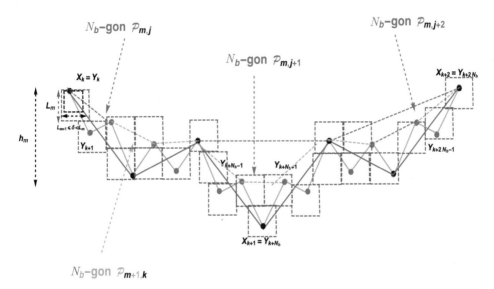

A square of side $\delta \in [L_{m+1}, L_m[$ intersecting polygons $\mathcal{P}_{m+1,j}, 0 \leq j \leq N_b^{m+1} - 1$.

There are N_b^{m+1} such polygons. One has to consider the vertical amplitude, taking account that the N_b^{m+1} polygons $\mathcal{P}_{m+1,j}, 0 \leq j \leq N_b^{m+1} - 1$, with N_b sides, are related to the elementary height h_{m+1}. This brings in a required number (related to the vertical amplitude) at least of

$$N_b \frac{h_{m+1}}{\delta}.$$

Thus

$$N_\delta(\Gamma_{\mathcal{W}}) \geq \frac{1}{\delta} \times N_b \frac{h_{m+1}}{\delta} \geq \frac{N_b \, h_{m+1}}{L_m^2}$$

i.e.

$$N_\delta(\Gamma_{\mathcal{W}}) \geq \frac{N_b \, h_{m+1}}{L_m^2} \geq \frac{N_b \, C_1(N_b) \, L_{m+1}^{2-D_{\mathcal{W}}}}{L_m^2}$$

i.e.

$$N_\delta(\Gamma_{\mathcal{W}}) \geq \frac{N_b \, C_1(N_b) \, L_m^{2-D_{\mathcal{W}}}}{L_m^2 \, N_b^{2-D_{\mathcal{W}}}}$$

i.e.

$$N_\delta\left(\Gamma_{\mathcal{W}}\right) \geq \frac{C_1(N_b)\,L_m^{-D_{\mathcal{W}}}}{N_b^{1-D_{\mathcal{W}}}}.$$

So, at least

$$\frac{C_1(N_b)\,L_m^{-D_{\mathcal{W}}}}{N_b^{1-D_{\mathcal{W}}}}$$

squares of side δ are required to cover $\Gamma_{\mathcal{W}}$ on $[0,1[$.

    Hence

$$N_\delta\left(\Gamma_{\mathcal{W}}\right) \geq \frac{C_1(N_b)\,L_m^{-D_{\mathcal{W}}}}{N_b^{1-D_{\mathcal{W}}}}$$

which yields

$$-\liminf_{\delta\to 0^+}\frac{\ln N_\delta\left(\Gamma_{\mathcal{W}}\right)}{\ln\delta} \geq -\liminf_{m\to+\infty}\frac{\ln\dfrac{C_1(N_b)\,L_m^{-D_{\mathcal{W}}}}{N_b^{1-D_{\mathcal{W}}}}}{\ln L_m} = -\liminf_{m\to+\infty}\frac{\ln L_m^{-D_{\mathcal{W}}}}{\ln L_m} = D_{\mathcal{W}}.$$

$\square$

Corollary 3.3. *The sequence of polygonal domains $(\mathcal{D}\left(\Gamma_{\mathcal{W}_m}\right))_{m\in\mathbb{N}}$ converges towards $\Gamma_{\mathcal{W}}$.*

Proof. For any natural integer m, the aforementioned squares, the side length of which is at most equal to L_m, that can cover the graph $\Gamma_{\mathcal{W}_m}$ on $[0,1[$, also cover the polygonal domain $\mathcal{D}\left(\Gamma_{\mathcal{W}_m}\right)$. Since

$$\lim_{m\to+\infty}L_m = 0$$

the convergence is obvious. $\square$

3.2   Non-differentiability of the Weierstrass function

The original proof of the non-differentiability of the Weierstrass function was given by K. Weierstrass in the case where $\lambda\,N_b > 1 + \frac{3\pi}{2}$, N_b being an odd positive integer (see [Tit39], pages 351–354). It is rather technical (two pages), and consists in proving that the $\mathcal{W}$ function has no finite derivative for any value of $x \in \mathbb{R}$, since the quantity

$$\left|\frac{\mathcal{W}\left(x+h\right) - \mathcal{W}\left(x\right)}{h}\right|$$

takes arbitrary large values when $h \to 0^+$.

    A slight improvement was given by T. J. I'A. Bromwich [Bro08], in the case where

$$\lambda\,N_b > 1 + \frac{3\pi}{2}\left(1-\lambda\right)$$

T. J. Bromwich seemed very proud of his result, and did not hesitate to qualify the seminal condition of Weierstrass of "unnecessarily narrow" ...

In [Har16], G. H. Hardy showed that all those conditions were artificial ones, which "arose in consequence of the methods employed", and, as it could have been expected, did not correspond to "any essential feature of the function". G. H. Hardy proved that in the general case, i.e. not depending on the fact that b was or was not an integer, and under the condition:

$$\lambda b > 1$$

the $\mathcal{W}$ function is not differentiable. Again, it is very technical, the aim being to obtain estimates that enable to get the expected result. Following the above remark of Hardy himself, one may say that, at the times, one did not have enough appropriate tools.

To get the expected limit, one simply requires a lower bound for the absolute value of the average rate of change

$$\left| \frac{\mathcal{W}(x+h) - \mathcal{W}(x)}{h} \right|$$

where h denotes a positive real number that tends to 0, which happens to be given by Theorem 3.1.

Corollary 3.4. (*of Theorem 3.1*) *In the case where*

$$0 < \lambda < 1, \quad b = N_b \in \mathbb{N} \quad and \quad \lambda N_b > 1$$

the $\mathcal{W}$ function is non-differentiable.

Proof. Given a point $X = (x, \mathcal{W}(x))$ of $\Gamma_{\mathcal{W}}$, and a natural integer m, one may note that, for

$$k_0 = \sup \left\{ k \in \{0, \ldots, N_b - 1\}, x \left(T_{\mathcal{M}_{m,j}} (P_k) \right) \leq x \right\}$$

where $\mathcal{M}_{m,j}, 0 \leq j \leq N_b^m - 1$ denotes a word of length m, one has

$$x \left(T_{\mathcal{M}_{m,j}} (P_{k_0}) \right) \leq x \leq x \left(T_{\mathcal{M}_{m,j}} (P_{k_0+1}) \right).$$

In the same time

$$\left| x \left(T_{\mathcal{M}_{m,j}} (P_{k_0}) \right) - x \left(T_{\mathcal{M}_{m,j}} (P_{k_0+1}) \right) \right| = \frac{1}{(N_b - 1) N_b^m} = L_m \xrightarrow[m \to +\infty]{} 0.$$

Thus

$$\left| \mathcal{W} \left(x \left(T_{\mathcal{M}_{m,j}} (P_{k_0}) \right) \right) - \mathcal{W} \left(x \left(T_{\mathcal{M}_{m,j}} (P_{k_0+1}) \right) \right) \right| \geq C_1(N_b) L_m^{2-D_{\mathcal{W}}}$$
$$= C_1(N_b) \left| x \left(T_{\mathcal{M}_{m,j}} (P_{k_0}) \right) - x \left(T_{\mathcal{M}_{m,j}} (P_{k_0+1}) \right) \right|^{2-D_{\mathcal{W}}}$$

122   Claire David

which leads to

$$\left| \frac{\mathcal{W}\left(x\left(T_{\mathcal{M}_{m,j}}(P_{k_0})\right)\right) - \mathcal{W}\left(x\left(T_{\mathcal{M}_{m,j}}(P_{k_0+1})\right)\right)}{\underbrace{x\left(T_{\mathcal{M}_{m,j}}(P_{k_0})\right) - x\left(T_{\mathcal{M}_{m,j}}(P_{k_0+1})\right)}_{L_m}} \right|$$

$$\geq C_1(N_b)\left|x\left(T_{\mathcal{M}_{m,j}}(P_{k_0})\right) - x\left(T_{\mathcal{M}_{m,j}}(P_{k_0+1})\right)\right|^{1-D_{\mathcal{W}}}$$
$$= C_1(N_b)\,L_m^{1-D_{\mathcal{W}}}$$

where

$$1 - D_{\mathcal{W}} = -1 - \frac{\ln \lambda}{\ln N_b} = -\frac{\ln(\lambda N_b)}{\ln N_b} < 0.$$

By passing to the limit when the integer m tends towards infinity, one gets the non-differentiability expected result:

$$\lim_{m\to+\infty} \left| \frac{\mathcal{W}\left(x\left(T_{\mathcal{M}_{m,j}}(P_{k_0})\right)\right) - \mathcal{W}\left(x\left(T_{\mathcal{M}_{m,j}}(P_{k_0+1})\right)\right)}{L_m} \right| = +\infty.$$

The key point of this proof is that the points of the mth order prefractal graph approximation, in particular, $T_{\mathcal{M}_{m,j}}(P_{k_0})$ and $T_{\mathcal{M}_{m,j}}(P_{k_0+1})$, are **also** on the Weierstrass Curve. One thus naturally falls on the limit position of the secant. □

Thanks
The author would like to thank JPG, FD, AK, for their very pertinent suggestions and advices, which helped a lot improving the original work.

References

[Amp06]   A.-M. Ampère. Recherches sur quelques points de la théorie des fonctions dérivées qui conduisent à une nouvelle démonstration de la série de Taylor, et à l'expression finie des termes qu'on néglige lorsqu'on arrête cette série à un terme quelconque. *J. École Polytech.*, 6: 148–181, 1806.

[BBR14]   K. Barański, B. Bárány, and J. Romanowska. On the dimension of the graph of the classical Weierstrass function. *Advances in Math.*, 265: 791–800, 2014.

[BD85]   M. F. Barnsley and S. Demko. Iterated function systems and the global construction of fractals. *The Proceedings of the Royal Society of London*, A(399): 243–275, 1985.

[BR75]   P.-G. du Bois-Reymond. Versuch eine Classification der willkürlichen Funktionen reeller Argumente nach ihren Aenderungen in den kleinsten Intervallen. *Journal für die reine und angewandte Mathematik*, 79: 21–37, 1875.

[Bro08]   T. J. I'A. Bromwich. *An Introduction to the Theory of Infinite Series.* MacMillan and Co., 1908.

[BU37]   A. S. Besicovitch and H. D. Ursell. Sets of Fractional dimensions (V): on dimensional numbers of some continuous curves. *J. London Math. Soc.*, 12(1): 18–25, 1937.

[Cor05]   *Correspondance d'Hermite et de Stieltjes, Tome II.* Gauthier-Villars, 1905.

[Dar75]   G. Darboux. Mémoire sur les fonctions discontinues. *Ann. Sci. École Norm. Sup. Sér. 2*, 4: 57–112, 1875.

[Dar79]   G. Darboux. Addition au mémoire sur les fonctions discontinues. *Ann. Sci. École Norm. Sup. Sér. 2*, 8: 195–202, 1879.

[Dav18]   Cl. David. Bypassing dynamical systems: A simple way to get the box-counting dimension of the graph of the Weierstrass function. *Proceedings of the International Geometry Center*, 11(2): 1–16, 2018.

[Din77]   U. Dini. Su alcune funzioni che in tutto un intervallo non hanno mai derivata. *Annali di Matematica*, 8: 122–137, 1877.

[Din78]   U. Dini. *Fondamenti per la teorica delle funzioni di variabili reali.* Tipografia T. Nistri e C., 1878.

[Fal86]   K. Falconer. *The Geometry of Fractal Sets.* Cambridge University Press, 1986.

[Ger70]   J. Gerver. The differentiability of the Riemann function at certain rational multiples of π. *American Journal Math.*, 82: 33–55, 1970.

[Har11]   G. Hardy. Theorems connected with Maclaurin's test for the convergence of series. *The Proceedings of the Royal Society of London*, s2-9(1): 126–144, 1911.

[Har16]   G. H. Hardy. Weierstrass's non-differentiable function. *Transactions of the American Mathematical Society*, 17(3): 301–325, 1916.

[HL93]   T.-Y. Hu and K.-S. Lau. Fractal dimensions and singularities of the Weierstrass type functions. *Transactions of the American Mathematical Society*, 335(2): 649–665, 1993.

[HSR92]   J. A. C. Humphrey, C. A. Schuler, and B. Rubinsky. On the use of the Weierstrass-Mandelbrot function to describe the fractal component of turbulent velocity. *Fluid Dynamics Research*, 9: 81–95, 1992.

[Hun98]   B. Hunt. The Hausdorff dimension of graphs of Weierstrass functions. *Proc. Amer. Math. Soc.*, 12(1): 791–800, 1998.

[JP15]   M. Jarnicki and P. Pflug. *Continuous Nowhere Differentiable Functions, The Monsters of Analysis.* Springer, 2015.

[Kah64]   J.-P. Kahane. Lacunary Taylor and Fourier series. *Bull. Amer. Math. Soc.*, 70(2): 169–181, 1964.

[Kel17]   G. Keller. A simpler proof for the dimension of the graph of the classical Weierstrass function. *Ann. Inst. Poincaré*, 53(1): 169–181, 2017.

[KH16]   Wolfgang König and Jürgen Sprekels Hrsg. *Karl Weierstrass (1815-1897): Aspekte seines Lebens und Werkes.* Springer, 2016.

[KMPY84]   J. L. Kaplan, J. Mallet-Paret, and J. A. Yorke. The Lyapunov dimension of a nowhere differentiable attracting torus. *Ergodic Theory Dynam. Systems*, 4: 261–281, 1984.

[Man77]   B. B. Mandelbrot. *Fractals: Form, Chance, and Dimension.* W. H. Freeman & Co., 1977.

[Man82]   B. B. Mandelbrot. *The Fractal Geometry of Nature.* W. H. Freeman & Co., 1982.

[Poi98]   H. Poincaré. L'œuvre mathématique de Weierstrass. *Acta mathematica*, 22, 1898.

[Poi99]   H. Poincaré. La logique et l'intuition dans la science mathématique et dans l'enseignement. *Enseign. Math.*, 12(1): 157–162, 1899.

[Poi90]   H. Poincaré. *Science et méthode.* E. Flammarion, 1908.

[PU89]   F. Przytycki and M. Urbański. On the Hausdorff dimension of some fractal sets. *Studia Math.*, 93(2): 155–186, 1989.

[Sag86]   H. Sagan. Approximating polygons for Lebesgue's and Schoenberg's space filling curves. *The American Mathematical Monthly*, 93(5): 361–368, 1986.

[Sag94]   H. Sagan. *Space Filling Curves.* Springer, 1994.

[She18]   W. Shen. Hausdorff dimension of the graphs of the classical Weierstrass functions. *Math. Z.*, 289: 223–266, 2018.

[Tit39]   E. C. Titschmarsh. *The Theory of Functions, Second edition.* Oxford University Press, 1939.

[Wei54]   K. Weierstrass. Zur Theorie der Abelschen Functionen. *J. Reine Angew. Math.*, 47: 289–306, 1854.

[Wei56]   K. Weierstrass. Theorie der Abel'schen Functionen. *J. Reine Angew. Math.*, 52: 285–380, 1856.

[Wei73]   K. Weierstrass. Briefe von K. Weierstrass an Paul du Bois-Reymond, 1873.

[Wei75]   K. Weierstrass. Über continuirliche Funktionen eines reellen Arguments, die für keinen Werth des letzteren einen bestimmten Differential quotienten besitzen. *Journal für die reine und angewandte Mathematik*, 79: 29–31, 1875.

[Wun73]   W. Wunderlich. Über Peano-Kurven. *Elemente Der Mathematik*, 28(1): 1–10, 1973.

[Zyg02]   A. Zygmund. *Trigonometric Series, Third edition. With a Foreword by Robert A. Fefferman*, volume I, II. Cambridge University Press, 2002.

Lectures on the Analysis of Nonlinear
Partial Differential Equations Vol. 6
MLM 6, pp. 125–170

L_p Estimates for Stokes Systems[*]

Hongjie Dong[†]

Abstract

These lecture notes are based on the several lectures which I gave in the Academy of Mathematics and Systems Science (AMSS) of the Chinese Academy of Sciences (CAS) in June, 2019. We give a self-contained approach to the L_p estimates for stationary and time-dependent Stokes systems with variable coefficients. In the stationary case, we consider the systems in the whole space, a half space, and on bounded Lipschitz domains when the coefficients are merely measurable functions in one direction and have locally small mean oscillations in the orthogonal directions in each small ball, where the direction is allowed to depend on the ball. In the time-dependent case, we obtain a priori estimates in a cylindrical domain for both divergence and non-divergence systems when the coefficients are measurable in the time variable and have small (bounded) mean oscillations with respect to the space variables.

1 Steady Stokes systems

We will first discuss stationary Stokes systems with variable coefficients in the whole space, a half space, and on bounded Lipschitz domains, following the argument in [14]. The regularity theory for the linear Stokes system has been extensively studied over the last fifty years by many authors. The classical Stokes system with the Laplace operator in smooth domains has the form

$$\begin{cases} \Delta u + \nabla p = f & \text{in } \Omega, \\ \operatorname{div} u = g & \text{in } \Omega \end{cases} \tag{1.1}$$

with the non-homogeneous Dirichlet boundary condition $u = \varphi$ on $\partial\Omega$. We refer the reader to Ladyženskaya [31], Sobolevskiĭ [41], Cattabriga [8], Vorovič and Judovič [44], and Amrouche and Girault [4]. In particular, Cattabriga [8] obtained

---

[*]Key words and phrases: Stokes systems, boundary value problem, measurable coefficients.

[†]Division of Applied Mathematics, Brown University, 182 George Street, Providence, RI 02912, USA. E-mail: Hongjie_Dong@brown.edu

the following W_q^1-estimate for the system in a bounded C^2 domain $\Omega \subset \mathbb{R}^3$: for any $q \in (1, \infty)$,

$$\|Du\|_{L_q(\Omega)} + \|p\|_{L_q(\Omega)} \leq N\|f\|_{W_q^{-1}(\Omega)} + N\|g\|_{L_q(\Omega)} + N\|\varphi\|_{W_q^{1-1/q}(\Omega)}.$$

The proof is based on the explicit representation of solutions using fundamental solutions. By using a result by Agmon, Douglas, and Nirenberg [2] for elliptic systems together with an interpolation argument, Cattabriga's result was later extended by Amrouche and Girault [4] to a bounded $C^{1,1}$ domain $\Omega \subset \mathbb{R}^d$, for any $d \geq 2$. The system (1.1) on a bounded Lipschitz domain was first studied by Galdi, Simader, and Sohr [21]. They proved W_q^1-estimates and solvability under the assumption that the Lipschitz constant of the domain is sufficiently small. The problem was also studied by Fabes, Kenig, and Verchota [19] in the case of arbitrary Lipschitz domains with the range of q restricted, using the layer potential method and Rellich identities. For this line of research, see also [40, 6, 36, 37, 22] and references therein, some of which obtain estimates in Besov spaces.

In this lecture series, we first consider the stationary Stokes system with variable coefficients:

$$\begin{cases} \mathcal{L}u + \nabla p = f + D_\alpha f_\alpha & \text{in } \Omega, \\ \operatorname{div} u = g & \text{in } \Omega, \end{cases} \tag{1.2}$$

where $\Omega \subseteq \mathbb{R}^d$ and $\mathcal{L}$ is a strongly elliptic operator, given by

$$\mathcal{L}u = D_\alpha \left(A^{\alpha\beta} D_\beta u \right), \quad A^{\alpha\beta} = [A_{ij}^{\alpha\beta}]_{i,j=1}^d$$

for $\alpha, \beta = 1, \ldots, d$. Here and throughout the notes, we use the Einstein summation convention on repeated indices. Such type of systems were considered by Giaquinta and Modica [23], where they obtained various regularity results for both linear and nonlinear Stokes systems when the coefficients are sufficiently regular. Besides its mathematical interests, the system (1.2) is also partly motivated by the study of inhomogeneous fluids with density dependent viscosity (see, for instance, [32, 33, 1]), as well as equations describing flows of shear thinning and shear thickening fluids with viscosity depending on pressure (see, for instance, [20, 7]).

We allow coefficients $A^{\alpha\beta}$ to be merely measurable in one direction. In particular, they may have jump discontinuities, so that the system can be used to model, for example, the motion of two fluids with interfacial boundaries. We will consider the system (1.2) in the whole space, a half space, and on bounded Lipschitz domains. In the whole and half spaces, we obtain a priori $\dot{W}_q^1$-estimates for any $q \in [2, \infty)$ in the case that the coefficients are merely measurable functions in one fixed direction (see Theorem 1.1). For the system on bounded Lipschitz domains with a small Lipschitz constant, we prove a W_q^1-estimate and solvability for (1.2) with any $q \in (1, \infty)$, when the coefficients are merely measurable functions in one direction and have locally small mean oscillations in the orthogonal directions in each small ball, with the direction depending on the ball (see Theorem 1.18). We note that the class of coefficients considered here was first introduced by Kim and Krylov [27] and Krylov [29], where they established the W_p^2-estimate for non-divergence form second-order elliptic equations in the whole space. Subsequently, such coefficients were also treated in [16, 15] for second- and higher-order elliptic and parabolic systems in regular and irregular domains.

For the Stokes system, the presence of the pressure term p gives an added difficulty, because in the usual L_2-estimate, instead of p one can only bound $p-(p)$ by Du. See, for instance, Lemma 1.7. Nevertheless, in Lemma 1.11 we show that for the homogeneous Stokes system and any integer $k \geq 1$, the L_2-norm of $D_{x'}^k p$ in a smaller ball can be controlled by that of Du in a larger ball. Finally, in order to deal with the system (1.2) in a Lipschitz domain, we apply a version of the Fefferman-Stein sharp function theorem for spaces of homogeneous type, which was recently proved in [12] (cf. Lemma 1.21). Furthermore, we employ a delicate cut-off argument, together with Hardy's inequality, which was first used in [15] and also in the recent paper [10].

We conclude by introducing some notation and definitions. We fix a half space to be $\mathbb{R}_+^d$, defined by

$$\mathbb{R}_+^d = \{x = (x_1, x') \in \mathbb{R}^d : x_1 > 0, \, x' \in \mathbb{R}^{d-1}\}.$$

Throughout the notes, $D_{x'} u$ denotes the partial derivative of u in the x_i direction, $i = 2, \ldots, d$.

Let $B_r(x_0)$ be the Euclidean ball of radius r in $\mathbb{R}^d$ centered at $x_0 \in \mathbb{R}^d$, and let $B_r^+(x_0)$ be the half ball

$$B_r^+(x_0) = B_r(x_0) \cap \mathbb{R}_+^d.$$

A ball in $\mathbb{R}^{d-1}$ is denoted by

$$B_r'(x') = \{y' \in \mathbb{R}^{d-1} : |x' - y'| < r\}.$$

We use the abbreviations $B_r := B_r(0), B_r^+ := B_r^+(0)$ where $0 \in \mathbb{R}^d$, and $B_r' := B_r'(0)$ where $0 \in \mathbb{R}^{d-1}$.

For a locally integrable function f, we define its average on Ω by

$$(f)_\Omega = \frac{1}{|\Omega|} \int_\Omega f \, dx = \fint_\Omega f \, dx.$$

We shall use the following function spaces:

$$W_q^1(\Omega) := \{f \in L_q(\Omega) : Df \in L_q(\Omega)\}, \quad W_q^1(\Omega)^d = \left(W_q^1(\Omega)\right)^d.$$

Finally, let $\mathring{W}_q^1(\Omega)$ be the completion of $C_0^\infty(\Omega)$ in $W_q^1(\Omega)$, and

$$\mathring{W}_q^1(\Omega)^d = \left(\mathring{W}_q^1(\Omega)\right)^d.$$

Throughout the notes, the coefficients $A^{\alpha\beta}$ are bounded and satisfy the strong ellipticity condition, i.e., there exists a constant $\delta \in (0,1)$ such that

$$|A^{\alpha\beta}| \leq \delta^{-1}, \quad \sum_{\alpha,\beta=1}^d \xi_\alpha \cdot A^{\alpha\beta} \xi_\beta \geq \delta \sum_{\alpha=1}^d |\xi_\alpha|^2$$

for any $\xi_\alpha \in \mathbb{R}^d$, $\alpha = 1, \ldots, d$. By the trace-extension theorem on Lipschitz domains, in the sequel we only consider the homogeneous boundary condition $u = 0$ on $\partial\Omega$, without loss of generality.

We say that $(u, p) \in W_q^1(\Omega)^d \times L_q(\Omega)$ is a solution to (1.2) if for any $\psi = (\psi_1, \ldots, \psi_d) \in C_0^\infty(\Omega)^d$, we have that

$$-\int_\Omega D_\alpha \psi \cdot A^{\alpha\beta} D_\beta u \, dx - \int_\Omega p \operatorname{div} \psi \, dx = \int_\Omega f \cdot \psi \, dx - \int_\Omega f_\alpha \cdot D_\alpha \psi \, dx,$$

where

$$\operatorname{div} \psi = D_1 \psi_1 + \cdots + D_d \psi_d \quad \text{and} \quad D_\alpha \psi = (D_\alpha \psi_1, D_\alpha \psi_2, \ldots, D_\alpha \psi_d).$$

1.1   Stokes systems in the whole space or the half space

Our first two results concern a priori L_q-estimates of the Stokes system defined in $\mathbb{R}^d$ or $\mathbb{R}_+^d$, when the coefficients $A^{\alpha\beta}$ are merely measurable functions of only x_1. In this case, we set

$$\mathcal{L}_0 u = D_\alpha \left(A^{\alpha\beta}(x_1) D_\beta u \right), \tag{1.3}$$

where $A^{\alpha\beta}(x_1) = [A_{ij}^{\alpha\beta}(x_1)]_{i,j=1}^d$. Note that we do not impose any regularity assumptions on $A^{\alpha\beta}(x_1)$.

Theorem 1.1. *Let* $q \in [2, \infty)$, *and let* Ω *be either* $\mathbb{R}^d$ *or* $\mathbb{R}_+^d$ *and* $A^{\alpha\beta} = A^{\alpha\beta}(x_1)$, *i.e.,* $\mathcal{L} = \mathcal{L}_0$. *If* $(u, p) \in W_q^1(\Omega)^d \times L_q(\Omega)$ *satisfies*

$$\begin{cases} \mathcal{L}_0 u + \nabla p = D_\alpha f_\alpha & \text{in } \Omega, \\ \operatorname{div} u = g & \text{in } \Omega, \\ u = 0 & \text{on } \partial\Omega \quad \text{in case } \Omega = \mathbb{R}_+^d, \end{cases}$$

where $f_\alpha, g \in L_q(\Omega)$, *then we have that*

$$\|Du\|_{L_q(\Omega)} + \|p\|_{L_q(\Omega)} \le N \left(\|f_\alpha\|_{L_q(\Omega)} + \|g\|_{L_q(\Omega)} \right), \tag{1.4}$$

where $N = N(d, \delta, q)$.

Remark 1.2. In Theorem 1.1 we only consider the case that $q \in [2, \infty)$ to simplify the exposition and to present our approach in the most transparent way. Indeed, if $q = 2$, the theorem holds even with measurable $A^{\alpha\beta}(x)$. See Theorem 1.6. Thus, in the proof of Theorem 1.1 we focus on the case $q \in (2, \infty)$, the proof of which well illustrates, in the simplest setting, our arguments based on mean oscillation estimates together with the sharp function and the maximal function theorems. One can prove the other case, with $q \in (1, 2)$, by using Theorem 1.18 below.

1.2 L_2 estimates

In this subsection, we assume that the coefficients $A^{\alpha\beta}$ are measurable functions of $x \in \mathbb{R}^d$. That is, no regularity assumptions are imposed on $A^{\alpha\beta}$.

We impose the following assumption on a bounded domain $\Omega \subset \mathbb{R}^d$ in Lemma 1.5 below.

Assumption 1.3. For any $g \in L_2(\Omega)$ such that $\int_\Omega g \, dx = 0$, there exist $Bg \in \mathring{W}_2^1(\Omega)^d$ and a constant $K_1 > 0$ depending only on d and Ω such that

$$\text{div } Bg = g \quad \text{in } \Omega, \quad \|D(Bg)\|_{L_2(\Omega)} \le K_1 \|g\|_{L_2(\Omega)}.$$

Remark 1.4. If $\Omega = B_R$ or $\Omega = B_R^+$, it follows from a scaling argument that the constant K_1 depends only on the dimension d. If Ω is a bounded Lipschitz domain satisfying Assumption 1.16, then Assumption 1.3 is satisfied with K_1 depending only on d, R_0, and $\text{diam } \Omega$.

Lemma 1.5. *Let $\Omega \subset \mathbb{R}^d$ be a bounded domain which satisfies Assumption 1.3, and $f, f_\alpha, g \in L_2(\Omega)$ with $(g)_\Omega = 0$. Then, there exists a unique $(u, p) \in W_2^1(\Omega)^d \times L_2(\Omega)$ with $(p)_\Omega = 0$ satisfying*

$$\begin{cases} \mathcal{L}u + \nabla p = f + D_\alpha f_\alpha & \text{in } \Omega, \\ \text{div } u = g & \text{in } \Omega, \\ u = 0 & \text{on } \partial\Omega. \end{cases}$$

Moreover, we have that

$$\|Du\|_{L_2(\Omega)} + \|p\|_{L_2(\Omega)} \le N \left(\|f\|_{L_2(\Omega)} + \|f_\alpha\|_{L_2(\Omega)} + \|g\|_{L_2(\Omega)} \right),$$

where $N = N(d, \delta, K_1)$. If $\Omega = B_R(x_0)$, $x_0 \in \mathbb{R}^d$, or $\Omega = B_R^+(x_0)$, $x_0 \in \partial\mathbb{R}_+^d$, then we have that

$$\|Du\|_{L_2(\Omega)} + \|p\|_{L_2(\Omega)} \le N \left(R\|f\|_{L_2(\Omega)} + \|f_\alpha\|_{L_2(\Omega)} + \|g\|_{L_2(\Omega)} \right),$$

where $N = N(d, \delta)$.

Proof. The lemma follows from the Lax-Milgram theorem and the Riesz representation theorem, by using the orthogonal decomposition

$$H_0^1(\Omega) = H_{0,\text{div}}^1 \oplus (H_{0,\text{div}}^1)^\perp,$$

where $H_{0,\text{div}}^1$ is a closed subspace of H_0^1 consisting of all functions which are divergence free. We leave this as an exercise. $\qquad\square$

As far as a priori estimates are concerned, one can have $\Omega = \mathbb{R}^d$ or $\Omega = \mathbb{R}_+^d$ in Lemma 1.5 if $f \equiv 0$. In this case we do not necessarily need that the integral of p over Ω is zero. For completeness and later reference, we state and prove this result in the theorem below.

Recall that we say that $(u, p) \in W_2^1(\Omega)^d \times L_2(\Omega)$ satisfies, for instance, $\mathcal{L}u + \nabla p = 0$ in Ω, if

$$\int_\Omega D_\alpha \psi \cdot A^{\alpha\beta} D_\beta u \, dx + \int_\Omega p \, \text{div } \psi \, dx = 0$$

for all $\psi \in C_0^\infty(\Omega)^d$. One can easily see that $C_0^\infty(\Omega)^d$ can be replaced by $\mathring{W}_2^1(\Omega)^d$.

Theorem 1.6. *Let Ω be either $\mathbb{R}^d$ or $\mathbb{R}^d_+$. If $(u,p) \in W_2^1(\Omega)^d \times L_2(\Omega)$ satisfies*

$$\begin{cases} \mathcal{L}u + \nabla p = D_\alpha f_\alpha & in\ \Omega, \\ \operatorname{div} u = g & in\ \Omega, \\ u = 0 & on\ \partial\Omega \quad when\ \Omega = \mathbb{R}^d_+, \end{cases}$$

where $f_\alpha, g \in L_2(\Omega)$, then we have that

$$\|Du\|_{L_2(\Omega)} + \|p\|_{L_2(\Omega)} \le N\left(\|f_\alpha\|_{L_2(\Omega)} + \|g\|_{L_2(\Omega)}\right),$$

where $N = N(d,\delta)$.

Proof. Since $u \in \mathring{W}_2^1(\Omega)^d$, we can use u as a test function to obtain that

$$\int_\Omega D_\alpha u \cdot A^{\alpha\beta} D_\beta u \, dx + \int_\Omega p g \, dx = \int_\Omega f_\alpha \cdot D_\alpha u \, dx.$$

From this, the ellipticity condition, and Young's inequality, we have that

$$\|Du\|_{L_2(\Omega)}^2 \le \varepsilon_0 \|p\|_{L_2(\Omega)}^2 + \frac{N}{\varepsilon_0}\|g\|_{L_2(\Omega)}^2 + N\|f_\alpha\|_{L_2(\Omega)}^2 \qquad (1.5)$$

for any $\varepsilon_0 > 0$, where $N = N(d,\delta)$.

Now, for any $\varepsilon > 0$, we can find $R > 0$ and $p_\varepsilon \in L_2(\Omega)$ such that $\operatorname{supp} p_\varepsilon \subset B_R$, $\int_\Omega p_\varepsilon \, dx = 0$, and $\|p - p_\varepsilon\|_{L_2(\Omega)} < \varepsilon$. To do this, because $p \in L_2(\Omega)$, we first find a function p_1 with a compact support in B_{R_1} such that

$$\|p - p_1\|_{L_2(\Omega)} < \varepsilon/2.$$

If $\int_\Omega p_1 \, dx = 0$, we set $R = R_1$ and $p_\varepsilon = p_1$. Otherwise, set

$$g = \frac{\varepsilon^2}{4C} I_{B_{R_2}}, \quad where \quad C = \int_\Omega p_1 \, dx,$$

and R_2 is a positive number satisfying

$$|B_{R_2}| = 4\frac{C^2}{\varepsilon^2} \ \ if \ \ \Omega = \mathbb{R}^d, \quad |B_{R_2}^+| = 4\frac{C^2}{\varepsilon^2} \ \ if \ \ \Omega = \mathbb{R}^d_+.$$

Then, we see that

$$\int_\Omega (p_1 - g) \, dx = 0 \quad and \quad \int_\Omega g^2 \, dx = \varepsilon^2/4.$$

Thus, it suffices to take $R = \max\{R_1, R_2\}$ and $p_\varepsilon = p_1 - g$.

Thanks to the fact that $(p_\varepsilon)_{B_R} = 0$ or $(p_\varepsilon)_{B_R^+} = 0$, there exists a solution $\psi \in \mathring{W}_2^1(B_R)^d$ or $\psi \in \mathring{W}_2^1(B_R^+)^d$ satisfying the divergence equation

$$\operatorname{div}\psi = p_\varepsilon$$

in B_R or B_R^+. Extend ψ to be zero on $\mathbb{R}^d \setminus B_R$ or $\mathbb{R}_+^d \setminus B_R^+$. Then we have $\psi \in \mathring{W}_2^1(\Omega)^d$ and

$$\|D\psi\|_{L_2(\Omega)} \leq N(d)\|p_\varepsilon\|_{L_2(\Omega)}. \tag{1.6}$$

By applying ψ as a test function to the system, we have that

$$\int_\Omega D_\alpha \psi \cdot A^{\alpha\beta} D_\beta u \, dx + \int_\Omega p\, p_\varepsilon \, dx = \int_\Omega f_\alpha \cdot D_\alpha \psi \, dx.$$

From this, Young's inequality, and (1.6), it follows that

$$\|p\|_{L_2(\Omega)}^2 \leq \|p p_\varepsilon\|_{L_1(\Omega)} + \|p(p - p_\varepsilon)\|_{L_1(\Omega)}$$

$$\leq \varepsilon_1 \|p_\varepsilon\|_{L_2(\Omega)}^2 + \frac{N}{\varepsilon_1}\|f_\alpha\|_{L_2(\Omega)}^2 + \frac{N}{\varepsilon_1}\|Du\|_{L_2(\Omega)}^2 + \varepsilon_1\|p\|_{L_2(\Omega)}^2 + \frac{N}{\varepsilon_1}\|p - p_\varepsilon\|_{L_2(\Omega)}^2$$

for any $\varepsilon_1 > 0$, where $N = N(d, \delta)$. Since $\varepsilon > 0$ is arbitrary, from the above inequality we obtain that

$$\|p\|_{L_2(\Omega)}^2 \leq N\|f_\alpha\|_{L_2(\Omega)}^2 + N\|Du\|_{L_2(\Omega)}^2,$$

which combined with (1.5) proves the desired inequality. $\qquad\square$

In the lemmas below, we do not assume that $(p)_{B_R} = 0$ or $(p)_{B_R^+} = 0$ unless specified.

Lemma 1.7. *Let $R > 0$. If $(u, p) \in W_2^1(B_R)^d \times L_2(B_R)$ satisfies*

$$\mathcal{L}u + \nabla p = 0 \quad in \; B_R, \tag{1.7}$$

then

$$\int_{B_R} |p - (p)_{B_R}|^2 \, dx \leq N \int_{B_R} |Du|^2 \, dx, \tag{1.8}$$

where $N = N(d, \delta)$. The same estimate holds if B_R is replaced by B_R^+.

Proof. We only prove the case with B_R^+, because the other case is similar. By Remark 1.4, one can find $\psi \in \mathring{W}_2^1(B_R^+)^d$ satisfying

$$\text{div } \psi = p - (p)_{B_R^+} \quad in \; B_R^+$$

and

$$\|D\psi\|_{L_2(B_R^+)} \leq N\|p - (p)_{B_R^+}\|_{L_2(B_R^+)}, \tag{1.9}$$

where $N = N(d)$. Then, apply ψ to (1.7) as a test function, and use Young's inequality and (1.9), to get (1.8) with B_R^+ in place of B_R. $\qquad\square$

Lemma 1.8. *Let $0 < r < R$.*

(1) If $(u,p) \in W_2^1(B_R)^d \times L_2(B_R)$ satisfies

$$\begin{cases} \mathcal{L}u + \nabla p = 0 & \text{in } B_R, \\ \operatorname{div} u = 0 & \text{in } B_R, \end{cases} \tag{1.10}$$

then for any $\varepsilon > 0$, we have that

$$\int_{B_r} |Du|^2 \, dx \le N(R-r)^{-2} \int_{B_R} |u|^2 \, dx + \varepsilon \int_{B_R} |Du|^2 \, dx, \tag{1.11}$$

where $N = N(d, \delta, \varepsilon)$.

(2) If $(u,p) \in W_2^1(B_R^+)^d \times L_2(B_R^+)$ satisfies

$$\begin{cases} \mathcal{L}u + \nabla p = 0 & \text{in } B_R^+, \\ \operatorname{div} u = 0 & \text{in } B_R^+, \\ u = 0 & \text{on } B_R \cap \partial \mathbb{R}_+^d, \end{cases} \tag{1.12}$$

then for any ε, we have that (1.11) holds with B_r^+ and B_R^+ replacing B_r and B_R, respectively, where $N = N(d, \delta, \varepsilon)$.

Proof. We only prove the second assertion of the lemma. The first is the same with obvious modifications, and is left as an exercise.

Set η to be an infinitely differentiable function on $\mathbb{R}^d$, such that

$$0 \le \eta \le 1, \quad \eta = 1 \text{ on } B_r, \quad \eta = 0 \text{ on } \mathbb{R}^d \setminus B_R, \quad |D\eta| \le N(d)(R-r)^{-1}. \tag{1.13}$$

Then, we apply $\eta^2 u$ to (1.12) as a test function (because $(\eta^2 u)|_{\partial B_R^+} = 0$), to obtain that

$$\int_{B_R^+} D_\alpha (\eta^2 u) \cdot A^{\alpha\beta} D_\beta u \, dx + \int_{B_R^+} p \, \operatorname{div}(\eta^2 u) \, dx = 0.$$

From this and the fact that $\int_{B_R^+} \operatorname{div}(\eta^2 u) \, dx = 0$, we have that

$$\int_{B_R^+} \eta D_\alpha u \cdot A^{\alpha\beta} \eta D_\beta u \, dx$$

$$= -2 \int_{B_R^+} (D_\alpha \eta) u \cdot A^{\alpha\beta} \eta D_\beta u \, dx - \int_{B_R^+} \left(p - (p)_{B_R^+} \right) \operatorname{div}(\eta^2 u) \, dx.$$

Together with (1.13), the ellipticity condition, Young's inequality, and the fact that $\operatorname{div} u = 0$, this shows that

$$\int_{B_R^+} \eta^2 |Du|^2 \, dx \le N(d, \delta, \varepsilon)(R-r)^{-2} \int_{B_R^+} |u|^2 \, dx + \varepsilon \int_{B_R^+} |p - (p)_{B_R^+}|^2 \, dx$$

for any $\varepsilon > 0$. The desired estimate follows by combining this with Lemma 1.7, and the fact that $\eta = 1$ on B_r.   □

Lemma 1.9. *Let $0 < r < R$.*

(1) If $(u,p) \in W_2^1(B_R)^d \times L_2(B_R)$ satisfies (1.10), then we have that

$$\int_{B_r} |Du|^2 \, dx \le N(R-r)^{-2} \int_{B_R} |u|^2 \, dx, \qquad (1.14)$$

where $N = N(d, \delta)$.

(2) If $(u,p) \in W_2^1(B_R^+)^d \times L_2(B_R^+)$ satisfies (1.12), then we have that (1.14) holds, with B_r^+ and B_R^+ replacing B_r and B_R, respectively.

Proof. Set

$$R_0 = r, \quad R_k = r + (R-r)(1 - 2^{-k}), \quad k = 1, 2, \dots.$$

Then, by Lemma 1.8 we have that

$$\int_{B_{R_k}} |Du|^2 \, dx \le N \frac{4^k}{(R-r)^2} \int_{B_{R_{k+1}}} |u|^2 \, dx + \varepsilon \int_{B_{R_{k+1}}} |Du|^2 \, dx, \quad k = 0, 1, 2, \dots,$$

for any $\varepsilon > 0$, where $N = N(d, \delta, \varepsilon)$. By multiplying both sides of the above inequality by ε^k and summing the terms with respect to $k = 0, 1, \dots$, we obtain that

$$\sum_{k=0}^{\infty} \varepsilon^k \int_{B_{R_k}} |Du|^2 \, dx \le \frac{N}{(R-r)^2} \sum_{k=0}^{\infty} (4\varepsilon)^k \int_{B_{R_{k+1}}} |u|^2 \, dx + \sum_{k=1}^{\infty} \varepsilon^k \int_{B_{R_k}} |Du|^2 \, dx,$$

where each summation is finite upon choosing, for instance, $\varepsilon = 1/8$. Since the first summation on the right-hand side of the above inequality is bounded by $\int_{B_R} |u|^2 \, dx$, we can arrive at (1.14) by subtracting $\sum_{k=1}^{\infty} \varepsilon^k \int_{B_{R_k}} |Du|^2 \, dx$ from both sides of the above inequality. The other case for half balls is proved in the same way. □

1.3 L_∞ and Hölder estimates

Now we prove L_∞ and Hölder estimates of certain linear combinations of Du and p, which are crucial for proving our main results. Recall the operator $\mathcal{L}_0$ given in (1.3), where the coefficients are functions of x_1 only. In this case, if a sufficiently smooth (u,p) satisfies $\mathcal{L}_0 u + \nabla p = 0$ in $\Omega \subset \mathbb{R}^d$, we see that

$$D_1 \left(A^{1\beta} D_\beta u + \begin{bmatrix} p \\ 0 \\ \vdots \\ 0 \end{bmatrix} \right) = - \sum_{\alpha \neq 1} A^{\alpha\beta} D_{\alpha\beta} u - \begin{bmatrix} 0 \\ D_2 p \\ \vdots \\ D_d p \end{bmatrix} \qquad (1.15)$$

in Ω. Set $U = (U_1, U_2, \dots, U_d)^{\mathrm{tr}}$, where

$$U_1 = \sum_{j=1}^{d} \sum_{\beta=1}^{d} A_{1j}^{1\beta} D_\beta u_j + p, \quad U_i = \sum_{j=1}^{d} \sum_{\beta=1}^{d} A_{ij}^{1\beta} D_\beta u_j, \quad i = 2, \dots, d. \qquad (1.16)$$

That is,
$$U = A^{1\beta}D_\beta u + (p, 0, \ldots, 0)^{\mathrm{tr}}.$$

Here and throughout we write $DD^k_{x'}u$, $k = 0, 1, \ldots$, to denote $D^\vartheta u$, where ϑ is a multi-index such that $\vartheta = (\vartheta_1, \ldots, \vartheta_d)$ with $\vartheta_1 = 0, 1$ and $|\vartheta| = k + 1$.

Lemma 1.10. *Let* $0 < r < R$, *and let* ℓ *be a constant.*

(1) *If* $(u, p) \in W_2^1(B_R)^d \times L_2(B_R)$ *satisfies*
$$\begin{cases} \mathcal{L}_0 u + \nabla p = 0 & \text{in } B_R, \\ \operatorname{div} u = \ell & \text{in } B_R, \end{cases} \tag{1.17}$$

then $DD_{x'}u \in L_2(B_r)$, *and*
$$\int_{B_r} |DD_{x'}u|^2 \, dx \le N(R - r)^{-2} \int_{B_R} |Du|^2 \, dx, \tag{1.18}$$

where $N = N(d, \delta)$.

(2) *If* $(u, p) \in W_2^1(B_R^+)^d \times L_2(B_R^+)$ *satisfies*
$$\begin{cases} \mathcal{L}_0 u + \nabla p = 0 & \text{in } B_R^+, \\ \operatorname{div} u = \ell & \text{in } B_R^+, \\ u = 0 & \text{on } B_R \cap \partial\mathbb{R}_+^d, \end{cases} \tag{1.19}$$

then $DD_{x'}u \in L_2(B_r^+)$, $D_{x'}u = 0$ *on* $B_r \cap \partial\mathbb{R}_+^d$, *and* (1.18) *is satisfied with* B_r^+ *and* B_R^+ *replacing* B_r *and* B_R, *respectively.*

Proof. We only deal with the second assertion here. Set $\delta_{j,h}f$ to be the difference quotient of f with respect to x_j, i.e.,
$$\delta_{j,h}f(x) = \frac{f(x + e_j h) - f(x)}{h},$$

and let $R_1 = (R + r)/2$. Then, since the coefficients are functions of x_1 only, we have for $0 < h < (R - r)/2$ that
$$\begin{cases} \mathcal{L}_0(\delta_{j,h}u) + \nabla(\delta_{j,h}p) = 0 & \text{in } B_{R_1}^+, \\ \operatorname{div}(\delta_{j,h}u) = 0 & \text{in } B_{R_1}^+, \\ \delta_{j,h}u = 0 & \text{on } B_{R_1} \cap \partial\mathbb{R}_+^d, \end{cases} \tag{1.20}$$

where $j = 2, \ldots, d$. By applying Lemma 1.9 to (1.20), we have that
$$\int_{B_r^+} |D(\delta_{j,h}u)|^2 \, dx \le N(R - r)^{-2} \int_{B_{R_1}^+} |\delta_{j,h}u|^2 \, dx,$$

which we can combine with the standard finite difference argument to imply the desired conclusion. $\square$

To estimate U, we also need to bound $D_{x'}p \in L_2(B_r)$, as in the following key lemma.

Lemma 1.11. *Let $0 < r < R$, and let ℓ be a constant.*

(1) *If $(u, p) \in W_2^1(B_R)^d \times L_2(B_R)$ satisfies (1.17) in B_R, then $D_{x'}p \in L_2(B_r)$ and*

$$\int_{B_r} |D_{x'}p|^2 \, dx \le N(R-r)^{-2} \int_{B_R} |Du|^2 \, dx, \qquad (1.21)$$

where $N = N(d, \delta) > 0$.

(2) *If $(u, p) \in W_2^1(B_R^+)^d \times L_2(B_R^+)$ satisfies (1.19) in B_R^+, then $D_{x'}p \in L_2(B_r^+)$, and (1.21) is satisfied with B_r^+ and B_R^+ replacing B_r and B_R, respectively.*

Proof. We only prove the second assertion here. Define $\delta_{j,h}f$ as in the proof of Lemma 1.10, and set $R_1 = (2R+r)/3$. Then, for $0 < h < (R-r)/3$, we have that

$$\mathcal{L}_0 (\delta_{j,h}u) + \nabla (\delta_{j,h}p) = 0 \qquad (1.22)$$

in $B_{R_1}^+$, where $j = 2, \ldots, d$. Set $R_2 = (R + 2r)/3$, and let η be an infinitely differentiable function on $\mathbb{R}^d$ such that

$$0 \le \eta \le 1, \quad \eta = 1 \text{ on } B_r, \quad \eta = 0 \text{ on } \mathbb{R}^d \setminus B_{R_2},$$
$$\text{and} \quad |D\eta| \le N(d)(R_2 - r)^{-1} = N(d)(R - r)^{-1}.$$

Find a function $\psi \in \mathring{W}_2^1(B_{R_1}^+)^d$ such that

$$\text{div } \psi = \delta_{j,h} \left((p - c)\eta^2 \right)$$

in $B_{R_1}^+$, where $c := (p)_{B_R^+}$. Note that

$$\int_{B_{R_1}^+} \delta_{j,h} \left((p - c)\eta^2 \right) \, dx = 0.$$

Then, by Remark 1.4 we have that

$$\|D\psi\|_{L_2(B_{R_1}^+)} \le N(d)\|\delta_{j,h} \left((p - c)\eta^2 \right)\|_{L_2(B_{R_1}^+)}.$$

Since

$$\delta_{j,h} \left((p - c)\eta^2 \right) = \eta^2 (x)(\delta_{j,h}p)(x) + (p - c)(x + h)(\delta_{j,h}\eta^2)(x),$$

it follows that

$$\|D\psi\|_{L_2(B_{R_1}^+)} \le N\|(\delta_{j,h}p)\eta\|_{L_2(B_{R_1}^+)} + N(R - r)^{-1}\|p - c\|_{L_2(B_R^+)}. \qquad (1.23)$$

Then, applying ψ to (1.22) as a test function, we have that

$$\int_{B_{R_1}^+} (\delta_{j,h}p) \, \delta_{j,h} \left((p - c)\eta^2 \right) \, dx = - \int_{B_{R_1}^+} D_\alpha \psi \cdot A^{\alpha\beta} D_\beta (\delta_{j,h}u) \, dx.$$

Thus, we have that

$$\int_{B_{R_1}^+} (\delta_{j,h}p)^2 \eta^2 \, dx$$

$$= -\int_{B_{R_1}^+} D_\alpha \psi \cdot A^{\alpha\beta} D_\beta (\delta_{j,h}u) \, dx$$

$$- \int_{B_{R_1}^+} (\delta_{j,h}p)(x)(p-c)(x+h)(\delta_{j,h}\eta)(x) \, (\eta(x+h) + \eta(x)) \, dx.$$

By Young's inequality and (1.23), we have for any $\varepsilon \in (0,1)$ that

$$\|(\delta_{j,h}p)\eta\|_{L_2(B_{R_1}^+)}^2$$

$$\leq \varepsilon\|D\psi\|_{L_2(B_{R_1}^+)}^2 + N(\varepsilon,\delta)\|D(\delta_{j,h}u)\|_{L_2(B_{R_1}^+)}^2 + \varepsilon\|(\delta_{j,h}p)\eta\|_{L_2(B_{R_1}^+)}^2$$

$$+\varepsilon\|(\delta_{j,h}p)\eta(\cdot+h)\|_{L_2(B_{R_1}^+)}^2 + N(\varepsilon)\|(p-c)(\cdot+h)(\delta_{j,h}\eta)\|_{L_2(B_{R_1}^+)}^2. \quad (1.24)$$

Here, we note that

$$\|(\delta_{j,h}p)\eta(\cdot+h)\|_{L_2(B_{R_1}^+)} \leq \|(\delta_{j,h}p)\eta\|_{L_2(B_{R_1}^+)} + \|(\delta_{j,h}p)\,(\eta(\cdot+h)-\eta(\cdot))\|_{L_2(B_{R_1}^+)},$$

where the last term is estimated by

$$\|(\delta_{j,h}p)\,(\eta(\cdot+h)-\eta(\cdot))\|_{L_2(B_{R_1}^+)}^2 = \int_{B_{R_1}^+} |(p(x+h)-p(x))\,(\delta_{j,h}\eta)(x)|^2 \, dx$$

$$\leq N(R-r)^{-2}\|p-c\|_{L_2(B_R^+)}^2.$$

In addition, note that

$$\|(p-c)(\cdot+h)(\delta_{j,h}\eta)\|_{L_2(B_{R_1}^+)}^2 \leq N(R-r)^{-2}\|p-c\|_{L_2(B_R^+)}^2,$$

and by Lemma 1.10 and the properties of $\delta_{j,h}$, it holds that

$$\|D(\delta_{j,h}u)\|_{L_2(B_{R_1}^+)} \leq \|DD_{x'}u\|_{L_2(B_{R'}^+)} \leq N(R-r)^{-2}\|Du\|_{L_2(B_R^+)},$$

where $R_1 < R' < R$. By using the above inequalities combined with (1.24), (1.23), and Lemma 1.7, and choosing a sufficiently small $\varepsilon > 0$, we obtain that

$$\|(\delta_{j,h}p)\eta\|_{L_2(B_{R_1}^+)} \leq N(d,\delta)(R-r)^{-2}\|Du\|_{L_2(B_R^+)}.$$

Together with the properties of the finite difference operator, this proves the desired inequality. $\qquad\square$

As usual, by $[u]_{C^\tau(\Omega)}$, $\tau \in (0,1)$, we denote the Hölder semi-norm of u defined by

$$[u]_{C^\tau(\Omega)} = \sup_{\substack{x,y\in\Omega \\ x\neq y}} \frac{|u(x)-u(y)|}{|x-y|^\tau}.$$

The following interior and boundary L_∞ and Hölder estimates are the key lemma of our proofs.

Lemma 1.12. *Let ℓ be a constant, and let $(u,p) \in W_2^1(B_2)^d \times L_2(B_2)$ satisfy (1.17) with $R = 2$. Then, we have that*

$$\|D_{x'}u\|_{L_\infty(B_1)} + [D_{x'}u]_{C^{1/2}(B_1)} + [U]_{C^{1/2}(B_1)} \leq N\|Du\|_{L_2(B_2)},$$

and

$$\|U_i\|_{L_\infty(B_1)} \leq N\|Du\|_{L_2(B_2)}, \quad i = 2, \ldots, d,$$

where $N = N(d, \delta)$.

Proof. See the proof of Lemma 1.13 below, with obvious modifications. $\qquad\square$

Lemma 1.13. *Let ℓ be a constant, and let $(u,p) \in W_2^1(B_2^+)^d \times L_2(B_2^+)$ satisfy (1.19) with $R = 2$. Then, we have that*

$$\|D_{x'}u\|_{L_\infty(B_1^+)} + [D_{x'}u]_{C^{1/2}(B_1^+)} + [U]_{C^{1/2}(B_1^+)} \leq N\|Du\|_{L_2(B_2^+)},$$

and

$$\|U_i\|_{L_\infty(B_1^+)} \leq N\|Du\|_{L_2(B_2^+)}, \quad i = 2, \ldots, d,$$

where $N = N(d, \delta)$.

Proof. From Lemmas 1.10 and 1.11, we have that $(D_{x'}u, D_{x'}p) \in W_2^1(B_{r_1}^+)^d \times L_2(B_{r_1}^+)$ and

$$\|DD_{x'}u\|_{L_2(B_{r_1}^+)}^2 + \|D_{x'}p\|_{L_2(B_{r_1}^+)}^2 \leq N(d, \delta, r_1) \int_{B_2^+} |Du|^2 \, dx,$$

where $1 < r_1 < 2$. Moreover, $(D_{x'}u, D_{x'}p)$ satisfies

$$\begin{cases} \mathcal{L}_0(D_{x'}u) + \nabla(D_{x'}p) = 0 & \text{in } B_{r_1}^+, \\ \operatorname{div}(D_{x'}u) = 0 & \text{in } B_{r_1}^+, \\ D_{x'}u = 0 & \text{on } B_{r_1} \cap \partial\mathbb{R}_+^d. \end{cases}$$

Then, we apply Lemmas 1.10 and 1.11 again as above, with r_2 in place of r_1 and with r_1 in place of 1, where $1 < r_2 < r_1 < 2$. By repeating this process, we see that $(D_{x'}^k u, D_{x'}^k p)$ belongs to $W_2^1(B_r^+)^d \times L_2(B_r^+)$ with $D_{x'}^k u = 0$ on $B_r \cap \partial\mathbb{R}_+^d$, and satisfies

$$\int_{B_r^+} |DD_{x'}^k u|^2 \, dx + \int_{B_r^+} |D_{x'}^k p|^2 \, dx \leq N(d, \delta, r, k) \int_{B_2^+} |Du|^2 \, dx \qquad (1.25)$$

for any $r \in [1, 2)$ and $k = 1, 2, \ldots$. In particular, this estimate means that $D_{x'}u$ has one derivative in x_1 and sufficiently many derivatives in x_i, $i = 2, \ldots, d$, the $L_2(B_1^+)$ norms of which are bounded by $\|Du\|_{L_2(B_2^+)}$. Then, by the anisotropic Sobolev embedding theorem with $k > (d-1)/2$ (see, for instance, the proof [16, Lemma 3.5]), we get that

$$\|D_{x'}u\|_{L_\infty(B_1^+)} + [D_{x'}u]_{C^{1/2}(B_1^+)} \leq N(d, \delta)\|Du\|_{L_2(B_2^+)}.$$

Now, we prove the Hölder semi-norm estimate of U and the sup-norm estimate of U_i, $i = 2, \ldots, d$. Set

$$\tilde{U} = A^{1\beta} D_\beta u + \left(p - (p)_{B_1^+}, 0, \ldots, 0 \right)^{\mathrm{tr}}.$$

By the definitions of U and $\tilde{U}$, we have that

$$\|D_{x'}^k \tilde{U}\|_{L_2(B_1^+)} = \|D_{x'}^k U\|_{L_2(B_1^+)} \le N(d,\delta) \|DD_{x'}^k u\|_{L_2(B_1^+)} + \|D_{x'}^k p\|_{L_2(B_1^+)},$$

where $k = 1, 2, \ldots$. In combination with (1.25), this shows that

$$\|D_{x'}^k \tilde{U}\|_{L_2(B_1^+)} \le N(d,\delta,k) \|Du\|_{L_2(B_2^+)}. \tag{1.26}$$

Similarly, since the $D_{\alpha\beta} u$ terms on the right-hand side of (1.15) are of the form $DD_{x'} u$, we have that

$$\|D_1 D_{x'}^k \tilde{U}\|_{L_2(B_1^+)} = \|D_1 D_{x'}^k U\|_{L_2(B_1^+)} \le N(d,\delta,k) \|Du\|_{L_2(B_2^+)}, \tag{1.27}$$

where $k = 0, 1, 2, \ldots$. To estimate $\|\tilde{U}\|_{L_2(B_1^+)}$, we apply Lemma 1.7 with $R = 1$ to obtain that

$$\|p - (p)_{B_1^+}\|_{L_2(B_1^+)} \le N \|Du\|_{L_2(B_1^+)} \le N \|Du\|_{L_2(B_2^+)},$$

where $N = N(d,\delta)$. Together with the definition of $\tilde{U}$, this shows that

$$\|\tilde{U}\|_{L_2(B_1^+)} \le N(d,\delta) \|Du\|_{L_2(B_2^+)}.$$

Together with (1.26) and (1.27), and using the anisotropic Sobolev embedding as above with $k > (d-1)/2$, this gives that

$$\|\tilde{U}\|_{L_\infty(B_1^+)} + [\tilde{U}]_{C^{1/2}(B_1^+)} \le N(d,\delta) \|Du\|_{L_2(B_2^+)}.$$

Since $[U]_{C^{1/2}(B_1^+)} = [\tilde{U}]_{C^{1/2}(B_1^+)}$ and $\tilde{U}_i = U_i$, $i = 2, \ldots, d$, we have obtained the desired inequalities. Thus, the lemma is proved. $\square$

1.4   Mean oscillation estimates

Now we prove mean oscillation estimates using the Hölder estimates developed in Subsection 1.3 and the L_2-estimates of the Stokes system given in Lemma 1.5. Throughout this subsection, we consider the operator $\mathcal{L}_0$, i.e., the coefficients $A^{\alpha\beta}$ are measurable functions of x_1 only.

Lemma 1.14. *Let* $r \in (0,\infty)$, $\kappa \ge 2$, $x_0 \in \mathbb{R}^d$, *and* $f_\alpha, g \in L_2(B_{\kappa r}(x_0))$. *If* $(u,p) \in W_2^1(B_{\kappa r}(x_0))^d \times L_2(B_{\kappa r}(x_0))$ *satisfies*

$$\begin{cases} \mathcal{L}_0 u + \nabla p = D_\alpha f_\alpha & \text{in } B_{\kappa r}(x_0), \\ \operatorname{div} u = g & \text{in } B_{\kappa r}(x_0), \end{cases}$$

then

$$\left(|D_{x'}u - (D_{x'}u)_{B_r(x_0)}|\right)_{B_r(x_0)} + \left(|U - (U)_{B_r(x_0)}|\right)_{B_r(x_0)}$$
$$\le N\kappa^{-1/2} \left(|Du|^2\right)_{B_{\kappa r}(x_0)}^{1/2} + N\kappa^{d/2} \left(|f_\alpha|^2 + |g|^2\right)_{B_{\kappa r}(x_0)}^{1/2},$$

where $N = N(d, \delta)$.

Proof. This lemma follows as a consequence of Lemmas 1.5 and 1.12. See Case 2 in the proof of Lemma 1.15 below. □

Lemma 1.15. *Let $r \in (0, \infty)$, $\kappa \ge 16$, $x_0 \in \overline{\mathbb{R}^d_+}$, and $f_\alpha, g \in L_2(B^+_{\kappa r}(x_0))$. If $(u, p) \in W^1_2(B^+_{\kappa r}(x_0))^d \times L_2(B^+_{\kappa r}(x_0))$ satisfies*

$$\begin{cases} \mathcal{L}_0 u + \nabla p = D_\alpha f_\alpha & \text{in } B^+_{\kappa r}(x_0), \\ \operatorname{div} u = g & \text{in } B^+_{\kappa r}(x_0), \\ u = 0 & \text{on } B_{\kappa r} \cap \partial \mathbb{R}^d_+, \end{cases}$$

then

$$\left(|D_{x'}u - (D_{x'}u)_{B^+_r(x_0)}|\right)_{B^+_r(x_0)} + \left(|U - (U)_{B^+_r(x_0)}|\right)_{B^+_r(x_0)}$$
$$\le N\kappa^{-1/2} \left(|Du|^2\right)_{B^+_{\kappa r}(x_0)}^{1/2} + N\kappa^{d/2} \left(|f_\alpha|^2 + |g|^2\right)_{B^+_{\kappa r}(x_0)}^{1/2}, \quad (1.28)$$

where $N = N(d, \delta)$.

Proof. Denote the first coordinate of x_0 by x_{01}. We consider the following two cases.

Case 1: $x_{01} \ge \kappa r/8$. In this case, we have that

$$B^+_r(x_0) = B_r(x_0) \subset B_{\kappa r/8}(x_0) \subset \mathbb{R}^d_+$$

and $\kappa/8 \ge 2$. Then, the estimate (1.28) follows from Lemma 1.14.

Case 2: $x_{01} < \kappa r/8$. Set $y_0 = (0, x'_0)$. Then, we have that

$$B^+_r(x_0) \subset B^+_{\kappa r/4}(y_0) \subset B^+_{\kappa r/2}(y_0) \subset B^+_{\kappa r}(x_0). \quad (1.29)$$

Considering dilation, it suffices to prove (1.28) when $r = 4/\kappa \le 1/4$ and $x_{01} < 1/2$. Furthermore, we assume that $y_0 = 0$. By Lemma 1.5, there exists $(w, p_1) \in \mathring{W}^1_2(B^+_2)^d \times L_2(B^+_2)$ such that $(p_1)_{B^+_2} = 0$,

$$\begin{cases} \mathcal{L}_0 w + \nabla p_1 = D_\alpha f_\alpha & \text{in } B^+_2, \\ \operatorname{div} w = g - (g)_{B^+_2} & \text{in } B^+_2, \end{cases}$$

and

$$\|Dw\|_{L_2(B^+_2)} + \|p_1\|_{L_2(B^+_2)} \le N \left(\|f_\alpha\|_{L_2(B^+_2)} + \|g\|_{L_2(B^+_2)}\right), \quad (1.30)$$

where $N = N(d, \delta)$. In particular, we have that $w = 0$ on $B_2 \cap \partial \mathbb{R}^d_+$. The estimate (1.30) clearly implies that

$$\|Dw\|_{L_2(B^+_r(x_0))} + \|p_1\|_{L_2(B^+_r(x_0))} \leq N \left(\|f_\alpha\|_{L_2(B^+_2)} + \|g\|_{L_2(B^+_2)} \right), \qquad (1.31)$$

where $N = N(d, \delta)$.

Now, we set $(v, p_2) = (u, p) - (w, p_1)$, which satisfies

$$\begin{cases} \mathcal{L}_0 v + \nabla p_2 = 0 & \text{in } B^+_2, \\ \operatorname{div} v = (g)_{B^+_2} & \text{in } B^+_2, \\ v = 0 & \text{on } B_2 \cap \partial \mathbb{R}^d_+. \end{cases}$$

Then, by Lemma 1.13,

$$\left(|D_{x'}v - (D_{x'}v)_{B^+_r(x_0)}| \right)_{B^+_r(x_0)} \leq (2r)^{1/2} [D_{x'}v]_{C^{1/2}(B^+_r(x_0))}$$
$$\leq (2r)^{1/2} [D_{x'}v]_{C^{1/2}(B^+_1)} \leq N\kappa^{-1/2} \|Dv\|_{L_2(B^+_2)},$$

where $N = N(d, \delta)$. Similarly, we have that

$$\left(|V - (V)_{B^+_r(x_0)}| \right)_{B^+_r(x_0)} \leq N\kappa^{-1/2} \|Dv\|_{L_2(B^+_2)},$$

where V is defined in exactly the same way as U in (1.16) with v in place of u. Then, it follows from the triangle inequality that

$$\left(|D_{x'}u - (D_{x'}u)_{B^+_r(x_0)}| \right)_{B^+_r(x_0)}$$
$$\leq \left(|D_{x'}v - (D_{x'}v)_{B^+_r(x_0)}| \right)_{B^+_r(x_0)} + 2 \left(|D_{x'}w| \right)_{B^+_r(x_0)}$$
$$\leq N\kappa^{-1/2} \left(|Dv|^2 \right)^{1/2}_{B^+_2} + N\kappa^{d/2} \|D_{x'}w\|_{L_2(B^+_r)},$$

where $N = N(d, \delta)$. Together with the estimates (1.30) and (1.31), and the fact that $u = v + w$, this shows that

$$\left(|D_{x'}u - (D_{x'}u)_{B^+_r(x_0)}| \right)_{B^+_r(x_0)}$$
$$\leq N\kappa^{-1/2} \left(|Du|^2 \right)^{1/2}_{B^+_2} + N\kappa^{d/2} \left(|f_\alpha|^2 + |g|^2 \right)^{1/2}_{B^+_2}.$$

It only remains to observe that the right-hand side is bounded by that of (1.28), because of (1.29).

We can similarly obtain the desired estimate for U. Thus, the lemma is proved. $\qquad \square$

1.5 Proof of Theorem 1.1

We use the following filtration of partitions of $\mathbb{R}^d$:

$$\mathbb{C}_n := \{C_n = C_n(i_1, \ldots, i_d) : (i_1, \ldots, i_d) \in \mathbb{Z}^d\},$$

where $n \in \mathbb{Z}$ and

$$C_n(i_1, \ldots, i_d) = [i_1 2^{-n}, (i_1 + 1)2^{-n}) \times \cdots \times [i_d 2^{-n}, (i_d + 1)2^{-n}).$$

For a filtration of partitions of $\mathbb{R}^d_+$, we replace $i_1 \in \mathbb{Z}$ by $i_1 \in \{0, 1, 2, \ldots\}$. Using these filtrations, we define the sharp function of $f \in L_{1,\text{loc}}(\Omega)$, where $\Omega = \mathbb{R}^d$ or $\Omega = \mathbb{R}^d_+$, by

$$f^\#(x) = \sup_{n < \infty} \fint_{C_n \ni x} |f(y) - (f)_{C_n}| \, dy,$$

where the supremum is taken with respect to all $C_n \in \mathbb{C}_n$ containing x, where $n \in \mathbb{Z}$. The maximal function of f in $\mathbb{R}^d$ or $\mathbb{R}^+$ is defined by

$$\mathcal{M}f(x) = \sup_{x_0 \in \bar{\Omega}, \Omega \cap B_r(x_0) \ni x} \fint_{\Omega \cap B_r(x_0)} |f(y)| \, dy, \qquad (1.32)$$

where $\Omega = \mathbb{R}^d$ or $\Omega = \mathbb{R}^d_+$, and the supremum is taken with respect to all $B_r(x_0)$ containing x with $r > 0$, where $x_0 \in \bar{\Omega}$.

Proof of Theorem 1.1. Because Theorem 1.6 covers the case with $q = 2$, we assume that $q \in (2, \infty)$. We prove the case when $\Omega = \mathbb{R}^d_+$. The other case is simpler.

For $x \in \mathbb{R}^d_+$ and $C_n \in \mathbb{C}_n$ such that $x \in C_n$, find $x_0 \in \mathbb{R}^d_+$ and the smallest $r \in (0, \infty)$ (indeed, $r = 2^{-n-1}\sqrt{d}$) satisfying $C_n \subset B_r(x_0)$ and

$$\fint_{C_n} |h(x) - (h)_{C_n}| \, dx \le N(d) \fint_{B_r(x_0)} |h(x) - (h)_{B_r(x_0)}| \, dx. \qquad (1.33)$$

Since $(u, p) \in W_2^1(B_{\kappa r}^+(x_0))^d \times L_2(B_{\kappa r}^+(x_0))$, it follows from Lemma 1.15 that we have the mean oscillation estimate (1.28) for $\kappa \ge 16$. Moreover, each term in the right-hand side of (1.28) is bounded by its maximal function at x. From this and (1.33), we have that

$$(|D_{x'}u - (D_{x'}u)_{C_n}|)_{C_n} + (|U - (U)_{C_n}|)_{C_n}$$
$$\le N\kappa^{-1/2} \left(\mathcal{M}(|Du|^2)(x)\right)^{1/2} + N\kappa^{d/2} \left(\mathcal{M}(|f_\alpha|^2)(x)\right)^{1/2} + \left(\mathcal{M}(|g|^2)(x)\right)^{1/2}$$

for $x \in C_n$ and $\kappa \ge 16$, where $N = N(d, \delta)$. By taking the supremum of the left-hand side of the above inequality with respect to all $C_n \ni x$, $n \in \mathbb{Z}$, we obtain that

$$(D_{x'}u)^\# (x) + U^\#(x) \le N\kappa^{-1/2} \left(\mathcal{M}(|Du|^2)(x)\right)^{1/2}$$
$$+ N\kappa^{d/2} \left(\mathcal{M}(|f_\alpha|^2)(x)\right)^{1/2} + N\kappa^{d/2} \left(\mathcal{M}(|g|^2)(x)\right)^{1/2}$$

for $x \in \mathbb{R}^d_+$ and $\kappa \ge 16$. Then, we employ the Fefferman-Stein theorem on sharp functions (see, for instance, [30, Theorem 3.2.10]) and the maximal function theorem (see, for instance, [30, Theorem 3.3.2] or Lemma 1.20, which also holds when $\Omega = \mathbb{R}^d_+$ with $N = N(d, q)$) on the above pointwise estimate, to obtain that

$$\|D_{x'}u\|_{L_q} + \|U\|_{L_q} \le N\kappa^{-1/2}\|Du\|_{L_q} + N\kappa^{d/2} \left(\|f_\alpha\|_{L_q} + \|g\|_{L_q}\right), \qquad (1.34)$$

where $L_q = L_q(\mathbb{R}^d_+)$ and $N = N(d, \delta, p)$. Note that on the left-hand side of the above inequality we do not yet have L_q-norms of $D_1 u_i$, $i = 1, 2, \ldots, d$, and p. To obtain L_q-estimates of such terms, we first note the relation

$$D_1 u_1 + \cdots + D_d u_d = g.$$

Using this and (1.34), we have that

$$\|D_1 u_1\|_{L_q} + \|D_{x'} u\|_{L_q} + \|U\|_{L_q} \leq N\kappa^{-1/2}\|Du\|_{L_q} + N\kappa^{d/2}\left(\|f_\alpha\|_{L_q} + \|g\|_{L_q}\right). \tag{1.35}$$

Then, we use the relation

$$\sum_{j=2}^{d} A_{ij}^{11} D_1 u_j = U_i - \sum_{j=1}^{d}\sum_{\beta=2}^{d} A_{ij}^{1\beta} D_\beta u_j - A_{i1}^{11} D_1 u_1, \quad i = 2, \ldots, d, \tag{1.36}$$

which follows from the definition of U_i, $i = 2, \ldots, d$. By the ellipticity condition on $A^{\alpha\beta}$, it follows that the $(d-1) \times (d-1)$ matrix $[A_{ij}^{11}]_{i,j=2}^{d}$ is invertible. Thus, from (1.36) and (1.35) we have that

$$\|Du\|_{L_q} + \|U\|_{L_q} \leq N\kappa^{-1/2}\|Du\|_{L_q} + N\kappa^{d/2}\left(\|f_\alpha\|_{L_q} + \|g\|_{L_q}\right).$$

Upon taking a sufficiently large $\kappa \geq 16$, which depends only on d, δ, and q, such that $N\kappa^{-1/2} \leq 1/2$, we arrive at

$$\|Du\|_{L_q} + \|U\|_{L_q} \leq N\left(\|f_\alpha\|_{L_q} + \|g\|_{L_q}\right). \tag{1.37}$$

Finally, from this estimate and the definition of U_1, we see that the L_q-norm of p is bounded by the right-hand side of (1.4). By this and (1.37), we can conclude that the estimate (1.4) holds, and the theorem is proved. $\qquad\square$

1.6   Stokes systems in a bounded Lipschitz domain

When the Stokes system is defined in a bounded Lipschitz domain Ω with a small Lipschitz constant, we show that the system is uniquely solvable in $L_q(\Omega)$ spaces. In this case, we allow coefficients not only to be measurable locally in one direction (almost perpendicular to the boundary of the domain), but also to have small mean oscillations in the other directions. To present this result, we impose the following assumptions.

Assumption 1.16. For any $x_0 \in \partial\Omega$ and $0 < r \leq R_0$, there is a coordinate system depending on x_0 and r such that in the new coordinate system we have

$$\Omega \cap B_r(x_0) = \{x \in B_r(x_0) : x_1 > \phi(x')\},$$

where $\phi : \mathbb{R}^{d-1} \to \mathbb{R}$ is a Lipschitz function with

$$\sup_{\substack{x', y' \in B_r'(x_0') \\ x' \neq y'}} \frac{|\phi(y') - \phi(x')|}{|y' - x'|} \leq \frac{1}{16}.$$

Assumption 1.17 (γ, ρ). Let $\gamma, \rho \in (0, 1/16)$. There exists $R_1 \in (0, R_0]$ satisfying the following.

(1) For $x_0 \in \Omega$ and $0 < r \le \min\{R_1, \text{dist}(x_0, \partial\Omega)\}$, there is a coordinate system depending on x_0 and r such that in this new coordinate system we have that

$$\fint_{B_r(x_0)} \left| A^{\alpha\beta}(y_1, y') - \fint_{B'_r(x'_0)} A^{\alpha\beta}(y_1, z') \, dz' \right| dx \le \gamma. \tag{1.38}$$

(2) For any $x_0 \in \partial\Omega$ and $0 < r \le R_1$, there is a coordinate system depending on x_0 and r such that in the new coordinate system we have that (1.38) holds, and

$$\Omega \cap B_r(x_0) = \{x \in B_r(x_0) : x_1 > \phi(x')\},$$

where $\phi : \mathbb{R}^{d-1} \to \mathbb{R}$ is a Lipschitz function with

$$\sup_{\substack{x', y' \in B'_r(x'_0) \\ x' \ne y'}} \frac{|\phi(y') - \phi(x')|}{|y' - x'|} \le \rho.$$

Theorem 1.18. Let $q, q_1 \in (1, \infty)$ satisfying $q_1 \ge qd/(q+d)$, $K > 0$, and let Ω be bounded $(\text{diam}\,\Omega \le K)$. Then, there exist constants $(\gamma, \rho) = (\gamma, \rho)(d, \delta, R_0, K, q) \in (0, 1/16)$ such that, under Assumptions 1.16 and 1.17 (γ, ρ), for $(u, p) \in W^1_q(\Omega)^d \times L_q(\Omega)$ satisfying $(p)_\Omega = 0$ and

$$\begin{cases} \mathcal{L}u + \nabla p = f + D_\alpha f_\alpha & \text{in } \Omega, \\ \text{div}\, u = g & \text{in } \Omega, \\ u = 0 & \text{on } \partial\Omega, \end{cases} \tag{1.39}$$

where $f \in L_{q_1}(\Omega)$, $f_\alpha, g \in L_q(\Omega)$, we have that

$$\|Du\|_{L_q(\Omega)} + \|p\|_{L_q(\Omega)} \le N \left(\|f\|_{L_{q_1}(\Omega)} + \|f_\alpha\|_{L_q(\Omega)} + \|g\|_{L_q(\Omega)} \right), \tag{1.40}$$

where $N > 0$ is a constant depending only on d, δ, R_0, R_1, K, q, and q_1. Moreover, for $f \in L_{q_1}(\Omega)$, $f_\alpha, g \in L_q(\Omega)$ with $(g)_\Omega = 0$, there exists a unique $(u, p) \in W^1_q(\Omega)^d \times L_q(\Omega)$ satisfying $(p)_\Omega = 0$ and (1.39).

For any $x_0 \in \mathbb{R}^d$ and $r > 0$, denote

$$\Omega_r(x_0) = \Omega \cap B_r(x_0).$$

We first derive the following mean oscillation estimate.

Lemma 1.19. Let $\mu, \nu \in (1, \infty)$ be such that $1/\mu + 1/\nu = 1$ and $\kappa \ge 32$. Then, under Assumption 1.17 (γ, ρ) such that $\rho\kappa \le 1/4$, for any $r \in (0, R_1/\kappa]$, $x_0 \in \overline{\Omega}$, and

$$(u, p) \in W^1_{2\mu}(\Omega_{\kappa r}(x_0))^d \times L_2(\Omega_{\kappa r}(x_0))$$

satisfying

$$\begin{cases} \mathcal{L}u + \nabla p = D_\alpha f_\alpha & \text{in } \Omega_{\kappa r}(x_0), \\ \text{div}\, u = g & \text{in } \Omega_{\kappa r}(x_0), \\ u = 0 & \text{on } \partial\Omega \cap B_{\kappa r}(x_0), \end{cases} \tag{1.41}$$

where $f_\alpha \in L_2(\Omega_{\kappa r}(x_0))$, there exists a d^2-dimensional vector-valued function $\mathcal{U}$ on $\Omega_{\kappa r}(x_0)$ such that on $\Omega_{\kappa r}(x_0)$,

$$N^{-1}|Du| \le |\mathcal{U}| \le N|Du| \tag{1.42}$$

and

$$\left(|\mathcal{U} - (\mathcal{U})_{\Omega_r(x_0)}|\right)_{\Omega_r(x_0)} \le N(\kappa^{-\frac{1}{2}} + \kappa\rho)\left(|Du|^2\right)^{\frac{1}{2}}_{\Omega_{\kappa r}(x_0)}$$
$$+ N\kappa^{\frac{d}{2}}\left(f_\alpha^2\right)^{\frac{1}{2}}_{\Omega_{\kappa r}(x_0)} + N\kappa^{\frac{d}{2}}\left(g^2\right)^{\frac{1}{2}}_{\Omega_{\kappa r}(x_0)}$$
$$+ N\kappa^{\frac{d}{2}}(\gamma+\rho)^{\frac{1}{2\nu}}\left(|Du|^{2\mu}\right)^{\frac{1}{2\mu}}_{\Omega_{\kappa r}(x_0)}, \tag{1.43}$$

where $N = N(d, \delta, \mu)$.

Proof. Let $\tilde{x} \in \partial\Omega$ be such that $|x_0 - \tilde{x}| = \text{dist}(x_0, \partial\Omega)$. As in the proof of Lemma 1.15, we consider two cases.
Case 1: $|x_0 - \tilde{x}| \ge \kappa r/16$. In this case, we have that

$$\Omega_r(x_0) = B_r(x_0) \subset B_{\kappa r/16}(x_0) \subset \Omega.$$

Since $\kappa/16 \ge 2$, (1.43) follows from Lemma 1.14, by using a rotation of coordinates and setting
$$\mathcal{U} = (D_{x'}u, \text{div }u, U_2, \ldots, U_d),$$

where for $i = 2, \ldots, d$, U_i are given as in (1.56) below. See the proof for Case 2. As in the proof of Theorem 1.1, by using the definition of U and the equality $\text{div }u = g$, we see that (1.42) is satisfied.
Case 2: $|x_0 - \tilde{x}| < \kappa r/16$. Without loss of generality, one may assume that $\tilde{x}$ is the origin. Note that

$$\Omega_r(x_0) \subset \Omega_{\kappa r/4} \subset \Omega_{\kappa r/2} \subset \Omega_{\kappa r}(x_0).$$

Denote $R = \kappa r/2 (\le R_1/2)$. Due to Assumption 1.17, we can take an orthogonal transformation to obtain that

$$\{(x_1, x') : \rho R < x_1\} \cap B_R \subset \Omega \cap B_R \subset \{(x_1, x') : -\rho R < x_1\} \cap B_R$$

and

$$\fint_{B_R} \left|A^{\alpha\beta}(x_1, x') - \bar{A}^{\alpha\beta}(x_1)\right| dx \le \gamma, \tag{1.44}$$

where

$$\bar{A}^{\alpha\beta}(x_1) = \fint_{B_R'} A^{\alpha\beta}(x_1, x') dx'. \tag{1.45}$$

Take a smooth function χ on $\mathbb{R}$ such that

$$\chi(x_1) \equiv 0 \quad \text{for } x_1 \le \rho R, \quad \chi(x_1) \equiv 1 \quad \text{for } x_1 \ge 2\rho R, \quad \text{and} \quad |\chi'| \le N(\rho R)^{-1}.$$

Denote $\mathcal{L}_0$ to be the elliptic operator with the coefficients $\bar{A}^{\alpha\beta}$ from (1.45). Let $\hat{u} = \chi u$, which vanishes on $B_R \cap \{x_1 \le \rho R\}$. From (1.41), it is easily seen that $(\hat{u}, p)$ satisfies

$$\begin{cases} \mathcal{L}_0\hat{u} + \nabla p = D_\alpha(\tilde{f}_\alpha + h_\alpha) & \text{in } B_R \cap \{x_1 > \rho R\}, \\ \operatorname{div} \hat{u} = \chi g + \chi' u_1 & \text{in } B_R \cap \{x_1 > \rho R\}, \\ \hat{u} = 0 & \text{on } B_R \cap \{x_1 = \rho R\}, \end{cases} \quad (1.46)$$

where

$$\tilde{f}_\alpha = f_\alpha + (\bar{A}^{\alpha\beta} - A^{\alpha\beta})D_\beta u \quad \text{and} \quad h_\alpha = \bar{A}^{\alpha\beta}D_\beta((\chi - 1)u).$$

For $\tau \in [0, \infty)$, set

$$\widetilde{B}_r^+(\tau, 0) = B_r(\tau, 0) \cap \{x_1 > \tau\},$$

where $0 \in \mathbb{R}^{d-1}$. Since $\rho \in (0, 1/16)$, we have that

$$\Omega_{R/2} \subset \Omega_{3R/4}(\rho R, 0) \quad \text{and} \quad \widetilde{B}_{3R/4}^+(\rho R, 0) \subset B_R \cap \{x_1 > \rho R\}.$$

By Lemma 1.5, there is a unique solution

$$(\hat{w}, p_1) \in W_2^1(\widetilde{B}_{3R/4}^+(\rho R, 0))^d \times L_2(\widetilde{B}_{3R/4}^+(\rho R, 0))$$

satisfying $(p_1)_{\widetilde{B}_{3R/4}^+(\rho R, 0)} = 0$ and

$$\begin{cases} \mathcal{L}_0\hat{w} + \nabla p_1 = D_\alpha(\tilde{f}_\alpha + h_\alpha) & \text{in } \widetilde{B}_{3R/4}^+(\rho R, 0), \\ \operatorname{div} \hat{w} = \chi g + \chi' u_1 - (\chi g + \chi' u_1)_{\widetilde{B}_{3R/4}^+(\rho R, 0)} & \text{in } \widetilde{B}_{3R/4}^+(\rho R, 0), \\ \hat{w} = 0 & \text{on } \partial\widetilde{B}_{3R/4}^+(\rho R, 0). \end{cases} \quad (1.47)$$

Moreover, it holds that

$$\begin{aligned} \|D\hat{w}\|_{L_2} &\leq N\big(\|\tilde{f}_\alpha\|_{L_2} + \|h_\alpha\|_{L_2} + \|\chi g + \chi' u\|_{L_2}\big) \\ &\leq N\big(\|f_\alpha\|_{L_2} + \|(\bar{A}^{\alpha\beta} - A^{\alpha\beta})D_\beta u\|_{L_2} \\ &\quad + \|D((\chi - 1)u)\|_{L_2} + \|g\|_{L_2} + \|\chi' u\|_{L_2}\big), \end{aligned} \quad (1.48)$$

where $\|\cdot\|_{L_2} = \|\cdot\|_{L_2(\widetilde{B}_{3R/4}^+(\rho R, 0))}$ and $N = N(d, \delta)$. Using the fact that $|A^{\alpha\beta}| \leq \delta^{-1}$, together with (1.44) and Hölder's inequality, it follows that

$$\|(\bar{A}^{\alpha\beta} - A^{\alpha\beta})D_\beta u\|_{L_2(\widetilde{B}_{3R/4}^+(\rho R, 0))} \leq N\gamma^{\frac{1}{2\nu}}R^{\frac{d}{2\nu}}\|Du\|_{L_{2\mu}(\widetilde{B}_{3R/4}^+(\rho R, 0))}. \quad (1.49)$$

Since $\chi - 1$ is supported on $\{x_1 \leq 2\rho R\}$, Hölder's inequality implies that

$$\|(\chi - 1)Du\|_{L_2(\widetilde{B}_{3R/4}^+(\rho R, 0))} \leq N\rho^{\frac{1}{2\nu}}R^{\frac{d}{2\nu}}\|Du\|_{L_{2\mu}(\Omega_R)}. \quad (1.50)$$

Using Hölder's inequality again, together with the fact that χ' is supported on $\{\rho R \leq x_1 \leq 2\rho R\}$, we have that

$$\begin{aligned} \|\chi' u\|_{L_2(\widetilde{B}_{3R/4}^+(\rho R, 0))} &\leq N\rho^{\frac{1}{2\nu}}R^{\frac{d}{2\nu}}\|\chi' u\|_{L_{2\mu}(\widetilde{B}_{3R/4}^+(\rho R, 0))} \\ &\leq N\rho^{\frac{1}{2\nu}}R^{\frac{d}{2\nu}}\|Du\|_{L_{2\mu}(\Omega_R)}, \end{aligned} \quad (1.51)$$

where the last inequality follows from Hardy's inequality, using the boundary condition $u = 0$ on $\partial\Omega$ and the observation that

$$|\chi'| \le N(x_1 - \phi(x'))^{-1}$$

for $(x_1, x') \in \Omega_R$, where Ω_R is given by $\{x \in B_R : x_1 > \phi(x')\}$. The inequalities (1.49), (1.50), and (1.51), together with (1.48), imply that

$$(|D\hat{w}|^2)^{\frac{1}{2}}_{\tilde{B}^+_{3R/4}(\rho R, 0)} \le N(\rho + \gamma)^{\frac{1}{2\nu}}(|Du|^{2\mu})^{\frac{1}{2\mu}}_{\Omega_R} + N(f_\alpha^2 + g^2)^{\frac{1}{2}}_{\Omega_R}. \tag{1.52}$$

We extend $\hat{w}$ to be zero in $\Omega_{3R/4}(\rho R, 0) \cap \{x_1 < \rho R\}$, so that $\hat{w} \in W_2^1(\Omega_{3R/4}(\rho R, 0))$, and we let

$$w = \hat{w} + (1 - \chi)u.$$

By the same reasoning as in (1.50) and (1.51), we have that

$$\|D((1-\chi)u)\|_{L_2(\Omega_{3R/4}(\rho R, 0))} \le N\rho^{\frac{1}{2\nu}} R^{\frac{d}{2\nu}} \|Du\|_{L_{2\mu}(\Omega_R)}.$$

From this and (1.52), we deduce that

$$(|Dw|^2)^{\frac{1}{2}}_{\Omega_{3R/4}(\rho R, 0)} \le N(\gamma + \rho)^{\frac{1}{2\nu}}(|Du|^{2\mu})^{\frac{1}{2\mu}}_{\Omega_R} + N(f_\alpha^2 + g^2)^{\frac{1}{2}}_{\Omega_R}. \tag{1.53}$$

Note that, because $\kappa\rho \le 1/4$, it holds that

$$\Omega_r(x_0) \subset \Omega_{3R/4}(\rho R, 0) \quad \text{and} \quad |\Omega_{3R/4}(\rho R, 0)|/|\Omega_r(x_0)| \le N(d)\kappa^d.$$

Thus, from (1.53) we also obtain that

$$(|Dw|^2)^{\frac{1}{2}}_{\Omega_r(x_0)} \le N\kappa^{\frac{d}{2}}(\gamma + \rho)^{\frac{1}{2\nu}}(|Du|^{2\mu})^{\frac{1}{2\mu}}_{\Omega_R} + N\kappa^{\frac{d}{2}}(f_\alpha^2 + g^2)^{\frac{1}{2}}_{\Omega_R}. \tag{1.54}$$

Next, we define $v = u - w$ ($= \chi u - \hat{w}$) in $\Omega_{3R/4}(\rho R, 0)$ and $p_2 = p - p_1$ in $\tilde{B}^+_{3R/4}(\rho R, 0)$. From (1.46) and (1.47), it is easily seen that (v, p_2) satisfies

$$\begin{cases} \mathcal{L}_0 v + \nabla p_2 = 0 & \text{in } \tilde{B}^+_{3R/4}(\rho R, 0), \\ \text{div } v = (\chi g + \chi' u_1)_{\tilde{B}^+_{3R/4}(\rho R, 0)} & \text{in } \tilde{B}^+_{3R/4}(\rho R, 0), \\ v = 0 & \text{on } B_{3R/4}(\rho R, 0) \cap \{x_1 = \rho R\}. \end{cases}$$

Denote

$$\mathcal{D}_1 = \Omega_r(x_0) \cap \{x_1 \le \rho R\}, \quad \mathcal{D}_2 = \Omega_r(x_0) \cap \{x_1 > \rho R\}, \quad \text{and } \mathcal{D}_3 = \tilde{B}^+_{R/4}(\rho R, 0).$$

We have that $\mathcal{D}_2 \subset \mathcal{D}_3$ and $|\mathcal{D}_1| \le N\kappa\rho|\Omega_r(x_0)|$, where the latter follows from the fact that $\mathcal{D}_1 = \Omega_r(x_0) \cap \{-\rho R \le x_1 \le \rho R\}$. We set

$$V_i = \sum_{j=1}^d \sum_{\beta=1}^d \bar{A}^{1\beta}_{ij} D_\beta v_j, \quad i = 2, \ldots, d,$$

where the coefficients $\bar{A}^{1\beta}(x_1)$ are taken from (1.45). Note that $v = V_i = 0$ in $\mathcal{D}_1$. Then, by applying Lemma 1.13 with a dilation, we get that

$$\left(|V_i - (V_i)_{\Omega_r(x_0)}|\right)_{\Omega_r(x_0)} + \left(|D_{x'}v - (D_{x'}v)_{\Omega_r(x_0)}|\right)_{\Omega_r(x_0)}$$

$$\leq Nr^{1/2}\left([V_i]_{C^{1/2}(\mathcal{D}_2)} + [D_{x'}v]_{C^{1/2}(\mathcal{D}_2)}\right) + N\kappa\rho\left(\|V_i\|_{L_\infty(\mathcal{D}_2)} + \|D_{x'}v\|_{L_\infty(\mathcal{D}_2)}\right)$$

$$\leq Nr^{1/2}\left([V_i]_{C^{1/2}(\mathcal{D}_3)} + [D_{x'}v]_{C^{1/2}(\mathcal{D}_3)}\right) + N\kappa\rho\left(\|V_i\|_{L_\infty(\mathcal{D}_3)} + \|D_{x'}v\|_{L_\infty(\mathcal{D}_3)}\right)$$

$$\leq N(\kappa^{-1/2} + \kappa\rho)(|Dv|^2)^{1/2}_{\Omega_{R/2}(\rho R,0)}. \tag{1.55}$$

Now, we set $\mathcal{U} = (D_{x'}u, \operatorname{div} u, U_2, \ldots, U_d)$, where

$$U_i = \sum_{j=1}^{d}\sum_{\beta=1}^{d} \bar{A}^{1\beta}_{ij} D_\beta u_j, \quad i = 2, \ldots, d. \tag{1.56}$$

Note that $\mathcal{U}$ satisfies (1.42). From the triangle inequality, (1.55), and the fact that $|\Omega_{2R}(x_0)| \leq N\kappa^d|\Omega_r(x_0)|$ by the condition that $\kappa\rho \leq 1/4$, we have that

$$\left(|\mathcal{U} - (\mathcal{U})_{\Omega_r(x_0)}|\right)_{\Omega_r(x_0)}$$

$$\leq N\left(|D_{x'}v - (D_{x'}v)_{\Omega_r(x_0)}|\right)_{\Omega_r(x_0)}$$

$$\quad + N\left(|V_i - (V_i)_{\Omega_r(x_0)}|\right)_{\Omega_r(x_0)} + N(|g|)_{\Omega_r(x_0)} + N\left(|Dw|\right)_{\Omega_r(x_0)}$$

$$\leq N(\kappa^{-1/2} + \kappa\rho)(|Dv|^2)^{1/2}_{\Omega_{R/2}(\rho R,0)} + N(|g|^2)^{1/2}_{\Omega_r(x_0)} + N\left(|Dw|^2\right)^{1/2}_{\Omega_r(x_0)}$$

$$\leq N(\kappa^{-1/2} + \kappa\rho)(|Du|^2)^{1/2}_{\Omega_{R/2}(\rho R,0)} + N(\kappa^{-1/2} + \kappa\rho)(|Dw|^2)^{1/2}_{\Omega_{R/2}(\rho R,0)}$$

$$\quad + N\kappa^{d/2}(|g|^2)^{1/2}_{\Omega_{2R}(x_0)} + N\left(|Dw|^2\right)^{1/2}_{\Omega_r(x_0)},$$

where we bound the second and last terms on the right-hand side of the last inequality by using (1.53) and (1.54). This completes the proof of Lemma 1.19. $\qquad\square$

Before we present the proof of Theorem 1.18, we note that a Lipschitz domain Ω in $\mathbb{R}^d$ satisfying Assumption 1.16 with $\operatorname{diam}\Omega \leq K$ is a space of homogeneous type, which is endowed with the Euclidean distance and a doubling measure μ that is naturally inherited from the Lebesgue measure. By a result by Christ [11, Theorem 11] (also see [25]), there exists a filtration of partitions of Ω in the following sense. For each $n \in \mathbb{Z}$, there exists a collection of disjoint open subsets $\mathbb{C}_n := \{Q^n_\alpha : \alpha \in I_n\}$ for some index set I_n, satisfying the following properties:

(1) for any $n \in \mathbb{Z}$, $\mu(\Omega \setminus \bigcup_\alpha Q^n_\alpha) = 0$;
(2) for each n and $\alpha \in I_n$, there exists a unique $\beta \in I_{n-1}$ such that $Q^n_\alpha \subset Q^{n-1}_\beta$;
(3) for each n and $\alpha \in I_n$, $\operatorname{diam}(Q^n_\alpha) \leq N_0\delta^n_0$;
(4) each Q^n_α contains some ball $\Omega_{\varepsilon_0\delta^n_0}(z^n_\alpha)$;

for some constants $\delta_0 \in (0,1)$, $\varepsilon_0 > 0$, and N_0 depending only on d, R_0, and K.

For any $f \in L_{1,\text{loc}}(\Omega)$, recall the definition of a maximal function in (1.32) with a bounded Lipschitz domain Ω in place of $\mathbb{R}^d$ or $\mathbb{R}^d_+$:

$$\mathcal{M}f(x) = \sup_{x_0 \in \bar{\Omega}, \Omega_r(x_0) \ni x} \fint_{\Omega_r(x_0)} |f(y)|\, dy.$$

Lemma 1.20. *Let $q \in (1, \infty)$ and Ω satisfy Assumption 1.16 with $\operatorname{diam} \Omega \leq K$. Then, for any $f \in L_q(\Omega)$, we have that*

$$\|\mathcal{M}f\|_{L_q(\Omega)} \leq N\|f\|_{L_q(\Omega)},$$

where $N > 0$ is a constant depending only on d, q, R_0, and K.

Proof. Since Ω is a space of homogeneous type, the lemma follows from the Hardy-Littlewood maximal function theorem for spaces of homogeneous type. See, for instance, [3]. Also see [12, Theorem 2.2].   □

Lemma 1.21. *Let $q \in (1, \infty)$ and Ω satisfy Assumption 1.16 with $\operatorname{diam} \Omega \leq K$. Suppose that*

$$F \in L_q(\Omega), \quad G \in L_q(\Omega), \quad H \in L_q(\Omega), \quad |F| \leq H,$$

and that for each $n \in \mathbb{Z}$ and $Q \in \mathbb{C}_n$, there exists a measurable function F^Q on Q such that

$$|F| \leq F^Q \leq H \quad on \ Q \quad and \quad \fint_Q |F^Q(x) - (F^Q)_Q| \, dx \leq N_0 G(y) \quad \forall y \in Q \quad (1.57)$$

for some constant $N_0 > 0$. Then, we have that

$$\|F\|^p_{L_q(\Omega)} \leq N N_0 \|G\|_{L_q(\Omega)} \|H\|^{p-1}_{L_q(\Omega)} + N\|H\|^p_{L_1(\Omega)}, \quad (1.58)$$

where $N > 0$ is a constant depending only on d, q, R_0, and K.

Proof. This lemma is a special case of Theorem 2.4 of [12], in which A_q weights are considered. When there is no weight as in the lemma, it is easily seen that β in that theorem is equal to 1.   □

We are now ready to present the proof of Theorem 1.18.

Proof of Theorem 1.18. We only derive the a priori estimate (1.40). The solvability then follows from (1.40), the Poincaré inequality, and the method of continuity. Furthermore, we assume that $f \equiv 0$. Otherwise, for $B_R \supseteq \Omega$, we find $w \in W^2_{q_1}(B_R)$ such that $\Delta w = f 1_\Omega$ in B_R and $w|_{\partial B_R} = 0$. Then, we consider

$$\mathcal{L}u + \nabla p = D_\alpha (D_\alpha w + f_\alpha),$$

where from the Sobolev embedding theorem and the well-known L_{q_1}-estimate for the Laplace equation we have that

$$\|D_\alpha w\|_{L_q(\Omega)} \leq \|D_\alpha w\|_{L_q(B_R)} \leq N\|w\|_{W^2_{q_1}(B_R)} \leq N\|f\|_{L_{q_1}(\Omega)}.$$

We consider the two cases with $q > 2$ and $q \in (1, 2)$. The case with $q = 2$ follows from Lemma 1.5.

Case 1: $q > 2$. We take $\mu \in (1, \infty)$, depending only on q, such that $2\mu < q$, and we let $\kappa \geq 32$ be a constant to be specified. By the properties (3) and (4) described above, for each Q in the partitions there exist $r \in (0, \infty)$ and $x_0 \in \bar{\Omega}$ such that

$$Q \subset \Omega_r(x_0) \quad \text{and} \quad |\Omega_r(x_0)| \leq N|Q|, \tag{1.59}$$

where N depends only on d, R_0, and K. In order to apply Lemma 1.21, we take $F = |Du|$, $H = N|Du|$, where $N = N(d, \delta, q) \geq 1$ from (1.42), and

$$G(y) = (\kappa^{-\frac{1}{2}} + \kappa\rho) \left[\mathcal{M}(|Du|^2)(y) \right]^{\frac{1}{2}} + \kappa^{\frac{d}{2}} \left[\mathcal{M}(f_\alpha^2)(y) \right]^{\frac{1}{2}}$$
$$+ \kappa^{\frac{d}{2}} \left[\mathcal{M}(g^2)(y) \right]^{\frac{1}{2}} + \kappa^{\frac{d}{2}} (\gamma + \rho)^{\frac{1}{2\nu}} \left[\mathcal{M}(|Du|^{2\mu})(y) \right]^{\frac{1}{2\mu}} + R_1^{-d} \kappa^d \|Du\|_{L_1(\Omega)}.$$

For F^Q, we consider two cases. When $\kappa r \leq R_1$, we choose $F^Q = \mathcal{U}$, where $\mathcal{U}$ is from Lemma 1.19. Thanks to (1.42), (1.43), and (1.59), we have that (1.57) holds with N_0 depending only on d, δ, R_0, K, and q. Otherwise, i.e., if $r > R_1/\kappa$ we take $F^Q = |Du|$. Then, by (1.59) we have

$$\fint_Q |F^Q(x) - (F^Q)_Q| \, dx \leq N \fint_{\Omega_r(x_0)} |Du| \, dx \leq N|\Omega_{R_1/\kappa}(x_0)|^{-1} \|Du\|_{L_1(\Omega)},$$

where $N = N(d, R_0, K)$. Since $|\Omega_{R_1/\kappa}(x_0)|^{-1} \leq NR_1^{-d}\kappa^d$, we still get that (1.57) holds with N_0 depending only on d, R_0, and K. Therefore, the conditions in Lemma 1.21 are satisfied. From (1.58), we obtain that

$$\|Du\|_{L_q(\Omega)} \leq N\|G\|_{L_q(\Omega)} + N\|Du\|_{L_1(\Omega)}$$
$$\leq NR_1^{-d}\kappa^d \|Du\|_{L_1(\Omega)} + N(\kappa^{-\frac{1}{2}} + \kappa\rho)\|\mathcal{M}(|Du|^2)^{\frac{1}{2}}\|_{L_q(\Omega)} + N\kappa^{\frac{d}{2}}\|\mathcal{M}(f_\alpha^2)^{\frac{1}{2}}\|_{L_q(\Omega)}$$
$$+ N\kappa^{\frac{d}{2}}\|\mathcal{M}(g^2)^{\frac{1}{2}}\|_{L_q(\Omega)} + N\kappa^{\frac{d}{2}}(\gamma + \rho)^{\frac{1}{2\nu}}\|\mathcal{M}(|Du|^{2\mu})^{\frac{1}{2\mu}}\|_{L_q(\Omega)}$$
$$= NR_1^{-d}\kappa^d \|Du\|_{L_1(\Omega)} + N(\kappa^{-\frac{1}{2}} + \kappa\rho)\|\mathcal{M}(|Du|^2)\|_{L_{q/2}(\Omega)}^{\frac{1}{2}} + N\kappa^{\frac{d}{2}}\|\mathcal{M}(f_\alpha^2)\|_{L_{q/2}(\Omega)}^{\frac{1}{2}}$$
$$+ N\kappa^{\frac{d}{2}}\|\mathcal{M}(g^2)\|_{L_{q/2}(\Omega)}^{\frac{1}{2}} + N\kappa^{\frac{d}{2}}(\gamma + \rho)^{\frac{1}{2\nu}}\|\mathcal{M}(|Du|^{2\mu})\|_{L_{q/2\mu}(\Omega)}^{\frac{1}{2\mu}},$$

where $N = N(d, \delta, R_0, K, q)$. By Lemma 1.20, the right-hand side above is bounded by

$$NR_1^{-d}\kappa^d \|Du\|_{L_1(\Omega)} + N(\kappa^{-\frac{1}{2}} + \kappa\rho)\||Du|^2\|_{L_{q/2}(\Omega)}^{\frac{1}{2}} + N\kappa^{\frac{d}{2}}\|f_\alpha^2\|_{L_{q/2}(\Omega)}^{\frac{1}{2}}$$
$$+ N\kappa^{\frac{d}{2}}\|g^2\|_{L_{q/2}(\Omega)}^{\frac{1}{2}} + N\kappa^{\frac{d}{2}}(\gamma + \rho)^{\frac{1}{2\nu}}\||Du|^{2\mu}\|_{L_{q/2\mu}(\Omega)}^{\frac{1}{2\mu}}$$
$$= NR_1^{-d}\kappa^d \|Du\|_{L_1(\Omega)} + N(\kappa^{-\frac{1}{2}} + \kappa\rho)\|Du\|_{L_q(\Omega)} + N\kappa^{\frac{d}{2}}\|f_\alpha\|_{L_q(\Omega)}$$
$$+ N\kappa^{\frac{d}{2}}\|g\|_{L_q(\Omega)} + N\kappa^{\frac{d}{2}}(\gamma + \rho)^{\frac{1}{2\nu}}\|Du\|_{L_q(\Omega)}.$$

Upon taking sufficiently large κ, then sufficiently small γ and ρ, depending on d, δ, R_0, K, and q (but independent of R_1), so that

$$N(\kappa^{-\frac{1}{2}} + \kappa\rho) + N\kappa^{\frac{d}{2}}(\gamma + \rho)^{\frac{1}{2\nu}} \leq 1/2,$$

we get that

$$\|Du\|_{L_q(\Omega)} \leq NR_1^{-d}\kappa^d \|Du\|_{L_1(\Omega)} + N\|f_\alpha\|_{L_q(\Omega)} + N\|g\|_{L_q(\Omega)}. \tag{1.60}$$

Since Ω is bounded, we have that $f_\alpha, g \in L_2(\Omega)$. Thus, by Lemma 1.5 (also see Remark 1.4) and Hölder's inequality,

$$\|Du\|_{L_2(\Omega)} + \|p\|_{L_2(\Omega)} \leq N \left(\|f_\alpha\|_{L_q(\Omega)} + \|g\|_{L_q(\Omega)} \right),$$

where $N = N(d, \delta, R_0, K, \rho, q)$. Combining this with (1.60) yields that

$$\|Du\|_{L_q(\Omega)} \leq N \left(\|f_\alpha\|_{L_q(\Omega)} + \|g\|_{L_q(\Omega)} \right), \tag{1.61}$$

where $N = N(d, \delta, R_0, R_1, K, q)$. Next, we estimate p. For any $\eta \in L_{q'}(\Omega)$ with $q' = q/(q-1)$, it follows from the solvability of the divergence equation in Lipschitz domains (cf. [5]) that there exists $\psi \in \mathring{W}_{q'}^1(\Omega)^d$ such that

$$\operatorname{div} \psi = \eta - (\eta)_\Omega \quad \text{in } \Omega \quad \text{and} \quad \|D\psi\|_{L_{q'}(\Omega)} \leq N \|\eta\|_{L_{q'}(\Omega)}, \tag{1.62}$$

where $N > 0$ is a constant depending only on d, R_0, K, and q'. We test the first equation of (1.39) by ψ, and using the fact that $(p)_\Omega = 0$ we obtain

$$\int_\Omega p\eta \, dx = \int_\Omega p(\eta - (\eta)_\Omega) \, dx = \int_\Omega D_\alpha \psi \cdot (f_\alpha - A^{\alpha\beta} D_\beta u) \, dx,$$

which combined with (1.61) and (1.62) yields

$$\left| \int_\Omega p\eta \, dx \right| \leq N \left(\|f_\alpha\|_{L_q(\Omega)} + \|g\|_{L_q(\Omega)} \right) \|\eta\|_{L_{q'}(\Omega)}.$$

Since $\eta \in L_{q'}(\Omega)$ is arbitrary, we can infer that

$$\|p\|_{L_q(\Omega)} \leq N \left(\|f_\alpha\|_{L_q(\Omega)} + \|g\|_{L_q(\Omega)} \right),$$

which together with (1.61) implies that (1.40) holds.

Case 2: $q \in (1, 2)$. We employ a duality argument. Let $q' = q/(q-1) \in (2, \infty)$ and $(\gamma, \rho) = (\gamma, \rho)(d, \delta, R_0, K, q')$ from Case 1. Then, for any $\eta = (\eta_\alpha)$, where $\eta_\alpha \in L_{q'}(\Omega)^d$ for $\alpha = 1, \ldots, d$, there exists a unique solution $(v, \pi) \in W_{q'}^1(\Omega)^d \times L_{q'}(\Omega)$ with $(\pi)_\Omega = 0$ satisfying

$$\begin{cases} D_\beta(A_{\operatorname{tr}}^{\alpha\beta} D_\alpha v) + \nabla \pi = D_\alpha \eta_\alpha & \text{in } \Omega, \\ \operatorname{div} v = 0 & \text{in } \Omega, \\ v = 0 & \text{on } \partial\Omega, \end{cases}$$

where $A_{\operatorname{tr}}^{\alpha\beta}$ is the transpose of the matrix $A^{\alpha\beta}$ for each $\alpha, \beta = 1, \ldots, d$. Moreover, we have that

$$\|Dv\|_{L_{q'}(\Omega)} + \|\pi\|_{L_{q'}(\Omega)} \leq N \|\eta_\alpha\|_{L_{q'}(\Omega)}, \tag{1.63}$$

where $N = N(d, \delta, R_0, R_1, K, q')$. Then, we test the equation of (v, π) by u, to obtain

$$\int_\Omega \eta_\alpha \cdot D_\alpha u \, dx = \int_\Omega \left(D_\beta u \cdot A_{\operatorname{tr}}^{\alpha\beta} D_\alpha v + \pi \operatorname{div} u \right) dx = \int_\Omega \left(f_\alpha \cdot D_\alpha v + \pi g \right) dx.$$

From this and (1.63), we get that

$$\left| \int_\Omega \eta_\alpha \cdot D_\alpha u \, dx \right| \le N \|\eta_\alpha\|_{L_{q'}(\Omega)} \left(\|f_\alpha\|_{L_q(\Omega)} + \|g\|_{L_q(\Omega)} \right).$$

Since $\eta \in \left(L_{q'}(\Omega) \right)^{d^2}$ is arbitrary, we obtain that (1.61) holds. The estimate of p is the same as in Case 1. Thus, the theorem is proved. $\square$

Remark 1.22. We refer the read to [13] for results about stationary Stokes systems in weighted Sobolev spaces and in more general domains.

2   Time-dependent Stokes systems

We now turn to non-stationary Stokes systems with singular measurable coefficients in divergence form:

$$u_t - D_i(a_{ij} D_j u) + \nabla p = \operatorname{div} f, \quad \operatorname{div} u = g, \tag{2.1}$$

where $a_{ij} = b_{ij}(t,x) + d_{ij}(t,x)$ is a given measurable matrix of viscosity coefficients, in which (b_{ij}) and (d_{ij}) are its symmetric and skew-symmetric parts, respectively. We assume that the matrix (a_{ij}) satisfies the following boundedness and ellipticity conditions with ellipticity constant $\nu \in (0,1)$:

$$\nu|\xi|^2 \le a_{ij}\xi_i\xi_j, \quad |b_{ij}| \le \nu^{-1}, \tag{2.2}$$

and

$$b_{ij} = b_{ji}, \quad d_{ij} \in L_{1,\mathrm{loc}}, \quad d_{ij} = -d_{ji}, \quad \forall\, i,j \in \{1,2,\ldots,d\}. \tag{2.3}$$

Note that in (2.2), the boundedness is not imposed on the skew-symmetric part $(d_{ij})_{i,j=1}^d$ of the coefficient matrix $(a_{ij})_{i,j=1}^d$. As a result, the viscosity coefficients in (2.1) can be singular. We also consider non-divergence form Stokes systems

$$u_t - a_{ij} D_{ij} u + \nabla p = f, \quad \operatorname{div} u = g, \tag{2.4}$$

and in this setting, the matrix $a_{ij} = b_{ij}$, i.e., $d_{ij} = 0$.

The interest in results concerning estimating solutions in mixed Sobolev norms arises, for example, when one wants to have better regularity of traces of solutions for each time slide while treating linear or nonlinear equations. See, for instance, [34, 43], where the initial-boundary value problem for the non-stationary Stokes system in mixed-norm Sobolev spaces was studied. Besides its mathematical interests, our motivation to study the Stokes systems (2.1) and (2.4) with variable coefficients comes from the study of inhomogeneous fluid with density dependent viscosity, see [1, 33]. Here we follow the approach developed in [18].

Let us introduce a few more notation. Let $Q_\rho(z_0)$ be the parabolic cylinder centered at $z_0 = (t_0, x_0) \in \mathbb{R}^{d+1}$ with radius $\rho > 0$:

$$Q_\rho(z_0) = (t_0 - \rho^2, t_0] \times B_\rho(x_0),$$

where $B_\rho(x_0)$ is the ball in $\mathbb{R}^d$ of radius ρ centered at $x_0 \in \mathbb{R}^d$. For abbreviation, when $z_0 = (0,0)$, we write $Q_\rho = Q_\rho(0,0)$ and $B_\rho = B_\rho(0)$.

For each integrable function f defined in a measurable set $Q \subset \mathbb{R}^{d+1}$, $(f)_Q$ is the average of f in Q, i.e.,

$$(f)_Q = \fint_Q f(t,x)\, dx\, dt.$$

For each $s, q \in [1, \infty)$ and each cylindrical domain $Q = \Gamma \times U \subset \mathbb{R} \times \mathbb{R}^d$, the mixed (s,q)-norm of a measurable function u defined in Q is

$$\|u\|_{L_{s,q}(Q)} = \left[\int_\Gamma \left(\int_U |u(t,x)|^q\, dx \right)^{s/q} dt \right]^{1/s}.$$

We define

$$L_{s,q}(Q) = \{\text{measurable } u : Q \to \mathbb{R} : \|u\|_{L_{s,q}(Q)} < \infty\} \quad \text{and} \quad L_q(Q) = L_{q,q}(Q).$$

We also define the parabolic Sobolev space

$$W^{1,2}_{s,q}(Q) = \{u : u, Du, D^2u \in L_{s,q}(Q),\ u_t \in L_1(Q)\},$$

and define

$$\mathbb{H}^{-1}_{s,q}(Q) = \{u = \operatorname{div} F + h \text{ in } Q : \|F\|_{L_{s,q}(Q)} + \|h\|_{L_{s,q}(Q)} < \infty\}.$$

Naturally, for any $u \in \mathbb{H}^{-1}_{s,q}(Q)$, we define the norm

$$\|u\|_{\mathbb{H}^{-1}_{s,q}(Q)} = \inf\{\|F\|_{L_{s,q}(Q)} + \|h\|_{L_{s,q}(Q)} \mid u = \operatorname{div} F + h\},$$

and it is easy to see that $\mathbb{H}^{-1}_{s,q}(Q)$ is a Banach space. Moreover, when $u \in \mathbb{H}^{-1}_{s,q}(Q)$ and $u = \operatorname{div} F + h$, we write

$$\langle u, \phi \rangle = \int_Q \left[-F \cdot \nabla \phi + h\phi \right] dx\, dt, \quad \text{for any} \quad \phi \in C^\infty_0(Q).$$

We also define

$$\mathcal{H}^1_{s,q}(Q) = \{u : u, Du \in L_{s,q}(Q), u_t \in \mathbb{H}^{-1}_{1,1}(Q)\}.$$

When $s = q$, we omit one of these two indices and write

$$L_q(Q) = L_{q,q}(Q), \quad W^{1,2}_q(Q) = W^{1,2}_{q,q}(Q),$$

$$\mathcal{H}^1_q(Q) = \mathcal{H}^1_{q,q}(Q), \quad \mathbb{H}^{-1}_q(Q) = \mathbb{H}^{-1}_{q,q}(Q).$$

Remark 2.1. In our definition of $W^{1,2}_{s,q}(Q)$ we only require that $u_t \in L_1(Q)$, not $u_t \in L_{s,q}(Q)$ as in the standard notation for the space $W^{1,2}_{s,q}(Q)$. Similarly in the definition of $\mathcal{H}^1_{s,q}(Q)$, we only require $u_t \in \mathbb{H}^{-1}_{1,1}(Q)$, not $u_t \in \mathbb{H}^{-1}_{s,q}(Q)$. This is because for the Stokes systems, local weak solutions may not possess good regularity in the time variable in view of Serrin's example [38]. It is possible to further relax the regularity assumptions of u in t and also p below, but we do not pursue in that direction.

For $s, q \in (1, \infty)$, we denote s' and q' to be the conjugates of s and q, i.e.,

$$1/s + 1/s' = 1, \quad 1/q + 1/q' = 1. \tag{2.5}$$

Then, under the assumption that $d_{kj} \in L_{s',q'}(Q_1)$ and $f_{kj} \in L_1(Q_1)$ for all $i, j = 1, 2, \ldots, d$, $g \in L_1(Q_1)$, we say that a vector field function $u = (u_1, u_2, \ldots, u_d) \in \mathcal{H}^1_{s,q}(Q_1)^d$ is a weak solution of the Stokes system (2.1) in Q_1 if for a.e. $t \in (-1, 0)$ and any $\varphi \in C_0^\infty(B_1)$,

$$\int_{B_1} u(t, x) \cdot \nabla \varphi(x) \, dx = - \int_{B_1} g(t, x) \varphi(x) \, dx$$

and

$$\langle \partial_t u_k, \phi_k \rangle + \int_{Q_1} a_{ij}(t, x) D_j u_k D_i \phi_k \, dx \, dt = - \int_{Q_1} f_{kj} D_j \phi_k \, dt dx,$$

for any $k = 1, 2, \ldots, d$ and $\phi = (\phi_1, \phi_2, \ldots, \phi_d) \in C_0^\infty(Q_1)^d$ such that $\mathrm{div}[\phi(t, \cdot)] = 0$ for $t \in (-1, 0)$. On the other hand, a vector field $u \in W^{1,2}_{1,1}(Q_1)^d$ is said to be a strong solution of (2.4) on Q_1 if (2.4) holds for a.e. $(t, x) \in Q_1$ for some $p \in L_1(Q_1)$ with $\nabla p \in L_1(Q_1)^d$.

In addition to the ellipticity condition (2.2), we need the following VMO$_x$ (vanishing mean oscillation in x) condition, first introduced in [28], with constants $\delta \in (0, 1)$ and $\alpha_0 \in [1, \infty)$ to be determined later. We note that such condition is weaker than the usual VMO condition in both t and x.

Assumption 2.2 (δ, α_0). There exists $R_0 \in (0, 1/4)$ such that for any $(t_0, x_0) \in Q_{2/3}$ and $r \in (0, R_0)$, there exists $\bar{a}_{ij}(t) = \bar{b}_{ij}(t) + \bar{d}_{ij}(t)$ for which $\bar{b}_{ij}(t)$ and $\bar{d}_{ij}(t)$ satisfy (2.2)–(2.3) and

$$\fint_{Q_r(t_0, x_0)} |a_{ij}(t, x) - \bar{a}_{ij}(t)|^{\alpha_0} \, dx \, dt \le \delta^{\alpha_0},$$

where $\delta \in (0, 1)$ and $\alpha_0 \in [1, \infty)$.

We are ready to state the main results for time-dependent Stokes systems. Our first theorem is about the $L_{s,q}$-estimate for gradients of weak solutions to (2.1).

Theorem 2.3. Let $s, q \in (1, \infty)$, $\nu \in (0, 1)$, $\alpha_0 \in (\min(s, q)/(\min(s, q) - 1), \infty)$, and $d_{ij} \in L_{s',q'}(Q_1)$ with s', q' being as in (2.5) and $i, j = 1, 2, \ldots, d$. There exists $\delta = \delta(d, \nu, s, q, \alpha_0) \in (0, 1)$ such that the following statement holds. Assume that (2.2)–(2.3) and Assumption 2.2 (δ, α_0) hold. Then, if $(u, p) \in \mathcal{H}^1_{s,q}(Q_1)^d \times L_1(Q_1)$ is a weak solution to (2.1) in Q_1, $f \in L_{s,q}(Q_1)^{d \times d}$, and $g \in L_{s,q}(Q_1)$, it holds that

$$\|Du\|_{L_{s,q}(Q_{1/2})} \le N(d, \nu, s, q, \alpha_0) \Big[\|f\|_{L_{s,q}(Q_1)} + \|g\|_{L_{s,q}(Q_1)} \Big] \tag{2.6}$$
$$+ N(d, \nu, s, q, R_0, \alpha_0) \|u\|_{L_{s,q}(Q_1)}.$$

For the non-divergence form Stokes system (2.4), we have the following mixed-norm estimates for the Hessian of solutions.

Theorem 2.4. *Let* $s, q \in (1, \infty)$ *and* $\nu \in (0, 1)$. *There exists* $\delta = \delta(d, \nu, s, q) \in$ $(0, 1)$ *such that the following statement holds. Suppose that* $d_{ij} = 0$, *the ellipticity condition* (2.2) *and Assumption 2.2* $(\delta, 1)$ *hold. Then, if* $u \in W_{s,q}^{1,2}(Q_1)^d$ *is a strong solution to* (2.1) *in* Q_1, $f \in L_{s,q}(Q_1)^{d \times d}$, *and* $Dg \in L_{s,q}(Q_1)^d$, *then it follows that*

$$
\|D^2 u\|_{L_{s,q}(Q_{1/2})} \le N(d, \nu, q) \Big[\|f\|_{L_{s,q}(Q_1)} + \|Dg\|_{L_{s,q}(Q_1)} \Big]
$$
$$
+ N(d, \nu, q, R_0) \|u\|_{L_{s,q}(Q_1)}.
\tag{2.7}
$$

Remark 2.5. By using interpolation and a standard iteration argument, (2.6) and (2.7) still hold if we replace the term $\|u\|_{L_{s,q}(Q_1)}$ on the right-hand sides with $\|u\|_{L_{s,1}(Q_1)}$.

Several remarks regarding our Theorems 2.3 and 2.4 are in order. The estimates in Theorems 2.3 and 2.4 do not contain any pressure term on the right-hand sides, which seem to be new even when the coefficients are constants. One can easily see that the estimates in Theorems 2.3 and 2.4 imply the available regularity estimates such as [39, Proposition 6.7, p. 84] in which the regularity for the pressure p is required. Even when $q = s = 2$ and $g \equiv 0$, the estimates (2.6) and (2.7) are already new for the non-stationary Stokes system with variable coefficients. These estimates are known as Caccioppoli type inequalities. When $a_{ij} = \delta_{ij}$, $f \equiv 0$, and $g \equiv 0$, Caccioppoli type inequalities for Stokes system were established in [26] by using special test functions. However, it is not so clear that this method can be extended to systems with variable coefficients and nonzero right-hand side. We also refer the reader to [45, 9] for Caccioppoli inequalities without the pressure term on the right-hand side for the Navier-Stokes equations.

Sobolev estimates for non-stationary Stokes system with constant coefficients were established in [42] many years ago, and recently in [24] with a different approach.

Finally, we mention that the smallness Assumption 2.2 (δ, α_0) is necessary for both Theorem 2.3 and Theorem 2.4. See an example in the well-known paper [35] for linear elliptic equations in which $d_{ij} = 0$. Our $L_{s,q}$-estimates for the Stokes system have applications to the Navier-Stokes equations. We refer the reader to [18] for some examples.

2.1 Preliminaries

The following result is a special case of [12, Theorem 2.3 (i)]. Let $\mathcal{X} \subset \mathbb{R}^{d+1}$ be a space of homogeneous type, which is endowed with the parabolic distance and a doubling measure μ that is naturally inherited from the Lebesgue measure. As in the previous section, we take a filtration of partitions of $\mathcal{X}$ and for any $f \in L_{1,\text{loc}}$ we define its dyadic sharp function $f_{\text{dy}}^{\#}$ in $\mathcal{X}$ associated with the filtration of partitions. Also for each $q \in [1, \infty]$, A_q is the Muckenhoupt class of weights.

Theorem 2.6. *Let* $s, q \in (1, \infty)$, $K_0 \ge 1$ *and* $\omega \in A_q$ *with* $[\omega]_{A_q} \le K_0$. *Suppose that* $f \in L_s(\omega d\mu)$. *Then,*

$$
\|f\|_{L_s(\omega d\mu)} \le N \Big[\|f_{\text{dy}}^{\#}\|_{L_s(\omega d\mu)} + \mu(\mathcal{X})^{-1} \omega(\text{supp}(f))^{\frac{1}{s}} \|f\|_{L_1(\mu)} \Big],
$$

where $N > 0$ is a constant depending only on s, q, K_0, and the doubling constant of μ.

The following lemma is a direct corollary of Theorem 2.6.

Lemma 2.7. *For any $s, q \in (1, \infty)$, there exists a constant $N = N(d, s, q) > 0$ such that*

$$\|f\|_{L_{s,q}(Q_R)} \leq N\left[\|f^\#_{\mathrm{dy}}\|_{L_{s,q}(Q_R)} + R^{\frac{2}{s}+\frac{d}{q}-d-2}\|f\|_{L_1(Q_R)}\right]$$

for any $R > 0$ and $f \in L_{s,q}(Q_R)$.

Proof. For $t \in (-R^2, 0)$, let

$$\psi(t) = \|f(t,\cdot)\|_{L_q(B_R)} \quad \text{and} \quad \phi(t) = \|f^\#_{\mathrm{dy}}(t,\cdot) + (|f|)_{Q_R}\|_{L_q(B_R)}.$$

Moreover, for any $\omega \in A_q((-R^2, 0))$ with $[\omega]_{A_q} \leq K_0$, we write $\tilde{\omega}(t, x) = \omega(t)$ for all $(t, x) \in Q_R$. Then, by applying Theorem 2.6 with $\mathcal{X} = Q_R$, we obtain

$$\|\psi\|_{L_q((-R^2,0),\omega)} = \|f\|_{L_q(Q_R,\tilde{\omega})} \leq N\|f^\#_{\mathrm{dy}} + (|f|)_{Q_R}\|_{L_q(Q_R,\tilde{\omega})} = N\|\phi\|_{L_q((-R^2,0),\omega)},$$

with $N = N(d, K_0, s)$. Then, by the extrapolation theorem (see, for instance, [12, Theorem 2.5]), we see that

$$\|\psi\|_{L_s((-R^2,0),\omega)} \leq 4N\|\phi\|_{L_s((-R^2,0),\omega)}, \quad \forall\, \omega \in A_s, \quad [\omega]_{A_s} \leq K_0.$$

Note that in the special case when $\omega \equiv 1$, $\|\psi\|_{L_s((-R^2,0),\omega)} = \|f\|_{L_{s,q}(Q_R)}$ and

$$\|\phi\|_{L_s((-R^2,0),\omega)} \leq \|f^\#_{\mathrm{dy}}\|_{L_{s,q}(Q_R)} + R^{2/s+d/q}(|f|)_{Q_R}.$$

Therefore, the desired estimate follows. $\qquad\square$

2.2 Stokes systems with simple coefficients

We first consider the time dependent Stokes system with coefficients that only depend on the time variable

$$u_t - D_i(a_{ij}(t)D_j u) + \nabla p = 0, \quad \operatorname{div} u = 0, \tag{2.8}$$

where $a_{ij} = b_{ij}(t) + d_{ij}(t)$ with $b_{ij} = b_{ji}$ and $d_{ij} = -d_{ji}$ for all $i, j = \{1, 2, \ldots, d\}$. Moreover, a_{ij} satisfies the ellipticity condition with ellipticity constant $\nu \in (0, 1)$: for any $\xi \in \mathbb{R}^d$,

$$\nu|\xi|^2 \leq b_{ij}\xi_i\xi_j, \quad |b_{ij}| \leq \nu^{-1}. \tag{2.9}$$

We have the following gradient estimate.

Lemma 2.8. *Assume that (2.9) holds. Let $q_0 \in (1, \infty)$, and $(u, p) \in \mathcal{H}^1_{q_0}(Q_1)^d \times L_1(Q_1)$ be a weak solution to (2.8) in Q_1. Then we have*

$$\|D^2 u\|_{L_{q_0}(Q_{1/2})} + \|Du\|_{L_{q_0}(Q_{1/2})} \leq N(d, \nu, q_0)\|u - [u]_{B_1}(t)\|_{L_{q_0}(Q_1)}, \tag{2.10}$$

where $[u]_{B_1}(t)$ is the average of $u(t, \cdot)$ in B_1.

Proof. By a mollification in x, we see that $\omega = \nabla \times u$ is a weak solution to the parabolic equation

$$\omega_t - D_i(a_{ij}(t)D_j\omega) = 0 \quad \text{in} \quad Q_1.$$

Observe that since the matrix $(d_{ij}(t))_{d \times d}$ is skew-symmetric, ω is indeed a weak solution of

$$\omega_t - D_i(b_{ij}(t)D_j\omega) = 0 \quad \text{in} \quad Q_1.$$

Since the matrix $(b_{ij})_{d \times d}$ satisfies the ellipticity condition as in (2.9), we can apply the local $\mathcal{H}_p^1$ estimate for linear parabolic equations with coefficients measurable in t (cf. [28, 30]) to obtain

$$\|D\omega\|_{L_{q_0}(Q_{2/3})} \leq N(d, \nu, q_0)\|\omega\|_{L_{q_0}(Q_{3/4})}. \tag{2.11}$$

Since u is divergence free, we have

$$\Delta u_i = -D_i \sum_{k=1}^{d} D_k u_k + \sum_{k=1}^{d} D_{kk} u_i = \sum_{k \neq i} D_k(D_k u_i - D_i u_k).$$

Thus by the local W_p^1 estimate for the Laplace operator,

$$\|Du\|_{L_{q_0}(Q_{1/2})} \leq N\|\omega\|_{L_{q_0}(Q_{2/3})} + N\|u\|_{L_{q_0}(Q_{2/3})}.$$

Similarly,

$$\|D^2 u\|_{L_{q_0}(Q_{1/2})} \leq N\|D\omega\|_{L_{q_0}(Q_{2/3})} + N\|Du\|_{L_{q_0}(Q_{2/3})} \leq N\|Du\|_{L_{q_0}(Q_{3/4})}$$
$$\leq \varepsilon\|D^2 u\|_{L_{q_0}(Q_{3/4})} + N\varepsilon^{-1}\|u - [u]_{B_1}(t)\|_{L_{q_0}(Q_{3/4})}$$

for any $\varepsilon \in (0, 1)$, where we used (2.11) in the second inequality, and multiplicative inequalities in the last inequality. It then follows from a standard iteration argument that

$$\|D^2 u\|_{L_{q_0}(Q_{1/2})} \leq N\|u - [u]_{B_1}(t)\|_{L_{q_0}(Q_1)},$$

from which and multiplicative inequalities we obtain (2.10). The lemma is proved. $\square$

Recall that for each $\alpha \in (0, 1]$, and each parabolic cylinder $Q \subset \mathbb{R}^{d+1}$, we write

$$[[u]]_{C^{\alpha/2,\alpha}(Q)} = \sup_{\substack{(t,x),(s,y) \in Q \\ (t,x) \neq (s,y)}} \frac{|u(t,x) - u(s,y)|}{|t - s|^{\alpha/2} + |x - y|^{\alpha}},$$

and

$$\|u\|_{C^{\alpha/2,\alpha}(Q)} = \|u\|_{L_\infty(Q)} + [[u]]_{C^{\alpha/2,\alpha}(Q)}.$$

Lemma 2.9. *Under the assumptions of Lemma 2.8, we have*

$$\|\omega\|_{C^{1/2,1}(Q_{1/2})} \leq N(d, \nu, q_0)\|\omega\|_{L_{q_0}(Q_1)},$$

where $\omega = \nabla \times u$.

Proof. The lemma follows by using mollifications in x, the interior estimate for parabolic equations with coefficients measurable in t (see [28, 30]), a bootstrap argument, and the parabolic embedding inequalities. $\square$

2.3 Divergence form Stokes systems

We need to establish several lemmas in order to prove Theorem 2.3. Our first lemma gives the control of $(|Du|^{q_0})^{1/q_0}_{Q_{r/2}}$ for weak solution u of the Stokes system (2.1).

Lemma 2.10. *Let* $\delta, \nu \in (0,1)$, $q_0 \in (1,\infty)$, $q \in (q_0,\infty)$. *Suppose that* (2.2)–(2.3) *hold, and Assumption 2.2* (δ, α_0) *holds with* $\alpha_0 \geq \frac{q_0 q}{q-q_0}$. *Then, for any* $r \in (0, R_0)$ *and weak solution* $(u,p) \in \mathcal{H}^1_{s,q}(Q_r)^d \times L_1(Q_r)$ *of* (2.1) *in* Q_r, *we have*

$$(|Du|^{q_0})^{1/q_0}_{Q_{r/2}} \leq N(d,\nu,q_0)\Big((|f|^{q_0})^{1/q_0}_{Q_r} + r^{-1}(|u - [u]_{B_r}(t)|^{q_0})^{1/q_0}_{Q_r}\Big)$$
$$+ N(d,\nu,q_0)\delta(|Du|^q)^{1/q}_{Q_r} + N(d,q_0)(|g|^{q_0})^{1/q_0}_{Q_r}.$$

Proof. Let (w, p_1) be a weak solution to

$$w_t - D_i(\bar{a}_{ij}(t)D_j w) + \nabla p_1 = \text{div}(I_{Q_r}f) + D_i(I_{Q_r}(a_{ij} - \bar{a}_{ij})D_j u), \quad \text{div } w = I_{Q_r}g$$

in $(-r^2, 0) \times \mathbb{R}^d$ with the zero initial condition on $\{t = -r^2\}$. Then $\nabla \times w$ is a so-called *adjoint solution* to the parabolic equation. By duality, we get

$$\|\nabla \times w\|_{L_{q_0}((-r^2,0)\times\mathbb{R}^d)} \leq N(d,\nu,q_0)\Big[\|f\|_{L_{q_0}(Q_r)} + \|(a_{ij} - \bar{a}_{ij})D_j u\|_{L_{q_0}(Q_r)}\Big]. \quad (2.12)$$

Then, from this and the equation $\text{div } w = I_{Q_r}g$, it follows that

$$\|Dw\|_{L_{q_0}((-r^2,0)\times\mathbb{R}^d)} \leq N(d,q_0)\Big[\|\nabla \times w\|_{L_{q_0}((-r^2,0)\times\mathbb{R}^d)} + \|I_{Q_r}g\|_{L_{q_0}((-r^2,0)\times\mathbb{R}^d)}\Big]$$
$$\leq N(d,\nu,q_0)\|f\|_{L_{q_0}(Q_r)} + N(d,\nu,q_0)\|(a_{ij} - \bar{a}_{ij})D_j u\|_{L_{q_0}(Q_r)} + N(d,q_0)\|g\|_{L_{q_0}(Q_r)}.$$

Thus, we have

$$(|Dw|^{q_0})^{1/q_0}_{Q_r}$$
$$\leq N(d,\nu,q_0)\Big[(|f|^{q_0})^{1/q_0}_{Q_r} + (|(a_{ij} - \bar{a}_{ij})D_j u|^{q_0})^{1/q_0}_{Q_r}\Big] + N(d,q_0)(|g|^{q_0})^{1/q_0}_{Q_r}$$
$$\leq N(d,\nu,q_0)\Big[(|f|^{q_0})^{1/q_0}_{Q_r} + \delta(|Du|^q)^{1/q}_{Q_r}\Big] + N(d,q_0)(|g|^{q_0})^{1/q_0}_{Q_r}, \quad (2.13)$$

where we used Assumption 2.2 with $\alpha_0 \geq \frac{q_0 q}{q-q_0}$ and Hölder's inequality for the middle term on the right-hand side in the last inequality. Now $(v, p_2) := (u - w, p - p_1)$ is a weak solution of

$$v_t - D_i(\bar{a}_{ij}(t)D_j v) + \nabla p_2 = 0, \quad \text{div } v = 0$$

in Q_r. By Lemma 2.8 with a scaling, we have

$$(|Dv|^{q_0})^{1/q_0}_{Q_{r/2}} \leq r^{-1}(|v - [v]_{B_r}(t)|^{q_0})^{1/q_0}_{Q_r}. \quad (2.14)$$

By (2.13), (2.14), the triangle inequality, and the Poincaré inequality, we get the desired inequality. $\qquad \square$

For a domain $\Omega \subset \mathbb{R}^d$ and $\rho > 0$, we denote

$$\Omega^\rho = \bigcup_{y \in \Omega} B_\rho(y).$$

We say that Ω satisfies the interior measure condition if there exists $\gamma \in (0,1)$ such that for any $x_0 \in \overline{\Omega}$ and $r \in (0, \text{diam}\,\Omega)$,

$$\frac{|B_r(x_0) \cap \Omega|}{|B_r(x_0)|} \geq \gamma. \tag{2.15}$$

Corollary 2.11. *Let* $\delta, \nu \in (0,1)$, $q_0 \in (1,\infty)$, $q \in (q_0, \infty)$, $r \in (0, R_0)$, $T > 0$, *and* $\Omega \subset \mathbb{R}^d$ *satisfy* (2.15) *for some* $\gamma > 0$ *and* $(-T, 0) \times \Omega \subset Q_{2/3}$. *Suppose that* (2.2)–(2.3) *hold, and Assumption 2.2* (δ, α_0) *holds with* $\alpha_0 \geq \frac{q_0 q}{q - q_0}$. *Then, for any weak solution* $(u,p) \in \mathcal{H}_q^1((-T - r^2, 0) \times \Omega^r)^d \times L_1((-T - r^2, 0) \times \Omega^r)$ *of* (2.1) *in* $(-T - r^2, 0) \times \Omega^r$, *we have*

$$(|Du|^{q_0})_{(-T,0) \times \Omega}^{\frac{1}{q_0}}$$

$$\leq N(d, \nu, q_0, \gamma) \frac{((T + r^2)|\Omega^r|)^{\frac{1}{q_0}}}{(T|\Omega|)^{\frac{1}{q_0}}} \left[(|f|^{q_0})_{(-T - r^2, 0) \times \Omega^r}^{\frac{1}{q_0}} + r^{-1}(|u|^{q_0})_{(-T - r^2, 0) \times \Omega^r}^{\frac{1}{q_0}} \right.$$

$$\left. + (|g|^{q_0})_{(-T - r^2, 0) \times \Omega^r}^{\frac{1}{q_0}} \right] + N(d, \nu, q_0, q, \gamma) \frac{((T + r^2)|\Omega^r|)^{\frac{1}{q}}}{(T|\Omega|)^{\frac{1}{q}}} \delta (|Du|^q)_{(-T - r^2, 0) \times \Omega^r}^{\frac{1}{q}}.$$

$$\tag{2.16}$$

Proof. We use a partition of unity argument. By Lemma 2.10, for any $x_0 \in \Omega$ and $t_0 \in (-T, 0)$, we have

$$(|Du|^{q_0})_{Q_{r/2}(t_0, x_0)} \leq N(d, \nu, q_0) \left[(|f|^{q_0})_{Q_r(t_0, x_0)} + r^{-q_0}(|u|^{q_0})_{Q_r(t_0, x_0)} \right]$$

$$+ N(d, \nu, q_0) \delta^{q_0} (|Du|^q)_{Q_r(t_0, x_0)}^{q_0/q} + N(d, q_0)(|g|^{q_0})_{Q_r(t_0, x_0)}.$$

Now to obtain (2.16), it suffices to integrate both sides of the above inequality with respect to $(t_0, x_0) \in (-T, 0) \times \Omega$ and use Hölder's inequality and the interior measure condition (2.15). $\qquad\square$

In the next lemma we prove a mean oscillation estimate of $\nabla \times u$.

Lemma 2.12. *Let* $q_1 \in (1, \infty)$, $q_0 \in (1, q_1)$, $\delta \in (0,1)$, $r \in (0, R_0)$, *and* $\kappa \in (0, 1/2)$. *Assume that* (2.2)–(2.3) *hold, and Assumption 2.2* (δ, α_0) *holds with* $\alpha_0 \geq \frac{q_0 q_1}{q_1 - q_0}$. *Suppose that* $(u,p) \in \mathcal{H}_{s_1, q_1}^1(Q_r)^d \times L_1(Q_r)$ *is a weak solution to* (2.1) *in* Q_r. *Then it holds that*

$$(|\omega - (\omega)_{Q_{\kappa r}}|)_{Q_{\kappa r}} \leq N(d, \nu, q_0) \kappa^{-\frac{d+2}{q_0}} (|f|^{q_0})_{Q_r}^{1/q_0}$$

$$+ N(n, \nu, q_0, q_1) \left(\kappa^{-\frac{d+2}{q_0}} \delta + \kappa \right) (|Du|^{q_1})_{Q_r}^{1/q_1},$$

where $\omega = \nabla \times u$.

Proof. Let (w, p_1) and (v, p_2) be as in the proof of Lemma 2.10. In particular, (w, p_1) is a weak solution of

$$w_t - D_i(\bar{a}_{ij}(t)D_j w) + \nabla p_1 = \mathrm{div}[I_{Q_r} f] + D_i(I_{Q_r}(a_{ij} - \bar{a}_{ij}(t))D_j u), \quad \mathrm{div}\, w = I_{Q_r} g$$

in $(-r^2, 0) \times \mathbb{R}^d$ with zero initial condition on $\{t = -r^2\}$. Also, $(v, p_2) = (u - w, p - p_1)$ is a weak solution of

$$v_t - D_i(\bar{a}_{ij}(t)D_j v) + \nabla p_2 = 0, \quad \mathrm{div}\, v = 0$$

in Q_r. Let $\omega_1 = \nabla \times w$ and $\omega_2 = \nabla \times v$. Observe that $\omega = \omega_1 + \omega_2$. Moreover, from (2.12),

$$(|\omega_1|^{q_0})_{Q_r}^{1/q_0} \le N(d, \nu, q_0)\left[(|f|^{q_0})_{Q_r}^{1/q_0} + \delta(|Du|^{q_1})_{Q_r}^{1/q_1}\right]. \tag{2.17}$$

On the other hand, by applying Lemma 2.9 to ω_2 with suitable scaling, we obtain

$$(|\omega_2 - (\omega_2)_{Q_{\kappa r}}|)_{Q_{\kappa r}} \le N\kappa r[[\omega_2]]_{C^{1/2,1}(Q_{r/2})} \le N(d, \nu, q_0)\kappa(|\omega_2|^{q_0})_{Q_r}^{1/q_0}$$
$$\le N(d, \nu, q_0)\kappa\left[(|\omega|^{q_0})_{Q_r}^{1/q_0} + (|\omega_1|^{q_0})_{Q_r}^{1/q_0}\right].$$

We then combine the last estimate with (2.17) and the fact that $\delta \in (0, 1)$ to deduce that

$$(|\omega_2 - (\omega_2)_{Q_{\kappa r}}|)_{Q_{\kappa r}} \le N(d, \nu, q_0)\kappa\left[(|f|^{q_0})_{Q_r}^{1/q_0} + (|Du|^{q_1})_{Q_r}^{1/q_1}\right]. \tag{2.18}$$

Moreover, by using the inequality

$$\fint_{Q_{\kappa r}} |\omega - (\omega)_{Q_{\kappa r}}|\, dx\, dt \le 2 \fint_{Q_{\kappa r}} |\omega - c|\, dx\, dt$$

with $c = (\omega_2)_{Q_{\kappa r}}$, and then applying the triangle inequality and Hölder's inequality, we have

$$\fint_{Q_{\kappa r}} |\omega - (\omega)_{Q_{\kappa r}}|\, dx\, dt \le 2 \fint_{Q_{\kappa r}} |\omega - (\omega_2)_{Q_{\kappa r}}|\, dx\, dt$$
$$\le 2 \fint_{Q_{\kappa r}} |\omega_2 - (\omega_2)_{Q_{\kappa r}}|\, dx\, dt + N(d, q_0)\kappa^{-\frac{d+2}{q_0}}\left(\fint_{Q_r} |\omega_1|^{q_0}\, dx\, dt\right)^{1/q_0}.$$

This last estimate together with (2.17) and (2.18) gives that

$$(|\omega - (\omega)_{Q_{\kappa r}}|)_{Q_{\kappa r}} \le N(d, \nu, q_0)\left(\kappa^{-\frac{d+2}{q_0}} + \kappa\right)(|f|^{q_0})_{Q_r}^{1/q_0}$$
$$+ N(d, \nu, q_0)\left(\kappa^{-\frac{d+2}{q_0}}\delta + \kappa\right)(|Du|^{q_1})_{Q_r}^{1/q_1},$$

which implies our desired estimate as $\kappa \in (0, 1/2)$. □

Our next lemma gives the key estimates of vorticity $\omega = \nabla \times u$ and Du in the mixed norm.

Lemma 2.13. *Let $R \in [1/2, 2/3]$, $R_1 \in (0, R_0)$, $\delta \in (0,1)$, $\kappa \in (0, 1/2)$, $s, q \in (1, \infty)$, and*

$$\alpha_0 \in (\min(s,q)/(\min(s,q) - 1), \infty).$$

Let $q_1 \in (1, \min(s,q))$ and $q_0 \in (1, q_1)$ such that $\alpha_0 \geq q_0 q_1 / (q_1 - q_0)$. Assume that (2.2)–(2.3) hold and Assumption 2.2 (δ, α_0) is satisfied. Suppose that $(u, p) \in \mathcal{H}^1_{s,q}(Q_{R+R_1})^d \times L_1(Q_{R+R_1})$ is a weak solution to (2.1) in Q_{R+R_1}, and $\omega = \nabla \times u$. Then we have

$$\|\omega\|_{L_{s,q}(Q_R)} \leq N\kappa^{-\frac{d+2}{q_0}} \|f\|_{L_{s,q}(Q_{R+R_1/2})} + N\kappa^{-\frac{d+2}{q_0}} \|g\|_{L_{s,q}(Q_{R+R_1/2})}$$
$$+ N\left(\kappa^{-\frac{d+2}{q_0}} \delta + \kappa\right) \|Du\|_{L_{s,q}(Q_{R+R_1/2})}$$
$$+ NR_1^{-1} \kappa^{-\frac{d+2}{q_0}} \|u\|_{L_{s,q}(Q_{R+R_1/2})}, \qquad (2.19)$$

and

$$\|Du\|_{L_{s,q}(Q_R)} \leq N\kappa^{-\frac{d+2}{q_0}} \|f\|_{L_{s,q}(Q_{R+R_1})} + N\kappa^{-\frac{d+2}{q_0}} \|g\|_{L_{s,q}(Q_{R+R_1})}$$
$$+ N\left(\kappa^{-\frac{d+2}{q_0}} \delta + \kappa\right) \|Du\|_{L_{s,q}(Q_{R+R_1})}$$
$$+ NR_1^{-1} \kappa^{-\frac{d+2}{q_0}} \|u\|_{L_{s,q}(Q_{R+R_1})}, \qquad (2.20)$$

where $N = N(d, \nu, s, q, q_0, q_1)$.

Proof. We consider two cases.
Case 1: $r \in (0, R_1/2)$. It follows from Lemma 2.12 that for all $z_0 \in Q_R$,

$$(|\omega - (\omega)_{Q_{\kappa r}(z_0)}|)_{Q_{\kappa r}(z_0)}$$
$$\leq N(d, \nu, q_0) \kappa^{-\frac{d+2}{q_0}} (|f|^{q_0})^{1/q_0}_{Q_r(z_0)} + N(d, \nu, q_0) \left(\kappa^{-\frac{d+2}{q_0}} \delta + \kappa\right) (|Du|^{q_1})^{1/q_1}_{Q_r(z_0)}.$$

Observe that because $r < R_1/2$, we have $Q_r(z_0) \subset Q_{R+R_1/2}$. Therefore,

$$(|f|^{q_0})^{1/q_0}_{Q_r(z_0)} \leq \mathcal{M}(I_{Q_{R+R_1/2}} |f|^{q_0})^{1/q_0}(z_0),$$
$$(|Du|^{q_1})^{1/q_1}_{Q_r(z_0)} \leq \mathcal{M}(I_{Q_{R+R_1/2}} |Du|^{q_1})^{1/q_1}(z_0),$$

which imply that

$$(|\omega - (\omega)_{Q_{\kappa r}(z_0)}|)_{Q_{\kappa r}(z_0)} \leq N\kappa^{-\frac{d+2}{q_0}} \mathcal{M}(I_{Q_{R+R_1/2}} |f|^{q_0})^{1/q_0}(z_0)$$
$$+ N\left(\kappa^{-\frac{d+2}{q_0}} \delta + \kappa\right) \mathcal{M}(I_{Q_{R+R_1/2}} |Du|^{q_1})^{1/q_1}(z_0).$$

Case 2: $r \in [R_1/2, 2R/\kappa)$ and $z_0 \in Q_R$ such that $t_0 \in [-R^2 + (\kappa r)^2/2, 0]$. In this case, we apply Corollary 2.11 to get

$$(|\omega - (\omega)_{Q_{\kappa r}(z_0) \cap Q_R}|)_{Q_{\kappa r}(z_0) \cap Q_R} \leq 2(|\omega|)_{Q_{\kappa r}(z_0) \cap Q_R} \leq 2(|\omega|^{q_0})^{\frac{1}{q_0}}_{Q_{\kappa r}(z_0) \cap Q_R}$$
$$\leq N\kappa^{-\frac{d+2}{q_0}} \left[(|f|^{q_0})^{\frac{1}{q_0}}_{Q_{\kappa r+R_1/2}(z_0) \cap Q_{R+R_1/2}} + R_1^{-1} (|u|^{q_0})^{\frac{1}{q_0}}_{Q_{\kappa r+R_1/2}(z_0) \cap Q_{R+R_1/2}} \right.$$
$$\left. + (|g|^{q_0})^{\frac{1}{q_0}}_{Q_{\kappa r+R_1/2}(z_0) \cap Q_{R+R_1/2}} \right] + N\kappa^{-\frac{d+2}{q_1}} \delta(|Du|^{q_1})^{\frac{1}{q_1}}_{Q_{\kappa r+R_1/2}(z_0) \cap Q_{R+R_1/2}},$$
$$(2.21)$$

where we used $R_1/2 \leq r$ in the last inequality.

Now we take $\mathcal{X} = Q_R$ and define the dyadic sharp function $\omega_{\mathrm{dy}}^{\#}$ of ω in $\mathcal{X}$. From the above two cases, we conclude that for any $z_0 \in \mathcal{X}$,

$$\omega_{\mathrm{dy}}^{\#}(z_0) \leq N(d,\nu,q_0)\kappa^{-\frac{d+2}{q_0}}\Big[\mathcal{M}(I_{Q_{R+R_1/2}}(|f|+|g|)^{q_0})^{1/q_0}(z_0)$$

$$+ R_1^{-1}\mathcal{M}(I_{Q_{R+R_1/2}}(|u|^{q_0}))^{1/q_0}(z_0)\Big]$$

$$+ N(d,\nu,q_0,q_1)\Big(\kappa^{-\frac{d+2}{q_1}}\delta + \kappa\Big)\mathcal{M}(I_{Q_{R+R_1/2}}|Du|^{q_1})^{1/q_1}(z_0).$$

Recalling that $1 < q_0 < q_1 < \min(s,q)$, by Lemma 2.7 and the Hardy-Littlewood maximum function theorem in mixed-norm spaces (see, for instance, [12, Corollary 2.6]),

$$\|\omega\|_{L_{s,q}(Q_R)} \leq N(d,s,q)\Big[\|\omega_{\mathrm{dy}}^{\#}\|_{L_{s,q}(Q_R)} + R^{\frac{2}{s}+\frac{d}{q}-d-2}\|\omega\|_{L_1(Q_R)}\Big]$$

$$\leq N\kappa^{-\frac{d+2}{q_0}}\|\mathcal{M}(I_{Q_{R+R_1/2}}(|f|+|g|)^{q_0})^{1/q_0}\|_{L_{s,q}(\mathbb{R}^{d+1})}$$

$$+ NR_1^{-1}\kappa^{-\frac{d+2}{q_0}}\|\mathcal{M}(I_{Q_{R+R_1/2}}|u|^{q_0})^{1/q_0}\|_{L_{s,q}(\mathbb{R}^{d+1})}$$

$$+ N\Big(\kappa^{-\frac{d+2}{q_0}}\delta + \kappa\Big)\|\mathcal{M}(I_{Q_{R+R_1/2}}|Du|^{q_1})^{1/q_1}\|_{L_{s,q}(\mathbb{R}^{d+1})} + NR^{\frac{2}{s}+\frac{d}{q}}(|\omega|)_{Q_R}$$

$$\leq N\Big[\kappa^{-\frac{d+2}{q_0}}(\|f\|_{L_{s,q}(Q_{R+R_1/2})} + \|g\|_{L_{s,q}(Q_{R+R_1/2})}) + R_1^{-1}\kappa^{-\frac{d+2}{q_0}}\|u\|_{L_{s,q}(Q_{R+R_1/2})}$$

$$+ \Big(\kappa^{-\frac{d+2}{q_0}}\delta + \kappa\Big)\|Du\|_{L_{s,q}(Q_{R+R_1/2})} + R^{\frac{2}{s}+\frac{d}{q}}(|\omega|)_{Q_R}\Big],$$

where $N = N(d,\nu,s,q,q_0,q_1)$. Similar to (2.21), by Corollary 2.11, the last term on the right-hand side above is bounded by

$$NR^{\frac{2}{s}+\frac{d}{q}}\Big[(|f|^{q_0})_{Q_{R+R_1/2}}^{\frac{1}{q_0}} + (|g|^{q_0})_{Q_{R+R_1/2}}^{\frac{1}{q_0}} + R_1^{-1}(|u|^{q_0})_{Q_{R+R_1/2}}^{\frac{1}{q_0}} + \delta(|Du|^{q_1})_{Q_{R+R_1/2}}^{\frac{1}{q_1}}\Big]$$

$$\leq N\Big[\|f\|_{L_{s,q}(Q_{R+R_1/2})} + \|g\|_{L_{s,q}(Q_{R+R_1/2})}$$

$$+ R_1^{-1}\|u\|_{L_{s,q}(Q_{R+R_1/2})} + \delta\|Du\|_{L_{s,q}(Q_{R+R_1/2})}\Big],$$

where we used Hölder's inequality in the last line. Combining the two inequalities above gives (2.19).

Next we show (2.20). Since $\operatorname{div} u = g$, as in the proof of Lemma 2.8, we have

$$\|Du\|_{L_{s,q}(Q_R)}$$
$$\leq N\|\omega\|_{L_{s,q}(Q_{R+R_1/2})} + N\|g\|_{L_{s,q}(Q_{R+R_1/2})} + NR_1^{-1}\|u\|_{L_{s,q}(Q_{R+R_1/2})}. \quad (2.22)$$

Combining (2.22) and (2.19) with $R + R_1/2$ in place of R, we reach (2.20). The lemma is proved. $\qquad\square$

Now we are ready to give the proof of Theorem 2.3.

Proof of Theorem 2.3. For $k = 1, 2, \ldots$, we denote $Q^k = (-(1-2^{-k})^2, 0) \times B_{1-2^{-k}}$. Let k_0 be the smallest positive integer such that $2^{-k_0-1} < R_0$. For $k \geq k_0$, we apply (2.20) with $R = 2/3 - 2^{-k}$ and $R_1 = 2^{-k-1}$ to get

$$\|Du\|_{L_{s,q}(Q^k)} \leq N\kappa^{-\frac{d+2}{q_0}} \|f\|_{L_{s,q}(Q^{k+1})} + N\kappa^{-\frac{d+2}{q_0}} \|g\|_{L_{s,q}(Q^{k+1})}$$
$$+ N\left(\kappa^{-\frac{d+2}{q_0}}\delta + \kappa\right)\|Du\|_{L_{s,q}(Q^{k+1})}$$
$$+ N\kappa^{-\frac{d+2}{q_0}} 2^k \|u\|_{L_{s,q}(Q^{k+1})}. \tag{2.23}$$

Note that the constants N above are independent of k. We then take κ sufficiently small and then δ sufficiently small so that $N\left(\kappa^{-\frac{d+2}{q_0}}\delta + \kappa\right) \leq 1/3$. Finally, we multiply both sides of (2.23) by 3^{-k} and sum in $k = k_0, k_0 + 1, \ldots$ to get the desired estimate. The theorem is proved. $\qquad\square$

2.4 Non-divergence form Stokes systems

Now we turn to the non-divergence form Stokes system and give the proof of Theorem 2.4. The following lemma is analogous to Lemma 2.10.

Lemma 2.14. *Let $q_0 \in (1, \infty)$, $q \in (q_0, \infty)$, $r \in (0, R_0)$, $\nu, \delta \in (0, 1)$, and $u \in W_q^{1,2}(Q_r)^d$ be a strong solution to (2.4) in Q_r. Suppose that (2.2) and Assumption 2.2 $(\delta, 1)$ hold. Then we have*

$$(|D^2 u|^{q_0})_{Q_{r/2}}^{1/q_0}$$
$$\leq N(d, q_0)(|Dg|^{q_0})_{Q_r}^{1/q_0} + N(d, \nu, q_0, q)(|f|^{q_0})_{Q_r}^{1/q_0}$$
$$+ N(d, \nu, q_0, q)\left[r^{-1}(|Du - [Du]_{B_r}(t)|^{q_0})_{Q_r}^{1/q_0} + \delta^{1/q_0 - 1/q}(|D^2 u|^q)_{Q_r}^{1/q}\right]. \tag{2.24}$$

Proof. The proof is similar to that of Lemma 2.10. Let (w, p_1) be a strong solution to

$$w_t - \bar{a}_{ij}(t)D_{ij}w + \nabla p_1 = I_{Q_r}(f + (a_{ij} - \bar{a}_{ij})D_{ij}u) \quad \text{div } w = \phi_r(g - [g]_{B_r}(t))$$

in $(-r^2, 0) \times \mathbb{R}^d$ with zero initial condition on $\{t = -r^2\}$, where $\phi_r \in C_0^\infty((-r^2, r^2) \times B_r)$ is a standard non-negative cut-off function, which satisfies $\phi_r = 1$ on $Q_{2r/3}$ and $|D\phi_r| \leq 4/r$. Observe that from the equation div $w = \phi_r(g - [g]_{B_r}(t))$ and the Poincaré inequality, we have

$$\|D^2 w\|_{L_{q_0}((-r^2, 0) \times \mathbb{R}^d)}$$
$$\leq N(d, q_0)\left[\|Dw\|_{L_{q_0}((-r^2,0)\times\mathbb{R}^d)} + \|D(\phi_r(g - [g]_{B_r}))\|_{L_{q_0}((-r^2,0)\times\mathbb{R}^d)}\right] \tag{2.25}$$
$$\leq N(d, q_0)\left[\|Dw\|_{L_{q_0}((-r^2,0)\times\mathbb{R}^d)} + \|Dg\|_{L_{q_0}(Q_r)}\right],$$

where $\omega = \nabla \times w$. Now ω is a weak solution to the divergence form parabolic equation

$$\omega_t - \bar{a}_{ij}(t)D_{ij}\omega = \nabla \times \left(I_{Q_r}(f + (a_{ij} - \bar{a}_{ij})D_{ij}u)\right).$$

By applying the $\mathcal{H}_p^1$ estimate for divergence form parabolic equations with coefficients measurable in t (see [28, 30]) and (2.25), we obtain

$$\|D\omega\|_{L_{q_0}((-r^2,0)\times\mathbb{R}^d)} \leq N(d,\nu,q_0)\Big[\|f\|_{L_{q_0}(Q_r)} + \|(a_{ij}-\bar{a}_{ij})D_{ij}u\|_{L_{q_0}(Q_r)}\Big]. \quad (2.26)$$

It then follows from (2.25) and (2.26) that

$$\|D^2w\|_{L_{q_0}((-r^2,0)\times\mathbb{R}^d)}$$
$$\leq N(d,\nu,q_0)\Big[\|f\|_{L_{q_0}(Q_r)} + \|(a_{ij}-\bar{a}_{ij})D_{ij}u\|_{L_{q_0}(Q_r)}\Big] + N(d,q_0)\|Dg\|_{L_{q_0}(Q_r)}.$$

From this and by using Assumption 2.2 (δ, 1) and Hölder's inequality for the middle term on the right hand side of the last estimate, we have

$$(|D^2w|^{q_0})_{Q_r}^{1/q_0} \leq N(d,q_0)(|Dg|^{q_0})_{Q_r}^{1/q_0}$$
$$+ N(d,\nu,q_0)\Big[(|f|^{q_0})_{Q_r}^{1/q_0} + \delta^{1/q_0-1/q}(|D^2u|^q)_{Q_r}^{1/q}\Big]. \quad (2.27)$$

Now $(v,p_2) := (u-w,p-p_1)$ satisfies

$$v_t - \bar{a}_{ij}(t)D_{ij}v + \nabla p_2 = 0, \quad \operatorname{div} v = [g]_{B_r}(t)$$

in $Q_{2r/3}$. By Lemma 2.8 applied to Dv with a scaling, we have

$$(|D^2v|^{q_0})_{Q_{r/2}}^{1/q_0} \leq r^{-1}(|Dv - [Dv]_{B_r}(t)|^{q_0})_{Q_r}^{1/q_0}. \quad (2.28)$$

By (2.27), (2.28), the triangle inequality, and the Poincaré inequality, we get the desired inequality. $\qquad\square$

Remark 2.15. By interpolation inequalities and iteration, we can replace the term $r^{-1}(|Du - [Du]_{B_r}(t)|^{q_0})_{Q_r}^{1/q_0}$ in (2.24) with $r^{-2}(|u - [u]_{B_r}(t)|^{q_0})_{Q_r}^{1/q_0}$.

Analogous to Corollary 2.11, from Lemma 2.14 we derive the following corollary.

Corollary 2.16. *Let $\delta,\nu \in (0,1)$, $q_0 \in (1,\infty)$, $q \in (q_0,\infty)$, $r \in (0,R_0)$, $T > 0$, and $\Omega \subset \mathbb{R}^d$ satisfy (2.15) for some $\gamma > 0$ and $(-T,0)\times\Omega \subset Q_{2/3}$. Suppose that (2.2) and Assumption 2.2 (δ,1) hold. Then, for any strong solution $(u,p) \in W_q^{1,2}((-T-r^2,0)\times\Omega^r)^d \times L_1((-T-r^2,0)\times\Omega^r)$ of (2.4) in $(-T-r^2,0)\times\Omega^r$, we have*

$$(|D^2u|^{q_0})_{(-T,0)\times\Omega}^{\frac{1}{q_0}}$$
$$\leq N(d,\nu,q_0,\gamma)\frac{((T+r^2)|\Omega^r|)^{\frac{1}{q_0}}}{(T|\Omega|)^{\frac{1}{q_0}}}\Big[(|f|^{q_0})_{(-T-r^2,0)\times\Omega^r}^{\frac{1}{q_0}}$$
$$+ r^{-1}(|Du|^{q_0})_{(-T-r^2,0)\times\Omega^r}^{\frac{1}{q_0}} + (|Dg|^{q_0})_{(-T-r^2,0)\times\Omega^r}^{\frac{1}{q_0}}\Big]$$
$$+ N(d,\nu,q_0,q,\gamma)\frac{((T+r^2)|\Omega^r|)^{\frac{1}{q}}}{(T|\Omega|)^{\frac{1}{q}}}\delta^{\frac{1}{q_0}-\frac{1}{q}}(|D^2u|^q)_{(-T-r^2,0)\times\Omega^r}^{\frac{1}{q}}.$$

In the next lemma we prove a mean oscillation estimate of $D\omega$.

Lemma 2.17. *Let $q_1 \in (1, \infty), q_0 \in (1, q_1), \delta \in (0, 1), R_0 \in (0, 1/4), r \in (0, R_0),$ $\kappa \in (0, 1/4)$. Suppose that (2.2) and Assumption 2.2 $(\delta, 1)$ hold. Suppose that $u \in W^{1,2}_{q_1}(Q_r)^d$ is a strong solution to (2.4) in Q_r. Then it holds that*

$$(|D\omega - (D\omega)_{Q_{\kappa r}}|)_{Q_{\kappa r}} \leq N(d, \nu, q_0)\kappa^{-\frac{d+2}{q_0}}(|f|^{q_0})^{1/q_0}_{Q_r}$$
$$+ N(n, \nu, q_0, q_1)\left(\kappa^{-\frac{d+2}{q_0}}\delta^{\frac{1}{q_0}-\frac{1}{q_1}} + \kappa\right)(|D^2u|^{q_1})^{1/q_1}_{Q_r},$$

where $\omega = \nabla \times u$.

Proof. The proof is similar to that of Lemma 2.12. Let (w, p_1) and (v, p_2) be as in the proof of Lemma 2.14. In particular, (w, p_1) is a strong solution of

$$w_t - \bar{a}_{ij}(t)D_{ij}w + \nabla p_1 = I_{Q_r}[f + (a_{ij} - \bar{a}_{ij}(t))D_{ij}u], \quad \text{div } w = \phi_r(g - [g]_{B_r}(t))$$

in $(-r^2, 0) \times \mathbb{R}^d$ with zero initial condition on $\{t = -r^2\}$. Moreover, $(v, p_2) = (u - w, p - p_1)$ is a strong solution of

$$v_t - \bar{a}_{ij}(t)D_{ij}v + \nabla p_2 = 0, \quad \text{div } v = [g]_{B_r}(t)$$

in $Q_{2r/3}$. Let $\omega_1 = \nabla \times w$ and $\omega_2 = \nabla \times v$ and we see that $\omega = \omega_1 + \omega_2$. Moreover, we can deduce from (2.26) that

$$(|D\omega_1|^{q_0})^{1/q_0}_{Q_r} \leq N(d, \nu, q_0)\left[(|f|^{q_0})^{1/q_0}_{Q_r} + \delta^{1/q_0-1/q}(|D^2u|^{q_1})^{1/q_1}_{Q_r}\right]. \qquad (2.29)$$

Also, by applying Lemma 2.9 to $D\omega_2$ with a suitable scaling, we obtain

$$(|D\omega_2 - (D\omega_2)_{Q_{\kappa r}}|)_{Q_{\kappa r}} \leq N\kappa r[[D\omega_2]]_{C^{1/2,1}(Q_{r/3})} \leq N(d, \nu, q_0)\kappa(|D\omega_2|^{q_0})^{1/q_0}_{Q_{2r/3}}$$
$$\leq N(d, \nu, q_0)\kappa\left[(|D\omega|^{q_0})^{1/q_0}_{Q_r} + (|D\omega_1|^{q_0})^{1/q_0}_{Q_r}\right].$$

Then, by combining this last estimate with (2.29) and the fact that $\delta \in (0, 1)$, we infer that

$$(|D\omega_2 - (D\omega_2)_{Q_{\kappa r}}|)_{Q_{\kappa r}} \leq N(d, \nu, q_0)\kappa\left[(|f|^{q_0})^{1/q_0}_{Q_r} + (|D^2u|^{q_1})^{1/q_1}_{Q_r}\right]. \qquad (2.30)$$

Now, by using the inequality

$$\fint_{Q_{\kappa r}} |D\omega - (D\omega)_{Q_{\kappa r}}| \, dx \, dt \leq 2 \fint_{Q_{\kappa r}} |D\omega - c| \, dx \, dt$$

with $c = (D\omega_2)_{Q_{\kappa r}}$, and then applying the triangle inequality, and Hölder's inequality, we have

$$\fint_{Q_{\kappa r}} |D\omega - (D\omega)_{Q_{\kappa r}}| \, dx \, dt \leq 2 \fint_{Q_{\kappa r}} |D\omega - (D\omega_2)_{Q_{\kappa r}}| \, dx \, dt$$

$$\leq 2 \fint_{Q_{\kappa r}} |D\omega_2 - (D\omega_2)_{Q_{\kappa r}}| \, dx \, dt + N(d, q_0)\kappa^{-\frac{d+2}{q_0}}\left(\fint_{Q_r} |D\omega_1|^{q_0} \, dx \, dt\right)^{1/q_0}.$$

This last estimate together with (2.29) and (2.30) implies that

$$(|D\omega - (D\omega)_{Q_{\kappa r}}|)_{Q_{\kappa r}} \le N(d,\nu,q_0)\kappa^{-\frac{d+2}{q_0}}(|f|^{q_0})_{Q_r}^{1/q_0}$$
$$+ N(d,\nu,q_0,q_1)\Big(\kappa^{-\frac{d+2}{q_0}}\delta^{1/q_0-1/q} + \kappa\Big)(|D^2u|^{q_1})_{Q_r}^{1/q_1}.$$

The proof is then complete. $\square$

Our next lemma gives the key estimates of $D\omega$ and D^2u in the mixed norm.

Lemma 2.18. *Let* $R \in [1/2,1]$, $R_1 \in (0,R_0)$, $\delta \in (0,1)$, $\kappa \in (0,1/4)$, $s,q \in (1,\infty)$, $q_1 \in (1,\min\{s,q\})$, *and* $q_0 \in (1,q_1)$. *Assume that (2.2) and Assumption 2.2 $(\delta,1)$ hold. Suppose that* $u \in W^{1,2}_{s,q}(Q_{R+R_1})^d$ *is a strong solution to (2.4) in* Q_{R+R_1}, *and* $\omega = \nabla \times u$. *Then we have*

$$\|D\omega\|_{L_{s,q}(Q_R)} \le N\kappa^{-\frac{d+2}{q_0}}\|f\|_{L_{s,q}(Q_{R+R_1/2})} + N\kappa^{-\frac{d+2}{q_0}}\|Dg\|_{L_{s,q}(Q_{R+R_1/2})}$$
$$+ N\Big(\kappa^{-\frac{d+2}{q_0}}\delta^{\frac{1}{q_0}-\frac{1}{q_1}} + \kappa\Big)\|D^2u\|_{L_{s,q}(Q_{R+R_1/2})}$$
$$+ N\kappa^{-\frac{d+2}{q_0}}R_1^{-1}\|Du\|_{L_{s,q}(Q_{R+R_1/2})} \tag{2.31}$$

and

$$\|D^2u\|_{L_{s,q}(Q_{R/2})} \le N\kappa^{-\frac{d+2}{q_0}}\|f\|_{L_{s,q}(Q_{R+R_1})} + N\kappa^{-\frac{d+2}{q_0}}\|Dg\|_{L_{s,q}(Q_{R+R_1})}$$
$$+ N\Big(\kappa^{-\frac{d+2}{q_0}}\delta^{\frac{1}{q_0}-\frac{1}{q_1}} + \kappa\Big)\|D^2u\|_{L_{s,q}(Q_{R+R_1})}$$
$$+ N\kappa^{-\frac{d+2}{q_0}}R_1^{-1}\|Du\|_{L_{s,q}(Q_{R+R_1})}. \tag{2.32}$$

Proof. As in the proof of Lemma 2.13, we discuss two cases.
Case 1: $r \in (0, R_1/2)$. It follows from Lemma 2.17 that for all $z_0 \in Q_R$,

$$(|D\omega - (D\omega)_{Q_{\kappa r}(z_0)}|)_{Q_{\kappa r}(z_0)}$$
$$\le N(d,\nu,q_0)\kappa^{-\frac{d+2}{q_0}}(|f|^{q_0})_{Q_r(z_0)}^{1/q_0}$$
$$+ N(d,\nu,q_0,q_1)\Big(\kappa^{-\frac{d+2}{q_0}}\delta^{\frac{1}{q_0}-\frac{1}{q_1}} + \kappa\Big)(|D^2u|^{q_1})_{Q_r(z_0)}^{1/q_1}.$$

Observe that because $r < R_1/2$, we have $Q_r(z_0) \subset Q_{R+R_1/2}$. Therefore,

$$(|f|^{q_0})_{Q_r(z_0)}^{1/q_0} \le \mathcal{M}(I_{Q_{R+R_1/2}}|f|^{q_0})^{1/q_0}(z_0), \quad \text{and}$$
$$(|D^2u|^{q_1})_{Q_r(z_0)}^{1/q_1} \le \mathcal{M}(I_{Q_{R+R_1/2}}|D^2u|^{q_1})^{1/q_1}(z_0),$$

where $\mathcal{M}$ is the Hardy-Littlewood maximal function. These estimates imply that

$$(|D\omega - (D\omega)_{Q_{\kappa r}(z_0)}|)_{Q_{\kappa r}(z_0)}$$
$$\le N\kappa^{-\frac{d+2}{q_0}}\mathcal{M}(I_{Q_{R+R_1/2}}|f|^{q_0})^{1/q_0}(z_0)$$
$$+ N\Big(\kappa^{-\frac{d+2}{q_0}}\delta^{\frac{1}{q_0}-\frac{1}{q_1}} + \kappa\Big)\mathcal{M}(I_{Q_{R+R_1/2}}|D^2u|^{q_1})^{1/q_1}(z_0).$$

Case 2: $r \in [R_1/2, 2R/\kappa)$ and $z_0 \in Q_R$ such that $t_0 \in [-R^2 + (\kappa r)^2/2, 0]$. In this case, we apply Corollary 2.16 to get

$$(|D\omega - (D\omega)_{Q_{\kappa r}(z_0) \cap Q_R}|)_{Q_{\kappa r}(z_0) \cap Q_R}$$

$$\leq 2(|D\omega|)_{Q_{\kappa r}(z_0) \cap Q_R} \leq 2(|D\omega|^{q_0})^{\frac{1}{q_0}}_{Q_{\kappa r}(z_0) \cap Q_R}$$

$$\leq N\kappa^{-\frac{d+2}{q_0}} \left[(|f|^{q_0})^{\frac{1}{q_0}}_{Q_{\kappa r + R_1/2}(z_0) \cap Q_{R+R_1/2}} + R_1^{-1}(|Du|^{q_0})^{\frac{1}{q_0}}_{Q_{\kappa r + R_1/2}(z_0) \cap Q_{R+R_1/2}} \right.$$

$$\left. + (|Dg|^{q_0})^{\frac{1}{q_0}}_{Q_{\kappa r + R_1/2}(z_0) \cap Q_{R+R_1/2}} \right]$$

$$+ N\kappa^{-\frac{d+2}{q_1}} \delta^{\frac{1}{q_0} - \frac{1}{q_1}} (|D^2 u|^{q_1})^{\frac{1}{q_1}}_{Q_{\kappa r + R_1/2}(z_0) \cap Q_{R+R_1/2}}, \tag{2.33}$$

where we used $R_1/2 \leq r$ in the last inequality.

Now, we take $\mathcal{X} = Q_R$ and define the dyadic sharp function $(D\omega)^\#_{\text{dy}}$ of $D\omega$ in $\mathcal{X}$. From the above two cases, we conclude that for any $z_0 \in \mathcal{X}$,

$$(D\omega)^\#_{\text{dy}}(z_0) \leq N \left[\kappa^{-\frac{d+2}{q_0}} \mathcal{M}(I_{Q_{R+R_1/2}}(|f| + |Dg|)^{q_0})^{1/q_0}(z_0) \right.$$

$$\left. + R_1^{-1} \mathcal{M}(I_{Q_{R+R_1/2}} |Du|^{q_0})^{1/q_0}(z_0) \right]$$

$$+ N \left(\kappa^{-\frac{d+2}{q_0}} \delta^{\frac{1}{q_0} - \frac{1}{q_1}} + \kappa \right) \mathcal{M}(I_{Q_{R+R_1/2}} |D^2 u|^{q_1})^{1/q_1}(z_0).$$

Recalling that $1 < q_0 < q_1 < \min\{s, q\}$, by Lemma 2.7 and the Hardy-Littlewood maximum function theorem in mixed-norm spaces (see, for instance, [12, Corollary 2.6]),

$$\|D\omega\|_{L_{s,q}(Q_R)} \leq N \left[\|(D\omega)^\#_{\text{dy}}\|_{L_{s,q}(Q_R)} + R^{2/s + d/q}(|D\omega|)_{Q_R} \right]$$

$$\leq N\kappa^{-\frac{d+2}{q_0}} \|\mathcal{M}(I_{Q_{R+R_1/2}} |f|^{q_0})^{1/q_0}\|_{L_{s,q}(\mathbb{R}^{d+1})}$$

$$+ N\kappa^{-\frac{d+2}{q_0}} \|\mathcal{M}(I_{Q_{R+R_1/2}} |Dg|^{q_0})^{1/q_0}\|_{L_{s,q}(\mathbb{R}^{d+1})}$$

$$+ N \left(\kappa^{-\frac{d+2}{q_0}} \delta^{\frac{1}{q_0} - \frac{1}{q_1}} + \kappa \right) \|\mathcal{M}(I_{Q_{R+R_1/2}} |D^2 u|^{q_1})^{1/q_1}\|_{L_{s,q}(\mathbb{R}^{d+1})}$$

$$+ NR^{2/s + d/q}(|D\omega|)_{Q_R}$$

$$\leq N \left[\kappa^{-\frac{d+2}{q_0}} \|f\|_{L_{s,q}(Q_{R+R_1/2})} + \kappa^{-\frac{d+2}{q_0}} \|Dg\|_{L_{s,q}(Q_{R+R_1/2})} \right.$$

$$\left. + \left(\kappa^{-\frac{d+2}{q_0}} \delta^{\frac{1}{q_0} - \frac{1}{q_1}} + \kappa \right) \|D^2 u\|_{L_{s,q}(Q_{R+R_1/2})} + R^{2/s + d/q}(|D\omega|)_{Q_R} \right].$$

Similar to (2.33), by Corollary 2.16, the last term on the right-hand side above is bounded by

$$NR^{\frac{2}{s} + \frac{d}{q}} \left[(|f|^{q_0})^{\frac{1}{q_0}}_{Q_{R+R_1/2}} + (|Dg|^{q_0})^{\frac{1}{q_0}}_{Q_{R+R_1/2}} + R_1^{-1}(|Du|^{q_0})^{\frac{1}{q_0}}_{Q_{R+R_1/2}} \right.$$

$$\left. + \delta^{\frac{1}{q_0} - \frac{1}{q_1}} (|D^2 u|^{q_1})^{\frac{1}{q_1}}_{Q_{R+R_1/2}} \right]$$

$$\leq N \left[\|f\|_{L_{s,q}(Q_{R+R_1/2})} + \|Dg\|_{L_{s,q}(Q_{R+R_1/2})} + R_1^{-1}\|Du\|_{L_{s,q}(Q_{R+R_1/2})} \right.$$

$$\left. + \delta^{\frac{1}{q_0} - \frac{1}{q_1}} \|D^2 u\|_{L_{s,q}(Q_{R+R_1/2})} \right].$$

Combining the above two inequalities, we reach (2.31).

Next we show (2.32). Since div $u = g$, as in the proof of Lemma 2.8, we have

$$\|D^2 u\|_{L_{s,q}(Q_R)}$$
$$\leq N\|D\omega\|_{L_{s,q}(Q_{R+R_1/2})} + N\|Dg\|_{L_{s,q}(Q_{R+R_1/2})}$$
$$+ NR^{-1}\|Du\|_{L_{s,q}(Q_{R+R_1/2})}. \tag{2.34}$$

Combining (2.34) and (2.31) with $R + R_1/2$ in place of R, we reach (2.32). The lemma is proved. □

Now we are ready to give

Proof of Theorem 2.4. As in the proof of Lemma 2.3, for $k = 1, 2, \ldots$, we denote $Q^k = (-(1-2^{-k})^2, 0) \times B_{1-2^{-k}}$. Let k_0 be the smallest positive integer such that $2^{-k_0-1} < R_0$. For $k \geq k_0$, we apply (2.32) with $R = 2/3 - 2^{-k}$ and $R_1 = 2^{-k-1}$ to get

$$\|D^2 u\|_{L_{s,q}(Q^k)} \leq N\kappa^{-\frac{d+2}{q_0}}\|f\|_{L_{s,q}(Q^{k+1})} + N\kappa^{-\frac{d+2}{q_0}}\|Dg\|_{L_{s,q}(Q^{k+1})}$$
$$+ N\left(\kappa^{-\frac{d+2}{q_0}}\delta^{\frac{1}{q_0}-\frac{1}{q_1}} + \kappa\right)\|D^2 u\|_{L_{s,q}(Q^{k+1})}$$
$$+ N\kappa^{-\frac{d+2}{q_0}}2^k\|Du\|_{L_{s,q}(Q^{k+1})}. \tag{2.35}$$

From (2.35) and interpolation inequalities, we get

$$\|D^2 u\|_{L_{s,q}(Q^k)} \leq N\kappa^{-\frac{d+2}{q_0}}\|f\|_{L_{s,q}(Q^{k+1})} + N\kappa^{-\frac{d+2}{q_0}}\|Dg\|_{L_{s,q}(Q^{k+1})}$$
$$+ N\left(\kappa^{-\frac{d+2}{q_0}}\delta^{\frac{1}{q_0}-\frac{1}{q_1}} + \kappa\right)\|D^2 u\|_{L_{s,q}(Q^{k+1})}$$
$$+ N\kappa^{-1-\frac{2(d+2)}{q_0}}2^{2k}\|u\|_{L_{s,q}(Q^{k+1})}. \tag{2.36}$$

Note that the constants N above are independent of k. We then take κ sufficiently small and then δ sufficiently small so that

$$N\left(\kappa^{-\frac{d+2}{q_0}}\delta^{\frac{1}{q_0}-\frac{1}{q_1}} + \kappa\right) \leq 1/5.$$

Finally, we multiply both sides of (2.36) by 5^{-k} and sum in $k = k_0, k_0 + 1, \ldots$ to get the desired estimate. The theorem is proved. □

Remark 2.19. We refer the reader to [17] for the corresponding results in the half space.

References

[1] H. Abidi, G. Gui, and P. Zhang. On the decay and stability of global solutions to the 3D inhomogeneous Navier-Stokes equations. *Comm. Pure Appl. Math.* 64(6): 832–881, 2011.

[2] S. Agmon, A. Douglis, and L. Nirenberg. Estimates near the boundary for solutions of elliptic partial differential equations satisfying general boundary conditions. II. *Comm. Pure Appl. Math.* 17: 35–92, 1964.

[3] H. Aimar and R. A. Macías. Weighted norm inequalities for the Hardy-Littlewood maximal operator on spaces of homogeneous type. *Proc. Amer. Math. Soc.* 91(2): 213–216, 1984.

[4] C. Amrouche and V. Girault. On the existence and regularity of the solution of Stokes problem in arbitrary dimension. *Proc. Japan Acad. Ser. A Math. Sci.* 67(5): 171–175, 1991.

[5] P. Auscher, E. Russ, and P. Tchamitchian. Hardy Sobolev spaces on strongly Lipschitz domains of $\mathbb{R}^n$. *J. Funct. Anal.* 218(1): 54–109, 2005.

[6] R. M. Brown and Z. Shen. Estimates for the Stokes operator in Lipschitz domains. *Indiana Univ. Math. J.* 44(4): 1183–1206, 1995.

[7] M. Bulíček, J. Málek, and K. R. Rajagopal. Navier's slip and evolutionary Navier-Stokes-like systems with pressure and shear-rate dependent viscosity. *Indiana Univ. Math. J.* 56(1): 51–85, 2007.

[8] L. Cattabriga. Su un problema al contorno relativo al sistema di equazioni di Stokes. *Rend. Sem. Mat. Univ. Padova* 31: 308–340, 1961.

[9] D. Chae and J. Wolf. On the Liouville type theorems for self-similar solutions to the Navier-Stokes equations. *Arch. Ration. Mech. Anal.* 225(1): 549–572, 2017.

[10] J. Choi and K. Lee. The Green function for the Stokes system with measurable coefficients. *Commun. Pure Appl. Anal.* 16(6): 1989–2022, 2017.

[11] M. Christ. A $T(b)$ theorem with remarks on analytic capacity and the Cauchy integral. *Colloq. Math.* 60/61(2): 601–628, 1990.

[12] H. Dong and D. Kim. On L_p-estimates for elliptic and parabolic equations with A_p weights. *Trans. Amer. Math. Soc.* 370(7): 5081–5130, 2018.

[13] H. Dong and D. Kim. Weighted L_q-estimates for stationary Stokes system with partially BMO coefficients. *J. Differential Equations* 264(7): 4603–4649, 2018.

[14] H. Dong and D. Kim. L_q-estimates for stationary Stokes system with coefficients measurable in one direction. *Bull. Math. Sci.* 9(1): 1950004, 30 pp, 2019.

[15] H. Dong and D. Kim. Higher order elliptic and parabolic systems with variably partially BMO coefficients in regular and irregular domains. *J. Funct. Anal.* 261(11): 3279–3327, 2011.

[16] H. Dong and D. Kim. Parabolic and elliptic systems in divergence form with variably partially BMO coefficients. *SIAM J. Math. Anal.* 43(3): 1075–1098, 2011.

[17] H. Dong, D. Kim, and T. Phan. Boundary Lebesgue mixed-norm estimates for non-stationary Stokes systems with VMO coefficients. arXiv:1910.00380.

[18] H. Dong and T. Phan. Mixed-norm L_p-estimates for non-stationary Stokes systems with singular VMO coefficients and applications. arXiv:1805.04143.

[19] E. B. Fabes, C. E. Kenig, and G. C. Verchota. The Dirichlet problem for the Stokes system on Lipschitz domains. *Duke Math. J.* 57(3): 769–793, 1988.

[20] M. Franta, J. Málek, and K. R. Rajagopal. On steady flows of fluids with pressure- and shear-dependent viscosities. *Proc. R. Soc. Lond. Ser. A Math. Phys. Eng. Sci.* 461(2055): 651–670, 2005.

[21] G. P. Galdi, C. G. Simader, and H. Sohr. On the Stokes problem in Lipschitz domains. *Ann. Math. Pure Appl. (4)* 167: 147–163, 1994.

[22] J. Geng and J. Kilty. The L^p regularity problem for the Stokes system on Lipschitz domains. *J. Differential Equations* 259(4): 1275–1296, 2015.

[23] M. Giaquinta and G. Modica. Nonlinear systems of the type of the stationary Navier-Stokes system. *J. Reine Angew. Math.* 330: 173–214, 1982.

[24] F. Hu, D. Li, and L. Wang. A new proof of L_p estimates of Stokes equations. *J. Math. Anal. Appl.* 420(2): 1251–1264, 2014.

[25] T. Hytönen and A. Kairema. Systems of dyadic cubes in a doubling metric space. *Colloq. Math.* 126(1): 1–33, 2012.

[26] B. J. Jin. On the Caccioppoli inequality of the unsteady Stokes system. *Int. J. Numer. Anal. Model.* Ser. B 4(3): 215–223, 2013.

[27] D. Kim and N. V. Krylov. Elliptic differential equations with coefficients measurable with respect to one variable and VMO with respect to the others. *SIAM J. Math. Anal.* 39(2): 489–506, 2007.

[28] N. V. Krylov. Parabolic and elliptic equations with VMO coefficients. *Comm. Partial Differential Equations* 32(1–3): 453–475, 2007.

[29] N. V. Krylov. Second-order elliptic equations with variably partially VMO coefficients. *J. Funct. Anal.* 257(6): 1695–1712, 2009.

[30] N. V. Krylov. *Lectures on Elliptic and Parabolic Equations in Sobolev Spaces*, volume 96 of *Graduate Studies in Mathematics*. American Mathematical Society, 2008.

[31] O. A. Ladyženskaya. Investigation of the Navier-Stokes equation for stationary motion of an incompressible fluid. *Uspehi Mat. Nauk*, 14(3): 75–97, 1959.

[32] O. A. Ladyženskaya and V. A. Solonnikov. The unique solvability of an initial-boundary value problem for viscous incompressible inhomogeneous fluids. Boundary value problems of mathematical physics, and related questions of the theory of functions, 8. *Zap. Naučn. Sem. Leningrad. Otdel. Mat. Inst. Steklov. (LOMI)*, 52: 52–109, 218–219, 1975.

[33] P.-L. Lions, *Mathematical Topics in Fluid Mechanics. Vol. 1. Incompressible Models*, Oxford Lecture Series in Mathematics and its Applications, 3. Oxford Science Publications. The Clarendon Press, Oxford University Press, 1996.

[34] P. Maremonti and V. A. Solonnikov. On estimates for the solutions of the nonstationary Stokes problem in S. L. Sobolev anisotropic spaces with a mixed norm. *Zap. Nauchn. Sem. S.-Peterburg. Otdel. Mat. Inst. Steklov. (POMI)* 222 (1995), *translation in J. Math. Sci. (New York)* 87(5): 3859–3877, 1997.

[35] N. G. Meyers. An L^p-estimate for the gradient of solutions of second order elliptic divergence equations. *Ann. Sc. Norm. Super. Pisa Cl. Sci. (3)* 17: 189–206, 1963.

[36] M. Mitrea and M. Taylor. Navier-Stokes equations on Lipschitz domains in Riemannian manifolds. *Math. Ann.* 321(4): 955–987, 2001.

[37] M. Mitrea and M. Wright. *Boundary Value Problems for the Stokes System in Arbitrary Lipschitz Domains.* Astérisque 344, 2012, viii+241 pp.

[38] J. Serrin. On the interior regularity of weak solutions of the Navier-Stokes equations. *Arch. Rat. Mech. Anal.* 9: 187–195, 1962.

[39] G. Seregin. *Lecture Notes on Regularity Theory for the Navier-Stokes Equations.* World Scientific Publishing Co. Pte. Ltd., 2015.

[40] Z. Shen. A note on the Dirichlet problem for the Stokes system in Lipschitz domains. *Proc. Amer. Math. Soc.* 123(3): 801–811, 1995.

[41] P. E. Sobolevskiĭ. On the smoothness of generalized solutions of the Navier-Stokes equations. *Dokl. Akad. Nauk SSSR* 131: 758–760 (Russian); *translation in Soviet Math. Dokl.* 1: 341–343, 1960.

[42] V. A. Solonnikov. Estimates for solutions of a non-stationary linearized system of Navier-Stokes equations. *Trudy Mat. Inst. Steklov.* 70: 213–317, 1964.

[43] V. A. Solonnikov. Estimates of solutions of the Stokes equations in S. L. Sobolev spaces with a mixed norm. *Zap. Nauchn. Sem. S.-Peterburg. Otdel. Mat. Inst. Steklov. (POMI)* 288 (2002), Kraev. Zadachi Mat. Fiz. i Smezh. Vopr. Teor. Funkts. 32: 204–231, 273–274; *translation in J. Math. Sci. (N.Y.)* 123(6): 4637–4653, 2004.

[44] I. I. Vorovič and V. I. Judovič. Steady flow of a viscous incompressible fluid. (Russian) *Mat. Sb. (N.S.)* 53(95): 393–428, 1961.

[45] J. Wolf, A new criterion for partial regularity of suitable weak solutions to the Navier-Stokes equations. *Advances in mathematical fluid mechanics*, 613–630, Springer, 2010.

Lectures on the Analysis of Nonlinear
Partial Differential Equations Vol. 6
MLM 6, pp. 171–226

Optimal and Almost Optimal Initial Values for Weak Solutions to the Navier-Stokes Equations[*]

Reinhard Farwig[†]

Abstract

Consider a global-in-time weak solution u of the instationary Navier-Stokes equations in a bounded domain $\Omega \subset \mathbb{R}^3$ with Dirichlet boundary data $u = 0$ on $\partial\Omega$ and a solenoidal initial value $u_0 \in L^2(\Omega)$. Since the pioneering work of J. Leray in 1933/34 it is an open problem whether weak solutions are unique and smooth. The main step—to nowadays knowledge—is to show that the given weak solution is a strong one in the sense of J. Serrin, *i.e.*, $u \in L^s\big(0, T; L^q(\Omega)\big)$ where $s \geq 2$, $q \geq 3$ and $\frac{2}{s} + \frac{3}{q} = 1$. This property should hold at least in some initial interval $[0, T)$ and requires further conditions on u_0 beyond $u_0 \in L^2(\Omega)$.

These lecture notes concern the optimal initial value condition and its consequences on local or global regularity, uniqueness and the validity of the energy equality. Moreover, we discuss slightly weaker, so-called almost optimal conditions on u_0 in a family of scale invariant Besov spaces leading to an analysis in L^s-spaces with algebraic weights in time.

1 Introduction

The Navier-Stokes system is the most classical model to describe the flow of a viscous incompressible fluid of so-called Newtonian type. Despite of more than 80 years of mathematical analysis since the seminal paper of J. Leray [61] on the existence of global weak solutions in the whole space $\mathbb{R}^3$ and corresponding results of E. Hopf [48] for domains, basic questions on uniqueness and regularity of weak solutions are still open in the three-dimensional case. These fundamental problems

---

[*]Key words and phrases: instationary Navier-Stokes equations, energy (in-)equality, optimal initial values, almost optimal initial values, weak solutions, strong solutions, one-sided regularity, blow up, uniqueness, admissible approximation schemes, scale invariant spaces, Besov spaces.

AMS Subject Classification: 35Q30, 35B65, 76D05, 76D03.

[†]Fachbereich Mathematik, Technische Universität Darmstadt, 64289 Darmstadt, Germany. E-mail: farwig@mathematik.tu-darmstadt.de

brought Clay Mathematics Institute in 2000 to classify the issue of regularity as one of the seven Millennium Prize Problem, see C. Fefferman [38].

Given a domain $\Omega \subset \mathbb{R}^3$ and a time interval $[0, T)$, $0 < T \leq \infty$, consider an external force $f : \Omega \times (0, T) \to \mathbb{R}^3$ and a solenoidal (divergence-free) initial value $u_0 : \Omega \to \mathbb{R}^3$ with vanishing normal trace on $\partial\Omega$. Then we are looking for a velocity field $u(t, x) = (u_1, u_2, u_3)(t, x)$ and an associated pressure function $p(t, x)$ such that

$$u_t - \nu\Delta u + u \cdot \nabla u + \nabla p = f, \quad \text{div } u = 0 \text{ in } \Omega \times (0, T),$$
$$u(0) = u_0, \quad u = 0 \text{ on } \partial\Omega \times (0, T). \tag{1.1}$$

For simplicity the coefficient of viscosity $\nu > 0$ will be set to $\nu = 1$ in this exposition. In the case $\Omega = \mathbb{R}^3$ the Dirichlet boundary condition $u = 0$ on $\partial\Omega$ has to be omitted. The nonlinear transport term $u \cdot \nabla u$ is defined by $\sum_{j=1}^{3} u_j \partial_j u$ and can also be written in the form

$$u \cdot \nabla u = \text{div } (u \otimes u)$$

since u is solenoidal; recall that $u \otimes v = (u_i v_j)_{i,j=1}^3$ and, for a matrix field $F = (F_{ij})$, that $\text{div } F = \left(\sum_{i=1}^{3} \partial_i F_{ij} \right)_{j=1}^{3}$.

In this article we use standard notation for Lebesgue, Sobolev and Bochner spaces, i.e. $\left(L^q(\Omega) = L^q, \| \cdot \|_q \right)$, $\left(W^{k,q}(\Omega) = W^{k,q}, \| \cdot \|_{W^{k,q}} \right)$ for $k \in \mathbb{N}$, and $\left(L^s(0, T; L^q(\Omega)) = L^s(0, T; L^q) = L^s(L^q); \| \cdot \|_{L^s(L^q)} \right)$, $1 \leq s, q \leq \infty$, respectively. We do not differ between spaces of scalar-, vector- and matrix-valued functions. The index σ will denote a subspace of solenoidal vector fields u with vanishing normal trace $u \cdot N = 0$ on $\partial\Omega$, the subscript 0 refers to a subspace of functions with vanishing trace. The duality product of functions on Ω and $\Omega \times (0, T)$ is denoted by $\langle \cdot, \cdot \rangle_\Omega$ and $\langle \cdot, \cdot \rangle_{\Omega, T}$, respectively.

Crucial in our analysis are *scale invariant* function spaces with respect to the scaling of solutions u of the Navier-Stokes system: If u (with associated pressure p) is a solution of (1.1) in $[0, \infty) \times \mathbb{R}^3$, then for each $\lambda > 0$ the vector field

$$u_\lambda(t, x) = \lambda u(\lambda^2 t, \lambda x)$$

(together with $p_\lambda(t, x) = \lambda^2 p(\lambda^2 t, \lambda x)$) is a solution of (1.1) with the same coefficient of viscosity. Then a Banach space $(\mathcal{X}, \| \cdot \|_\mathcal{X})$ of functions u on $(0, \infty) \times \mathbb{R}^3$ is said to be scale invariant iff $\|u\|_\mathcal{X} = \|u_\lambda\|_\mathcal{X}$ for all $\lambda > 0$ and $u \in \mathcal{X}$. Although for a finite time interval and a domain $\Omega \neq \mathbb{R}^3$ this scaling also modifies the time interval and the domain, its impact on regularity theory and questions of uniqueness is the same. It is easily seen that the spaces $L^s(0, T; L^q(\Omega))$ are scale invariant with respect to the above scaling if and only if $\frac{2}{s} + \frac{3}{q} = 1$. Looking at $s = \infty$ or ignoring the time variable, the space $L^3(\Omega)$ is scale invariant. The same holds when $L^q(\Omega)$ is replaced by a Lorentz space $L^{q,r}(\Omega)$ for any $1 \leq r \leq \infty$.

The (almost) optimal and scale invariant spaces we are looking for are Besov spaces $B_{q,s}^r(\Omega)$ with order of differentiability $r = -1 + \frac{3}{q}$. Since we are mainly considering domains $\Omega \neq \mathbb{R}^3$ there is no need to introduce carefully the definition and characterization of Besov spaces $B_{q,s}^r(\mathbb{R}^3)$ by Littlewood-Paley theory (see

[6], [79]), although those spaces are one of the starting points for the definition of Besov spaces on domains. The latter Besov spaces are introduced either by restriction of functions or distributions in $B^r_{q,s}(\mathbb{R}^3)$ to Ω ([79]) or, equivalently, by real interpolation of Sobolev spaces on Ω and duality arguments. This relation is analyzed in details by Amann in [2, 3, 4] for the spaces $B^r_{q,s}(\Omega)$. Subspaces of solenoidal vector fields in $B^r_{q,s}(\Omega)$ are denoted by $\mathbb{B}^r_{q,s}(\Omega)$. In the limit case $s = \infty$, i.e., with the second parameter of real interpolation $s = \infty$, the space $\mathbb{B}^{-1+3/q}_{q,\infty}(\Omega)$ will no longer be separable. The corresponding closed and separable subspace of $\mathbb{B}^{-1+3/q}_{q,\infty}(\Omega)$ defined by the *continuous interpolation method* is denoted by $\mathring{\mathbb{B}}^{-1+3/q}_{q,\infty}(\Omega)$ as in [2, 3, 4]. For details, in particular, on equivalent norms for these Besov spaces, we refer to Sect. 4.

To consider (1.1) as an abstract nonlinear evolution equation we introduce the Helmholtz projection on $L^q(\Omega)$,

$$P_q : L^q(\Omega) \to L^q_\sigma(\Omega) = \overline{C^\infty_{0,\sigma}(\Omega)}^{\|\cdot\|_q},$$

where $1 < q < \infty$ and $C^\infty_{0,\sigma}(\Omega) = \{u \in C^\infty_0(\Omega) : \operatorname{div} u = 0\}$. We recall that P_q is a well-defined bounded projection for bounded and exterior C^1-domains and defines an algebraic and topological decomposition

$$L^q(\Omega) = L^q_\sigma(\Omega) \oplus G_q(\Omega)$$

with range $L^q_\sigma(\Omega)$ and kernel $G_q(\Omega) = \{\nabla p \in L^q(\Omega) : p \in L^q_{\mathrm{loc}}(\overline{\Omega})\}$ of P_q; for details see [40], [72], [73].

Then we define the Stokes operator $A_q = -P_q\Delta$ on $L^q_\sigma(\Omega)$, $1 < q < \infty$, by

$$\mathcal{D}(A_q) = W^{2,q}(\Omega) \cap W^{1,q}_0(\Omega) \cap L^q_\sigma(\Omega),$$
$$A_q : \mathcal{D}(A_q) \subset L^q_\sigma(\Omega) \to L^q_\sigma(\Omega), \quad u \mapsto A_q u = -P_q\Delta u. \tag{1.2}$$

It is well-known, see *e.g.* [30], [43], that $-A_q$ generates a bounded analytic semigroup $\{e^{-tA_q}; t \geq 0\}$ for bounded and exterior domains of class $C^{1,1}$. Since A_q coincides with A_r on $\mathcal{D}(A_q) \cap \mathcal{D}(A_r)$, $1 < r, q < \infty$, we often simply write A; by analogy, since P_q and P_r coincide on $L^q(\Omega) \cap L^r(\Omega)$, we may write only P. For further details we refer to Subsect. 4.1 below. Note that in general non-smooth or general unbounded domains P and A may fail to exist, see [8], [64]. However, for $q = 2$ and any domain $\Omega \subset \mathbb{R}^3$, Hilbert space methods can be used to define P_2 and $A_2 = -P_2\Delta$ with the properties mentioned above.

Applying the Helmholtz projection P we get rid of the pressure term ∇p in (1.1) and rewrite the Navier-Stokes system as the abstract nonlinear evolution equation

$$u_t + Au + P(u \cdot \nabla u) = Pf \text{ in } (0, T), \quad u(0) = u_0 \tag{1.3}$$

in $L^q_\sigma(\Omega)$. Here we assume the condition $u_0 \in L^q_\sigma(\Omega)$ leading to $P_q u_0 = u_0$. Now we are in the position to introduce several notions of solutions u to (1.1).

Definition 1.1. (i) *A weak solution u of* (1.1) *is a solution in the sense of distributions*, i.e.

$$-\langle u, w_t \rangle_{\Omega,T} + \langle \nabla u, \nabla w \rangle_{\Omega,T} + \langle u \cdot \nabla u, w \rangle_{\Omega,T} = \langle f, w \rangle_{\Omega,T} + \langle u_0, w(0) \rangle_\Omega \tag{1.4}$$

for all test functions $w \in C_0^\infty\big([0,T); C_{0,\sigma}^\infty(\Omega)\big)$, *and if it is contained in the* Leray-Hopf *class*

$$u \in \mathcal{LH}_T = L^\infty(0,T; L_\sigma^2(\Omega)) \cap L_{\mathrm{loc}}^2\big([0,T); H_0^1(\Omega)\big). \tag{1.5}$$

Moreover, u *is called a* weak solution *in the sense of Leray-Hopf if it satisfies the energy inequality* (EI)

$$\frac{1}{2}\|u(t)\|_2^2 + \int_0^t \|\nabla u\|_2^2\,\mathrm{d}\tau \le \frac{1}{2}\|u_0\|_2^2 + \int_0^t \langle f, u\rangle_\Omega\,\mathrm{d}\tau \tag{1.6}$$

for all $t \in [0,T)$.

(ii) *A weak solution* $u \in \mathcal{LH}_T$ *is called a* strong solution (in the sense of Serrin) *if there are exponents* s, q *such that*

$$u \in L^s(0,T; L^q(\Omega)), \quad s > 2,\ q > 3, \quad \frac{2}{s} + \frac{3}{q} = 1. \tag{1.7}$$

Under the assumption $\frac{2}{s} + \frac{3}{q} = 1$ *the space* $L^s\big(0,T; L^q(\Omega)\big)$ *is called a* Serrin class. *Note that by this definition a strong solution is a weak one.*

(iii) *A* mild solution u *is a solution of the nonlinear integral equation*

$$u(t) = e^{-tA}u_0 + \int_0^t e^{-(t-\tau)A}P\big(f(\tau) - \mathrm{div}\,(u \otimes u)(\tau)\big)\,\mathrm{d}\tau \tag{1.8}$$

which originates from the variation of constants formula applied to the evolution problem (1.3).

Let us recall several important results on these different notions of solutions.

Remark 1.2. (1) The existence of weak solutions is known for arbitrary domains $\Omega \subset \mathbb{R}^3$, initial values in $u_0 \in L_\sigma^2(\Omega)$ and forces $f \in L^1(0,T; L^2(\Omega))$. More generally, f may be a functional of the form $f = \mathrm{div}\,F$, $F \in L^2(0,T; L^2(\Omega))$ so that in (1.4)

$$\langle f, w\rangle_{\Omega,T} := -\langle F, \nabla w\rangle_{\Omega,T}.$$

Redefining the weak solution u on a null set of $(0,T)$ we may assume that

$$u : [0,T) \to L_\sigma^2(\Omega) \text{ is weakly continuous.} \tag{1.9}$$

(2) Weak Leray-Hopf solutions can be constructed by several methods, *e.g.* by Galerkin's approximation method, approximation by time discretization or by approximation of the term $u \cdot \nabla u$ by $J_k u \cdot \nabla u$ via Yosida operators $J_k = \big(I + \frac{1}{k}A^{1/2}\big)^{-1}$.

(3) Weak Leray-Hopf solutions constructed as in (2) satisfy the energy inequality (EI) in (1.6). The reason for the inequality rather than an equality is the use of approximation techniques and the lower semi-continuity of the norm $\|\cdot\|_2$ with respect to weak convergence. When Ω is bounded, as a consequence of compact embedding theorems and the resulting strong convergence of the sequence of

approximate solutions in $L^2(0,T;L^2(\Omega))$, weak solutions can be shown to satisfy the *strong energy inequality* (SEI)

$$\frac{1}{2}\|u(t)\|_2^2 + \int_{t_0}^t \|\nabla u\|_2^2\,d\tau \leq \frac{1}{2}\|u(t_0)\|_2^2 + \int_{t_0}^t \langle f,u\rangle_\Omega\,d\tau \qquad (1.10)$$

for *almost all* $t_0 \in [0,T)$ (including $t_0 = 0$) and for all $t \in (t_0,T)$. In the case of an exterior domain maximal regularity estimates yielding $L^s(L^q)$-estimates of the associated pressure gradient ∇p are needed to construct a weak solution satisfying (SEI), see [67].

When a weak solution satisfies (1.10) for a specific t_0, we will say that u satisfies $(EI)_{t_0}$. Hence

$$u \text{ satisfies } (SEI) \quad \Leftrightarrow \quad u \text{ satisfies } (EI)_{t_0} \text{ for a.a. } t_0 \in [0,T].$$

It is an open problem whether (EI) for a weak solution implies (SEI). Moreover, it is even open whether every weak solution satisfying (1.4) and (1.5) also satisfies the energy inequality (EI).

(4) One of the fundamental problems on weak solutions is the question of uniqueness. A classical theorem, the so-called *Serrin uniqueness theorem*, states that a weak solution u lying in a Serrin class $L^s(0,T;L^q(\Omega))$, see (1.7), is unique in the class of all Leray-Hopf weak solutions. Note that by definition a Leray-Hopf type weak solution satisfies (EI) and that a weak solution in a Serrin class even satisfies the energy equality (EE), see (5) below. Uniqueness for a weak solution in the limit class $L^\infty(0,T;L^3(\Omega))$ was proved by Kozono and Sohr [57] and in [35], see also Remark 3.10 below.

(5) Less restrictive assumptions than in (4) are needed to prove that a weak solution u satisfies the energy equality (EE), *i.e.*

$$\frac{1}{2}\|u(t)\|_2^2 + \int_0^t \|\nabla u\|_2^2\,d\tau = \frac{1}{2}\|u_0\|_2^2 + \int_0^t \langle f,u\rangle_\Omega\,d\tau \qquad (1.11)$$

for all $t \in (0,T)$. A classical theorem requires the condition

$$u \in L^4(0,T;L^4(\Omega)) \qquad (1.12)$$

or equivalently $u \otimes u \in L^2(0,T;L^2)$. In this case, the term $u \cdot \nabla u = \text{div}\,(u \otimes u)$ has the same properties as the external force $\text{div}\,F$ with $F \in L^2(0,T;L^2)$. In particular, u may be used as test function w in (1.4), $(u \cdot \nabla u)u \in L^1(0,T;L^1)$, and the fact that $\langle u \cdot \nabla u, u\rangle_\Omega(t) = 0$ for a.a. $t \in (0,T)$ implies that $\langle u \cdot \nabla u, w\rangle_{\Omega,T} = 0$ for $w = u$ in (1.4). However, in the general three-dimensional case a weak solution $u \in \mathcal{LH}_T$ is not an admissible test function in (1.4).

Assumptions different to (1.12) were discussed by Shinbrot [71]: if $u \in L^r(0,T;L^q)$ where $\frac{2}{r} + \frac{3}{q} \leq 1 + \frac{1}{q}$, $q \geq 4$, then u satisfies (EE). Actually, Shinbrot's condition together with the Leray-Hopf integrability $u \in L^\infty(0,T;L^2)$ implies by Hölder's inequality that $u \in L^4(0,T;L^4)$. A similar argument can be applied when $u \in L^r(0,T;L^q)$, where $\frac{2}{r} + \frac{3}{q} \leq 1 + \frac{1}{r}$, $r \geq 4$, together with the fact that $u \in L^2(0,T;L^6)$. Weaker assumptions than (1.12) will be discussed in §3.2.

(6) The main open problem for weak solutions is the question of regularity, see [38]: is every weak solution u (and an associated pressure p) of class C^∞ in space and time provided that f, u_0 and $\partial\Omega$ are of class C^∞? The classical result requires that u lies in Serrin's class $L^s(0,T;L^q)$ as in (1.7); the proof of the implication $u \in L^s(0,T;L^q) \Rightarrow u \in C^\infty$ is based on mathematical induction, $cf.$ [74, Ch. V, Theorems 1.8.1 and 1.8.2].

(7) In the definition of mild solutions, see (1.8), the term $\operatorname{div}(u\otimes u) = u\cdot\nabla u$ is assumed to be at least integrable in space-time such that $e^{-(t-\tau)A}P\operatorname{div}(u\otimes u)(\tau)$ is well defined. However, if u is not weakly differentiable or if the nonlinear term $u\cdot\nabla u$ has only "low" integrability properties in space-time, a smoothing operator such as $A^{-1/2}$ will be used in front of this term in (1.8). Moreover, in general the force term is assumed to have the form $f = \operatorname{div} F$ with a matrix-valued function $F \in L^2(0,T;L^2(\Omega))$. In this case the variation of constants formula (1.8) has to be rewritten in the form

$$u(t) = e^{-tA}u_0 + \int_0^t A^{-1/2}e^{-(t-\tau)A}\big(A^{-1/2}P\operatorname{div}\big)\big(-F(\tau) - (u\otimes u)(\tau)\big)\,\mathrm{d}\tau. \quad (1.13)$$

The meaning of this formula will be explained in Subsect. 4.1 and used frequently throughout this article.

(8) Problems described as in (7) are inherent in the definition of *very weak solutions* which are solenoidal vector fields in Serrin's class $L^s(0,T;L^q(\Omega))$ but without any kind of differentiability except for the solenoidality $\operatorname{div} u = 0$. In this case all derivatives in space and time are applied to the test functions. These solutions can also be defined via duality arguments from strong solutions as in [26]. The theory of very weak solutions is used to construct solutions with low regularity data and to prove regularity results for weak solutions by identifying a weak solution with a very weak one, see [27, 28, 29]. For a result in exterior domains we refer to [23]. In the framework of Besov spaces very weak solutions are investigated by Amann ([2, 3, 4]).

These and further related results, including also unbounded domains and the Boussinesq system, can be found in the articles [23, 24], [25, 27, 28, 29], [31, 32], [33, 34, 35, 36], [52, 53, 54].

This article is organized as follows. In Sect. 2 we present the main results on existence of weak solutions with optimal and almost optimal initial values, postponing the proofs to Sect. 5. §2.1 starts with the discussion of the optimal initial value condition on u_0 to guarantee the existence of a local strong solution u. The main Theorem 2.1 is the basis for most of the following results and uses a Besov space characterization to be introduced in the beginning of §2.1 and, more thoroughly, in Sect. 4. The next subsection, §2.2, considers weaker assumptions on u_0 and needs algebraic weights in time for the initial value, the solution u and the external force field F. §2.3 deals with the continuity of u in time with values in the Besov spaces used in §2.1–§2.2. Whereas §2.1 directly leads to Serrin's uniqueness theorem, a corresponding result under the assumptions of §2.2 for domains $\Omega \neq \mathbb{R}^3$ is not known; conditional uniqueness results are presented in §2.4. The optimal initial value condition of §2.1 is used in §3.1 to improve known regularity criteria to some extent beyond Serrin's condition. Similar results are not available for almost optimal initial values since an analogue of Serrin's uniqueness theorem is not proved. The question of the validity of the energy equality and of uniqueness of weak solutions is addressed in §3.2. For the case $\Omega = \mathbb{R}^3$ which admits the use of special techniques of harmonic analysis in Besov spaces some classical and also recent results will be shortly discussed in §3.3. The longer proofs of results of §2.1–§2.2 are presented in Sect. 5. Next, details of the technical proof of continuity in time, see §2.3, are found in Sect. 6. The conditional uniqueness results of §2.4 are proved in Sect. 7. Finally, proofs of results from Sect. 3 are collected in Sect. 8.

2 Main results

2.1 Optimal initial values

The aim is to find an optimal condition on initial values $u_0 \in L^2_\sigma(\Omega)$ to allow for a local-in-time strong solution $u \in L^{s_q}(0, T; L^q(\Omega))$ of Serrin type $(\frac{2}{s_q} + \frac{3}{q} = 1)$ of the Navier-Stokes system (1.1). It is natural to require that at least the solution

$$E_0(t) := e^{-tA} u_0$$

of the corresponding linear Stokes system with vanishing external force has the property $E_0 \in L^{s_q}(0, T; L^q(\Omega))$. Actually, this condition which is well-known for the case of the whole space $\Omega = \mathbb{R}^3$ yields both a necessary and sufficient condition for smooth bounded and exterior domains, see [33, 34], [24, 52]. The integrability condition on E_0, say

$$\int_0^\infty \|e^{-\tau A} u_0\|_q^{s_q} \, d\tau < \infty, \tag{2.1}$$

can be considered in terms of Besov spaces. Starting with the classical Besov space $B^{2/s_q}_{q',s'_q}(\Omega)$ for a domain $\Omega \subset \mathbb{R}^3$ (cf. [79, Ch. 4]) where q', s'_q are the conjugate exponents to q, s_q, respectively, and $\frac{2}{s_q} + \frac{3}{q} = 1$, solenoidal subspaces

$$\mathbb{B}^{2/s_q}_{q',s'_q}(\Omega) = B^{2/s_q}_{q',s'_q}(\Omega) \cap L^{q'}_\sigma(\Omega) = \{v \in B^{2/s_q}_{q',s'_q}(\Omega) : \text{ div } v = 0, \ N \cdot v|_{\partial\Omega} = 0\}$$

were defined in [3]. Actually, $B^{2/s_q}_{q',s'_q}(\Omega)$ is the optimal space of initial values v_0 such that the solution $e_0(t) = e^{t\Delta} v_0$ of the scalar heat equation with Dirichlet boundary condition on $\partial\Omega$ satisfies $(e_0)_t, \Delta e_0(\cdot) \in L^{s_q}(0, T; L^{q'}(\Omega))$. By analogy, $\mathbb{B}^{2/s_q}_{q',s'_q}(\Omega)$ defines the optimal space of solenoidal vector fields u_0 such that $E_0(t) = e^{-tA} u_0$ is a classical strong solution of the homogeneous Stokes problem with initial value u_0 satisfying $(E_0)_t, AE_0(\cdot) \in L^{s_q}(0, T; L^{q'}_\sigma(\Omega))$.

To pass from strong solutions to weak ones we need the dual space

$$\mathbb{B}^{-1+3/q}_{q,s_q}(\Omega) = \mathbb{B}^{-2/s_q}_{q,s_q}(\Omega) := \left(\mathbb{B}^{2/s_q}_{q',s'_q}(\Omega)\right)^* \tag{2.2}$$

which has the equivalent norm

$$\|u_0\|_{\mathbb{B}_{q,s_q}^{-1+3/q}} = \left(\int_0^\infty \|e^{-\tau A} u_0\|_q^{s_q} \, d\tau \right)^{1/s_q}. \tag{2.3}$$

We note that $\left(\int_0^\infty \|e^{-\tau A} u_0\|_q^{s_q} \, d\tau \right)^{1/q}$ and $\left(\int_0^T \|e^{-\tau A} u_0\|_q^{s_q} \, d\tau \right)^{1/q}$ define equivalent norms on $\mathbb{B}_{q,s_q}^{-1+3/q}(\Omega)$ for any $0 < T \le \infty$ since the semigroup $\{e^{-tA}\}$ decays exponentially fast, see (4.3). For later use it is reasonable to work with such an equivalent norm, and we introduce the notation

$$\|u_0\|_{\mathbb{B}_{q,s_q;T}^{-1+3/q}} = \left(\int_0^T \|e^{-\tau A} u_0\|_q^{s_q} \, d\tau \right)^{1/s_q}. \tag{2.4}$$

If $u_0 \in L_\sigma^2(\Omega)$ as in our application, the norms in (2.3) and (2.4) are well defined (or equal to infinity) since $e^{-\tau A} u_0 \in L_\sigma^q(\Omega)$ for each $\tau > 0$. Without the assumption $u_0 \in L_\sigma^2(\Omega)$ the semigroup $\{e^{-tA}\}$ applied to a functional u_0 must be defined by an extrapolation method, see §4.3.

For further details on these Besov spaces and norms we refer to Sect. 4.

Theorem 2.1. *Let $\Omega \subset \mathbb{R}^3$ be a bounded domain with boundary $\partial\Omega \in C^{1,1}$, let $0 < T \le \infty$, $2 < s_q < \infty$, $3 < q < \infty$ with $\frac{2}{s_q} + \frac{3}{q} = 1$ be given, and consider the Navier-Stokes system with initial value $u_0 \in L_\sigma^2(\Omega)$ and an external force $f = \operatorname{div} F$ where $F \in L^2(0,T; L^2(\Omega)) \cap L^{s_q/2}(0,T; L^{q/2}(\Omega))$.*

(i) There exists an absolute constant $\varepsilon_* = \varepsilon_*(q,\Omega) > 0$ with the following property: If $u_0 \in \mathbb{B}_{q,s_q}^{-1+3/q}(\Omega)$ and

$$\|u_0\|_{\mathbb{B}_{q,s_q;T}^{-1+3/q}} + \|F\|_{L^{s_q/2}(0,T;L^{q/2})} \le \varepsilon_*, \tag{2.5}$$

then the Navier-Stokes system (1.1) has a unique strong solution $u \in L^{s_q}(0,T; L^q(\Omega))$.

(ii) The condition $u_0 \in \mathbb{B}_{q,s_q}^{-1+3/q}(\Omega)$ is necessary and sufficient for the existence and uniqueness of a local-in-time strong solution $u \in L^{s_q}(0,T'; L^q(\Omega))$ of (1.1) for some $0 < T' \le T$.

We note that the second integrability condition on F in Theorem 2.1, *i.e.*, $F \in L^{s_q/2}(0,T; L^{q/2}(\Omega))$ as used in [31, 33], can be generalized to $F \in L^{s_*/2}(0,T; L^{q_*/2}(\Omega))$ where $\frac{2}{s_*} + \frac{3}{q_*} = 1$ and $0 \le \frac{3}{q_*/2} - \frac{3}{q} \le 1$. The proof of Theorem 2.1 in Sect. 5 will include this more general assumption.

Next let us recall further results and extensions of Theorem 2.1.

Remark 2.2. (1) A classical result, see *e.g.* [50], states that an initial value $u_0 \in \mathcal{D}(A_2)$ allows for a local strong solution. The much weaker condition $u_0 \in \mathcal{D}(A_2^{1/4})$ is due to Fujita and Kato [39] for a smooth bounded domain. Note that the condition $u_0 \in \mathcal{D}(A_2^{1/4})$ has the important property of being scale invariant; to be more precise, the leading part $\|A^{1/4} u_0\|_2$ of the graph norm of $\mathcal{D}(A_2^{1/4})$ is scale invariant. Since this result is based on L^2-theory, it can be generalized to arbitrary bounded and unbounded domains, see [74, Ch. V, Theorem 4.2.2].

(2) Fabes, Jones and Rivière [19] as well as Miyakawa [66] proved that the condition $u_0 \in L^q_\sigma(\Omega)$, $q > 3$, yields a local strong solution. An extension to Lorentz spaces $L^{q,r}_\sigma(\Omega)$, $q > 3$, $r \le \infty$, and solutions in $L^{s_q,\infty}(0,T;L^{q,\infty})$ is found in an article by Sohr [75].

(3) The crucial step from $\mathcal{D}(A_2^{1/4})$ to the condition $u_0 \in L^3_\sigma(\Omega)$ was performed by Kato [49] for $\Omega = \mathbb{R}^3$, introducing the famous Kato iteration scheme; Giga and Miyakawa [46] and Giga [45] extended the results to domains $\Omega \subset \mathbb{R}^3$.

(4) This shows that the spaces for initial values to yield a local in time strong solution were getting larger and larger during the last decades. The optimal and largest space is the Besov space $\mathbb{B}_{q,s_q}^{-1+3/q}$. Here we mention the continuous embeddings

$$\mathcal{D}(A_2^{1/4}) \hookrightarrow L^3_\sigma(\Omega) \hookrightarrow L^{3,s}_\sigma(\Omega) \hookrightarrow \mathbb{B}_{q,s_q}^{-1+3/q}(\Omega) \qquad (2.6)$$

where each space can be equipped with an equivalent and scale invariant norm; for the latter embedding which holds when $3 < q \le s \le s_q \le \infty$ we refer to [3, (0.16)].

(5) For a smooth exterior domain $\Omega \subset \mathbb{R}^3$ results similar to Theorem 2.1 were obtained in [24], [52]. Since in this case Sobolev embeddings are more restricted than for bounded domains, also a smallness assumption on F in $L^2(L^2)$ is needed.

(6) For a general bounded or unbounded domain $\Omega \subset \mathbb{R}^3$ only L^2-theory for the Stokes operator and the Helmholtz projection is available. In this case Theorem 2.1(i) holds with the exponents $q = 4$, $s = 8$, cf. [33, Sect. 4]. The exponent $q = 4$ occurs since due to the nonlinear term $u \otimes u$ the analysis must be performed in the space $L^{q/2}(\Omega) = L^2(\Omega)$.

2.2 Almost optimal initial values

In view of the optimal space $\mathbb{B}_{q,s_q}^{-1+3/q}(\Omega)$ it makes sense to analyze the less optimal situation when replacing s_q by $s \in (s_q, \infty]$ and $\mathbb{B}_{q,s_q}^{-1+3/q}(\Omega)$ by the slightly larger space $\mathbb{B}_{q,s}^{-1+3/q}(\Omega)$; note that s acts as the second parameter in the real interpolation of two Sobolev spaces. Since $\frac{2}{s_q} + \frac{3}{q} = 1$, the term s will satisfy the condition

$$\frac{2}{s} + \frac{3}{q} = 1 - 2\alpha \quad \text{where } 0 < \alpha < \frac{1}{2}. \qquad (2.7)$$

Actually, as shown in §4 the space $\mathbb{B}_{q,s}^{-1+3/q}(\Omega)$ has the equivalent weighted norm

$$\|u\|_{\mathbb{B}_{q,s}^{-1+3/q}} = \begin{cases} \left(\displaystyle\int_0^\infty \left(\tau^\alpha \|e^{-\tau A} u\|_q \right)^s \, d\tau \right)^{1/s} & \text{if } s < \infty, \\[2mm] \sup_{0 < \tau < \infty} \tau^\alpha \|e^{-\tau A} u\|_q & \text{if } s = \infty. \end{cases} \qquad (2.8)$$

Since Ω is bounded and $\tau^\alpha \|e^{-\tau A} u\|_q$ tends to 0 exponentially fast as $\tau \to \infty$, see (4.3) below, the behaviour of $\tau \mapsto \tau^\alpha e^{-\tau A} u$ in (2.8) is crucial only for small $\tau > 0$. Restricting the range of τ to $(0,T)$, $0 < T < \infty$, we get equivalent norms on

$\mathbb{B}_{q,s}^{-1+3/q}$ which will be denoted by $\|u_0\|_{\mathbb{B}_{q,s;T}^{-1+3/q}}$. In this context, it is important to note that the corresponding subspace of the non-separable space $\mathbb{B}_{q,\infty}^{-1+3/q}$ defined by the continuous interpolation method, $\mathring{\mathbb{B}}_{q,\infty}^{-1+3/q}$, is characterized by the criterion

$$\mathring{\mathbb{B}}_{q,\infty}^{-1+3/q}(\Omega) = \{u \in \mathbb{B}_{q,\infty}^{-1+3/q}(\Omega) : \tau^\alpha \|e^{-\tau A}u\|_q \to 0 \text{ as } \tau \to 0\}. \quad (2.9)$$

The first theorem concerns the case when $s < \infty$, see Theorem 2.3 below. Here we also need Bochner spaces with weights τ^α in time: For $1 \le s, q < \infty$ and $\alpha \in \mathbb{R}$ let $L_\alpha^s(0,T;L^q(\Omega))$ denote the space of measurable functions $u(\tau,x)$ such that $\|u\|_{L_\alpha^s(0,T;L^q)} := \|\tau^\alpha u\|_{L^s(0,T;L^q)}$ is finite. The modification in the case $s = \infty$ is obvious. However, in this case, in Theorem 2.4 and Corollary 2.5 we differ between the Besov space $\mathbb{B}_{q,\infty}^{-1+3/q}$ and its closed subspace $\mathring{\mathbb{B}}_{q,\infty}^{-1+3/q}$. To get consistent results also the weighted space $L_{2\alpha}^\infty(0,T;L^q(\Omega))$ for the external force F on $(0,T)$ is to be replaced by the closed subspace

$$\mathring{L}_{2\alpha}^\infty(0,T;L^q(\Omega)) = \Big\{F \in L_{2\alpha}^\infty(0,T;L^q(\Omega)) : \operatorname*{ess\,sup}_{\tau\in(0,t)} \tau^{2\alpha}\|F(\tau)\|_q \to 0 \text{ as } t \to 0\Big\}.$$

The following results with slightly stronger assumptions on F can be found in [20, 21]. Note that the proofs are very similar in all cases and will be carried out together.

Theorem 2.3. *Let* $0 < T \le \infty$, $u_0 \in L_\sigma^2(\Omega)$ *and* $f = \operatorname{div} F$ *where* $F \in L^2(0,T;L^2(\Omega)) \cap L_{2\alpha_*}^{s_*/2}(0,T;L^{q_*/2}(\Omega))$; *here,* $2 < s, s_* < \infty$, $3 < q, q_* < \infty$ *and* $0 < \alpha, \alpha_* < \frac{1}{2}$ *satisfying*

$$\frac{2}{s} + \frac{3}{q} = 1 - 2\alpha, \quad \frac{2}{s_*} + \frac{3}{q_*} = 1 - 2\alpha_* \quad s \ge \frac{s_*}{2}, \ \alpha \le 2\alpha_*, \quad (2.10)$$

$$0 \le \frac{3}{q_*/2} - \frac{3}{q} \le 1 \quad or, \ equivalently, \quad \frac{1}{s} + \alpha \le 2\Big(\frac{1}{s_*} + \alpha_*\Big) \le \frac{1}{2} + \frac{1}{s} + \alpha.$$
$$\quad (2.11)$$

(i) *There exists a constant* $\epsilon_* = \epsilon_*(q, q_*, s, s_*, \alpha, \alpha_*, \Omega) > 0$ *with the following property: If* $u_0 \in \mathbb{B}_{q,s}^{-1+3/q}(\Omega)$ *and*

$$\|u_0\|_{\mathbb{B}_{q,s;T}^{-1+3/q}} + \|F\|_{L_{2\alpha_*}^{s_*/2}(0,T;L^{q_*/2})} \le \varepsilon_*, \quad (2.12)$$

then the Navier-Stokes system (1.1) *has a unique strong* $L_\alpha^s(L^q)$-*solution with data* u_0, f *on the interval* $[0,T)$. *This solution even satisfies the energy equality* (EE) *on* $[0,T)$.

(ii) *The condition* $u_0 \in \mathbb{B}_{q,s}^{-1+3/q}(\Omega)$ *is sufficient and necessary for the existence of a* (unique) *local-in-time strong* $L_\alpha^s(L^q)$-*solution of the Navier-Stokes system* (1.1).

Theorem 2.4. *Let* $0 < T \le \infty$, $3 < q < \infty$, $s = \infty$, $2 < s_* \le \infty$ *and* $0 < \alpha, \alpha_* < \frac{1}{2}$ *be given. Consider the Navier-Stokes equation* (1.1) *with initial value* $u_0 \in L_\sigma^2(\Omega)$

and an external force $f = \operatorname{div} F$ *where* $F \in L^2(0, T; L^2(\Omega)) \cap L_{2\alpha_*}^{s_*/2}(0, T; L^{q_*/2}(\Omega))$
and

$$\frac{3}{q} = 1 - 2\alpha, \quad \frac{2}{s_*} + \frac{3}{q_*} = 1 - 2\alpha_* \quad and \quad 0 \le \frac{3}{q_*/2} - \frac{3}{q} < 1, \ \alpha < 2\alpha_*. \quad (2.13)$$

(i) *There exists a constant* $\epsilon_* = \epsilon_*(q, q_*, s_*, \alpha, \alpha_*, \Omega) > 0$ *such that if* $u_0 \in \mathbb{B}_{q,\infty}^{-1+3/q}(\Omega)$ *and*

$$\|e^{-\tau A} u_0\|_{L_\alpha^\infty(0,T;L^q)} + \|F\|_{L_{2\alpha_*}^{s_*/2}(0,T;L^{q_*/2})} \le \epsilon_*, \quad (2.14)$$

then there exists a strong $L_\alpha^\infty(L^q)$-*solution with data* u_0, f *on the interval* $[0, T)$. *This solution is unique in the class of all strong* $L_\alpha^\infty(L^q)$-*solutions on* $(0, T)$ *with sufficiently small norm in* $L_\alpha^\infty(L^q(\Omega))$ *and satisfies the energy equality* (EE) *on* $[0, T)$.

(ii) *If* $u_0 \in L_\sigma^2(\Omega) \cap \mathring{\mathbb{B}}_{q,\infty}^{-1+3/q}(\Omega)$ *and* $F \in L^2(0, T; L^2(\Omega)) \cap \mathring{L}_{2\alpha_*}^\infty(0, T; L^{q_*/2}(\Omega))$, *then* (1.1) *has a unique strong* $L_\alpha^\infty(L^q)$-*solution with data* u_0, f *on some interval* $[0, T') \subset [0, T)$.

Concerning necessity of the conditions in Theorem 2.4 we mention the following results.

Corollary 2.5. *Suppose that the assumptions of Theorem 2.4 are fulfilled.*
(i) *The condition*

$$\operatorname*{ess\,sup}_{\tau \in (0,\infty)} \tau^\alpha \|e^{-\tau A} u_0\|_q < \infty \quad (2.15)$$

is necessary for the existence of a unique strong $L_\alpha^\infty(L^q)$-*solution* $u \in L_\alpha^\infty(0, T; L^q(\Omega))$ *of* (1.1) *with data* u_0, f *in some interval* $[0, T)$, $0 < T \le \infty$.
(ii) *If additionally* $F \in \mathring{L}_{2\alpha_*}^\infty(0, T; L^{q_*/2}(\Omega))$, *then the condition*

$$u_0 \in \mathring{\mathbb{B}}_{q,\infty}^{-1+\frac{3}{q}}(\Omega) \quad (2.16)$$

is even necessary and sufficient for the existence of a unique strong $L_\alpha^\infty(L^q)$-*solution* $u \in \mathring{L}_\alpha^\infty(0, T; L^q(\Omega))$ *of* (1.1).

2.3 Continuity in time

In addition to the weak L^2-continuity of weak solutions, see (1.9), the solutions constructed in Theorems 2.1, 2.3 and 2.4, are continuous in time with values in the corresponding Besov spaces $\mathbb{B}_{q,s}^{-1+3/q}(\Omega)$, $q \le s \le \infty$, and $\mathring{\mathbb{B}}_{q,\infty}^{-1+3/q}(\Omega)$. Actually, this continuity is a property of mild $\mathbb{B}_{q,s}^{-1+3/q}(\Omega)$-valued solutions and does not refer to L^2-theory; in particular, it is not assumed for the initial value that $u_0 \in L_\sigma^2(\Omega)$. The results of this subsection are found in [22].

Theorem 2.6. *Let* $\Omega \subset \mathbb{R}^3$ *be a bounded domain with* $C^{2,1}$ *boundary,* $0 < T \le \infty$, *and let the exponents* s, q, α *and* s_*, q_*, α_* *be given as in Theorem 2.3 or Theorem 2.4.*

(i) *Assume $s < \infty$ and let u be a strong $L_\alpha^s(L^q)$-solution with initial data $u_0 \in \mathbb{B}_{q,s}^{-1+3/q}$ and external force $f = \operatorname{div} F$ with $F \in L_{2\alpha_*}^{s_*/2}(0,T;L^{q_*/2})$. Then*

$$u \in C\big([0,T];\mathbb{B}_{q,s}^{-1+3/q}\big). \tag{2.17}$$

(ii) *Assume $s = s_* = \infty$ and let u be a strong $L_\alpha^\infty(L^q)$-solution with initial data $u_0 \in \mathring{\mathbb{B}}_{q,\infty}^{-1+3/q}$ and $f = \operatorname{div} F$ satisfying $F \in \mathring{L}_{2\alpha_*}^\infty(0,T;L^{q_*/2})$. Then*

$$u \in C\big([0,T];\mathring{\mathbb{B}}_{q,\infty}^{-1+3/q}\big). \tag{2.18}$$

We further observe that solutions are continuous with respect to initial data and external forces.

Theorem 2.7. *Under the assumptions of Theorem 2.6(i) or (ii) with data u_0, F, and solution u, let v be a strong $L_\alpha^s(L^q)$-solution with initial data $v_0 \in \mathbb{B}_{q,s}^{-1+3/q}$ and external force $G \in L_{2\alpha_*}^{s_*/2}(0,T;L^{q_*/2})$.*

(i) *If $s < \infty$ there are constants ε_* and $C > 0$ depending only on $q, q_*, s, s_*, \alpha, \alpha_*$ and Ω with the following property: If $\|u\|_{L_\alpha^s(0,T_0;L^q)} \le \varepsilon_*$ and $\|v\|_{L_\alpha^s(0,T_0;L^q)} \le \varepsilon_*$ for some $T_0 \le T$ then for all $t \in (0,T_0)$*

$$\|(u-v)(t)\|_{\mathbb{B}_{q,s}^{-1+3/q}} \le C\Big(\|u_0 - v_0\|_{\mathbb{B}_{q,s}^{-1+3/q}} + \|F - G\|_{L_{2\alpha_*}^{s_*/2}(0,T_0;L^{q_*/2})}\Big). \tag{2.19}$$

(ii) *If $s = s_* = \infty$ there are constants ε_* and C with the following property: If even $u_0, v_0 \in \mathring{\mathbb{B}}_{q,\infty}^{-1+3/q}$ and $F, G \in \mathring{L}_{2\alpha_*}^\infty(0,T;L^{q_*/2})$ and $\|u\|_{L_\alpha^\infty(0,T_0;L^q)} \le \varepsilon_*$, $\|v\|_{L_\alpha^\infty(0,T_0;L^q)} \le \varepsilon_*$ for some $T_0 \le T$, then for all $t \in (0,T_0)$*

$$\|(u-v)(t)\|_{\mathbb{B}_{q,\infty}^{-1+3/q}} \le C\Big(\|u_0 - v_0\|_{\mathbb{B}_{q,\infty}^{-1+3/q}} + \|F - G\|_{L_{2\alpha_*}^\infty(0,T_0;L^{q_*/2})}\Big). \tag{2.20}$$

2.4 Conditional uniqueness

It is an open problem whether any weak solution of the Navier-Stokes system (1.1) with initial value $u_0 \in L_\sigma^2(\Omega)$ is uniquely determined by u_0; the same holds for weak solutions in the sense of Leray-Hopf. By *Serrin's uniqueness theorem*, see Remark 1.2, the existence of a strong $L^{sq}(0,T;L^q)$-solution u implies that all weak solutions of Leray-Hopf type with the same initial value are unique; here it is crucial that Leray-Hopf weak solutions satisfy the energy inequality (EI). However, the classical proof does not seem to work for solutions with almost optimal initial values. On the other hand, uniqueness was proved by Barker [5] in the whole space case $\Omega = \mathbb{R}^3$ crucially exploiting Littlewood-Paley theory for Besov spaces $\mathbb{B}_{q,s}^{-1+3/q}(\mathbb{R}^3)$.

Therefore, we restrict ourselves to weak solutions constructed by an admissible approximation procedure. For such well-chosen weak solutions the weak-strong uniqueness theorem still holds.

In the following, for simplicity, we choose $q_* = q$, $s_* = s$, and $\alpha_* = \alpha$.

Definition 2.8. *Given $u_0 \in L^2_\sigma(\Omega)$ a well-chosen weak solution v is a weak solution of the Navier-Stokes system (1.1) with $v(0) = u_0$ satisfying the strong energy inequality (SEI), see (1.10), and defined by a concrete, so-called admissible approximation procedure, compatible with the notion of strong $L^s_\alpha(L^q)$-solutions in the following sense:*

(1) *There exists a sequence of initial data $(u_{0n}) \subset L^2_\sigma(\Omega) \cap \mathbb{B}_{q,s}^{-1+3/q}(\Omega)$ converging to u_0 in $L^2_\sigma(\Omega) \cap \mathbb{B}_{q,s}^{-1+3/q}(\Omega)$ as $n \to \infty$.*

(2) *The external force $F \in L^2(0,T;L^2) \cap L^{s/2}_{2\alpha}(0,T;L^{q/2})$ is approximated by a sequence $(F_n) \subset L^2(0,T;L^2) \cap L^{s/2}_{2\alpha}(0,T;L^{q/2})$ such that $F_n \to F$ in $L^2(0,T;L^2) \cap L^{s/2}_{2\alpha}(0,T;L^{q/2})$ as $n \to \infty$.*

(3) *The admissible approximation method yields approximate weak solutions u_n on $(0,T)$ uniformly bounded in $\mathcal{LH}_T$ and containing a subsequence (u_{n_k}) such that $u_{n_k} \rightharpoonup v$ in the sense of $\mathcal{LH}_T$, i.e., $u_{n_k} \rightharpoonup v$ in $L^2(0,T;H^1)$ and $u_{n_k} \overset{*}{\rightharpoonup} v$ in $L^\infty(0,T;L^2_\sigma)$ as $k \to \infty$.*

(4) *(u_n) is uniformly bounded in $L^s_\alpha(0,T';L^q)$ for some $T' \in (0,T]$.*

Remark 2.9. (1) The crucial part of Definition 2.8 is the assumption (4) on (u_n).

(2) The strong convergence $u_{0n} \to u_0$ in $L^2(\Omega)$ in Definition 2.8 (1) can be replaced by the corresponding weak convergence. By analogy, the strong convergence $F_n \to F$ in $L^2(0,T;L^2)$ may be replaced by a weak one.

(3) However, the strong convergence $u_{0n} \to u_0$ in $\mathbb{B}_{q,s}^{-1+3/q}(\Omega)$ is crucial to find $0 < T' \leq T$ independent of $n \in \mathbb{N}$ such that, as in (2.12), (2.14),

$$\int_0^{T'} \left(\tau^\alpha \|e^{-\tau A} u_{0n}\|_q\right)^s d\tau \leq \varepsilon_*.$$

A similar remark applies to the strong convergence $F_n \to F$ in $L^{s/2}_{2\alpha}(0,T;L^{q/2})$.

Now our main theorem reads as follows.

Theorem 2.10. *Let $2 < s < \infty$, $3 < q < \infty$, $0 < \alpha < \frac{1}{2}$ with $\frac{2}{s} + \frac{3}{q} = 1 - 2\alpha$ and suppose that $u_0 \in L^2_\sigma(\Omega) \cap \mathbb{B}_{q,s}^{-1+3/q}(\Omega)$ and an external force $f = \operatorname{div} F$ with $F \in L^2(0,T;L^2) \cap L^{s/2}_{2\alpha}(0,T;L^{q/2})$ are given. Furthermore, let $u \in L^s_\alpha(0,T;L^q)$ be the unique strong $L^s_\alpha(L^q)$-solution of (1.1) with data u_0, F.*

(i) *Then u is unique within the class of all well-chosen weak solutions of (1.1) in the sense of Definition 2.8.*

(ii) *Given an admissible approximation scheme yielding an approximate sequence (u_n) as in Definition 2.8 assume that each subsequence of (u_n) converging weakly in the sense of $\mathcal{LH}_T$ defines a limit which again is a weak Leray-Hopf solution of (1.1).*

Then the whole sequence (u_n) converges to u. Moreover, in this case, for any sequence of initial values (u_{0n}) and external forces (F_n) approximating u_0 and F in the sense of Definition 2.8(1), (2), respectively, and generating approximate solutions with a subsequence weakly convergent to a weak solution of (1.1) in $\mathcal{LH}_T$, the whole sequence of approximate solutions converges weakly in $\mathcal{LH}_T$ to u.

Note that in general a weak limit in the sense of $\mathcal{LH}_T$, due to a lack of strong convergence, is not necessarily a weak solution. This disadvantage is excluded, however, by the assumption in Theorem 2.10 (ii).

The crucial point in the application of Theorem 2.10 is to show that a concrete approximation procedure for the construction of weak solutions is compatible with the notion of strong $L_\alpha^s(L^q)$-solutions as required in Definition 2.8.

Theorem 2.11. *Let* $2 < s < \infty$, $3 < q < \infty$, $0 < \alpha < \frac{1}{2}$ *and* $\frac{2}{s} + \frac{3}{q} = 1 - 2\alpha$.

(i) The Yosida approximation scheme and, if $3 < q \leq 4$, *the Galerkin approximation scheme are admissible, i.e., they define well-chosen weak solutions in* $L_\alpha^s(L^q)$. *To be more precise, we consider the following approximation methods:*

<u>Case 1</u> *(Yosida approximation scheme) Let* $J_n = \left(I + \frac{1}{n}A^{1/2}\right)^{-1}$ *denote the Yosida operator, let* $u_{0n} = J_n u_0$, *and assume that* $F_n \to F$ *in* $L^2(0,T;L^2) \cap L_{2\alpha}^{s/2}(0,T^*;L^{q/2})$ *for some* $0 < T^* \leq T$. *Then the approximate solution* u_n *is defined as the solution of the approximate Navier-Stokes system*

$$\partial_t u_n - \Delta u_n + (J_n u_n) \cdot \nabla u_n + \nabla p_n = \operatorname{div} F_n, \quad \operatorname{div} u_n = 0,$$

$$u_n|_{\partial\Omega} = 0, \quad u_n(0) = u_{0n}. \tag{2.21}$$

<u>Case 2</u> *(Galerkin approximation scheme) Let* Π_n *denote the* L_σ^2-*projection onto the space of the first n eigenfunctions of the Stokes operator* A_2, *and suppose that* $u_{0n} \in \Pi_n L_\sigma^2(\Omega)$ *as well as* $F_n \in L^2(0,T;L^2)$ *satisfy the assumptions of Definition 2.8(1), (2). Then let* u_n *denote the Galerkin approximation of the Navier-Stokes system with data* u_{0n}, F_n.

(ii) In both cases the assumption in Theorem 2.10(ii) is satisfied. Hence the whole sequence given by the Yosida approximation scheme or the Galerkin approximation scheme converges to the well-chosen weak solution, irrespective of the sequences (u_{0n}) *and* (F_n).

3 Applications: Regularity

3.1 Regularity criteria

It suggests itself to exploit Theorem 2.1(i) not only at $t_0 = 0$, but at all or almost all $t_0 \in [0,T)$ to show that a weak solution is a strong one. For simplicity, let $F = 0$ in the following. Then the problem of global regularity splits into two subproblems: Show that

- u is *left-sided* $L^s(L^q)$-*regular at* $t_0 \in (0,T)$, i.e., there exists $\varepsilon = \varepsilon(t_0) \in (0,t_0)$ such that $u \in L^s((t_0 - \varepsilon, t_0); L^q(\Omega))$,
- u is *right-sided* $L^s(L^q)$-*regular at* $t_0 \in (0,T)$, i.e., there exists $\varepsilon = \varepsilon(t_0) \in (0, T - t_0)$ such that $u \in L^s((t_0, t_0 + \varepsilon); L^q(\Omega))$.

For the following detailed analysis recall the definition of the equivalent norm $\|\cdot\|_{\mathbb{B}_{q,s;T}^{-1+3/q}}$ in (2.4) of the Besov space $\mathbb{B}_{q,s}^{-1+3/q}(\Omega)$. Since for almost optimal initial values only a conditional uniqueness result is available and Serrin's uniqueness theorem will be a crucial tool, in the following the exponent s equals s_q such that $\frac{2}{s_q} + \frac{3}{q} = 1$.

Theorem 3.1. *Let $u \in \mathcal{LH}_T$ be a weak solution of (1.1) in a bounded smooth domain $\Omega \subset \mathbb{R}^3$ satisfying the strong energy inequality (SEI), and let $2 < s_q < \infty$, $3 < q < \infty$, $\frac{2}{s_q} + \frac{3}{q} = 1$.*

   (i) Let u satisfy $u(t) \in \mathbb{B}_{q,s_q}^{-1+3/q}(\Omega)$ for all $t \in [0,T)$. Given $t_1 \in (0,T)$ assume that

$$\limsup_{t \nearrow t_1} \|u(t)\|_{\mathbb{B}_{q,s_q;t_1-t}^{-1+3/q}} < \varepsilon_*, \tag{3.1}$$

where ε_* is the constant from (2.5). Then u is left-sided $L^{s_q}(L^q)$-regular at t_1.

   (ii) Given $t_1 \in (0,T)$ assume that $u(t_1) \in \mathbb{B}_{q,s_q}^{-1+3/q}(\Omega)$ and that u satisfies $(EI)_{t_1}$, i.e., the strong energy inequality in t_1. Then u is right-sided $L^{s_q}(L^q)$-regular at t_1.

   (iii) Let u satisfy $u(t) \in \mathbb{B}_{q,s_q}^{-1+3/q}(\Omega)$ for all $t \in [0,T)$ and condition (3.1) at every $t_1 \in (0,T)$. Then $u \in L_{\mathrm{loc}}^{s_q}([0,T); L^q(\Omega))$.

Remark 3.2. (1) Actually, the proof will show that condition (3.1) can be weakened to the following assumption: There exists $t_0 \in (0, t_1)$ such that $\|u(t_0)\|_{\mathbb{B}_{q,s_q;t_1-t_0}^{-1+3/q}} \leq \varepsilon_*$ and $(EI)_{t_0}$ is satisfied.

(2) Note that in Theorem 3.1 (i) the condition $u(t) \in \mathbb{B}_{q,s_q}^{-1+3/q}$ has been required for *all* $t \in [0,T)$; actually, it suffices to use this assumption in a left-sided neighborhood of t_1. This assumption does not occur in (ii) where we need the much weaker condition $(EI)_{t_1}$.

The regularity criterion of Theorem 3.1 has the disadvantage that the norm of $u(t)$ in $\mathbb{B}_{q,s_q;t_1-t}^{-1+3/q}(\Omega)$ can not be controlled directly for the solution u. However, there are many easy applications of Theorem 3.1 yielding more concrete criteria.

Corollary 3.3. *Let $u \in \mathcal{LH}_T$ be a weak solution of (1.1) on a bounded smooth domain $\Omega \subset \mathbb{R}^3$ satisfying (SEI).*

   (i) If there exists $\delta \in (0, \infty]$ such that $u \in C^0([0,T); \mathbb{B}_{q,s_q;\delta}^{-1+3/q}(\Omega))$, then u is a strong solution on $[0,T)$.

   (ii) If there exists $\delta \in (0, \infty]$ such that $\|u\|_{L^\infty(0,T;\mathbb{B}_{q,s_q;\delta}^{-1+3/q})} \leq \varepsilon_*$, then u is a strong solution on $[0,T)$.

   (iii) In (i) and (ii) the space $\mathbb{B}_{q,s_q;\delta}^{-1+3/q}(\Omega)$ can be replaced by any of the spaces $\mathcal{D}(A^{1/4})$, $L_\sigma^3(\Omega)$ and $L_\sigma^{3,s}(\Omega)(s \geq q > 3)$.

It is easy to see that for all $0 < T \leq \infty$ the Bochner spaces $L^\infty\left(0,T; \mathbb{B}_{q,s_q}^{-1+3/q}\right)$ and $L^\infty\left(0,T; \mathbb{B}_{q,s_q;\infty}^{-1+3/q}\right)$ (respectively $C^0\left([0,T); \mathbb{B}_{q,s_q;\delta}^{-1+3/q}\right)$ and $C^0\left([0,T); \mathbb{B}_{q,s_q;\infty}^{-1+3/q}\right)$) coincide with equivalent norms for any $\delta \in (0, \infty]$.

The next criteria are based on Theorem 2.1 and, in a certain sense, hold beyond Serrin's condition at the expense of a further smallness assumption.

Theorem 3.4. *Let* $u \in \mathcal{LH}_T$ *be a weak solution of* (1.1) *satisfying* (SEI) *as in Corollary 3.3. Further let the exponents* $2 < s_q < \infty$, $3 < q < \infty$ *satisfy* $\frac{2}{s_q} + \frac{3}{q} = 1$ *and* $1 \le r \le s_q$. *Then there exists a constant* $\tilde{\varepsilon}_* > 0$ *related to* ε_* *in* (3.1) *with the following properties:*

(i) *If at* $t_1 \in (0, T)$

$$\liminf_{\delta \to 0} \frac{1}{\delta^{1-r/s_q}} \int_{t_1 - \delta}^{t_1} \|u(\tau)\|_q^r \, d\tau \le \tilde{\varepsilon}_*, \tag{3.2}$$

then u *is regular at* t_1, *i.e.* $u \in L^{s_q}(t_1 - \varepsilon, t_1 + \varepsilon; L^q(\Omega))$ *for some* $\varepsilon > 0$.

(ii) *Assume that for* $0 \le t_0 < t_1 < T$ *and some* $t_1 < T' \le T$

$$\frac{1}{t_1 - t_0} \int_{t_0}^{t_1} (T' - \tau)^{r/s_q} \|u(\tau)\|_q^r \, d\tau \le \tilde{\varepsilon}_*. \tag{3.3}$$

Then u *is regular at* t_1.

An easy consequence of Theorem 3.4 (i) is the well-known fact that a weak solution $u \in \mathcal{LH}_T$ is regular almost everywhere (even everywhere in $(0, T)$ except for a possible set $S \subset (0, T)$ of vanishing Hausdorff measure $\mathcal{H}^{1/2}(S) = 0$). Actually, since $u \in L^2(0, T; L^6)$, Lebesgue's differentiation theorem implies that $\frac{1}{\delta} \int_{t_1 - \delta}^{t_1} \|u\|_6^2 \, d\tau \to \|u(t_1)\|_6^2$ t_1-a.e. Hence the term in (3.2) (with $q = 6$, $r = 2$ and $s_q = 4$) vanishes t-a.e.

Finally, we describe a regularity criterion based on the *kinetic energy*

$$E_k(t) = \frac{1}{2}\|u(t)\|_2^2. \tag{3.4}$$

Theorem 3.5. *Let* $u \in \mathcal{LH}_T$ *be a weak solution of* (1.1) *satisfying* (SEI) *as in Corollary 3.3. Assume that at* $t_1 \in (0, T)$ *for* $\alpha \in (\frac{1}{2}, 1]$ *the left-sided Hölder* α-*semi-norm*

$$[E_k(t_1-)]_\alpha = \sup_{\delta > 0} \frac{|E_k(t_1 - \delta) - E_k(t_1)|}{\delta^\alpha}$$

is finite (with the supremum taken only for small $\delta > 0$*) or that*

$$[E_k(t_1-)]_{1/2} \le \tilde{\varepsilon}_*. \tag{3.5}$$

Then u *is regular at* t_1.

Since this result is an immediate corollary of Theorem 3.4 we will present its short proof directly.

Proof. Obviously the condition $[E_k(t_1-)]_\alpha < \infty$ for $\alpha \in (\frac{1}{2}, 1]$ implies that $[E_k(t_1-)]_{1/2} \le \varepsilon_*$ (with the supremum taken only for small $\delta > 0$). Hence it suffices to assume the second condition where $\alpha = \frac{1}{2}$. With $r = 2$, $q = 6$ and $s_q = 4$ we get that

$$\int_{t_1 - \delta}^{t_1} \|u(\tau)\|_q^r \, d\tau \le c \int_{t_1 - \delta}^{t_1} \|\nabla u(\tau)\|_2^2 \, d\tau$$

$$\le c\big(E_k(t_1 - \delta) - E_k(t_1)\big) \tag{3.6}$$

$$\le c\tilde{\varepsilon}_* \delta^{1/2}$$

provided we choose only those $\delta > 0$ such that $(EI)_{t_1-\delta}$ holds. We conclude that (3.2) is satisfied and consequently that u is regular at t_1. $\square$

Note that for the Hölder exponent $\alpha = \frac{1}{2}$ we do need a smallness condition on the local left-sided Hölder seminorm. Actually, it is known, see [61, p. 227] for the case when $\Omega = \mathbb{R}^3$ and [41, Theorem 6.4] for bounded domains, that if $(t_0, t_1) \subset [0, T]$ is a maximal interval of regularity of a weak solution u, then necessarily

$$\|\nabla u(\tau)\|_2 \geq c(t_1 - \tau)^{-1/4}, \quad 0 < \tau < t_1,$$

with some $c = c(\Omega) > 0$. Hence the estimate

$$2c^2 \leq \frac{1}{\delta^{1/2}} \int_{t_1-\delta}^{t_1} \|\nabla u(\tau)\|_2^2 \, d\tau \leq \frac{1}{\delta^{1/2}}\big(E_k(t_1 - \delta) - E_k(t_1)\big)$$

for a.a. $\delta \in (t_1 - t_0, t_1)$ shows in this case that the condition (3.5) with an arbitrary (not sufficiently small) $\tilde{\varepsilon}_* > 0$ does not imply regularity. For the case including an external force we refer to [29].

For further regularity criteria beyond Serrin's condition we refer to [27].

3.2   The energy equality and uniqueness

As already mentioned in Remark 1.2 (5), a weak solution $u \in \mathcal{LH}_T$ satisfies (EE) if $u \in L^4(0, T; L^4)$ or if related $L^s(L^q)$-conditions are fulfilled which imply $u \in L^4(L^4)$ by Hölder's inequality and the assumption $u \in \mathcal{LH}_T$. In other words, those $L^s(L^q)$-conditions are closer to Serrin's class as to the class of weak solutions: actually, for those q, s we have $\frac{2}{s} + \frac{3}{q} \leq \frac{5}{4} < \frac{3}{2}$. Recently, Cheskidov, Friedlander and Shvydkoy [15] and jointly with Constantin [16] found conditions which concerning their scaling behavior are of the type $L^3(0, T; L^{9/2})$ where

$$\frac{2}{3} + \frac{3}{9/2} = \frac{4}{3} > \frac{5}{4}.$$

Theorem 3.6. ([15], [16])
    (i) Let $\Omega \subset \mathbb{R}^3$ be a C^2-domain and $u \in \mathcal{LH}_T$ be a weak solution of (1.1) such that

$$u \in L^3\big(0, T; \mathcal{D}\big(A_2^{5/12}\big)\big).$$

Then u satisfies the energy equality (EE).
    (ii) Let $\Omega = \mathbb{R}^3$ and $u \in \mathcal{LH}_T$ be a weak solution of (1.1) such that

$$u \in L^3\big(0, T; B_{3,\infty}^{1/3}(\mathbb{R}^3)\big).$$

Then u satisfies (EE).

We note that $\mathcal{D}\big(A_2^{5/12}\big) \subset L^{9/2}(\Omega)$ equals a subspace of the Bessel potential space $H^{5/6}(\Omega)$. The second result is available only for the whole space case since the characterization of the Besov space $B_{3,\infty}^{1/3}$ via Littlewood-Paley decomposition (*cf.* Remark 3.11 (3) below) is crucially used. Formally, the order $\frac{1}{3}$ of fractional derivatives in $B^{1/3}_{3,\infty}$ is optimal in view of the term $\int_{\mathbb{R}^3}(u \cdot \nabla u)u\,dx$ which must be shown to vanish.

   An intermediate result for domains exploiting derivatives of fractional order $\frac{1}{2}$ was proved in [37]. To meet the same scaling as in Theorem 3.6 the space $\mathcal{D}\big(A_2^{5/12}\big)$ is replaced by $\mathcal{D}\big(A_{18/7}^{1/4}\big) \subset L^{9/2}(\Omega)$.

Theorem 3.7. *Let $\Omega \subset \mathbb{R}^3$ be a bounded domain with $\partial\Omega \in C^{1,1}$ and let $u \in \mathcal{L}\mathcal{H}_T$ be a weak solution of* (1.1) *with $u(0) = u_0 \in L^2_\sigma(\Omega)$ satisfying*

$$u \in L^3\big(0,T,\mathcal{D}\big(A_{18/7}^{1/4}\big)\big). \tag{3.7}$$

Then u satisfies the energy equality (EE).

   This theorem will be proved in Section 8. It can be extended to general unbounded domains $\Omega \subset \mathbb{R}^3$ of uniform $C^{1,1}$-type where $\mathcal{D}\big(A_{18/7}^{1/4}\big)$ has to be replaced by the domain $\mathcal{D}\big((I + \tilde{A}_{18/7})^{1/4}\big)$ where $\tilde{A}_q$ denotes the Stokes operator on the space $\tilde{L}^q = L^q \cap L^2$, $q > 2$, rather than on L^q. For further details we refer to [37] where also a generalization to domains of fractional powers of the Stokes operator on Lorentz spaces can be found.

   The final problem to be addressed is the question of uniqueness of weak solutions.

Theorem 3.8. *Let $u \in \mathcal{L}\mathcal{H}_T$ be a weak solution of the Navier-Stokes system* (1.1) *in a bounded smooth domain $\Omega \subset \mathbb{R}^3$ with $u_0 \in L^2_\sigma(\Omega)$ such that for some $2 < s_q < \infty$, $3 < q < \infty$, $\frac{2}{s_q} + \frac{3}{q} = 1$ there holds $u(t_0) \in \mathbb{B}^{-1+3/q}_{q,s_q}(\Omega)$ for all $t_0 \in [0,T)$. Moreover, assume that u satisfies $(EI)_{t_0}$ for all $t_0 \in [0,T)$, i.e.,*

$$\frac{1}{2}\|u(t)\|_2^2 + \int_{t_0}^t \|\nabla u\|_2^2\,d\tau \le \frac{1}{2}\|u(t_0)\|_2^2 \tag{3.8}$$

for each $t \in [t_0,T)$. Then u is uniquely determined by the initial value u_0.

   To apply Theorem 3.8 we need conditions to guarantee the validity of the energy inequality $(EI)_{t_0}$ for all $t_0 \in [0,T)$. Let $E_k(\cdot)$ denote the kinetic energy, see (3.4).

Proposition 3.9. *Let $u \in \mathcal{L}\mathcal{H}_T$ be a weak Leray-Hopf solution.*
   (i) *If u satisfies the strong energy inquality (SEI) and $E_k(\cdot) \in C^0([0,T))$, then u satisfies $(EI)_{t_0}$ for all $t_0 \in [0,T)$. Actually, it suffices that at each $t_0 \in [0,T)$ the function E_k is either right-continuous or left-continuous.*
   (ii) *If $u \in L^4_{\mathrm{loc}}([0,T);L^4(\Omega))$ or*

$$u \in L^\infty_{\mathrm{loc}}\big([0,T);L^{3,\infty}_\sigma(\Omega)\big), \tag{3.9}$$

then u even satisfies the energy equality (EE).

Remark 3.10. Let u be a weak solution of (1.1) in a bounded smooth domain $\Omega \subset \mathbb{R}^3$ with $u_0 \in L_\sigma^2$. It is interesting to discuss uniqueness and regularity properties of u satisfying Serrin's condition $u \in L_{\mathrm{loc}}^s([0,T); L^q)$ in the limit case $s = s_3 = \infty, q = 3$. In this case, a classical argument using the reflexivity of the space L^3 and the weak $L^2(\Omega)$-continuity of $u(\cdot)$ imply that $u(t) \in L^3$ for *each* $t \in [0,T)$, cf. the proof of Lemma 8.1 below.

Since $L^3 \subset L^{3,\infty} \cap \mathbb{B}_{q,s_q}^{-1+3/q}$ Proposition 3.9 (ii) and Theorem 3.8 yield a short proof of the uniqueness property for weak solutions $u \in L^\infty([0,T); L_\sigma^3)$; cf. [55]. Furthermore, for $u \in L_{\mathrm{loc}}^\infty([0,T); L_\sigma^3)$ we get the local right-sided regularity property for each $t \in [0,T)$, see Theorem 3.1 (ii). On the other hand, from Corollary 3.3 (iii) we see that the stronger assumption $u \in C([0,T); L_\sigma^3)$ is sufficient to get the regularity $u \in L_{\mathrm{loc}}^{s_q}([0,T); L^q)$ with certain $2 < s_q < \infty, 3 < q < \infty, \frac{2}{s_q} + \frac{3}{q} = 1$.

Hence Theorem 3.1 is a slightly weaker result than that in a series of papers on the celebrated $L^\infty(0,T; L^3(\Omega))$-regularity result of Escauriaza, Seregin and Šverák [18] when $\Omega = \mathbb{R}^3$. We refer to Seregin [69] in domains up to a flat part of the boundary, and to Mikhailov and Shilkin [65] where in domains with curved boundaries an additional condition on the pressure has to be assumed.

3.3   The Case $\Omega = \mathbb{R}^n$

The case when $\Omega = \mathbb{R}^n$ (partly also when Ω equals the periodic cell or torus $\mathbb{T}^n = (\mathbb{R}/2\pi\mathbb{Z})^n$) admits the use of tools from harmonic analysis and yields interesting and deep results in homogeneous and nonhomogeneous Besov as well as Triebel-Lizorkin spaces, $\dot{B}_{q,r}^s(\mathbb{R}^n), B_{q,r}^s(\mathbb{R}^n)$ and $\dot{F}_{q,r}^s(\mathbb{R}^n), F_{q,r}^s(\mathbb{R}^n)$, respectively, $s \in \mathbb{R}$, $1 \le q,r \le \infty$; in particular the space $BMO(\mathbb{R}^n) = \dot{F}_{\infty,2}^0(\mathbb{R}^n)$ of functions with bounded mean oscillation is of great interest. Another advantage of the whole space case is the fact that the Helmholtz projection commutes with the Laplacian so that properties of the heat semigroup can be exploited.

In the following we will only describe some results selected from the numerous and deep results of the last decades concerning the whole space case $\Omega = \mathbb{R}^3$. As basic reference for these tools and function spaces involved we mention the monographs of Bahouri, Chemin and Danchin [6], Cannone [11], Lemarié-Rieusset [59, 60] and Triebel [79]. We note that the weak solutions to be considered in the following are solutions in the sense of distributions but not necessarily of Leray-Hopf type.

Remark 3.11. (1) The largest space of initial values to yield local- or global-in-time solutions in scale invariant function spaces known so far was found by Koch and Tataru [51]. Assume that u_0 lies in the scale invariant space BMO^{-1}, *i.e.*, u_0 can be written in the form $u_0 = \mathrm{div}\,\Phi$ with some $\Phi \in BMO$, and has small norm in this space. Then there exists a unique local/global solution $u \in L_{\mathrm{loc}}^2(\mathbb{R}^3 \times [0,\infty))$ which is equipped with a scale invariant norm.

(2) A BMO-result in the spirit of conditional regularity is based on the

vorticity $\omega = \mathrm{rot}\, u$: If

$$\int_0^T \|\mathrm{rot}\, u\|_{BMO}\, d\tau < \infty,$$

then u is regular, see Kozono and Taniuchi ([58]). This condition should be compared with the celebrated work by Beale, Kato and Majda [7] who used for the Euler equations the integral $\int_0^T \|\mathrm{rot}\, u\|_\infty\, d\tau$ with the $L^\infty(\mathbb{R}^3)$-norm instead of the BMO-norm of $\mathrm{rot}\, u$. Since $BMO \subset \dot{B}_{q,\infty}^0$ for $q = \infty$, the result (*cf.* [56] by Kozono, Ogawa and Taniuchi)

$$\int_0^T \|\mathrm{rot}\, u\|_{\dot{B}_{q,\infty}^0}^s\, d\tau < \infty, \qquad \frac{2}{s} + \frac{3}{q} = 2,\ 3 \le q \le \infty$$

contains the previous one. Here the strong energy inequality (*SEI*) had to be assumed.

(3) The largest scale invariant space for the Navier-Stokes system is the homogeneous Besov space $\dot{B}_{\infty,\infty}^{-1}$ (for a proof see [11, Proposition 7]) containing the space BMO^{-1} from [51]; hence conditions formulated in this space are among the most interesting. Since $B_{\infty,\infty}^{-1}$ is larger than $\dot{B}_{\infty,\infty}^{-1}$, it will not make sense to consider initial values in $B_{\infty,\infty}^{-1}$. However, regularity question can be posed in this space. By a recent result of Cheskidov and Shvydkoy [14] a weak solution $u \in C^0((0,T]; B_{\infty,\infty}^{-1})$ is regular on $(0,T]$. In fact, $u(t)$ may be even discontinuous in time with values in $B_{\infty,\infty}^{-1}$ to get the same conclusion as long as

$$\sup_{t_0 \in (0,T]} \limsup_{t \to t_0} \|u(t) - u(t_0)\|_{B_{\infty,\infty}^{-1}} < c\nu,$$

where $c > 0$ is an absolute constant and $\nu > 0$ is the viscosity of the fluid.

(4) The Navier-Stokes system is *ill-posed* in the space $\dot{B}_{\infty,\infty}^{-1}$, see Bourgain-Pavlović [10]. Actually, the authors proved that for any $\delta > 0$ there exists an initial value u_0 in the Schwartz class $\mathcal{S}(\mathbb{R}^3)$, a $T \in (0,\delta)$ and a corresponding solution (u,p) to the Navier-Stokes system such that

$$\|u_0\|_{\dot{B}_{\infty,\infty}^{-1}} \le \delta \quad \text{and} \quad \|u(T)\|_{\dot{B}_{\infty,\infty}^{-1}} \ge \frac{1}{\delta}.$$

This phenomenon of norm inflation was improved by Yoneda [81] who replaced the space $\dot{B}_{\infty,\infty}^{-1}$ by a slightly smaller space $\mathcal{B}$ satisfying $\dot{B}_{\infty,2}^{-1} \subset \mathcal{B} \subset \dot{B}_{\infty,r}^{-1}$ for any $r > 2$ or by any of the spaces $\dot{B}_{\infty,r}^{-1}$ or $\dot{F}_{\infty,r}^{-1}$ ($r > 2$); recall that $BMO = \dot{F}_{\infty,2}^0$. The result was extended by B. Wang [80] to the homogeneous Besov spaces $\dot{B}_{\infty,r}^{-1}$ for $1 \le r \le 2$.

(5) Another striking phenomenon is described by Cheskidov and Shvydkoy [13]: For any initial value $u_0 \in L_\sigma^2(\mathbb{T}^3)$ there exists a weak solution u such that

$$\limsup_{t \to 0} \|u(t) - u_0\|_{B_{\infty,\infty}^{-1}} \ge \delta_0$$

for some $\delta_0 = \delta_0(u) > 0$. The lower bound δ_0 can be chosen uniformly in the class of weak Leray-Hopf solution. This also holds in $\mathbb{B}_{q,\infty}^{-1+3/q}$ for all $1 < q < \infty$ ([13, p.

1060]). In other words, $u(t)$ is discontinuous as $t \to 0$ in the topology of $\mathbb{B}_{q,\infty}^{-1+3/q}$ although it is continuous in L_σ^2 at $t = 0$. On the other hand, note that by Theorem 2.6 the space $\mathring{\mathbb{B}}_{q,\infty}^{-1+3/q}$, $3 < q < \infty$, admits continuity in time and does not lead to norm inflation, see (2.18).

(6) Nevertheless, there exist initial values of arbitrarily large norm in $\mathring{B}_{\infty,\infty}^{-1}$ allowing for global smooth solutions of the Navier-Stokes system. To this aim Chemin, Gallagher and Paicu [12] construct for the case of the torus $\mathbb{T}^3$ special initial values with different behaviour in the first two components compared to the third one.

For further results on the well-posedness of the Navier-Stokes system in critical function spaces over $\mathbb{R}^3$ and $\mathbb{T}^3$ we refer to a recent review by Gallagher [42].

4 Stokes operator and Besov spaces

4.1 The Stokes operator

Recall the definition of the Helmholtz projection $P = P_q$ and of the Stokes operator $A = A_q = -P\Delta$ with domain $\mathcal{D}(A_q) \subset L_\sigma^q(\Omega)$, see (1.2). Here we focus on the case of bounded domains $\Omega \subset \mathbb{R}^3$ and leave aside some restrictions necessary for exterior and other types of unbounded domains.

The operator A_q, $1 < q < \infty$, is sectorial and generates an analytic C^0-semigroup. Moreover, it admits fractional powers A_q^α, $-1 \leq \alpha \leq 1$ such that A_q^α, $0 \leq \alpha \leq 1$, is an injective, closed and densely defined operator with domain $\mathcal{D}(A_q^\alpha) \subset L_\sigma^q(\Omega)$, range $\mathcal{R}(A_q^\alpha) = L_\sigma^q(\Omega)$, and such that $\mathcal{D}(A_q) \subset \mathcal{D}(A_q^\alpha) \subset \mathcal{D}(A_q^\beta) \subset L_\sigma^q(\Omega)$ for $0 \leq \beta \leq \alpha \leq 1$. Hence, for $-1 \leq \alpha < 0$, we may define the bounded operators $A_q^\alpha = (A_q^{-\alpha})^{-1} : L_\sigma^q(\Omega) \to \mathcal{R}(A_q^\alpha) = \mathcal{D}(A_q^{-\alpha})$. Important properties are the embeddings

$$\|v\|_q \leq c\|A_r^\alpha v\|_r, \quad v \in \mathcal{D}(A_r^\alpha), \quad 1 < r \leq q < \infty, \quad 2\alpha + \frac{3}{q} = \frac{3}{r}, \qquad (4.1)$$

$$\|\nabla v\|_q \sim \|A_q^{1/2} v\|_q, \quad v \in \mathcal{D}(A_q^{1/2}) = W_{0,\sigma}^{1,q}(\Omega), \quad 1 < q < \infty, \qquad (4.2)$$

and $\|\nabla v\|_2 = \|A_2^{1/2} v\|_2$ for $v \in W_{0,\sigma}^{1,2}(\Omega)$; in (4.1) we assume $0 \leq \alpha \leq 1$.

The inverse $A_q^{-1} : L_\sigma^q(\Omega) \to L_\sigma^q(\Omega)$ is compact and has a purely discrete point spectrum $\sigma(A_q) \subset (0, \infty)$ independent of q. Moreover, $-A_q$ is the infinitesimal generator of an analytic and exponentially decaying semigroup e^{-tA}, $t \geq 0$, i.e., there exists $\delta = \delta(q, \Omega) > 0$ and $c = c(q, \delta, \Omega) > 0$ such that $\|e^{-tA_q} v\|_q \leq ce^{-\delta t}\|v\|_q$ for $v \in L_\sigma^q(\Omega)$ and $t \geq 0$. It even holds

$$\|A_q^\alpha e^{-tA_q} v\|_q \leq ct^{-\alpha} e^{-\delta t}\|v\|_q, \quad v \in L_\sigma^q(\Omega), \quad 0 \leq \alpha \leq 1. \qquad (4.3)$$

Summarizing (4.1)–(4.3) we obtain for any $1 < q \leq p < \infty$ the following important

L^p-L^q-estimates of the Stokes semigroup:[1]

$$\|e^{-tA_q}v\|_p \leq ct^{-\frac{3}{2}(\frac{1}{q}-\frac{1}{p})}\|v\|_q, \qquad v \in L^q_\sigma(\Omega), \tag{4.4}$$

$$\|\nabla e^{-tA_q}v\|_p \leq ct^{-\frac{1}{2}-\frac{3}{2}(\frac{1}{q}-\frac{1}{p})}\|v\|_q, \qquad v \in L^q_\sigma(\Omega). \tag{4.5}$$

An important property of the Stokes operator is the maximal regularity. Consider the abstract inhomogeneous linear Cauchy problem

$$u_t + Au = f \quad \text{in } (0,T), \quad u(0) = u_0 \tag{4.6}$$

where $-A$ is the generator of an analytic semigroup e^{-tA} on a Banach space X. Then A is said to admit *maximal L^s-regularity* on $[0,T)$, $1 < s < \infty$, $0 < T \leq \infty$, if for suitable u_0 and all $f \in L^s(0,T;X)$ the solution of (4.6) given by the variation of constants formula

$$u(t) = e^{-tA}u_0 + \int_0^t e^{-(t-\tau)A}f(\tau)\,d\tau, \quad t \in [0,T) \tag{4.7}$$

is differentiable a.e., $u(t) \in \mathcal{D}(A)$ a.e. and u_t, $Au \in L^s(0,T;X)$. In this case the closed graph theorem yields a constant $C = C(s,T) > 0$ such that

$$\|u_t; Au\|_{L^s(0,T;X)} \leq C\left(\int_0^T \|Ae^{-tA}u_0\|_X^s\,dt\right)^{1/s} + C\|f\|_{L^s(0,T;X)}. \tag{4.8}$$

It is well-known that maximal L^s-regularity for one $s \in (1,\infty)$ implies maximal L^s-regularity for all $s \in (1,\infty)$.

The Stokes operator $A = A_q$ on $X = L^q_\sigma(\Omega)$, $1 < q < \infty$, has maximal L^s-regularity on $[0,\infty)$ for bounded (and other classes of) domains $\Omega \subset \mathbb{R}^3$ of class $C^{1,1}$; e.g., see [47], [63], [76]. In particular, for any $f \in L^s(0,\infty;L^q)$ the Cauchy problem (4.6) with $A_q = -P_q\Delta$, $u_0 = 0$ and f replaced by $P_q f$ has a unique strong solution $u \in C^0_{\text{loc}}([0,\infty);L^q_\sigma)$ such that u_t, $Au \in L^s(0,\infty;L^q)$, $1 < s < \infty$, and

$$\|u_t; Au\|_{L^s(0,\infty,L^q)} \leq c\|f\|_{L^s(0,\infty;L^q)}. \tag{4.9}$$

Moreover, there exists an associated pressure p such that

$$\|u_t; \nabla^2 u; \nabla p\|_{L^s(0,\infty;L^q)} \leq c\|f\|_{L^s(0,\infty;L^q)}. \tag{4.10}$$

The notion of maximal L^s-regularity can be extended to L^s-integrals with weights in time t^α where $0 \leq \alpha < 1 - \frac{1}{s}$; for a related result see Lemma 4.3 below.

For more information on the Stokes operator and maximal regularity we refer to Borchers and Miyakawa [9], Denk, Hieber and Prüß [17], Giga [43], [44] and Sohr [74].

---

[1]For an exterior domain (4.3) holds only for $\delta = 0$; moreover, (4.1) is valid for $\alpha \leq \frac{1}{2}$, $1 < r < 3$ and $0 \leq \alpha \leq 1$, $1 < r < \frac{3}{2}$. The norm equivalence (4.2) is restricted to $1 < q < 3$ although $\|A_q^{1/2}v\|_q \leq c\|\nabla v\|_q$ holds for all $1 < q < \infty$. Finally, (4.5) holds for $p \leq 3$ only.

The inhomogeneous Cauchy problem (4.6) with initial value u_0 in the real interpolation space $\left(L_\sigma^q, \mathcal{D}(A_q)\right)_{1-1/s,s} \subset L_\sigma^q$ with equivalent norm

$$\|u_0\|_{(L_\sigma^q, \mathcal{D}(A_q))_{1-1/s,s}} := \|u_0\|_q + \left(\int_0^\infty \|Ae^{-tA}u_0\|_q^s \,d\tau\right)^{1/s}$$

admits a unique solution as well; however, in the *a priori* estimates (4.9), (4.10) the additional term $\|u_0\|_{(L_\sigma^q,\mathcal{D}(A_q))_{1-1/s,s}}$ is needed on the right-hand side. For details on the above real interpolation space, even for an abstract Cauchy problem $u_t + Au = f$ on an abstract Banach space X and closed sectorial operator A with $\mathcal{D}(A) \subset X$, we refer to [79, Chapters 1.13, 1.14]. The integration over $(0,\infty)$ in the norm $\|\cdot\|_{(L_\sigma^q,\mathcal{D}(A_q))_{1-1/s,s}}$ may be replaced by an integration over $(0,\delta)$ for any $\delta > 0$, and, since Ω is bounded, the term $\|u_0\|_q$ may be omitted, see §4.2 below and [79, Theorem 1.14.5].

The relation between a weak solution u of the instationary Stokes system with force term $f = \operatorname{div} F$ and initial value u_0 and the notion of a mild solution, *cf.* (1.13) when omitting the nonlinear term $\operatorname{div}(u \otimes u)$, is given by the representation

$$u(t) = e^{-tA}u_0 + A^{1/2}\int_0^t e^{-(t-\tau)A}(A^{-1/2}P\operatorname{div})F(\tau)\,d\tau. \tag{4.11}$$

The formal operator $A^{-1/2}P\operatorname{div}$ applied to matrix-valued $L^q(\Omega)$-functions F is defined by duality, see Giga and Miyakawa [46] or [74]. Actually, we generalize (4.11) by replacing $A^{\pm\frac{1}{2}}$ by $A^{\pm\gamma}$ where $\frac{1}{2} \le \gamma < 1$, *i.e.*, we consider the representation

$$u(t) - e^{-tA}u_0 = A^\gamma \int_0^t e^{-(t-\tau)A}(A^{-\gamma}P\operatorname{div})F(\tau)\,d\tau \tag{4.12}$$

for suitable vector fields F. Ignoring the time variable for a moment, we fix $1 < p_* \le r < \infty$, let $F \in L^{p_*}(\Omega)$ and $\gamma \in [\frac{1}{2},1)$ be given such that

$$\gamma = \frac{3}{2p_*} - \frac{3}{2r} + \frac{1}{2}. \tag{4.13}$$

Then for $\varphi \in C_{0,\sigma}^\infty(\Omega)$

$$\langle A^{-\gamma}P\operatorname{div}F, \varphi\rangle := -\langle F, \nabla A_{p_*'}^{-\gamma}\varphi\rangle_\Omega \tag{4.14}$$

so that (4.1), (4.2) yield the estimate

$$|\langle F, \nabla A_{p_*'}^{-\gamma}\varphi\rangle| \le \|F\|_{p_*}\|A_{p_*'}^{1/2-\gamma}\varphi\|_{p_*'} \le c_{p_*'}\|F\|_{p_*}\|\varphi\|_{r'}.$$

Here $A_{p_*'}$ denotes the Stokes operator on $L_\sigma^{p_*'}$ and coincides with the Banach space adjoint $(A_{p_*})' : \mathcal{D}((A_{p_*})') = \mathcal{D}(A_{p_*'}) \subset L_\sigma^{p_*'} \to L_\sigma^{p_*'}$ of A_{p_*}. Since $C_{0,\sigma}^\infty(\Omega)$ is dense in $L_\sigma^r(\Omega)$, a duality argument implies that

$$A^{-\gamma}P\operatorname{div} \in \mathcal{L}\left(L^{p_*}(\Omega), L_\sigma^r(\Omega)\right), \quad \|A^{-\gamma}P\operatorname{div}F\|_r \le c_{p_*'}(\Omega)\|F\|_{p_*}. \tag{4.15}$$

Obviously, the closed operator A^γ commutes with the integral in (4.12). The corresponding formal integral equation leads due to (4.3) to a (weakly) singular integral which converges under suitable assumptions on F.

Due to [74, Ch. IV, Theorem 2.4.1, Lemma 2.4.2] for $u_0 \in L^2_\sigma$ and $F \in L^2(0,T;L^2)$, the vector field u in (4.11) is the unique weak solution $u \in \mathcal{LH}_T$ of the instationary Stokes system (4.6) with data u_0 and $f = \operatorname{div} F$. Moreover, u satisfies the energy equality (EE), cf. (1.11), leading to the estimate

$$\|u\|^2_{L^\infty(0,T;L^2)} + \|\nabla u\|^2_{L^2(0,T;L^2)} \le \|u_0\|^2_2 + \|F\|^2_{L^2(0,T;L^2)}. \tag{4.16}$$

4.2 Besov spaces

In this subsection we use methods of real interpolation, see [62, 79]. Let $(\cdot,\cdot)_{\theta,r}$, $1 \le r < \infty$, denote the general real interpolation functor and $(\cdot,\cdot)^0_{\theta,\infty}$ denote the continuous interpolation functor. For $1 < q < \infty$, $1 \le r \le \infty$ and $t \in \mathbb{R}$ the usual family of Besov spaces on $\mathbb{R}^3$ is denoted by $B^t_{q,r}(\mathbb{R}^3)$, see [79, 2.3.1]. Then we define for a bounded domain $\Omega \subset \mathbb{R}^3$ the space $B^t_{q,r}(\Omega)$ by restriction of elements in $B^t_{q,r}(\mathbb{R}^3)$ in the sense of distributions to Ω; the norm of $u \in B^t_{q,r}(\Omega)$ is given by

$$\|u\|_{B^t_{q,r}(\Omega)} = \inf \left\{ \|v\|_{B^t_{q,r}(\mathbb{R}^3)} : v \in B^t_{q,r}(\mathbb{R}^3),\ v\big|_\Omega = u \right\}.$$

Concerning Besov spaces on Ω with vanishing trace (if possible), the definition is modified as follows: Considering only vector fields rather than scalar-valued functions and the range $t \in [-2,2]$ we follow Amann [2, 4] and define

$$\mathbf{B}^t_{q,r}(\Omega) = \begin{cases} \{u \in B^t_{q,r}(\Omega)^3;\ u_{|\partial\Omega} = 0\}, & 1/q < t \le 2, \\ \{u \in B^{1/q}_{q,r}(\mathbb{R}^3)^3;\ \operatorname{supp}(u) \subset \overline{\Omega}\}, & 1/q = t, \\ B^t_{q,r}(\Omega)^3, & 0 \le t < 1/q, \\ \left(\mathbf{B}^{-t}_{q',r'}(\Omega)\right)' \ (1 < r \le \infty), & -2 \le t < 0. \end{cases} \tag{4.17}$$

For spaces of solenoidal vector fields on Ω let

$$\mathbb{B}^t_{q,r}(\Omega) = \begin{cases} \mathbf{B}^t_{q,r}(\Omega) \cap L^q_\sigma(\Omega), & 0 < t \le 2, \\ \operatorname{cl}\left(C^\infty_{c,\sigma}(\Omega)\right) \text{ in } \mathbf{B}^0_{q,r}(\Omega), & t = 0, \\ \left(\mathbb{B}^{-t}_{q',r'}(\Omega)\right)' \ (1 < r \le \infty), & -2 \le t < 0, \end{cases} \tag{4.18}$$

where "cl" denotes the closure. Note that $u \in \mathbb{B}^t_{q,r}(\Omega)$ with $\frac{1}{q} < t \le 2$ vanishes on $\partial\Omega$ by (4.17), but that only the normal component of u vanishes on $\partial\Omega$ when $0 < t \le \frac{1}{q}$ since $u \in L^q_\sigma(\Omega)$. Moreover, we need the spaces

$$\mathring{\mathbb{B}}^t_{q,\infty}(\Omega) := \operatorname{cl}\left(\mathbf{H}^t_q(\Omega) \cap L^q_\sigma(\Omega)\right) \text{ in } \mathbb{B}^t_{q,\infty}(\Omega),$$

where $\mathbf{H}^t_q(\Omega)$ is a Bessel potential space defined by restriction of the usual Bessel potential space $H^t_q(\mathbb{R}^3)^3$ to vector fields on Ω (and vanishing on $\partial\Omega$ as in (4.17)), cf. [4, pp. 3–4]. Obviously, $\mathring{\mathbb{B}}^t_{q,\infty}(\Omega)$ is a closed, separable subspace of $\mathbb{B}^t_{q,\infty}(\Omega)$.

The spaces $\mathbb{B}_{q,r}^t(\Omega)$ and $\mathring{\mathbb{B}}_{q,\infty}^t(\Omega)$ are real interpolation spaces; indeed, by [2, Theorem 3.4], for $0 < \theta < 1$

$$(L_\sigma^q(\Omega), \mathcal{D}(A_q))_{\theta,r} = \mathbb{B}_{q,r}^{2\theta}(\Omega), \tag{4.19}$$

$$(L_\sigma^q(\Omega), \mathcal{D}(A_q))_{\theta,\infty}^0 = \mathring{\mathbb{B}}_{q,\infty}^{2\theta}(\Omega). \tag{4.20}$$

Note that $\mathcal{D}(A_q)$ is equipped with its graph norm, and that for a bounded domain this graph norm can be simplified to $\|A_q \cdot\|_q$. As is well-known ([62, Proposition 6.2, Exercise 6.1.1 (1)]), equivalent norms on the spaces $(L_\sigma^q(\Omega), \mathcal{D}(A_q))_{\theta,r}$, $1 \leq r \leq \infty$, are given by

$$\|u\|_{(L_\sigma^q(\Omega), \mathcal{D}(A_q))_{\theta,r}} \sim \begin{cases} \left(\int_0^T \left(\tau^{1-\theta} \|A_q e^{-\tau A_q} u\|_q \right)^r \frac{d\tau}{\tau} \right)^{1/r} & \text{if } 1 \leq r < \infty, \\ \sup_{(0,T)} \tau^{1-\theta} \|A_q e^{-\tau A_q} u\|_q & \text{if } r = \infty, \end{cases}$$
$$\tag{4.21}$$

where $T \in (0,\infty)$ can be chosen arbitrarily. The space $\mathring{\mathbb{B}}_{q,\infty}^{2\theta}(\Omega)$ is equipped with the norm of $(L_\sigma^q(\Omega), \mathcal{D}(A_q))_{\theta,\infty}$, but its elements enjoy the further property that

$$\lim_{\tau \to 0} \tau^{1-\theta} \|A_q e^{-\tau A_q} u\|_q = 0. \tag{4.22}$$

We note that the identity (4.20) for the usual interpolation space $(L_\sigma^q(\Omega), \mathcal{D}(A_q))_{\theta,\infty}$ and $\mathbb{B}_{q,\infty}^{2\theta}(\Omega)$ is not found in the literature.

To find similar representations for negative exponents of regularity as well we use the duality theorem of real interpolation to get that for $-1 < \theta < 0$ and $1 < r < \infty$ by (4.18), (4.19)

$$\left(L_\sigma^q(\Omega), \mathcal{D}(A_{q'})' \right)_{-\theta,r} = \left((L_\sigma^{q'}(\Omega), \mathcal{D}(A_{q'}))_{-\theta,r'} \right)' = \left(\mathbb{B}_{q',r'}^{-2\theta}(\Omega) \right)' = \mathbb{B}_{q,r}^{2\theta}(\Omega).$$

For the cases $r = 1$ and $r = \infty$ we note that A_q is an isomorphism from $\mathcal{D}(A_q)$ to $L_\sigma^q(\Omega)$ and by duality also from $L_\sigma^q(\Omega)$ to $\mathcal{D}(A_{q'})'$, see also Sect. 4.3. Hence, for $1 \leq r \leq \infty$ and $-1 < \theta < 0$

$$\left(\mathcal{D}(A_{q'})', L_\sigma^q(\Omega) \right)_{1+\theta,r} = A\left((L_\sigma^q(\Omega), \mathcal{D}(A_q))_{1+\theta,r} \right), \tag{4.23}$$

with a similar result for the continuous interpolation functor $(\cdot, \cdot)_{\theta,\infty}^0$. Then we get the characterizations (here $-1 < \theta < 0$):

$$\left(\mathcal{D}(A_{q'})', L_\sigma^q(\Omega) \right)_{1+\theta,r} = \mathbb{B}_{q,r}^{2\theta}(\Omega), \quad 1 \leq r < \infty, \tag{4.24}$$

$$\left(\mathcal{D}(A_{q'})', L_\sigma^q(\Omega) \right)_{1+\theta,\infty} = \mathbb{B}_{q,\infty}^{2\theta}(\Omega) \cong \mathbb{B}_{q,\infty}^{2\theta}(\Omega) / \left(\mathbb{B}_{q',1}^{-2\theta}(\Omega) \right)^\perp, \tag{4.25}$$

$$\left(\mathcal{D}(A_{q'})', L_\sigma^q(\Omega) \right)_{1+\theta,\infty}^0 = \mathring{\mathbb{B}}_{q,\infty}^{2\theta}(\Omega) = \mathrm{cl}\left(\mathbf{H}_q^2(\Omega) \right) \text{ in } \left(\mathbb{B}_{q',1}^{-2\theta}(\Omega) \right)'. \tag{4.26}$$

Actually, (4.24) for $r = 1$ and (4.26) follow from [2, Theorem 3.4], [4, p. 4], for all $-1 < \theta < 0$; the space $\mathring{\mathbb{B}}_{q,\infty}^{2\theta}(\Omega)$ also coincides with the closure $\mathrm{cl}\left(L_\sigma^q(\Omega) \right)$ in

$\mathbb{B}^{2\theta}_{q,\infty}(\Omega)$. To prove (4.25) we use (4.19), the duality theorem of real interpolation and (4.18) to get that

$$\left(\mathcal{D}(A_{q'})', L^q_\sigma(\Omega)\right)_{1+\theta,\infty} = \left(\left(L^{q'}_\sigma(\Omega), \mathcal{D}(A_{q'})\right)_{-\theta,1}\right)' = \left(\mathbb{B}^{-2\theta}_{q',1}(\Omega)\right)' = \mathbb{B}^{2\theta}_{q,\infty}(\Omega).$$

The second part of (4.25) is based on the isomorphism

$$\mathbb{B}^{2\theta}_{q,\infty}(\Omega) = \left(\mathbb{B}^{-2\theta}_{q',1}(\Omega)\right)' \cong \mathbb{B}^{2\theta}_{q,\infty}(\Omega) / (\mathbb{B}^{-2\theta}_{q',1}(\Omega))^\perp,$$

see also [2, Remark 3.6] and its proof; here $\mathbb{B}^{2\theta}_{q,\infty}(\Omega) = \left(\mathbb{B}^{-2\theta}_{q',1}(\Omega)\right)'$ by definition.

Thus for any $1 \le r \le \infty$ and $-1 < \theta < 0$, by (4.23), (4.24), (4.25) and (4.19), we conclude that $\left(\mathcal{D}(A_{q'})', L^q_\sigma(\Omega)\right)_{1+\theta,r} = A\left(\left(L^q_\sigma(\Omega), \mathcal{D}(A_q)\right)_{1+\theta,r}\right) = \mathbb{B}^{2\theta}_{q,r}(\Omega)$ with equivalent norm

$$\|u\|_{A(L^q_\sigma(\Omega), \mathcal{D}(A_q))_{1+\theta,r}} \sim \begin{cases} \left(\displaystyle\int_0^T \left(\tau^{-\theta}\|e^{-\tau A_q}u\|_q\right)^r \dfrac{d\tau}{\tau}\right)^{1/r} & \text{if } 1 \le r < \infty, \\[2mm] \sup_{\tau\in(0,T)} \tau^{-\theta}\|e^{-\tau A_q}u\|_q & \text{if } r = \infty; \end{cases}$$

(4.27)

see [33] when $\frac{2}{r} + \frac{3}{q} = 1$, $\theta = 0$, $2 < r < \infty$. For the continuous interpolation space $\left(\mathcal{D}(A_{q'})', L^q_\sigma(\Omega)\right)^0_{1+\theta,\infty} = \mathring{\mathbb{B}}^{2\theta}_{q,\infty}(\Omega)$ we have the norm defined in (4.27), with the additional property that

$$\lim_{\tau\to 0} \tau^{-\theta}\|e^{-\tau A_q}u\|_q = 0.$$

Summarizing the previous arguments we get the following theorem.

Theorem 4.1. *Let $\Omega \subset \mathbb{R}^3$ be a bounded smooth domain, and let $T \in (0,\infty)$.*

(i) Let $2 < s < \infty$, $3 < q < \infty$ and $0 < \alpha < \frac{1}{2}$ such that $\frac{2}{s} + \frac{3}{q} = 1 - 2\alpha$. Then the real interpolation space $\left(\mathcal{D}(A_{q'})', L^q_\sigma(\Omega)\right)_{1-\alpha,s}$ coincides with the Besov space $\mathbb{B}^{-1+3/q}_{q,s}(\Omega)$ and has the equivalent norm

$$\left(\int_0^T \left(\tau^\alpha \|e^{-\tau A_q}u\|_q\right)^s d\tau\right)^{1/s}.$$

(ii) If $3 < q < \infty$ and $0 < \alpha < \frac{1}{2}$ such that $\frac{3}{q} = 1 - 2\alpha$, the real interpolation space $\left(\mathcal{D}(A_{q'})', L^q_\sigma(\Omega)\right)_{1-\alpha,\infty}$ coincides with the space of Besov-type $\mathbb{B}^{-1+3/q}_{q,\infty}(\Omega)$ and has the equivalent norm

$$\sup_{\tau\in(0,T)} \tau^\alpha \|e^{-\tau A_q}u\|_q.$$

(iii) The interpolation space $\left(\mathcal{D}(A_{q'})', L^q_\sigma(\Omega)\right)^0_{1-\alpha,\infty}$ equals the Besov space $\mathring{\mathbb{B}}^{-1+3/q}_{q,\infty}(\Omega)$, equipped with the norm of $\mathbb{B}^{-1+3/q}_{q,\infty}(\Omega)$ in (ii) such that the following property additionally holds for $u \in \mathring{\mathbb{B}}^{-1+3/q}_{q,\infty}(\Omega)$:

$$\lim_{\tau\to 0} \tau^\alpha \|e^{-\tau A_q}u\|_q = 0.$$

4.3 Abstract spaces and analytic semigroups

In the final subsection we put the results of §4.1–§4.2 into a more abstract setting so that they can easily be applied to the proof of continuity in Sect. 6. In the application in Sect. 6 we let $X = L^q_\sigma(\Omega)$, $1 < q < \infty$, and $Z = \big(\mathcal{D}(A')\big)'$ where $A = A_q$ is the Stokes operator.

Let X be a Banach space equipped with the norm $\|\cdot\|_X$, and let $-A$ denote the infinitesimal generator of an analytic and bounded C^0-semigroup e^{-tA} in X. Moreover, assume that 0 lies in the resolvent set of A. Then $A^{-1} : X \to \mathcal{D}(A)$ is bounded and $A : \mathcal{D}(A) \to X$ is an isometry when $\mathcal{D}(A)$ is equipped with the homogeneous graph norm $\|A \cdot \|_X$. Furthermore, the semigroup e^{-tA} decays exponentially in time, i.e., $\big\|e^{-tA}\big\|_{\mathcal{L}(X)} \le C_0 e^{-\nu t}$ with some constants C_0 and $\nu > 0$; here $\|\cdot\|_{\mathcal{L}(X)}$ denotes the operator norm on X.

Following [1, Chapter V] we define the extrapolation space $Z = X_{-1}$ with norm $\|z\|_Z = \|A^{-1}z\|_X$ as the completion $\overline{(X, \|\cdot\|_Z)}$. Then A_{-1}, defined as the closure of A in X_{-1}, is the unique continuous extension of the isometry $A : \mathcal{D}(A) \to X$ and yields an isometry $A_{-1} : X = \mathcal{D}(A_{-1}) \subset X_{-1} \to X_{-1}$. The semigroup operators e^{-tA} possess continuous extensions from X to X_{-1} defining an exponentially decaying analytic semigroup with infinitesimal generator A_{-1}, see Proposition 4.2 below. For simplicity we will denote this semigroup by e^{-tA} again. For details we refer to [1, Chapter V, p. 262]. If X is reflexive, then Z is isomorphic to $\big(\mathcal{D}(A')\big)'$, see [1, Theorem V.1.4.6].

Hence, with an abuse of notation, we will write

$$A : X \to Z = AX = \big(\mathcal{D}(A')\big)'$$

defining the isometry $\|Ax\|_Z = \|x\|_X$ for $x \in X$.

For $1 \le s \le \infty$ and $\alpha \in \mathbb{R}$ such that $0 < \alpha + \frac{1}{s} < 1$ we define the space

$$\mathcal{X}_{s,\alpha} = \{f \in Z : \|f\|_{\mathcal{X}_{s,\alpha}} < \infty\}, \qquad (4.28)$$

$$\|f\|_{\mathcal{X}_{s,\alpha}} := \|A^{-1}f\|_X + \begin{cases} \left(\displaystyle\int_0^\infty \tau^{\alpha s}\, \big\|e^{-\tau A}f\big\|_X^s \, d\tau\right)^{1/s} & \text{when } s < \infty, \\[2mm] \sup_{0<\tau<\infty} \tau^\alpha \big\|e^{-\tau A}f\big\|_X & \text{when } s = \infty. \end{cases}$$

We note that due to the spectral properties of A the term $\|A^{-1}f\|_X$ in (4.28) can be omitted. Hence we obtain that for any $0 < T < \infty$ the norm in (4.28) can be replaced by the equivalent norm

$$\|f\|_{\mathcal{X}^T_{s,\alpha}} = \begin{cases} \left(\displaystyle\int_0^T \tau^{\alpha s}\, \big\|e^{-\tau A}f\big\|_X^s \, d\tau\right)^{1/s} & \text{when } s < \infty, \\[2mm] \sup_{0<\tau<T} \tau^\alpha \big\|e^{-\tau A}f\big\|_X & \text{when } s = \infty. \end{cases} \qquad (4.29)$$

To indicate which equivalent norm is used in $\mathcal{X}_{s,\alpha}$ we also use the notation $\mathcal{X}^T_{s,\alpha}$ instead of $\mathcal{X}_{s,\alpha}$ when $\mathcal{X}_{s,\alpha}$ is equipped with the norm $\|\cdot\|_{\mathcal{X}^T_{s,\alpha}}$.

From real interpolation theory applied to the spaces Z and $X = \mathcal{D}(A_{-1})$, e.g., see [62, Proposition 6.2], we conclude that $\mathcal{X}^T_{s,\alpha} = \mathcal{X}_{s,\alpha}$ also coincides with the real interpolation space $(Z, X)_{1-\alpha-1/s,s}$ endowed with the equivalent norm

$$\|f\|_{(Z,X)_{1-\alpha-1/s,s}} = \left(\int_0^\infty \left(\tau^{\alpha+1/s} \|Ae^{-\tau A} f\|_Z \right)^s \frac{d\tau}{\tau} \right)^{1/s} \qquad (4.30)$$

where the term $\|A^{-1}f\|_X$ is omitted. In the limit case when $s = \infty$, [62, Exercise 6.1.1 (1)] implies that $\mathcal{X}_{\infty,\alpha} = \mathcal{X}_{\infty,\alpha}^T = (Z, X)_{1-\alpha,\infty}$ for all $0 < T \le \infty$ with equivalent norms. Thus, for fixed $\theta = 1-\alpha-\frac{1}{s} \in (0,1)$, i.e., $\alpha = \alpha(s) := 1-\theta-\frac{1}{s} \in [0, 1-\theta]$, we get the scale of interpolation spaces $(Z, X)_{\theta,s}$ for $\frac{1}{1-\theta} =: s_1 \le s \le \infty$ with continuous embeddings

$$X \subset (Z, X)_{\theta,s_1} \subset (Z, X)_{\theta,s} \subset (Z, X)_{\theta,\infty} \subset Z, \qquad (4.31)$$

or, since $(Z, X)_{\theta,s} = \mathcal{X}_{s,\alpha(s)}^T$, the scale $X \subset \mathcal{X}_{s_1,\alpha(s_1)}^T \subset \mathcal{X}_{s,\alpha(s)}^T \subset \mathcal{X}_{\infty,\alpha(\infty)}^T \subset Z$.

Proposition 4.2. (i) *For $t > 0$ and $f \in Z$ we have that $e^{-tA} f \in Z$ such that*

$$\|e^{-tA} f\|_Z \le \|e^{-tA}\|_{\mathcal{L}(X)} \|f\|_Z.$$

Moreover, e^{-tA} extends to a bounded linear operator from Z to X. To be more precise, there exists a constant $c > 0$ independent of t and $f \in Z$ such that

$$\|e^{-tA} f\|_X \le ct^{-1} \|f\|_Z, \quad t > 0.$$

(ii) *The space X is continuously embedded into $\mathcal{X}_{s,\alpha}^T$ for all $\alpha \ge 0$, $1 \le s \le \infty$, and $0 < T \le \infty$.*

Proof. (i) By analyticity we observe that for $f \in Z = X_{-1}$ and for $t > 0$

$$\|e^{-tA} f\|_Z = \|A_{-1}^{-1} e^{-tA} f\|_X = \|e^{-tA} A_{-1}^{-1} f\|_X$$
$$\le \|e^{-tA}\|_{\mathcal{L}(X)} \|A^{-1} f\|_X \le \|e^{-tA}\|_{\mathcal{L}(X)} \|f\|_Z.$$

If $f = A_{-1} x \in Z$ with $x \in X$, then

$$\|e^{-tA} f\|_X = \|A_{-1} e^{-tA} x\|_X \le ct^{-1} \|x\|_X = ct^{-1} \|f\|_Z,$$

with some constant c independent of f and t.

(ii) is well-known from real interpolation theory since $X \subset Z$, see also (4.31).   □

Besides the spaces $\mathcal{X}_{\infty,\alpha}$ we also need the closed subspace $\mathring{\mathcal{X}}_{\infty,\alpha}$ defined by

$$\mathring{\mathcal{X}}_{\infty,\alpha} = \left\{ f \in \mathcal{X}_{\infty,\alpha} : \sup_{0 < \tau < \tau_0} \tau^\alpha \|e^{-\tau A} f\|_X \to 0 \quad \text{as } \tau_0 \to 0 \right\}.$$

By [62, Exercise 6.1.1 (1)] it is known that $\mathring{\mathcal{X}}_{\infty,\alpha}$ coincides with the continuous interpolation space $(Z, X)_{1-\alpha,\infty}^0$ with equivalent norms. Thus obviously $X \subset \mathring{\mathcal{X}}_{\infty,\alpha} \subset \mathcal{X}_{\infty,\alpha} \subset Z$.

In view of (4.28), (4.29) we often suppress the T-dependence of $\mathcal{X}_{s,\alpha}^T$ and assume that $0 < T < \infty$ is fixed. The notation $\mathcal{X}_{s,\alpha}^T$ is important in the construction of strong $L_\alpha^s(L^q)$-solution where T yields a control of the interval of existence.

Finally, we note that in the case where $X = L_\sigma^q(\Omega)$, $q > 3$, A equals the Stokes operator A_q, and $Z = (\mathcal{D}(A_{q'}))'$, by §4.1–§4.2

$$\mathcal{X}_{s,\alpha} = \mathbb{B}_{q,s}^{-1+3/q}(\Omega) \quad \text{and} \quad \mathring{\mathcal{X}}_{\infty,\alpha} = \mathring{\mathbb{B}}_{q,\infty}^{-1+3/q}(\Omega), \qquad \frac{2}{s} + \frac{3}{q} = 1 - 2\alpha.$$

Additionally, we need a weighted maximal regularity estimate extending (4.8) to the case with weights in time.

Lemma 4.3. [68] *Let the operator A on X have the property of maximal regularity for some $s \in (1,\infty)$ (and hence for all $s \in (0,\infty)$). Then for any $1 < s < \infty$ and $\alpha \in [0, 1 - \frac{1}{s})$ the operator A has the property of weighted in time maximal regularity, i.e., the Cauchy problem (4.6) with $u_0 = 0$ and $f \in L_\alpha^s(0,\infty; X)$ has a unique strong (mild) solution u such that*

$$\|u_t\|_{L_\alpha^s(0,\infty;X)} + \|Au\|_{L_\alpha^s(0,\infty;X)} \le c\|f\|_{L_\alpha^s(0,\infty;X)}.$$

5 Proof of existence

As an important technical tool for the forthcoming proofs we need a version of the Hardy-Littlewood-Sobolev inequality on weighted L^s-spaces on $\mathbb{R}$, cf. [77, 78],

$$L_\alpha^s(\mathbb{R}) = \left\{ u : \|u\|_{L_\alpha^s} = \left(\int_{\mathbb{R}} (|\tau|^\alpha |u(\tau)|)^s \, d\tau \right)^{1/s} < \infty \right\}, \qquad \alpha \in \mathbb{R}, \ s \ge 1.$$

Lemma 5.1. *Let $0 < \lambda < 1$, $1 \le s_1 \le s_2 \le \infty$, $0 < \alpha_1 + \frac{1}{s_1}$, $\alpha_2 + \frac{1}{s_2} < 1$, and $\lambda + (\alpha_1 + \frac{1}{s_1}) = 1 + (\alpha_2 + \frac{1}{s_2})$, $\alpha_2 \le \alpha_1$, where $\alpha_2 = \alpha_1$ is admitted only if $1 < s_1 \le s_2 < \infty$. Then the linear integral operator*

$$I_\lambda f(t) = \int_{\mathbb{R}} |t - \tau|^{-\lambda} f(\tau) \, d\tau$$

is bounded as operator $I_\lambda : L_{\alpha_1}^{s_1}(\mathbb{R}) \to L_{\alpha_2}^{s_2}(\mathbb{R})$.

Note that Lemma 5.1 reduces to the classical Hardy-Littlewood-Sobolev inequality when $\alpha_1 = \alpha_2 = 0$ and $\lambda + \frac{1}{s_1} = 1 + \frac{1}{s_2}$; this will be needed for the proof of Theorem 2.1. In the weighted case, the exponents s_1 and s_2 formally are replaced by $(\alpha_1 + \frac{1}{s_1})^{-1}$ and $(\alpha_2 + \frac{1}{s_2})^{-1}$, respectively; this is consistent with the modified Serrin condition $(\frac{2}{s_j} + 2\alpha_j) + \frac{3}{q_j} = 1$, cf. (2.7). Obviously, the weight functions $|\tau|^{\alpha_j}$ are Muckenhoupt weights.

Proof of Theorems 2.1, 2.3, 2.4 and Corollary 2.5. Let u be a weak solution of (1.1) with initial value $u_0 \in L_\sigma^2$ and external force $f = \operatorname{div} F$ where $F \in L^2(L^2) \cap$

$L_{2\alpha_*}^{s_*/2}(L^{q_*/2})$. Moreover, let $v = E_{f,u_0}$, $f = \operatorname{div} F$, denote the solution of the linear Stokes problem

$$\partial_t v - \Delta v + \nabla p = f, \quad \operatorname{div} v = 0,$$
$$v|_{\partial\Omega} = 0, \quad v(0) = u_0.$$

Recalling the definition $\gamma = \frac{3}{2p_*} - \frac{3}{2r} + \frac{1}{2} \in [\frac{1}{2}, 1)$, see (4.13), with $p_* = \frac{q_*}{2}$ we use (4.12) in the form

$$v(t) = E_{f,u_0}(t) = e^{-tA}u_0 + \int_0^t A^\gamma e^{-(t-\tau)A}(A^{-\gamma}P\operatorname{div})F(\tau)\,d\tau. \qquad (5.1)$$

As explained in §4.1 v satisfies the energy equality (1.11) and the energy estimate (4.16).

Case 1: $\frac{s_*}{2} \le s < \infty$. Assume $E_{0,u_0} \in L_\alpha^s(L^q)$; in other words, $\int_0^t \|\tau^\alpha e^{-\tau A}u_0\|_q^s\,d\tau < \infty$ or, by Theorem 4.1, $u_0 \in \mathbb{B}_{q,s}^{-1+3/q}$. Moreover, if $0 \le \beta := \frac{3}{2r} - \frac{3}{2q}$ is strictly less than $1 - \gamma$, (4.15) and (4.1) yield the estimate

$$\|E_{f,0}(t)\|_q \le c\int_0^t \left\|A^{\gamma+\beta}e^{-(t-\tau)A}(A^{-\gamma}P\operatorname{div})F(\tau)\right\|_r d\tau$$

$$\le c\int_0^t (t-\tau)^{-\beta-\gamma}\|F(\tau)\|_{\frac{q_*}{2}}\,d\tau.$$

Note that the necessary conditions $q \ge r \ge p_*$ in the definition of $\beta \ge 0$, $\gamma \ge \frac{1}{2}$, and $\beta + \gamma < 1$ can be fulfilled by choosing r sufficiently close to p_*. Then the weighted Hardy-Littlewood-Sobolev inequality (Lemma 5.1) with exponents $s_2 = s$, $\alpha_2 = \alpha$, $s_1 = s_*/2$, $\alpha_1 = 2\alpha_*$, $\lambda = \beta + \gamma \in (0,1)$, $0 < \alpha + \frac{1}{s} < 1$ and $0 < 2\alpha_* + \frac{1}{s_*/2} < 1$, see (2.10), implies that

$$\|E_{f,0}\|_{L_\alpha^s(L^q)} \le c\|F\|_{L_{2\alpha_*}^{s_*/2}(L^{q_*/2})} \qquad (5.2)$$

since $(\beta + \gamma) + \frac{2}{s_*} + 2\alpha_* = 1 + \frac{1}{s} + \alpha$ by assumption (2.10).

However, if $\beta + \gamma = 1$, we write A^γ in front of the integral $E_{f,0}$ in (5.1) to see that for almost all $t > 0$

$$\|E_{f,0}(t)\|_q \le c\left\|A^{\gamma+\beta}\int_0^t e^{-(t-\tau)A}(A^{-\gamma}P\operatorname{div})F(\tau)\,d\tau\right\|_r$$

where $A^{\gamma+\beta} = A$ and $w(t) := \int_0^t e^{-(t-\tau)A}(A^{-\gamma}P\operatorname{div})F(\tau)\,d\tau$ is the mild solution of the Stokes system $w_t + Aw = (A^{-\gamma}P\operatorname{div})F$, $w(0) = 0$. Then the weighted maximal regularity Lemma 4.3 applied to w implies that

$$\|E_{f,0}\|_{L_\alpha^s(L^q)} \le c\left\|(A^{-\gamma}P\operatorname{div})F\right\|_{L_\alpha^s(L^r)}$$

$$\le c\|F\|_{L_\alpha^s(L^{q_*/2})}.$$

Now we define $\tilde{u} := u - E_{f,u_0}$ which solves the (Navier-)Stokes system

$$\partial_t \tilde{u} - \Delta \tilde{u} + u \cdot \nabla u + \nabla p = 0, \quad \operatorname{div} \tilde{u} = 0,$$
$$\tilde{u}|_{\partial\Omega} = 0, \quad \tilde{u}(0) = 0.$$

So we can write at least formally

$$\tilde{u}(t) = \mathcal{F}(\tilde{u}) := -\int_0^t e^{-(t-\tau)A} P\operatorname{div}(u \otimes u)(\tau)\, d\tau \tag{5.3}$$

$$= -\int_0^t A^{1/2} e^{-(t-\tau)A}(A^{-1/2}P\operatorname{div})(u \otimes u)(\tau)\, d\tau.$$

For the estimates to follow we use the previous ideas with $s_* = s$, $q_* = q$, $\alpha_* = \alpha$, $\gamma = \frac{1}{2}$, and $\beta = \frac{3}{2q} < \frac{1}{2}$. Consequently,

$$\|\tilde{u}(t)\|_q \le c \int_0^t \left\|A^{\frac{1}{2}+\beta} e^{-(t-\tau)A}\right\| \left\|A^{-\frac{1}{2}}P\operatorname{div}\right\| \|(u \otimes u)(\tau)\|_{\frac{q}{2}}\, d\tau$$

$$\le c \int_0^t (t-\tau)^{-\frac{1}{2}-\beta} \|u(\tau)\|_q^2\, d\tau, \tag{5.4}$$

and, by the Hardy-Littlewood-Sobolev inequality,

$$\|\tilde{u}\|_{L_\alpha^s(L^q)} \le c\|(\|u\|_q^2)\|_{L_{2\alpha}^{s/2}} = c\|u\|_{L_\alpha^s(L^q)}^2. \tag{5.5}$$

Since $u = \tilde{u} + E_{f,u_0}$ we have

$$\|\tilde{u}\|_{L_\alpha^s(0,T;L^q)} \le c_0 \left(\|\tilde{u}\|_{L_\alpha^s(0,T;L^q)} + \|F\|_{L_{2\alpha_*}^{s_*/2}(0,T;L^{q_*}/2)} + \|e^{-\tau A}u_0\|_{L_\alpha^s(0,T;L^q)}\right)^2$$

$$= c_0 \left(\|\tilde{u}\|_{L_\alpha^s(0,T;L^q)} + b\right)^2 \tag{5.6}$$

where $c_0 > 0$ is independent of T and the data, and

$$b = b(T) = \|F\|_{L_{2\alpha_*}^{s_*}(0,T;L^{q_*}/2)} + \|e^{-\tau A}u_0\|_{L_\alpha^s(0,T;L^q)}.$$

Under the smallness assumption (2.5) or (2.12) we can guarantee that $4bc_0 < 1$ and get a closed ball of radius less than b, $\mathcal{B} \subset L_\alpha^s(0,T;L^q)$, such that $\mathcal{F}$ maps $\mathcal{B}$ into itself. Moreover, it is straightforward to modify the above estimates to show that for $\tilde{u}, \hat{u} \in \mathcal{B}$

$$\|\mathcal{F}\tilde{u} - \mathcal{F}\hat{u}\|_{L_\alpha^s(0,T;L^q)} \le 4bc_0\|\tilde{u} - \hat{u}\|_{L_\alpha^s(0,T;L^q)}, \tag{5.7}$$

i.e., $\mathcal{F}$ is a strict contraction on $\mathcal{B}$, cf. [33, p. 99f]. Hence Banach's Fixed Point Theorem yields the existence of a unique fixed point $\tilde{u} \in \mathcal{B} \subset L_\alpha^s(0,T;L^q)$ of $\mathcal{F}$, solving

$$\partial_t \tilde{u} - \Delta \tilde{u} + (\tilde{u} + E_{f,u_0}) \cdot \nabla(\tilde{u} + E_{f,u_0}) + \nabla p = 0, \quad \operatorname{div} \tilde{u} = 0,$$
$$\tilde{u}|_{\partial\Omega} = 0, \quad \tilde{u}(0) = 0$$

provided (2.5) or (2.12) are satisfied. Then $u = \tilde{u} + E_{f,u_0} \in L^s_\alpha(0,T;L^q)$ is the solution we are looking for. It satisfies the estimate $\|u\|_{L^s_\alpha(0,T;L^q)} \le 2bc_0 + b$, i.e.,

$$\|u\|_{L^s_\alpha(0,T;L^q)} \le C\big(\|F\|_{L^{s*/2}_{2\alpha_*}(0,T;L^{q*}/2)} + \|e^{-\tau A}u_0\|_{L^s_\alpha(0,T;L^q)}\big). \tag{5.8}$$

Case 2: $s_* \le s = \infty$. As in the previous case the Stokes problem with $u_0 \in L^2_\sigma(\Omega)$ and $F \in L^2(0,T;L^2)$ admits a weak solution $E_{f,u_0} \in C^0([0,T];L^2) \cap L^2(H^1)$ satisfying the energy equality (1.11) and the energy estimate (4.2). By assumption (2.14) there holds $E_{0,u_0} \in L^\infty_\alpha(0,T;L^q)$. With the help of (4.15), (4.1) and $2\beta + \frac{3}{q} = \frac{3}{r}$ as above, and hence $\beta + \gamma = 1 + \alpha - 2\alpha_* - \frac{2}{s_*}$,

$$\|E_{f,0}(t)\|_q \le c \int_0^t \big\|A^{\beta+\gamma} e^{-(t-\tau)A}(A^{-\gamma}P\mathrm{div})F(\tau)\big\|_r \, d\tau$$

$$\le c \int_0^t (t-\tau)^{-1-\alpha+2\alpha_*+2/s_*} \|F(\tau)\|_{\frac{q_*}{2}} \, d\tau.$$

Then the weighted Hardy-Littlewood-Sobolev inequality (Lemma 5.1) with exponents $s_2 = \infty$, $\alpha_2 = \alpha < \alpha_1 = 2\alpha_*$ (since $s_2 = \infty$), and $s_1 = 2s_* \le \infty$, implies that

$$\|E_{f,0}\|_{L^\infty_\alpha(0,T;L^q)} \le c\|F\|_{L^{s*/2}_{2\alpha_*}(0,T;L^{q*}/2)}. \tag{5.9}$$

Adding the estimate for E_{0,u_0} we get that

$$\|E_{f,u_0}\|_{L^\infty_\alpha(0,T;L^q)} \le c\big(\|e^{-\tau A}u_0\|_{L^\infty_\alpha(0,T;L^q)} + \|F\|_{L^\infty_{2\alpha_*}(0,T;L^{q*}/2)}\big) \tag{5.10}$$

with a constant $c > 0$ independent of $t > 0$.

Then $\tilde{u} = u - E_{f,u_0}$ is a fixed point of the nonlinear map $\mathcal{F}$, see (5.3), which with $q_* = q$, $\alpha_* = \alpha$, and hence $\beta = \frac{3}{2q} = \frac{1}{2} - \alpha$, satisfies the estimate (cf. (5.4))

$$\|\mathcal{F}(\tilde{u})(t)\|_q \le c \int_0^t \big\|A^{\frac{1}{2}+\beta} e^{-(t-\tau)A}\big\|\big\|A^{-\frac{1}{2}}P\mathrm{div}\big\|\|(u \otimes u)\|_{\frac{q}{2}} \, d\tau$$

$$\le c \int_0^t (t-\tau)^{-1+\alpha}\|u\|^2_q \, d\tau.$$

We proceed as before and conclude that

$$\|\mathcal{F}\tilde{u}\|_{L^\infty_\alpha(0,T;L^q)} \le c\|u\|^2_{L^\infty_\alpha(0,T;L^q)} \le c_0\big(\|\tilde{u}\|_{L^\infty_\alpha(0,T;L^q)} + b\big)^2 \tag{5.11}$$

where $c_0 > 0$ is independent of T and u_0, F, and

$$b = b(T) = \|F\|_{L^\infty_{2\alpha_*}(0,T;L^{q*}/2)} + \|e^{-\tau A}u_0\|_{L^\infty_\alpha(0,T;L^q)}.$$

As in Case 1 we have to find $b = b(T)$ sufficiently small so that Banach's Fixed Point Theorem yields the existence of a unique fixed point $\tilde{u}$ of $\mathcal{F}$ in a closed ball $\mathcal{B} \subset L^\infty_\alpha(0,T;L^q)$ of radius b. In contrast to Case 1 where $b(T) \to 0$ as $T \to 0$, in the present L^∞-case when even $s_* = \infty$ we either have to assume this smallness by Theorem 2.4 (i) explicitly or use the properties of the spaces $\overset{\circ}{\mathbb{B}}_{q,\infty}^{-1+3/q}(\Omega)$ and $\overset{\circ}{L}_{2\alpha_*}^{\infty}(0,T;L^{q_*/2})$ in Theorem 2.4 (ii). The fixed point $\tilde{u}$ solves (5.3), and the mild solution $u = \tilde{u} + E_{f,u_0}$ is contained in $L_\alpha^\infty(0,T;L^q)$ satisfying an estimate of the type $\|u\|_{L_\alpha^\infty(0,T;L^q)} \le 2bc_0 + b$, *i.e.,* (5.8) with s replaced by ∞.

Now we will prove that this mild solution u is indeed a weak solution under the following conditions, *cf.* the assumptions in Theorem 2.3 and some facts already proved:

$$u, \tilde{u}, e^{-\tau A}u_0 \in L_\alpha^s(L^q), \quad u_0 \in L_\sigma^2 \cap \mathbb{B}_{q,s}^{-1+3/q}, \quad F \in L^2(L^2) \cap L_{2\alpha_*}^{s_*/2}(L^{q_*/2}).$$

In Theorem 2.4 (i) we have the same conditions with s replaced by ∞. Moreover, in Theorem 2.4 (ii) there even holds

$$u, \tilde{u}, e^{-\tau A}u_0 \in \overset{\circ}{L}_\alpha^\infty(L^q), \quad u_0 \in L_\sigma^2 \cap \overset{\circ}{\mathbb{B}}_{q,\infty}^{-1+3/q}, \quad F \in L^2(L^2) \cap \overset{\circ}{L}_{2\alpha_*}^\infty(L^{q_*/2}).$$

To this aim we need three lemmata.

Lemma 5.2. *The mild solution u constructed above satisfies $\nabla u \in L^2(0,T;L^2(\Omega))$.*

Proof. We use a modification of the proof described in [33]. Since for the moment we have no differentiability property for the mild solution u, we apply the Yosida operator $J_n = (I + \frac{1}{n}A^{\frac{1}{2}})^{-1}, n \in \mathbb{N}$, to (5.3) and write $J_n P\mathrm{div}\, u \otimes u$ in the form $J_n P\mathrm{div}\,(u \otimes (\tilde{u} + E_{f,u_0})), \tilde{u} = (I + \frac{1}{n}A^{\frac{1}{2}})\tilde{u}_n$, where $\tilde{u}_n = J_n\tilde{u}$. Then we have

$$J_n P\mathrm{div}\, u \otimes u = J_n P(u \cdot \nabla E_{f,u_0}) + J_n P(u \cdot \nabla \tilde{u}_n) + \frac{1}{n}J_n P\mathrm{div}\,(u \otimes A^{\frac{1}{2}}\tilde{u}_n)$$

$$= J_n P(u \cdot \nabla E_{f,u_0}) + J_n P(u \cdot \nabla \tilde{u}_n) + \frac{1}{n}A^{\frac{1}{2}}J_n(A^{-\frac{1}{2}}P\mathrm{div}\,)(u \otimes A^{\frac{1}{2}}\tilde{u}_n).$$

We use Hölder's inequality with $\frac{1}{\gamma} = \frac{1}{2} + \frac{1}{q}$ to obtain the estimate

$$\|J_n P\mathrm{div}\,(u \otimes u)\|_\gamma \le c\|u\|_q(\|\nabla E_{f,u_0}\|_2 + \|\nabla \tilde{u}_n\|_2 + \|A^{\frac{1}{2}}\tilde{u}_n\|_2)$$

$$= c\|u\|_q(\|\nabla E_{f,u_0}\|_2 + 2\|A^{\frac{1}{2}}\tilde{u}_n\|_2)$$

since $\|J_n\|_{\mathcal{L}(L_\sigma^r)} \le c$ and $\|\frac{1}{n}A^{\frac{1}{2}}J_n\|_{\mathcal{L}(L_\sigma^r)} \le c$ uniformly in $n \in \mathbb{N}$ for each $1 < r < \infty$ with a constant $c = c(r) > 0$.

From (5.3) we get that

$$A^{\frac{1}{2}}\tilde{u}_n(t) = -\int_0^t A^{\frac{1}{2}}e^{-(t-\tau)A}J_n P\mathrm{div}\,(u \otimes u)(\tau)\,\mathrm{d}\tau$$

so that by the embedding estimate (4.1) with $2\beta + \frac{3}{2} = \frac{3}{\gamma}$, *i.e.* $\beta = \frac{3}{2q}$,

$$\|A^{\frac{1}{2}}\tilde{u}_n(t)\|_2 \le c\int_0^t \|A^{\frac{1}{2}+\beta}e^{-(t-\tau)A}\|\|J_n P\mathrm{div}\,(u \otimes u)(\tau)\|_\gamma\,\mathrm{d}\tau.$$

Applying Lemma 5.1 we have for $0 < T_1 < T$

$$\|A^{\frac{1}{2}}\tilde{u}_n(t)\|_{L^2(0,T_1;L^2)} \le c\left(\int_0^{T_1} \left(\tau^\alpha \|u\|_q (\|\nabla E_{f,u_0}\|_2 + \|A^{\frac{1}{2}}\tilde{u}_n\|_2)\right)^{s_1} d\tau\right)^{1/s_1}$$

where $\frac{1}{s_1} = \frac{1}{2} + \frac{1}{s}$, $\alpha_1 = \alpha$, $s_2 = 2$, $\alpha_2 = 0$, and $(\frac{1}{2} + \frac{1}{s}) + \frac{1}{2} + \beta + \alpha = 1 + \frac{1}{2} + 0$, which is equivalent to $\frac{2}{s} + \frac{3}{q} = 1 - 2\alpha$. Thus, by Hölder's inequality,

$$\|A^{\frac{1}{2}}\tilde{u}_n\|_{L^2(0,T_1;L^2)}$$
$$\le c\|u\|_{L^s_\alpha(0,T_1;L^q)}(\|\nabla E_{f,u_0}\|_{L^2(0,T_1;L^2)} + \|A^{\frac{1}{2}}\tilde{u}_n\|_{L^2(0,T_1;L^2)}). \qquad (5.12)$$

The same result obviously holds in case that $s = \infty$.

Case 1: $s, s_* < \infty$.   Choose $0 < T_1 < T$ so small such that $c\|u\|_{L^s_\alpha(0,T_1;L^q)} \le \frac{1}{2}$ holds. Then the absorption argument leads from (5.12) to the estimate

$$\|A^{\frac{1}{2}}\tilde{u}_n\|_{L^2(0,T_1;L^2)} \le 2c\|u\|_{L^s_\alpha(0,T_1;L^q)}\|\nabla E_{f,u_0}\|_{L^2(0,T_1;L^2)} < \infty \qquad (5.13)$$

independent of $n \in \mathbb{N}$. From (5.13) we conclude by a classical weak convergence and reflexivity argument that $A^{\frac{1}{2}}\tilde{u}, \nabla\tilde{u} \in L^2(0,T_1;L^2)$ and hence $\nabla u \in L^2(0,T_1;L^2)$. Since by (5.8) $u \in L^s_\alpha(0,T;L^q)$ we repeat this step with a fixed time length $T_2 > 0$ and achieve in finitely many steps that $\nabla\tilde{u}, \nabla u \in L^2(0,T;L^2)$.

Case 2: $s_* \le s = \infty$.   Choosing ε_* sufficiently small, (5.8) with $s = \infty$ and Theorem 2.4 (i) directly imply that $c\|u\|_{L^\infty_\alpha(0,T_1;L^q)} \le \frac{1}{2}$ and hence $\nabla\tilde{u}, \nabla u \in L^2(0,T;L^2)$ as in Case 1. In case of Theorem 2.4 (ii) we find due to (5.8) a sufficiently small $T_1 > 0$ such that $c\|u\|_{L^\infty_\alpha(0,T_1;L^q)} \le \frac{1}{2}$. Now we conclude that $\nabla\tilde{u}, \nabla u \in L^2(0,T;L^2)$.

This completes the proof. $\square$

Lemma 5.3. *Let $0 < \alpha < \frac{1}{2}$. Then the mild solution u satisfies $u \in L^{s_2}(0,T;L^{q_2}(\Omega))$ for all $2 \le s_2 \le \infty$, $2 \le q_2 \le 6$ satisfying $\frac{2}{s_2} + \frac{3}{q_2} = \frac{3}{2}$. Moreover,*

$$\tilde{u}(t) \to 0 \quad \text{and} \quad u(t) \to u_0 \quad \text{in } L^2(\Omega) \quad \text{as } t \to 0+ .$$

Finally, $u \in L^4_{\alpha/(2+8\alpha)}(0,T;L^4(\Omega))$.

Proof. The result for $s_2 = 2$, $q_2 = 6$ follows from Lemma 5.2. Next we consider the case $s_2 = \infty$, $q_2 = 2$, and choose as in the proof of Lemma 5.2 $\frac{1}{\gamma} = \frac{1}{2} + \frac{1}{q}$ and $\beta = \frac{3}{2q}$. Then (4.1) and Hölder's inequality directly imply that

$$\|\tilde{u}(t)\|_2 \le c\int_0^t \|A^\beta e^{-(t-\tau)A}\| \|P(u \cdot \nabla u)\|_\gamma \, d\tau$$
$$\le c\int_0^t (t-\tau)^{-\frac{3}{2q}} \tau^{-\alpha}(\tau^\alpha \|u\|_q)\|\nabla u\|_2 \, d\tau$$
$$\le C\|u\|_{L^s_\alpha(L^q)}\|\nabla u\|_{L^2(L^2)} \qquad (5.14)$$

since the integral $\int_0^t ((t-\tau)^{-\frac{3}{2q}} \tau^{-\alpha})^{(\frac{1}{2}-\frac{1}{s})^{-1}} \, d\tau$ is finite and independent of t; note that for the last argument $\alpha > 0$ was necessary and that Lemma 5.1 because of $s_2 = \infty$, $\alpha_2 = 0$ did not apply.

To be more precise, with a constant $C > 0$ independent of t,

$$\|\tilde{u}\|_{L^\infty(0,t;L^2)} \leq C\|u\|_{L_\alpha^s(0,t;L^q)} \|\nabla u\|_{L^2(0,t;L^2)} \to 0 \quad \text{as } t \to 0.$$

So $\|\tilde{u}(t)\|_2 \to 0$ as $t \to 0$. Hence $u(t) = \tilde{u}(t) + E_{f,u_0}(t) \to u_0$ in $L^2(\Omega)$ as $t \to 0$.

From $\nabla u \in L^2(L^2)$ which implies $u \in L^2(L^6)$ and from $u \in L^\infty(L^2)$, it follows immediately by Hölder's inequality in space and time that $u \in L^{s_2}(L^{q_2})$ for all remaining pairs (s_2, q_2) satisfying $\frac{2}{s_2} + \frac{3}{q_2} = \frac{3}{2}$.

For the L^4-result let $\beta = \frac{1}{2+8\alpha}$ and define q_1, s_1 by $\frac{1}{4} = \frac{\beta}{q} + \frac{1-\beta}{q_1}$ and $\frac{1}{4} = \frac{\beta}{s} + \frac{1-\beta}{s_1}$. From Hölder's inequality we know that $\|u(t)\|_4 \leq \|u\|_q^\beta \|u\|_{q_1}^{1-\beta}$. Hence

$$\int_0^T \tau^{4\alpha\beta} \|u\|_4^4 \, d\tau \leq \int_0^T (\tau^\alpha \|u\|_q)^{4\beta} \|u\|_{q_1}^{4(1-\beta)} \, d\tau$$
$$\leq \|u\|_{L_\alpha^s(0,T;L^q)}^{4\beta} \|u\|_{L^{s_1}(0,T;L^{q_1})}^{4(1-\beta)} < \infty$$

since $\frac{2}{s_1} + \frac{3}{q_1} = \frac{3}{2}$. This argument works also in the case $s = \infty$.

The proof is now complete. $\qquad\qquad\qquad\qquad\qquad\qquad\qquad\qquad\qquad\square$

It remains to discuss the more classical case when $\alpha = 0$. Although the direct argument to prove (5.14) does not work, the L^4-result from Lemma 5.3 should hold in the form $u \in L^4(L^4)$. Actually, this property will be proved first and help to deduce from the theory of weak solutions that $u \in L^\infty(L^2)$.

Lemma 5.4. *Let $\alpha = 0$. Then the solution u satisfies $u \in L^4(0,T;L^4(\Omega)) \cap L^\infty(0,T;L^2(\Omega))$ and the energy equality (EE) holds.*

Proof. To estimate $\tilde{u}$, see (5.3), choose $2\beta + \frac{3}{q_1} = \frac{3}{q_2}$ where $\frac{1}{q_1} = \frac{1}{2} - \frac{1}{q}$, $\frac{1}{q_2} = \frac{1}{2} + \frac{1}{q}$, and use (4.1) with $\beta = \frac{3}{q}$ to obtain that

$$\|\tilde{u}(t)\|_{q_1} \leq c \int_0^t \left\| A^\beta e^{-(t-\tau)A_{q_2}} P(u \cdot \nabla u) \right\|_{q_2} \, d\tau$$
$$\leq c \int_0^t (t-\tau)^{-\beta} \|u \cdot \nabla u\|_{q_2} \, d\tau.$$

Then we apply the Hardy-Littlewood estimate without weights and $\frac{1}{s_1} = \frac{1}{2} - \frac{1}{s}$, $\frac{1}{s_2} = \frac{1}{2} + \frac{1}{s}$ so that $1 + \frac{1}{s_1} = \beta + \frac{1}{s_2}$, and obtain with Hölder's inequality that

$$\|\tilde{u}\|_{L^{s_1}(L^{q_1})} \leq c\|u \cdot \nabla u\|_{L^{s_2}(L^{q_2})} \leq c\|u\|_{L^s(L^q)} \|\nabla u\|_{L^2(L^2)} < \infty. \qquad (5.15)$$

Since $\frac{2}{s_1} + \frac{3}{q_1} = \frac{3}{2}$, E_{f,u_0} as a weak solution of the instationary Stokes system, see [74, Ch. IV, Theorem 2.4.1], satisfies $E_{f,u_0} \in L^{s_1}(L^{q_1})$. Then (5.15) implies that $u = \tilde{u} + E_{f,u_0} \in L^{s_1}(L^{q_1})$. Hence Hölder's inequality yields

$$\|u\|_{L^4(L^4)} \leq c\|u\|_{L^s(L^q)}^{1/2} \|u\|_{L^{s_1}(L^{q_1})}^{1/2} < \infty.$$

Finally we recall that u solves the nonlinear integral equation (1.13). Since $-uu - F \in L^2(0, T; L^2(\Omega))$, u is the unique weak solution of the linear system with right-hand side $-uu - F$, in other words, u is a weak solution to the Navier-Stokes system (1.1) and satisfies the energy equality on $[0, T)$, see [74, Ch. IV, Theorems 2.3.1, 2.4.1]. □

Returning to the case $\alpha > 0$ and Lemma 5.3 we conclude from $u \in L^4_{\alpha/(2+8\alpha)}(L^4)$ that $u \in L^4(\epsilon, T; L^4)$ for all $0 < \epsilon < T$. So, by [74, Ch. IV, Theorem 2.3.1, Lemma 2.4.2] and for a.a. $\epsilon \in (0, T)$, u is the unique weak solution in $L^4(\epsilon, T; L^4)$ on (ϵ, T) of the linear Stokes problem

$$\partial_t u - \Delta u + \nabla p = \operatorname{div} \tilde{F}, \quad \operatorname{div} u = 0,$$
$$u|_{\partial\Omega} = 0, \quad u|_{t=\epsilon} = u(\epsilon)$$

with external force $\operatorname{div} \tilde{F}$, $\tilde{F} = -F - u \otimes u \in L^2(\epsilon, T; L^2)$ and initial value $u(\epsilon) \in L^4(\Omega) \subset L^2(\Omega)$. Therefore, u satisfies the energy equality on (ϵ, T), i.e.,

$$\frac{1}{2}\|u(t)\|_2^2 + \int_\epsilon^t \|\nabla u\|_2^2 \, d\tau = \frac{1}{2}\|u(\epsilon)\|_2^2 - \int_\epsilon^t (F, \nabla u) \, d\tau$$

for all $t \in (\epsilon, T)$ and a.a. $\epsilon \in (0, T)$. Moreover, $u \in C^0([\epsilon, T); L^2)$ and hence even $u \in C^0((0, T); L^2)$, see [74, Theorem IV.2.3.1]. Furthermore, since by Lemma 5.3 $u(t) \to u_0$ in $L^2(\Omega)$ as $t \to 0$, it satisfies the energy equality on $[0, T)$. Consequently, u is a weak solution on $[0, T)$ enjoying the energy equality (EE).

Now the proof of Theorem 2.3 is complete. □

Finally, we give a proof concerning the necessary conditions described in Theorem 2.3 (ii) and Corollary 2.5.

Proof of Theorem 2.3(ii). (1) Using the assumption $u_0 \in \mathbb{B}_{q,s}^{-1+3/q}$ and the condition on F we find $0 < T \le \infty$ such that (2.12) is satisfied. Then Theorem 2.3 yields the existence of a unique $L^s_\alpha(L^q)$-strong solution $u \in L^s_\alpha(0, T; L^q)$ of (1.1).

Conversely, assume that $u \in L^s_\alpha(0, T; L^q)$, $0 < T \le \infty$, is an $L^s_\alpha(L^q)$-strong solution of (1.1). Recall that $E_{0,u_0} = u - \tilde{u} - E_{f,0}$ where by (5.5) and (5.2) $\tilde{u} \in L^s_\alpha(L^q)$ and $E_{f,0} \in L^s_\alpha(L^q)$, respectively. Hence $E_{0,u_0} \in L^s_\alpha(L^q)$ as well, and hence $u_0 \in \mathbb{B}_{q,s}^{-1+3/q}$. This proves part (ii) of Theorem 2.3. □

Proof of Corollary 2.5. (i) Assume that $u \in L^\infty_\alpha(0, T; L^q)$. Since $E_{0,u_0} = u - \tilde{u} - E_{f,0}$ and $\tilde{u}, E_{f,0} \in L^\infty_\alpha(L^q)$ by (5.11) and (5.9), we get that $E_{0,u_0} \in L^\infty_\alpha(0, T; L^q)$. Then the exponential decay of e^{-tA} completes the proof.

(ii) Conversely, let $F \in \mathring{L}^\infty_{2\alpha_*}(0, T; L^{q_*/2})$ and assume $u \in \mathring{L}^\infty_\alpha(0, T; L^q)$. Then (5.11) implies that even $E_{0,u_0} = u - \tilde{u} - E_{f,0} \in \mathring{L}^\infty_\alpha(0, T; L^q)$; consequently, (2.16) holds.

The sufficiency of the condition $u_0 \in \mathring{\mathbb{B}}_{q,\infty}^{-1+3/q}$ is given by Theorem 2.4 (ii). □

6   Proof of continuity in time

6.1   Estimates of continuity

In the following we use the notation introduced in Subsect. 4.3 and recall the setting of the closed operator A on the Banach space X generating an analytic bounded C^0-semigroup e^{-tA} on X as well as on $Z = (\mathcal{D}(A'))'$. We fix $0 < T < \infty$ and simply write $\mathcal{X}_{s,\alpha}$ for $\mathcal{X}_{s,\alpha}^T$ where $\mathcal{X}_{s,\alpha} = A\,(X, \mathcal{D}(A))_{1-\alpha-\frac{1}{s},s} = (Z, X)_{1-\alpha-\frac{1}{s},s}$ is a real interpolation space equipped with the equivalent weighted integral norm $\|f\|_{\mathcal{X}_{s,\alpha}^T} = \left(\int_0^T \left(\tau^\alpha \|e^{-\tau A} f\|_X \right)^s \, d\tau \right)^{1/s}$, see (4.29).

The first continuity result considers the homogeneous part $e^{-tA} u_0$ in (1.8).

Proposition 6.1. *Let $s \in [1, \infty]$ and $\alpha \geq 0$. Assume that $u_0 \in \mathcal{X}_{s,\alpha}$.*

(i) *For $t \in (0, T]$ the estimate*

$$\left\| e^{-tA} u_0 \right\|_{\mathcal{X}_{s,\alpha}^T} \leq C_T \|u_0\|_{\mathcal{X}_{s,\alpha}^T}$$

holds with the constant $C_T = \sup_{t \in (0,T)} \left\| e^{-tA} \right\|_{\mathcal{L}(X)}$.

(ii) $e^{-tA} u_0 \in C\big([0, \infty); \mathcal{X}_{s,\alpha}\big)$ *if $u_0 \in \mathcal{X}_{s,\alpha}$ and $s < \infty$.*

(iii) $e^{-tA} u_0 \in C\big([0, \infty); \mathring{\mathcal{X}}_{\infty,\alpha}\big)$ *if $u_0 \in \mathring{\mathcal{X}}_{\infty,\alpha}$.*

(iv) *For $u_0 \in \mathcal{X}_{\infty,\alpha}$, continuity holds except at $t = 0$, i.e., $e^{-tA} \in C\big((0, \infty); \mathring{\mathcal{X}}_{\infty,\alpha}\big)$. Moreover, $e^{-tA} u_0 \overset{*}{\rightharpoonup} u_0$ as $t \to 0$ in $\mathcal{X}_{\infty,\alpha}$; for the latter result X is assumed to be reflexive.*

To prove Proposition 6.1, we use the strong continuity of the semigroup e^{-tA} on X and on $\mathcal{D}(A)$ near $t = 0$.

Lemma 6.2. (i) *For each $\beta \in (0, 1]$ and $f \in \mathcal{D}(A^\beta)$ there holds for all $t \in (0, T)$ the estimate $\left\| (e^{-tA} - I) f \right\|_X \leq c_{\beta,T} t^\beta \left\| A^\beta f \right\|_X$ with a constant $c_{\beta,T} > 0$ independent of f and $t > 0$.*

(ii) *There holds $\left\| (e^{-tA} - I) e^{-\tau A} f \right\|_X \leq c_{\beta,T} \left(\frac{t}{\tau} \right)^\beta \|f\|_X$ for each $\beta \in (0, 1]$, $t \in (0, T)$ and $f \in X$ with $c_{\beta,T}$ independent of t, τ and f.*

Proof of Lemma 6.2. (i) By the fundamental theorem of calculus,

$$e^{-tA} f - f = -\int_0^t A e^{-\tau A} f \, d\tau = -\int_0^t A^{1-\beta} e^{-\tau A} A^\beta f \, d\tau.$$

Since $\left\| A^\lambda e^{-tA} \right\|_{\mathcal{L}(X)} \leq c_\lambda \tau^{-\lambda} \ (\lambda > 0)$ by analyticity, we observe that

$$\left\| e^{-tA} f - f \right\|_X \leq c_{1-\beta} \int_0^t \tau^{\beta-1} \left\| A^\beta f \right\|_X \, d\tau = c'_\beta t^\beta \left\| A^\beta f \right\|_X.$$

(ii) This follows from (i) since $\left\| A^\beta e^{-\tau A} \right\|_{\mathcal{L}(X)} \leq c_\beta \tau^{-\beta}$. $\square$

Proof of Proposition 6.1. (i) If $s < \infty$ then

$$\left\| e^{-tA} u_0 \right\|_{\mathcal{X}_{s,\alpha}^T}^s = \int_0^T \tau^{\alpha s} \left\| e^{-(\tau+t)A} u_0 \right\|_X^s \, d\tau \leq C_T^s \|u_0\|_{\mathcal{X}_{s,\alpha}^T}^s.$$

The case $s = \infty$ is treated similarly.

(ii) Let $t_0, t \geq 0$. Then by Lebesgue's Theorem on Dominated Convergence

$$\left\| e^{-tA} u_0 - e^{-t_0 A} u_0 \right\|_{\mathcal{X}_{s,\alpha}^T}^s = \int_0^T \tau^{\alpha s} \left\| \left(e^{-tA} - e^{-t_0 A} \right) e^{-\tau A} u_0 \right\|_X^s \, d\tau \to 0$$

as $t \to t_0$ since the integrand is uniformly estimated by an integrable function in $(0, T)$ and converges to 0 in the pointwise sense.

(iii) Let $t, t_0 \geq 0$. We take $\delta \in (0, T)$ and divide the supremum into two parts:

$$\left\| e^{-tA} u_0 - e^{-t_0 A} u_0 \right\|_{\mathcal{X}_{\infty,\alpha}^T} \leq \left(\sup_{\delta < \tau < T} + \sup_{0 < \tau < \delta} \right) \tau^{\alpha} \left\| \left(e^{-tA} - e^{-t_0 A} \right) e^{-\tau A} u_0 \right\|_X$$

$$=: J_1 + J_2.$$

Similarly to the case $s < \infty$, we observe that $J_1 \to 0$ as $t \to t_0$. The second term is estimated as

$$J_2 \leq 2 C_0 \sup_{0 < \tau < \delta} \tau^{\alpha} \left\| e^{-\tau A} u_0 \right\|_X \qquad \text{uniformly in } t, t_0.$$

If $u_0 \in \mathring{\mathcal{X}}_{\infty,\alpha}$, the right-hand side tends to zero as $\delta \to 0$. Thus we conclude the continuity of $e^{-tA} u_0$ up to $t = 0$ with values in $\mathring{\mathcal{X}}_{\infty,\alpha}$.

(iv) If $u_0 \in \mathcal{X}_{\infty,\alpha}$, the function $e^{-tA} u_0$ may not be continuous at $t = 0$ with values in $\mathcal{X}_{\infty,\alpha}$. However, since $e^{-tA} u_0 \in X$ by Proposition 4.2 for $t > 0$ and $X \subset \mathring{\mathcal{X}}_{s,\alpha}$, the assertion $e^{-tA} u_0 \in C\big((0, \infty); \mathring{\mathcal{X}}_{s,\alpha} \big)$ holds.

For the analysis at $t = 0$ we apply the duality theorem of real interpolation, see [79, Theorem 1.11.2], and consider $\mathcal{X}_{\infty,\alpha} = (Z, X)_{1-\alpha,\infty}$ as the dual space of $(Z', X')_{1-\alpha,1} = (X', \mathcal{D}(A'))_{\alpha,1}$ with the weighted norm $\int_0^T \tau^{-\alpha} \left\| A' e^{-\tau A'} \varphi \right\|_{X'} d\tau$ for $\varphi \in (X', \mathcal{D}(A'))_{\alpha,1}$, cf. (4.30) with $s = 1$. Given φ we get that

$$\left| \langle e^{-tA} u_0 - u_0, \varphi \rangle \right| = \left| \langle u_0, e^{-tA'} \varphi - \varphi \rangle \right|$$

$$\leq \| u_0 \|_{\mathcal{X}_{\infty,\alpha}} \left\| e^{-tA'} \varphi - \varphi \right\|_{(X', \mathcal{D}(A'))_{\alpha,1}}.$$

To show that the latter term converges to 0 as $t \to 0$ we note that the argument in part (ii) of this proposition holds also in this case. $\square$

To estimate nonlinear terms as on the right-hand side of (1.8), we consider for $\mu > 0$ the integral operator

$$(Nf)(t) = \int_0^t A^{\mu} e^{-(t-\tau)A} f(\tau) \, d\tau \tag{6.1}$$

for $f \in L_{\alpha_1}^{s_1}(0, T; Y)$. Here Y is another Banach space containing X and e^{-tA} can be extended to Y having a regularizing estimate

$$\left\| e^{-tA} a \right\|_X \leq c_T t^{-\eta} \| a \|_Y, \quad a \in Y, \quad t \in (0, T) \tag{6.2}$$

for some $\eta > 0$ with c_T independent of a; in the application $X = L_\sigma^q(\Omega)$, $Y = L_\sigma^{q/2}(\Omega)$ and (4.1) yields the regularizing effect with $\eta = \frac{3}{2q}$.

Proposition 6.3. *Assume that* $\lambda = \mu + \eta \in (0,1)$ *for positive* μ, η *as in* (6.1), (6.2) *satisfies the scale balance* $\frac{1}{s_1} + \alpha_1 + \lambda = \frac{1}{s_2} + \alpha_2 + 1$ *where* $1 \leq s_1 \leq s_2 \leq \infty$, $\alpha_2 \leq \alpha_1$ *and* $0 < \alpha_1 + \frac{1}{s_1} < 1$, $0 < \alpha_2 + \frac{1}{s_2} < 1$. *Moreover, if* $\alpha_1 = \alpha_2$ *then* $1 < s_1 \leq s_2 < \infty$. *Then* N *defined by* (6.1) *is a bounded operator from* $L^{s_1}_{\alpha_1}(0,T;Y)$ *to* $L^{s_2}_{\alpha_2}(0,T;X)$ *with an operator norm independent of* T.

Proof. The proposition is a consequence of the semigroup estimate $\left\|A^\mu e^{-tA}\right\|_{\mathcal{L}(X)} \leq Ct^{-\mu}$ and (6.2) yielding

$$\|(Nf)(t)\|_X \leq C \int_0^t (t-\tau)^{-\mu-\eta} \|f(\tau)\|_Y \, d\tau. \tag{6.3}$$

Then Lemma 5.1 finishes the proof. $\qquad\qquad\qquad\qquad\qquad\qquad\qquad\square$

We claim that $Nf(\cdot)$ belongs to $C\left([0,T]; \mathcal{X}_{s_2,\alpha_2}\right)$ and start with the case $s_2 < \infty$.

Theorem 6.4. *Assume that* $\lambda = \mu + \eta \in (0,1)$ *for positive* μ, η *as in* (6.1), (6.2) *satisfies the scale balance* $\frac{1}{s_1} + \alpha_1 + \lambda = \frac{1}{s_2} + \alpha_2 + 1$ *for exponents* $1 < s_1 \leq s_2 < \infty$, $\alpha_2 \leq \alpha_1$ *where* $0 < \alpha_1 + \frac{1}{s_1} < 1$, $0 < \alpha_2 + \frac{1}{s_2} < 1$. *If* $f \in L^{s_1}_{\alpha_1}(0,T;Y)$, *then*

$$\|Nf(t)\|_{\mathcal{X}^T_{s_2,\alpha_2}} \leq C\|f\|_{L^{s_1}_{\alpha_1}(0,t;Y)}, \quad t \in [0,T]. \tag{6.4}$$

Moreover,

$$Nf \in C\left([0,T]; \mathcal{X}_{s_2,\alpha_2}\right).$$

Proof. By definition we get from (6.3) that

$$\|Nf(t)\|_{\mathcal{X}^T_{s_2,\alpha_2}} = \left\|e^{-\tau A}(Nf)(t)\right\|_{L^{s_2}_{\alpha_2}(0,t;X)}$$

$$\leq C \left\|\int_0^t (t + \cdot - \rho)^{-\mu-\eta} \|f(\rho)\|_Y \, d\rho\right\|_{L^{s_2}_{\alpha_2}(0,T)} \tag{6.5}$$

$$= C \left(\int_0^T \left(\tau^{\alpha_2} \int_{\mathbb{R}} |t + \tau - \rho|^{-\lambda} \|(f\chi)(\rho)\|_Y \, d\rho\right)^{s_2} d\tau\right)^{1/s_2}$$

with $\chi = \chi_{(0,t)}$, the characteristic function of the interval $(0,t)$. Using the change of variables $\tau' = \tau + t$ and that $0 \leq \tau' - t \leq \tau'$ Lemma 5.1 implies for each $t \in [0,T]$ that

$$\|Nf(t)\|_{\mathcal{X}^T_{s_2,\alpha_2}} \leq C \left(\int_t^{t+T} \left((\tau' - t)^{\alpha_2} \int_{\mathbb{R}} |\tau' - \rho|^{-\lambda} \|(f\chi)(\rho)\|_Y \, d\rho\right)^{s_2} d\tau'\right)^{1/s_2}$$

$$\leq C \left\|I_\lambda(\|f\chi\|_Y)\right\|_{L^{s_2}_{\alpha_2}(t,t+T)}$$

$$\leq C \left\|\|f\chi\|_Y\right\|_{L^{s_1}_{\alpha_1}} = C\|f\|_{L^{s_1}_{\alpha_1}(0,t;Y)}.$$

The proof of continuity is based on the previous estimates. By definition for $t_1 \geq t_2 \geq 0$, we observe that

$$(Nf)(t_1) - (Nf)(t_2)$$

$$= \int_{t_2}^{t_1} A^\mu e^{-(t_1-\rho)A} f(\rho)\,d\rho + \int_0^{t_2} \left(A^\mu e^{-(t_1-\rho)A} - A^\mu e^{-(t_2-\rho)A}\right) f(\rho)\,d\rho$$

$$=: \mathcal{I}_1 + \mathcal{I}_2.$$

The first term is easy to estimate. Replacing f by $f\chi_{(t_2,t_1)}$ and rewriting $\mathcal{I}_1$ as an integral for $f\chi_{(t_2,t_1)}(\rho)$ with $\rho \in (0,t_1)$, (6.4) proves that

$$\|\mathcal{I}_1\|_{\mathcal{X}^T_{s_2,\alpha_2}} \leq C \left\|\int_0^{t_1} (t_1+\tau-\rho)^{-\mu-\eta}\|f(\rho)\chi_{(t_2,t_1)}(\rho)\|_Y\,d\rho\right\|_{L^{s_2}_{\alpha_2}(0,T)}$$

$$\leq C\|f\|_{L^{s_1}_{\alpha_1}(t_2,t_1;Y)} \to 0$$

as $t_1 - t_2 \to 0$. The integral $\mathcal{I}_2$ is divided into two parts:

$$\|\mathcal{I}_2\|^{s_2}_{\mathcal{X}^T_{s_2,\alpha_2}} = \int_0^T \tau^{\alpha_2 s_2} \|e^{-\tau A}\mathcal{I}_2\|_X^{s_2}\,d\tau = \left(\int_0^\delta + \int_\delta^T\right) \tau^{\alpha_2 s_2} \|e^{-\tau A}\mathcal{I}_2\|_X^{s_2}\,d\tau.$$

The first part is estimated as follows: For $t_{1,2} = t_1$ and $t_{1,2} = t_2$

$$C \int_0^\delta \tau^{\alpha_2 s_2} \left\|\int_0^{t_2} A^\mu e^{-(t_{1,2}+\tau-\rho)A} f(\rho)\,d\rho\right\|_X^{s_2}\,d\tau$$

$$\leq C \int_0^\delta \tau^{\alpha_2 s_2} \left(\int_0^{t_2} (t_2+\tau-\rho)^{-\lambda}\|f(\rho)\|_Y\,d\rho\right)^{s_2}\,d\tau;$$

for t_1 we used that $(t_1+\tau-\rho)^{-\lambda} \leq (t_2+\tau-\rho)^{-\lambda}$ since $t_1 \geq t_2$. Replacing δ by T, we conclude—as for the estimate of $\|\mathcal{I}_1\|_{\mathcal{X}^T_{s_2,\alpha_2}}$—from Lemma 5.1 that the right-hand double integral is bounded by $C\|f\|^{s_2}_{L^{s_1}_{\alpha_1}(0,t_2;Y)}$. Hence, as a function of δ, the right-hand side converges to 0 as $\delta \to 0$, uniformly in $0 \leq t_2 \leq t_1 \leq T$.

To estimate the integral over (δ, T) in $\|\mathcal{I}_2\|^{s_2}_{\mathcal{X}^T_{s_2,\alpha_2}}$ we observe that

$$\int_\delta^T \tau^{\alpha_2 s_2} \|e^{-\tau A}\mathcal{I}_2\|_X^{s_2}\,d\tau = \int_\delta^T \tau^{\alpha_2 s_2} \varphi(\tau, t_1, t_2)\,d\tau$$

where by Lemma 6.2 (ii) for any $\nu_1 \in (0,1)$

$$\varphi(\tau, t_1, t_2) = \left\|\int_0^{t_2} \left(e^{-(t_1-t_2)A} - I\right) e^{-\tau A} A^\mu e^{-(t_2-\rho)A} f(\rho)\,d\rho\right\|_X^{s_2}$$

$$\leq C \left|\frac{t_2-t_1}{\tau}\right|^{\nu_1 s_2} \left(\int_0^{t_2} \|e^{-\tau A/2} A^\mu e^{-(t_2-\rho)A} f(\rho)\|_X\,d\rho\right)^{s_2}.$$

Thus

$$\int_{\delta}^{T} \tau^{\alpha_2 s_2} \varphi(\tau, t_1, t_2) \, d\tau$$

$$\leq C \left| \frac{t_2 - t_1}{\delta} \right|^{\nu_1 s_2} \int_0^T \left(\int_0^{t_2} \left(t_2 + \frac{\tau}{2} - \rho \right)^{-\lambda} \|f(\rho)\|_Y \, d\rho \right)^{s_2} d\tau$$

$$\leq C \left| \frac{t_2 - t_1}{\delta} \right|^{\nu_1 s_2} \|f\|_{L_{\alpha_1}^{s_1}(0, t_2; Y)}^{s_2}$$

converges to 0 as $t_2 - t_1 \to 0$ for fixed $\delta > 0$.

Now the proof of continuity in the case of finite s_1 and s_2 is complete.   □

Next we handle the case $\mathcal{X}_{\infty, \alpha}$ and set $s_1 = s_2 = \infty$.

Theorem 6.5. *Assume that* $\lambda = \mu + \eta \in (0, 1)$ *for positive* μ, η *as in* (6.1), (6.2), *and let* $0 < \alpha_1, \alpha_2 < 1$ *satisfy the balance condition* $0 < \alpha_2 + 1 = \lambda + \alpha_1$. *Moreover, suppose* $f \in \mathring{L}_{\alpha_1}^{\infty}(0, T; Y)$ *so that*

$$\|f\|_{L_{\alpha_1}^{\infty}(t)} := \|f\|_{L_{\alpha_1}^{\infty}(0, t; Y)} \to 0 \quad as \quad t \to 0. \tag{6.6}$$

(i) *For* $t \in (0, T)$

$$\|Nf(t)\|_{\mathcal{X}_{\infty, \alpha_2}^T} \leq C_T \|f\|_{L_{\alpha_1}^{\infty}(t)}. \tag{6.7}$$

Particularly, $Nf(t) \to 0$ *as* $t \to 0$ *in* $\mathcal{X}_{\infty, \alpha_2}$ *and* $Nf(t) \in \mathring{\mathcal{X}}_{\infty, \alpha_2}$.

(ii) $Nf \in C([0, T]; \mathring{\mathcal{X}}_{\infty, \alpha_2})$.

Proof. (i) We first observe, by (6.2) and the analyticity of e^{-tA}, that for $0 \leq \tau < T$

$$\tau^{\alpha_2} \left\| e^{-\tau A} Nf(t) \right\|_X \leq C \tau^{\alpha_2} \int_0^t (t + \tau - \rho)^{-\lambda} \rho^{-\alpha_1} \, d\rho \, \|f\|_{L_{\alpha_1}^{\infty}(t)}.$$

Thus, for $t \leq \tau < T$,

$$\sup_{t \leq \tau < T} \tau^{\alpha_2} \left\| e^{-\tau A} Nf(t) \right\|_X \leq C \sup_{t \leq \tau < T} \tau^{\alpha_2} \int_0^{\tau} (\tau - \rho)^{-\lambda} \rho^{-\alpha_1} \, d\rho \, \|f\|_{L_{\alpha_1}^{\infty}(t)}$$

$$\leq CB \|f\|_{L_{\alpha_1}^{\infty}(t)}$$

by the scale balance, where $B = B(1 - \lambda, 1 - \alpha_1)$ is the Beta function. For $\tau \leq t$ we have

$$\sup_{0 < \tau < t} \tau^{\alpha_2} \left\| e^{-\tau A} Nf(t) \right\|_X \leq C \sup_{0 < \tau < t} \tau^{\alpha_2} \int_0^t (t - \rho)^{-\lambda} \rho^{-\alpha_1} \, d\rho \, \|f\|_{L_{\alpha_1}^{\infty}(t)}$$

$$= CB \sup_{0 < \tau < t} \tau^{\alpha_2} t^{-\alpha_2} \|f\|_{L_{\alpha_1}^{\infty}(t)} \tag{6.8}$$

$$= CB \|f\|_{L_{\alpha_1}^{\infty}(t)}.$$

Hence, in view of (6.6),

$$\|Nf(t)\|_{\mathcal{X}^T_{\infty,\alpha_2}} \le CB\|f\|_{L^\infty_{\alpha_1}(t)} \to 0 \quad \text{as} \quad t \to 0.$$

For fixed $t > 0$, a modification of (6.8) also yields for $0 < \tau < \tau_0 < t$ the estimate

$$\sup_{0<\tau<\tau_0} \tau^{\alpha_2} \left\|e^{-\tau A}Nf(t)\right\|_X \le C(t) \sup_{0<\tau<\tau_0} \tau^{\alpha_2} \cdot \|f\|_{L^\infty_{\alpha_1}(t)},$$

i.e., $Nf(t) \in \overset{\circ}{\mathcal{X}}_{\infty,\alpha_2}$.

(ii) It remains to prove the continuity of $Nf(t)$ in $\mathcal{X}_{\infty,\alpha_2}$ for $t \ge \delta > 0$ for arbitrary $\delta > 0$. By definition for $t_1 \ge t_2 \ge \delta > 0$, we observe that

$$(Nf)(t_1) - (Nf)(t_2)$$
$$= \int_{t_2}^{t_1} A^\mu e^{-(t_1-\rho)A}f(\rho)\,d\rho + \int_0^{t_2} \left(e^{-(t_1-t_2)A} - I\right) A^\mu e^{-(t_2-\rho)A}f(\rho)\,d\rho$$
$$=: \mathcal{I}_1 + \mathcal{I}_2.$$

The term $\mathcal{I}_1$ is easy to treat. Due to the boundedness of the operator family $e^{-\tau A}$, $0 \le \tau \le T$, on X it suffices to consider $\|\mathcal{I}_1\|_X$ directly. If $0 < \tau < T$,

$$\|\mathcal{I}_1\|_X \le c \int_{t_2}^{t_1} (t_1 - \rho)^{-\lambda}\rho^{-\alpha_1}\,d\rho \,\|f\|_{L^\infty_{\alpha_1}(T)}$$
$$\le c_\delta \int_{t_2}^{t_1} (t_1 - \rho)^{-\lambda}\,d\rho \,\|f\|_{L^\infty_{\alpha_1}(T)}$$
$$\le C_\delta (t_1 - t_2)^{1-\lambda}\|f\|_{L^\infty_{\alpha_1}(T)}.$$

Thus

$$\limsup_{\substack{t_1-t_2\to 0 \\ t_1,t_2\ge\delta}} \sup_{0<\tau<T} \tau^{\alpha_2} \left\|e^{-\tau A}\mathcal{I}_1\right\|_X = 0.$$

Concerning the estimate of $\mathcal{I}_2$ Lemma 6.2 (ii) implies with $0 < \beta < 1 - \lambda$ that

$$\|\mathcal{I}_2\|_X \le c \int_0^{t_2} \left(\frac{t_1 - t_2}{t_2 - \rho}\right)^\beta (t_2 - \rho)^{-\lambda}\|f(\rho)\|_Y\,d\rho$$
$$\le c(t_1 - t_2)^\beta \int_0^{t_2} (t_2 - \rho)^{-\lambda-\beta}\rho^{-\alpha_1}\,d\rho \,\|f\|_{L^\infty_{\alpha_1}(T)}$$
$$\le c_\delta (t_1 - t_2)^\beta \|f\|_{L^\infty_{\alpha_1}(T)}$$

since $t_2 \ge \delta > 0$. We thus conclude that

$$\limsup_{\substack{t_1-t_2\to 0 \\ t_1,t_2\ge\delta}} \sup_{0<\tau<T} \left\|\tau^{\alpha_2}e^{-\tau A}\mathcal{I}_2\right\|_X = 0.$$

Now the assertion $Nf \in C\big([0,T];\overset{\circ}{\mathcal{X}}_{\infty,\alpha_2}\big)$ is proved. $\square$

6.2 Proof of Theorems 2.6 and 2.7

We shall prove Theorem 2.6 based on the abstract results given in the previous section.

Proof of Theorem 2.6(i). We first note that if $X = L^q_\sigma(\Omega)$, A is taken as the Stokes operator and, if $2/s + 3/q = 1 - 2\alpha$, then the Besov space $\mathbb{B}^{-1+3/q}_{q,s}$ coincides with the weighted space $\mathcal{X}^T_{s,\alpha} = X_{s,\alpha}$ (for all $0 < T \leq \infty$) with equivalent norms, see Theorem 4.1.

For $u_0 \in \mathbb{B}^{-1+3/q}_{q,s}$ the $L^s_\alpha(L^q)$-strong solution u satisfies the integral equation (1.13). From the assumptions $u \in L^s_\sigma(L^q)$ and $F \in L^{s/2}_{2\alpha}(L^{q/2})$ we conclude that

$$\widetilde{f} := (A^{-1/2}P\mathrm{div})\left((u \otimes u) - F\right) \in L^{s/2}_{2\alpha}(0, T; L^{q/2}_\sigma).$$

We take $Y = L^{q/2}_\sigma$ and $X = L^q_\sigma$ and rewrite (1.13) as

$$u(t) = e^{-tA}u_0 + Nf(t)$$

with $\mu = 1/2$. By Proposition 6.1, $e^{-tA}u_0 \in C\left([0, \infty); X_{s,\alpha}\right)$. Since due to (4.1), (4.3)

$$\left\|e^{-tA}v\right\|_X \leq C_T t^{-\eta}\|v\|_Y$$

with $\eta = 3/2q$, the operator N satisfies the assumptions (6.1), (6.3) with $\mu = 1/2, \eta = 3/2q$. Thus Theorem 6.4 implies that $N\widetilde{f} \in C\left([0, \infty); X_{s,\alpha}\right)$. $\quad\square$

Proof of Theorem 2.6(ii). We recall the construction of the solution of (1.13) by the fixed point iteration

$$u_1(t) = e^{-tA}u_0,$$

$$u_{m+1}(t) = e^{-tA}u_0 - \int_0^t A^{1/2}e^{-(t-\rho)A}\left(A^{-1/2}P\mathrm{div}\right)(u_m \otimes u_m - F)(\rho)\,\mathrm{d}\rho \quad (m \geq 1).$$

Since $u_0 \in \mathring{\mathbb{B}}^{-1+3/q}_{q,\infty} = \mathring{\mathcal{X}}_{\infty,\alpha}$ with $3/q = 1 - 2\alpha$, by Proposition 6.1 (iii) $u_1(t) = e^{-tA}u_0 \in C\left([0, T]; \mathring{\mathcal{X}}_{\infty,\alpha}\right)$ and even $\|u_1\|_{L^\infty_\alpha(0,t;X)} \to 0$ as $t \to 0$. By the assumption on F in Theorem 2.6 (ii) we conclude from Proposition 6.3 inductively that $\|u_{m+1}\|_{L^\infty_\alpha(0,t;X)} \to 0$ as $t \to 0$ for every $m \in \mathbb{N}$. Hence also the limit solution u which is the limit of (u_m) in $L^\infty_\alpha(0, T; X)$ has the same property at $t = 0$.

We now consider (1.13) and apply Proposition 6.1 and Theorem 6.5, where u, F satisfy (6.6), to get the desired continuity. $\quad\square$

Proof of Theorem 2.7(i). Let u, v be $L^s_\alpha(L^q)$-strong solutions of (1.1) with data $f = \mathrm{div}\, F, u_0$ and $g = \mathrm{div}\, G, v_0$. Being mild solutions of (1.13) the difference $w = u - v$ solves the integral equation

$$w(t) = e^{-tA}w_0 - \int_0^t A^{1/2}e^{-(t-\rho)A}\left(A^{-1/2}P\mathrm{div}\right)(w \otimes u + v \otimes w - (F - G))(\rho)\,\mathrm{d}\rho$$

$$\tag{6.9}$$

$$= e^{-tA}w_0 - (Ng)(t)$$

where N, $X = L^q_\sigma$, $Y = L^{q/2}_\sigma$, $\mu = \frac{1}{2}$, and $\eta = \frac{3}{2q}$ are defined as before, and

$$\widetilde{g} = (A^{-1/2}P\mathrm{div})(w \otimes u + v \otimes w - (F - G))$$

satisfies the elementary estimate

$$\|\widetilde{g}\|_{L^{s/2}_{2\alpha}(L^{q/2})} \leq c\Big(\|w\|_{L^s_\alpha(L^q)}\big(\|u\|_{L^s_\alpha(L^q)} + \|v\|_{L^s_\alpha(L^q)}\big) + \|F - G\|_{L^{s/2}_{2\alpha}(L^{q/2})}\Big). \tag{6.10}$$

Now Lemma 5.1 with exponents $s_1 = \frac{s}{2}$, $\alpha_1 = 2\alpha$, $s_2 = s$, $\alpha_2 = \alpha$ and $\lambda = \frac{1}{2} + \frac{3}{2q}$ implies that

$$\|N\widetilde{g}\|_{L^{s_2}_{\alpha_2}(L^q)} \leq C\|\widetilde{g}\|_{L^{s_1}_{\alpha_1}(L^{q/2})}.$$

Since $\|e^{-tA}w_0\|_{L^s_\alpha(L^q)} \leq \|w_0\|_{\mathcal{X}^T_{s,\alpha}}$, Proposition 6.3 and (6.10) lead to the estimate

$$\|w\|_{L^s_\alpha(L^q)}$$
$$\leq \|w_0\|_{\mathcal{X}^T_{s,\alpha}} + C\Big(\|w\|_{L^s_\alpha(L^q)}\big(\|u\|_{L^s_\alpha(L^q)} + \|v\|_{L^s_\alpha(L^q)}\big) + \|F - G\|_{L^{s/2}_{2\alpha}(L^{q/2})}\Big), \tag{6.11}$$

where the $L^s_\alpha(L^q)$-norm is considered on a time interval $(0, T)$. Choosing $T_0 \leq T$ such that, as in the assumption of Theorem 2.7, $\|u; v\|_{L^s_\alpha(0,T_0;L^q)} \leq \varepsilon_*$ with $2C\varepsilon_* \leq \frac{1}{2}$ the term involving w on the right-hand side of (6.11) can be absorbed. This proves the estimate (2.19), but still with the left-hand side replaced by $\|u - v\|_{L^s_\alpha(L^q)}$.

Next we apply (6.4) from Theorem 6.4 to (6.9) to get that

$$\|w(t)\|_{\mathcal{X}^T_{s,\alpha}} \leq \|u_0 - v_0\|_{\mathcal{X}^T_{s,\alpha}} + \|N\widetilde{g}(t)\|_{\mathcal{X}^T_{s,\alpha}}$$
$$\leq \|u_0 - v_0\|_{\mathcal{X}^T_{s,\alpha}} + C\|\widetilde{g}\|_{L^{s/2}_{2\alpha}(0,T;L^{q/2})} \tag{6.12}$$
$$\leq \|u_0 - v_0\|_{\mathcal{X}^T_{s,\alpha}} + C\|F - G\|_{L^{s/2}_{2\alpha}(0,T;L^{q/2})}$$
$$+ C\|w\|_{L^s_\alpha(0,T;L^q)}\|u; v\|_{L^s_\alpha(0,T;L^q)}.$$

Under the smallness assumption on $\|u; v\|_{L^s_\alpha(0,T_0;L^q)}$ for a suitable $0 < T_0 \leq T$ we can insert (6.11), i.e., (2.19) with the left-hand side $\|u - v\|_{L^s_\alpha(0,T_0;L^q)}$ instead of $\|u - v\|_{\mathcal{X}^T_{s,\alpha}}$, into (6.12) and conclude the proof of Theorem 2.7.   □

Proof of Theorem 2.7(ii). The proof is similar to the proof of Theorem 2.7 (i). Here $s'_1 = s'_2 = 1$ and (6.7) from Theorem 6.5 is applied to (6.11).   □

7   Proof of conditional uniqueness

Proof of Theorem 2.10. (i) By Definition 2.8. (3) there exists a sequence of approximate weak solutions (u_n) bounded in $\mathcal{LH}_T$ such that a subsequence (u_{n_k}) converges to a weak Leray-Hopf solution $v \in \mathcal{LH}_T$ of (1.1) satisfying (SEI).

Since (u_n) is uniformly bounded in $L^s_\alpha(0, T'; L^q)$ with T' in Definition 2.8 (4) we find a subsequence $(u_{n'_k})$ of (u_{n_k}) converging weakly in $L^s_\alpha(0, T'; L^q)$ to an element $v' \in L^s_\alpha(0, T'; L^q)$. Now, since $u_{n_k} \rightharpoonup v$ in $\mathcal{LH}_{T'}$, we may conclude that $v = v'$ on $(0, T')$; in particular, v' is a weak and even a strong $L^s_\alpha(L^q)$-solution of (1.1) on $(0, T')$. Since strong $L^s_\alpha(L^q)$-solutions are unique, $v = v' = u$ on $(0, T')$.

If $T' < T$, then we find due to (SEI) applied to v some $0 < T'' \le T'$ such that the weak solution v satisfies the energy inequality on $[T'', T)$ with initial time T''. Since $u \in L^s(T'', T; L^q(\Omega))$ with $\frac{2}{s} + \frac{2}{q} < 1$ is a "classical" strong solution, Serrin's uniqueness theorem implies that $u = v$ even on $[0, T)$.

(ii) Consider any subsequence (u_{m_k}) of (u_n) converging weakly in $\mathcal{LH}_T$. By the assumption in Theorem 2.10 (ii) this limit is a weak Leray-Hopf solution v'' on $(0, T)$ and even a well-chosen solution on an interval $(0, T')$. Now the assertion on uniqueness in (i) proves that $v'' = v = u$. Then an elementary argument implies that the whole sequence (u_n) converges weakly to v. Moreover, again due to uniqueness, this result will hold for any sequence (u_{0n}) and (F_n) with convergence properties as described in (ii).   □

Proof of Theorem 2.11.

(i) Case 1: *The Yosida approximation scheme:*
Given u_0, u_{0n} and F, F_n as in Definition 2.8 classical L^2-methods, see [74, Ch. V.2], prove the existence of a unique approximate solution $u_n \in \mathcal{LH}_T$ of (2.21) and the convergence of a subsequence of (u_n) to a weak solution $u \in \mathcal{LH}_T$ of (1.1). Indeed, u_n satisfies the energy equality (EE), see (1.9), and consequently the energy estimate

$$\|u_n(t)\|^2_2 + \int_0^t \|\nabla u_n\|^2_2 \, d\tau \le \|u_{0n}\|^2_2 + \int_0^t \|F_n\|^2_2 \, d\tau,$$

where the right-hand side is uniformly bounded with respect to $n \in \mathbb{N}$ and $0 < t < T$ due to the weak convergence properties in Definition 2.8. Finally, $(\partial_t u_n)$ is uniformly bounded in $L^{4/3}(0, T; H^1_{0,\sigma}(\Omega)')$, see [74, Lemma V. 2.6.1, Theorem V. 1.6.2]. Hence, by the Aubin-Lions-Simon compactness theorem for Bochner spaces, there exists a subsequence (u_{n_k}) of (u_n) and $v \in \mathcal{LH}_T$ such that

$$u_{n_k} \rightharpoonup v \text{ in } \mathcal{LH}_T, \quad u_{n_k} \to v \text{ in } L^2(0, T; L^2_\sigma(\Omega)) \qquad (7.1)$$

as $k \to \infty$. Furthermore,

$$u_{n_k}(t) \to v(t) \text{ in } L^2_\sigma(\Omega) \text{ for a.a. } t \in (0, T) \qquad (7.2)$$

as $k \to \infty$; this step needs the extraction of a further subsequence, as the case may be. Now (7.1) allows to pass to the limit in (2.21) and show that v is a weak Leray-Hopf solution of (1.1). In particular, v satisfies the energy inequality (EI) and, due to (7.2), even the strong energy inequality (SEI).

In the second step of the proof we improve the previous results by exploiting the properties of u_0 in $\mathbb{B}^{-1+3/q}_{q,s}$ and of F in $L^{s/2}_{2\alpha}(0, T; L^{q/2})$, see Definition 2.8. Since (u_{0n}) converges strongly to u_0 in $\mathbb{B}^{-1+3/q}_{q,s}$ we find some $T' \in (0, T]$ such that

$$\|u_{0n}\|_{\mathbb{B}^{-1+3/q}_{q,s,T'}} \le \frac{\varepsilon_*}{2}$$

for all $n \in \mathbb{N}$ where $\varepsilon_* > 0$ is the absolute constant from (1.5). Furthermore, since $F_n \to F$ in $L_{2\alpha}^{s/2}(0,T;L^{q/2})$ we may also assume that

$$\|F_n\|_{L_{2\alpha}^{s/2}(0,T';L^{q/2})} \le \frac{\varepsilon_*}{2}$$

for all $n \in \mathbb{N}$. We follow the construction of strong $L_\alpha^s(L^q)$-solutions as in Sect. 5 and decompose the solution u_n of (2.21) into $u_n = \tilde{u}_n + E_n$ where $E_n = E_{f_n,u_{0n}}$ solves the linear nonhomogeneous Stokes problem with data $u_{0n}, f_n = \operatorname{div} F_n$.

As in (5.3) $\tilde{u}_n = u_n - E_n$ has an integral representation based on the variation of constants formula and can be considered as solution of the fixed point problem $\tilde{u}_n = \mathcal{F}\tilde{u}_n$ in $L_\alpha^s(0,T';L^q)$ where

$$\mathcal{F}_n \tilde{u}_n(t) = - \int_0^t A^{1/2} e^{-(t-\tau)A} \big(A^{-1/2} P \operatorname{div}\big)\big(J_n(\tilde{u}_n + E_n) \otimes (\tilde{u}_n + E_n)\big)(\tau)\, \mathrm{d}\tau;$$

note that $\mathcal{F}_n$ differs from $\mathcal{F}$ in (5.3) only by the additional term J_n. Due to fundamental properties of the Yosida operators J_n, see the proof of Lemma 5.2, the fixed point of $\mathcal{F}_n$ can be constructed by Banach's Fixed Point Theorem in the same way as in Sect. 5. By the assumptions on u_n, F_n the solution $\tilde{u}_n$ and u_n satisfy the estimate

$$\|\tilde{u}_n;\, u_n\|_{L_\alpha^s(0,T';L^q)} \le C\varepsilon^* \tag{7.3}$$

with a constant $C > 0$ independent of $n \ge n_0(\varepsilon^*, T')$.

Case 2: *The Galerkin approximation scheme*:
It is well known that the Stokes operator A_2 on the bounded $C^{1,1}$-domain $\Omega \subset \mathbb{R}^3$ admits an orthonormal basis of eigenfunctions $\psi_k \in \mathcal{D}(A_2) = H^2 \cap H_{0,\sigma}^1$ with corresponding eigenvalues λ_k monotonically increasing to $+\infty$ as $k \to \infty$. For $n \in \mathbb{N}$ let

$$\Pi_n : L_\sigma^2 \to V_n := \operatorname{span}\{\psi_1,\dots,\psi_n\} \subset L_\sigma^2$$

denote the orthogonal projection. Obviously, $\|\Pi_n\|_{\mathcal{L}(L_\sigma^2)} = 1$ for all $n \in \mathbb{N}$. In the Galerkin method we are looking for a solution $u_n : [0,T) \to V_n$ of the ordinary differential $n \times n$-system

$$(\partial_t u_n, \psi_k) + (\nabla u_n, \nabla \psi_k) - (u_n \otimes u_n, \nabla \psi_k) = -(F_n, \nabla \psi_k),$$
$$u_n(0) = u_{0n} \in V_n \tag{7.4}$$

on $(0,T)$ for each $k = 1,\dots,n$. By the L^2-assumptions on u_{0n} and F_n we know that there exists a sequence of unique solutions (u_n) to (7.4) bounded in $\mathcal{LH}_T$. Moreover, $(\partial_t u_n)$ is uniformly bounded in $L^{4/3}(0,T;H_{0,\sigma}^1(\Omega)')$. As in the first part of the proof we find a subsequence (u_{n_k}) of (u_n) and a vector field v satisfying (7.1) and (7.2). In particular, $v \in \mathcal{LH}_T$ is a weak solution to (1.1) satisfying (SEI).

The crucial question is whether u_n is also a strong $L_\alpha^s(0,T';L^q)$-solution, uniformly bounded in n. To address this problem for fixed n we consider arbitrary linear combinations of $(7.4)_1$ to see that for all $w \in H_{0,\sigma}^1$

$$(\partial_t u_n, \Pi_n w) + (\nabla u_n, \nabla \Pi_n w) - (u_n \otimes u_n, \nabla \Pi_n w) = -(F_n, \nabla \Pi_n w),$$
$$u_n(0) = u_{0n} \in V_n. \tag{7.5}$$

Since $\Pi_n = P\Pi_n$, $P^* = P$, A commutes with Π_n, and $(\nabla u_n, \nabla \Pi_n w) = (Au_n, w)$, we may omit the test function $w \in H^1_{0,\sigma}$ and rewrite (7.5) in the form

$$\partial_t u_n + Au_n + \Pi_n P \operatorname{div}(u_n \otimes u_n) = \Pi_n P \operatorname{div} F_n, \quad u_n(0) = u_{0n} \in V_n. \quad (7.6)$$

Thus $u_n(t)$ can be considered as a solution in $W^{1,4/3}(0,T)$ (with respect to time) of an abstract Cauchy problem and as a mild solution with integral representation

$$u_n(t) = e^{-tA} u_{0n} - \int_0^t A^{1/2} e^{-(t-\tau)A} \big(A^{-1/2}\Pi_n P \operatorname{div}\big)(u_n \otimes u_n + F_n)(\tau)\,\mathrm{d}\tau. \quad (7.7)$$

Although $\|\Pi_n\|_{\mathcal{L}(L^2_\sigma)} = 1$ and $A^{-1/2}P \operatorname{div} \in \mathcal{L}(L^q)$ for each $1 < q < \infty$, similar estimates will not hold for Π_n on L^q_σ and for the operator $A^{-1/2}\Pi_n P \operatorname{div}$ on L^q uniformly in $n \in \mathbb{N}$ unless $q = 2$. The reason seems to be the non-uniform distribution of eigenvalues λ_k as $k \to \infty$ compared to a Fourier series setting. Therefore, the idea is to estimate Π_n in a step where only the L^2-norm is involved.

Applying (4.1) with $3 < q \le 4$, exploiting the uniform boundedness and commutator properties of Π_n on L^2_σ and finally (4.1), (4.3) we get with $\gamma = \frac{3}{4} - \frac{3}{2q}$ and $\gamma' = \frac{3}{q} - \frac{3}{4}$ the estimate

$$\big\|A^{1/2}e^{-(t-\tau)A}\big(A^{-1/2}\Pi_n P \operatorname{div}\big)(u_n \otimes u_n - F_n)\big\|_q$$
$$\le c\big\|A^{\gamma+1/2}e^{-(t-\tau)A}\big(A^{-1/2}\Pi_n P \operatorname{div}\big)(u_n \otimes u_n - F_n)\big\|_2$$
$$\le c\big\|A^{\gamma+1/2}e^{-(t-\tau)A}\big(A^{-1/2}P \operatorname{div}\big)(u_n \otimes u_n - F_n)\big\|_2$$
$$\le c\big\|A^{\gamma+\gamma'+1/2}e^{-(t-\tau)A}\big(A^{-1/2}P \operatorname{div}\big)(u_n \otimes u_n - F_n)\big\|_{q/2}$$
$$\le c(t-\tau)^{-1+\alpha+\frac{1}{s}}\|u_n \otimes u_n - F_n\|_{q/2}.$$

Now Lemma 5.1 implies with a constant $c > 0$ independent of $n \in \mathbb{N}$ and T that

$$\|u_n - e^{-tA}u_{0n}\|_{L^s_\alpha(0,T;L^q)} \le c\Big(\|u_n\|^2_{L^s_\alpha(0,T;L^q)} + \|F_n\|_{L^{s/2}_{2\alpha}(0,T;L^{q/2})}\Big).$$

Then by standard arguments we find $T' \in (0,T)$ independent of $n \in \mathbb{N}$ such that $(u_n) \subset L^s_\alpha(0,T';L^q)$ is uniformly bounded.

Now we complete the proof as in the previous case.

(ii) It is well-known that each sequence of approximate solutions, (u_n), given by the Yosida approximation scheme or the Galerkin method, respectively, allows for a uniform estimate of the time derivative $\partial_t u_n$. Hence by compactness arguments $(u_n) \subset L^2(0,T;L^2)$ is precompact. Consequently, there exists a subsequence converging strongly in $L^2(0,T;L^2)$ to a limit function which is a weak solution to (1.1). $\qquad\qquad\square$

8 Proof of regularity

Before proving Theorem 3.1 we consider an interesting property for weak solutions u in $L^\infty_{\mathrm{loc}}\big([0,T);\mathbb{B}^{-1+3/q}_{q,s}(\Omega)\big)$.

Lemma 8.1. *Let $u \in \mathcal{LH}_T$ be a Leray-Hopf type weak solution of* (1.1) *such that* $u \in L^\infty_{\mathrm{loc}}\big([0,T); \mathbb{B}^{-1+3/q}_{q,s}(\Omega)\big)$. *Then* $u(t) \in \mathbb{B}^{-1+3/q}_{q,s}(\Omega)$ *is well-defined for each* $t \in [0,T)$ *and*

$$\|u\|_{L^\infty(0,T';\mathbb{B}^{-1+3/q}_{q,s})} = \sup_{t \in [0,T']} \|u(t)\|_{\mathbb{B}^{-1+3/q}_{q,s}}, \quad 0 < T' < T. \tag{8.1}$$

The same results holds when $\|\cdot\|_{\mathbb{B}^{-1+3/q}_{q,s}}$ *is replaced by the equivalent norm* $\|\cdot\|_{\mathbb{B}^{-1+3/q}_{q,s;\delta}}$ *for any* $0 < \delta < \infty$.

Proof. Fix $0 < T' < T$. By definition of L^∞ there exists a Lebesgue null set $N \subset [0,T')$ such that $\|u\|_{L^\infty(0,T';\mathbb{B}^{-1+3/q}_{q,s})} = \sup_{t \in [0,T') \setminus N} \|u(t)\|_{\mathbb{B}^{-1+3/q}_{q,s}}$. Consider any $t_0 \in N$ and choose a sequence $(t_j) \subset (0,T') \setminus N$ such that $t_j \to t_0$ as $j \to \infty$. Obviously, $(u(t_j))$ is a sequence bounded in $\mathbb{B}^{-1+3/q}_{q,s}$ as well as in L^2_σ. Since u is weakly continuous on L^2_σ we get that $u(t_j) \rightharpoonup u(t_0) \in L^2_\sigma$. Moreover, by the reflexivity of $\mathbb{B}^{-1+3/q}_{q,s}(\Omega)$ we conclude by standard arguments that $u(t_j) \rightharpoonup u(t_0)$ even in $\mathbb{B}^{-1+3/q}_{q,s}$. Hence $u(t_0) \in \mathbb{B}^{-1+3/q}_{q,s}$ and $\|u(t_0)\|_{\mathbb{B}^{-1+3/q}_{q,s}} \leq \liminf_{t_j \to t_0} \|u(t_j)\|_{\mathbb{B}^{-1+3/q}_{q,s}}$. $\square$

Proof of Theorem 3.1. (i) Let $t_1 \in (0,T)$. To show that t_1 is a left-sided regular point of u we find due to the assumption (3.1) and Lemma 8.1 a $\delta > 0$ such that $\|u(t)\|_{\mathbb{B}^{-1+3/q}_{q,s_q;t_1-t}} \leq \varepsilon_*$ for all $t \in (t_1 - \delta, t_1)$. Here $\varepsilon_* > 0$ is the constant from Theorem 2.1, see (2.5). In particular, we can choose a $t_0 \in (t_1 - \delta, t_1)$ such that $(EI)_{t_0}$ holds. By Serrin's uniqueness theorem and Theorem 2.1 we conclude that $u \in L^{s_q}(t_0, t_1; L^q)$. Hence u is left-sided regular in t_1.

(ii) Since $u(t_1) \in \mathbb{B}^{-1+3/q}_{q,s_q}(\Omega)$, there exists a strong solution $v \in L^{s_q}(t_1, t_1 + \delta; L^q)$, $\delta > 0$, of (1.1) with initial value $v(t_1) = u(t_1)$. Moreover, by assumption the energy inequality for u holds with initial time t_1. Hence Serrin's uniqueness theorem implies that $v = u$ in $[t_1, t_1 + \delta)$, and u is right-sided regular in t_1.

(iii) To combine the results from (i) and (ii), in particular to apply (ii), it suffices to prove that u satisfies $(EI)_{t_1}$ at any t_1. Let $t_1 \in (0,T)$ be an instant where the validity of the energy inequality is not guaranteed by (SEI). By (i) t_1 is a left-sided regular point for u and, consequently, $u \in L^{s_q}(l_0, t_1; L^q)$ for some $0 < t_0 < t_1$. Therefore, u satisfies (EE) for all initial times $t'_0 \in (t_0, t_1)$; in particular, with $f = 0$,

$$\frac{1}{2}\|u(t_1)\|_2^2 + \int_{t'_0}^{t_1} \|\nabla u\|_2^2 \, d\tau = \frac{1}{2}\|u(t'_0)\|_2^2.$$

Thus $\lim_{t'_0 \nearrow t_1} \|u(t'_0)\|_2^2 = \|u(t_1)\|_2^2$ for $t'_0 \in (t_0, t_1)$. Moreover, by (SEI), there is a sequence $t_j \nearrow t_1$ such that for $t > t_1$

$$\frac{1}{2}\|u(t)\|_2^2 + \int_{t_j}^{t} \|\nabla u\|_2^2 \, d\tau \leq \frac{1}{2}\|u(t_j)\|_2^2, \quad t_j \leq t < T.$$

Passing to the limit $t_j \nearrow t_1$ we get that u satisfies $(EI)_{t_1}$.

Finally, since $u(0) = u_0 \in \mathbb{B}_{q,s_q}^{-1+3/q}(\Omega)$, we know that $u \in L^{s_q}(0,\varepsilon_0; L^q)$ for some $\varepsilon_0 > 0$. Now an elementary compactness argument proves that $u \in L_{\mathrm{loc}}^{s_q}([0,T); L^q)$. □

Proof of Corollary 3.3. (i) If $u \in C^0\big([0,T']; \mathbb{B}_{q,s_q;\delta}^{-1+3/q}\big)$, $0 < T' < T$, then u is uniformly continuous in $t \in [0,T']$ with values in $\mathbb{B}_{q,s_q;\delta}^{-1+3/q}$. Given ε_* from Theorem 2.1 the uniform continuity in t allows to find $\delta' \in (0,\delta]$ such that $\|u(t)\|_{\mathbb{B}_{q,s_q;\delta'}^{-1+3/q}} \leq \varepsilon_*$ for all $t \in [0,T']$. Then a compactness argument on $[0,T']$ implies that $u \in L^{s_q}\big([0,T']; L^q\big)$.

(ii) From Lemma 8.1 we know that $\|u(t)\|_{\mathbb{B}_{q,s_q;\delta}^{-1+3/q}} \leq \|u\|_{L^\infty(0,T;\mathbb{B}_{q,s_q;\delta}^{-1+3/q})} \leq \varepsilon_*$ for all $t \in [0,T)$. Now a compactness argument as in (i) completes the proof.

(iii) The continuous embeddings (2.6) and (i), (ii) yield (iii). □

Proof of Theorem 3.4. (i) Assuming (3.2) we find $\delta > 0$ such that with $t_0 = t_1 - \delta$ and $T' = t_1 + \delta \leq T$

$$\frac{1}{t_1 - t_0} \int_{t_0}^{t_1} (T' - \tau)^{r/s_q} \|u(\tau)\|_q^r \, d\tau \leq \frac{2^{r/s_q}}{\delta^{1-r/s_q}} \int_{t_0}^{t_1} \|u(\tau)\|_q^r \, d\tau \leq \tilde{\varepsilon}_*,$$

i.e., (3.3) is satisfied. Thus it suffices to prove (ii).

(ii) Let us check condition (3.1); to be more precise and as noted in Remark 3.2 (1), it suffices to find $t \in (0,t_1)$ such that $\|u(t)\|_{\mathbb{B}_{q,s_q;t_1-t}^{-1+3/q}} \leq \varepsilon_*$ and that $(EI)_t$ is satisfied. Indeed, due to (3.3) we find $t \in (t_0,t_1)$ such that

$$(T' - t)^{r/s} \|u(t)\|_q^r \leq \tilde{\varepsilon}_*$$

or, equivalently, $(T' - t)\|u(t)\|_q^s \leq \varepsilon_*^{s/r}$ and that $(EI)_t$ holds. Hence, employing the boundedness of the Stokes semigroup on L_σ^q,

$$\int_0^{T'-t} \|e^{-\tau A} u(t)\|_q^s \, d\tau \leq C(T' - t)\|u(t)\|_q^s \leq \varepsilon_*$$

with an appropriately chosen $\tilde{\varepsilon}_*$ in (3.3) so that $\|u(t)\|_{\mathbb{B}_{q,s_q;t_1-t}^{-1+3/q}} \leq \|u(t)\|_{\mathbb{B}_{q,s_q;T'-t}^{-1+3/q}} \leq \varepsilon_*$. □

Proof of Theorem 3.7. The main idea to analyze the term $\int_\Omega (u \cdot \nabla u) u \, dx$ is the splitting of the third factor u into a low frequency and a high frequency part, u_ℓ and u_h, respectively. For u_ℓ we take

$$u_\ell(t) = Su(t) := e^{-\delta A} u(t), \quad 0 < \delta < 1,$$

and use it as (admissible) test function in (1.4). By Lemma 8.2 below we get that for $0 < t < T$

$$\frac{1}{2}\langle u(t), e^{-\delta A} u(t)\rangle_\Omega + \int_0^t \langle A^{1/2} u, e^{-\delta A} A^{1/2} u\rangle_\Omega \, d\tau$$

$$= \frac{1}{2}\langle u_0, e^{-\delta A} u_0\rangle_\Omega + \int_0^t \langle u \cdot \nabla u, e^{-\delta A} u\rangle_\Omega \, d\tau.$$

Since $A^{1/2}u \in L^2(0,T; L^2_\sigma(\Omega))$ we pass to the limit $\delta \to 0$ to get that

$$\frac{1}{2}\|u(t)\|_2^2 + \int_0^t \|\nabla u\|_2^2 \,\mathrm{d}\tau = \frac{1}{2}\|u_0\|_2^2 + \lim_{\delta \to 0} \int_0^t \langle u \cdot \nabla u, e^{-\delta A} u \rangle_\Omega \,\mathrm{d}\tau.$$

To discuss the last term we write $\nabla u = \nabla u_\ell + \nabla u_h$, use that $\langle u \cdot \nabla u_\ell, u_\ell \rangle_\Omega = 0$ and obtain, after an integration by parts, for a.a. $\tau \in (0,t)$ the estimate

$$\begin{aligned}
|\langle u \cdot \nabla u, u_\ell \rangle_\Omega| &= |\langle u \cdot \nabla u_\ell, u_h \rangle_\Omega| \leq \|u\|_{9/2} \|\nabla u_\ell\|_{18/7} \|u_h\|_{18/7} \\
&\leq c\|u\|_{9/2} \|A^{1/2} u_\ell\|_{18/7} \|u_h\|_{18/7} \\
&\leq c\|A^{1/4} u\|_{18/7} \big(\delta^{-1/4}\|A^{1/4} u\|_{18/7}\big)\big(\delta^{1/4}\|A^{1/4} u\|_{18/7}\big) \\
&= c\|A^{1/4} u\|_{18/7}^3;
\end{aligned}$$

in the second last step we used the embedding (4.1) with $\alpha = \frac{1}{4}$, the estimate $\|A^{1/2} u_\ell\|_{18/7} \leq c\delta^{-1/4}\|A^{1/4} u\|_{18/7}$, see (4.3) with $\alpha = \frac{1}{4}$, and the identity $u_h = u - e^{-\delta A} u = \int_0^\delta A e^{-\tau A} u \,\mathrm{d}\tau$ for $u \in \mathcal{D}\big(A^{1/4}_{18/7}\big)$ together with (4.3). Moreover, since by (4.3) and (4.1)

$$\|u_h\|_3 \leq c\delta^{1/4}\|A^{1/4} u\|_3 \leq c\delta^{1/4}\|A^{1/2} u\|_2,$$

we get the pointwise convergence

$$|\langle u \cdot \nabla u, u_\ell \rangle_\Omega - \langle u \cdot \nabla u, u \rangle_\Omega| \leq \|u\|_6 \|\nabla u\|_2 \, c\delta^{1/4}\|\nabla u\|_2 \to 0 \quad \tau\text{-a.e.}$$

as $\delta \to 0$. Hence the assumption (3.7), i.e., $\|A^{1/4} u\|_{18/7}^3 \in L^1(0,T)$, and Lebesgue's convergence theorem yield the convergence

$$\int_0^t \langle u \cdot \nabla u, e^{-\delta A} u \rangle_\Omega \,\mathrm{d}\tau \to \int_0^t \langle u \cdot \nabla u, u \rangle_\Omega \,\mathrm{d}\tau = 0.$$

Now Theorem 3.7 is proved. $\qquad\qquad\qquad\qquad\qquad\qquad\qquad\qquad\qquad\qquad\quad \square$

Lemma 8.2. *Let $u \in \mathcal{LH}_T$ be a weak solution of (1.1) with $u_0 \in L^2_\sigma(\Omega)$ and let $S \in \mathcal{L}\big(L^2_\sigma(\Omega)\big)$ be a self-adjoint operator satisfying the estimate $\|Sv\|_{\mathcal{D}(A_2)} \leq c\|v\|_2$ for all $v \in L^2_\sigma(\Omega)$ and commuting with $A^{1/2}$ on $\mathcal{D}(A_2)$. Then for all $0 \leq t < T$*

$$\frac{1}{2}\langle u(t), S\, u(t) \rangle_\Omega + \int_0^t \langle S\, A^{1/2} u, A^{1/2} u \rangle_\Omega \,\mathrm{d}\tau = \frac{1}{2}\langle u_0, S\, u_0 \rangle_\Omega - \int_0^t \langle u \cdot \nabla u, S\, u \rangle_\Omega \,\mathrm{d}\tau.$$

Proof. Following an approximation argument of Serrin [70] let $0 \leq \varrho \in C_0^\infty(\mathbb{R})$ be an even cut-off function with $\int \varrho \,\mathrm{d}\tau = 1$ and let $\varrho_\varepsilon(\tau) = \frac{1}{\varepsilon}\varrho(\frac{\tau}{\varepsilon})$, $\varepsilon > 0$. Then fix $0 < t < T$, define the convolution

$$u_\varepsilon(\tau) := \int_0^t \varrho_\varepsilon(\tau - s) u(s) \,\mathrm{d}s$$

and note that $u_\varepsilon \to u$ and $A^{1/2}u_\varepsilon \to A^{1/2}u$ in $L^2(0,t;L^2)$ as $\varepsilon \to 0$. Since S is self-adjoint, an elementary calculation using symmetry arguments and a change of variables imply that

$$\int_0^t \langle u, \partial_\tau S\, u_\varepsilon\rangle_\Omega\, d\tau = \int_0^t \int_0^t \langle u(\tau), \partial_\tau \varrho_\varepsilon(\tau - s)S\, u(s)\rangle_\Omega\, d\tau\, ds = 0.$$

Moreover, the weak L^2-continuity of u with respect to time yields the convergence

$$\langle u(t), S\, u_\varepsilon(t)\rangle_\Omega \to \frac{1}{2}\langle u(t), S\, u(t)\rangle_\Omega, \quad \langle u_0, S(u_0)_\varepsilon\rangle_\Omega \to \frac{1}{2}\langle u_0, S\, u_0\rangle_\Omega$$

as $\varepsilon \to 0$. We also note that

$$\int_0^t \langle \nabla u, \nabla S\, u_\varepsilon\rangle_\Omega\, d\tau = \int_0^t \langle A^{1/2}u, A^{1/2}S\, u_\varepsilon\rangle_\Omega\, d\tau$$

$$= \int_0^t \langle S\, A^{1/2}u, A^{1/2}u_\varepsilon\rangle_\Omega\, d\tau \to \int_0^t \langle S\, A^{1/2}u, A^{1/2}u\rangle_\Omega\, d\tau$$

and that

$$\left| \int_0^t \langle u \cdot \nabla u, S(u_\varepsilon - u)\rangle_\Omega\, d\tau \right|$$

$$\leq c\|u\|_{L^\infty(0,T;L^2)}\|\nabla u\|_{L^2(0,T;L^2)} \left(\int_0^t \|S(u_\varepsilon - u)\|_{D(A_2)}^2\, d\tau \right)^{1/2}$$

$$\leq c\|u\|_{L^\infty(0,T;L^2)}\|\nabla u\|_{L^2(0,T;L^2)}\, \|u_\varepsilon - u\|_{L^2(0,t;L^2)}$$

converges to 0 as $\varepsilon \to 0$. Now the assertion follows from (1.4) with the admissible test function $w = S\, u_\varepsilon$. □

Proof of Theorem 3.8. Assume that u satisfies $u(t_0) \in \mathbb{B}_{q,s_q}^{-1+3/q}$ and $(EI)_{t_0}$ for all $t_0 \in [0,T)$. Moreover, let $\tilde{u}$ be another weak solution for the same initial value $u_0 \in L_\sigma^2(\Omega)$. Since $u \in L^{s_q}(0,\delta;L^q)$ with some $0 < \delta < T$, Serrin's uniqueness theorem implies that $u(t) = \tilde{u}(t)$ for $0 \leq t < \delta$.

Let $[0,T_1)$, $0 < T_1 \leq T$, be the largest half-open interval such that $u(t) = \tilde{u}(t)$ is satisfied for each $t \in [0,T_1)$. If $T_1 < T$, then the weak L^2-continuity in time, see (1.9), implies that $u(T_1) = \tilde{u}(T_1)$.

Now let $t_0 := T_1$. Since u satisfies $u(t_0) \in \mathbb{B}_{q,s_q}^{-1+3/q}$ and $(EI)_{t_0}$ we conclude that $u \in L^{s_q}(t_0, t_0 + \delta; L^q(\Omega))$. Moreover, the weak L^2-continuity in time and $(EI)_{t_0}$ for u imply that also $\tilde{u}$ satisfies $(EI)_{t_0}$. Consequently, by Serrin's uniqueness theorem, $u = \tilde{u}$ in $[0, T_1 + \delta)$. This is a contradiction to the construction of T_1. □

Proof of Proposition 3.9. (i) Since the given weak solution in $\mathcal{LH}_T$ satisfies the strong energy inequality, it is easy to see that the one-sided continuity of the kinetic energy E_k implies $(EI)_{t_0}$ for all $t_0 \in [0,T)$.

222 Reinhard Farwig

(ii) For $u \in L^4(L^4)$ Remark 1.2 (5) shows that u satisfies (EE). Now assume that u fulfills (3.9). By Hölder's inequality in Lorentz spaces ([55, Lemma 2.2]) and Sobolev's embedding $W_0^{1,2} \subset L^{6,2}$ ([55, Lemma 2.1 (1)]) we get that

$$\|uu\|_{L^2} \le c\|uu\|_{L^{2,2}} \le c\|u\|_{L^{3,\infty}}\|u\|_{L^{6,2}}$$
$$\le c\|u\|_{L^{3,\infty}}\|u\|_{W^{1,2}},$$

where $c = c(\Omega) > 0$. Hence $u \in L^4(L^4)$. Now we proceed as above. $\square$

Acknowledgements. These lecture notes are an extended version of a series of lectures given at the Institute of Mathematics, Academy of Mathematics and Systems Science at the Chinese Academy of Sciences, Beijing, in November 2017. We thank Prof. Dr. Ping Zhang for the invitation, his hospitality and the inspiring atmosphere.

References

[1] H. Amann: Linear and Quasilinear Parabolic Equations. Birkhäuser Verlag, Basel 1995

[2] H. Amann: On the strong solvability of the Navier-Stokes equations. J. Math. Fluid Mech. 2, 16–98 (2000)

[3] H. Amann: Nonhomogeneous Navier-Stokes equations with integrable low-regularity data. Int. Math. Ser., Kluwer Academic/Plenum Publishing, New York 2002, 1–28

[4] H. Amann: Navier-Stokes equations with nonhomogeneous Dirichlet data. J. Nonlinear Math. Physics 10, Supplement 1, 1–11 (2003)

[5] T. Barker: Uniqueness results for weak Leray-Hopf solutions of the Navier-Stokes system with initial values in critical spaces. J. Math. Fluid Mech. 20, 133–160 (2018)

[6] H. Bahouri, J.-Y. Chemin and R. Danchin: Fourier Analysis and Nonlinear Partial Differential Equations. Grundl. Math. Wiss. 343. Springer, Heidelberg 2011

[7] J. T. Beale, T. Kato and A. Majda: Remarks on the breakdown of smooth solutions for the 3-D Euler equations. Comm. Math. Phys. 94, 61–66 (1984)

[8] M. E. Bogovskiĭ: Decomposition of $L_p(\Omega, \mathbb{R}^n)$ into the direct sum of subspaces of solenoidal and potential vector fields. Soviet Math. Dokl. 33, 161–165 (1986)

[9] W. Borchers and T. Miyakawa: Algebraic L^2 decay for Navier-Stokes flows in exterior domains. Acta Math. 165, 189–227 (1990)

[10] J. Bourgain and N. Pavlović: Ill-posedness of the Navier-Stokes equations in a critical space in 3D. J. Funct. Anal. 255, 2233–2247 (2008)

[11] M. Cannone: Harmonic analysis tools for solving the incompressible Navier-Stokes equations. In: Handbook of Mathematical Fluid Dynamics, vol. III, North-Holland, Amsterdam 2004, 161–244

[12] J.-Y. Chemin, I. Gallagher and M. Paicu: Global regularity for some classes of large solutions to the Navier-Stokes equations. Ann. of Math. 173, 983–1012 (2011)

[13] A. Cheskidov and R. Shvydkoy: Ill-posedness of the basic equations of fluid mechanics in Besov spaces. Proc. Amer. Math. Soc 138, 1059–1067 (2010)

[14] A. Cheskidov and R. Shvydkoy: The regularity of weak solutions of the 3D Navier-Stokes equations in $B_{\infty,\infty}^{-1}$. Arch. Ration. Mech. Anal. 195, 159–169 (2010)

[15] A. Cheskidov, S. Friedlander and R. Shvydkoy: On the energy equality for weak solutions of the 3D Navier-Stokes equations. In: R. Rannacher and A. Sequeira (eds.), Advances in Mathematical Fluid Mechanics, Springer 2010, 171–175

[16] A. Cheskidov, P. Constantin, S. Friedlander and R. Shvydkoy: Energy conservation and Onsager's conjecture for the Euler equations. Nonlinearity 21, 1233–1252 (2008)

[17] R. Denk, M. Hieber and J. Prüß: $\mathcal{R}$-boundedness, Fourier multipliers and problems of elliptic and parabolic type. Mem. Amer. Math. Soc. 166 (2003), no. 788

[18] L. Escauriaza, G. A. Seregin and V. Šverák: $L_{3,\infty}$-solutions of Navier-Stokes equations and backward uniqueness. Uspekhi Mat. Nauk 58, 3–44 (2003) (Russian). Russian Math. Surveys 58, 211–250 (2003) (English)

[19] E. B. Fabes, B. F. Jones and N. M. Rivière: The initial value problem from the Navier-Stokes equations with data in L^p, Arch. Ration. Mech. Anal. 45, 222–240 (1972)

[20] R. Farwig, Y. Giga and P.-Y. Hsu: Initial values for the Navier-Stokes equations in spaces with weights in time. Funkcial. Ekvac. 59, 199–216 (2016)

[21] R. Farwig, Y. Giga and P.-Y. Hsu: The Navier-Stokes equations with initial values in Besov spaces of type $B_{q,\infty}^{-1+3/q}$. J. Korean Math. Soc. 54, 1483–1504 (2017)

[22] R. Farwig, Y. Giga and P.-Y. Hsu: On the continuity of the solutions to the Navier-Stokes equations with initial data in critical Besov spaces. Ann. Mat. Pura Appl. 198, 1495–1511 (2019)

[23] R. Farwig and C. Komo: Regularity of weak solutions to the Navier-Stokes equations in exterior domains. Nonlinear Differ. Equ. Appl. 17, 303–321 (2010)

[24] R. Farwig and C. Komo: Optimal initial value conditions for strong solutions of the Navier–Stokes equations in an exterior domain. Analysis (Munich) 33, 101–119 (2013)

[25] R. Farwig, H. Kozono and H. Sohr: An L^q–approach to Stokes and Navier-Stokes equations in general domains. Acta Math. 195, 21–53 (2005)

[26] R. Farwig, H. Kozono and H. Sohr: Very weak, weak and strong solutions to the instationary Navier-Stokes system. In: P. Kaplický and Š. Nečasová (eds.), Topics on Partial Differential Equations. J. Nečas Center Math. Model., Lecture Notes, Vol. 2, Praha 2007, 1–54

[27] R. Farwig, H. Kozono and H. Sohr: Local in time regularity properties of the Navier-Stokes equations. Indiana Univ. Math. J. 56, 2111–2131 (2007)

[28] R. Farwig, H. Kozono and H. Sohr: Energy-based regularity criteria for the Navier-Stokes equations. J. Math. Fluid Mech. 11, 1–14 (2008)

[29] R. Farwig, H. Kozono and H. Sohr: Regularity of weak solutions for the Navier-Stokes equations via energy criteria. In: R. Rannacher and A. Sequeira (eds.), Advances in Mathematical Fluid Mechanics, Springer 2010, 215–227

[30] R. Farwig and H. Sohr: Generalized resolvent estimates for the Stokes system in bounded and unbounded domains. J. Math. Soc. Japan 46, 607–643 (1994)

[31] R. Farwig and H. Sohr: Optimal initial value conditions for the existence of local strong solutions of the Navier-Stokes equations. Math. Ann. 345, 631–642 (2009)

[32] R. Farwig and H. Sohr: On the existence of local strong solutions for the Navier-Stokes equations in completely general domains. Nonlinear Anal. 73, 1459–1465 (2010)

[33] R. Farwig, H. Sohr and W. Varnhorn: On optimal initial value conditions for local strong solutions of the Navier-Stokes equations. Ann. Univ. Ferrara Sez. VII Sci. Mat. 55, 89–110 (2009)

[34] R. Farwig, H. Sohr and W. Varnhorn: Necessary and sufficient conditions on local strong solvability of the Navier-Stokes systems. Appl. Anal. 90, 47–58 (2011)

[35] R. Farwig, H. Sohr and W. Varnhorn: Extensions of Serrin's uniqueness and regularity conditions for the Navier-Stokes equations. J. Math. Fluid Mech. 14, 529–540 (2012)

[36] R. Farwig, H. Sohr and W. Varnhorn: Besov space regularity conditions for weak solutions of the Navier-Stokes equations. J. Math. Fluid Mech. 16, 307–320 (2014)

[37] R. Farwig and Y. Taniuchi: On the energy equality of Navier-Stokes equations in general unbounded domains. Arch. Math. 95, 447–456 (2010)

[38] C. Fefferman: Existence and Smoothness of the Navier–Stokes Equations. http://www.claymath.org/sites/default/files/navierstokes.pdf

[39] H. Fujita and T. Kato: On the Navier-Stokes initial value problem. Arch. Ration. Mech. Anal. 16, 269–315 (1964)

[40] D. Fujiwara and H. Morimoto: An L_r-theorem of the Helmholtz decomposition of vector fields. J. Fac. Sci. Univ. Tokyo (1A) 24, 685–700 (1977)

[41] G. P. Galdi: An introduction to the Navier-Stokes initial-boundary value problem. In: G. P. Galdi, J. G. Heywood, and R. Rannacher (eds.), Fundamental Directions in Mathematical Fluid Mechanics, Birkhäuser, Basel 2000, 1–70

[42] I. Gallagher: Critical function spaces for the well-posedness of the Navier-Stokes initial value problem. In: Y. Giga and A. Novotný A. (eds), Handbook of Mathematical Analysis in Mechanics of Viscous Fluids, Springer, Cham 2018

[43] Y. Giga: Analyticity of the semigroup generated by the Stokes operator in L_r-spaces. Math. Z. 178, 287–329 (1981)

[44] Y. Giga: Domains of fractional powers of the Stokes operator in L_r-spaces. Arch. Ration. Mech. Anal. 89, 251–265 (1985)

[45] Y. Giga: Solution for semilinear parabolic equations in L^p and regularity of weak solutions for the Navier-Stokes system. J. Differential Equations 61, 186–212 (1986)

[46] Y. Giga and T. Miyakawa: Solutions in L_r of the Navier-Stokes initial value problem. Arch. Ration. Mech. Anal. 89, 267–281 (1985)

[47] Y. Giga and H. Sohr: Abstract L^q-estimates for the Cauchy problem with applications to the Navier-Stokes equations in exterior domains. J. Funct. Anal. 102, 72–94 (1991)

[48] E. Hopf: Über die Anfangswertaufgabe für die hydrodynamischen Grundgleichungen. Math. Nachr. 4, 213–231 (1950–1951)

[49] T. Kato: Strong L^p-solutions of the Navier-Stokes equation in $\mathbb{R}^m$, with applications to weak solutions. Math. Z. 187, 471–480 (1984)

[50] A. A. Kiselev and O. A. Ladyzhenskaya: On the existence and uniqueness of solutions of the non-stationary problems for flows of non-compressible fluids. Amer. Math. Soc. Transl. Ser. 2, 24, 79–106 (1963)

[51] H. Koch and D. Tataru: Well-posedness for the Navier-Stokes equations. Adv. Math. 157, 22–35 (2001)

[52] C. Komo: Necessary and sufficient conditions for local strong solvability of the Navier-Stokes equations in exterior domains. J. Evol. Equ. 14, 713–725 (2014)

[53] C. Komo: Strong solutions of the Boussinesq system in exterior domains. Analysis 35, 161–175 (2015)

[54] C. Komo: Uniqueness criteria and strong solutions of the Boussinesq equations in completely general domains. Z. Anal. Anwend. 34, 147–164 (2015)

[55] H. Kozono: Uniqueness and regularity of weak solutions to the Navier-Stokes equations. Lecture Notes in Num. Appl. Anal., Kinokuniya, Tokyo, 16, 161–208 (1998)

[56] H. Kozono, T. Ogawa and Y. Taniuchi: The critical Sobolev inequalities in Besov spaces and regularity criterion to some semi-linear evolution equations. Math. Z. 242, 251–278 (2002)

[57] H. Kozono and H. Sohr: Remark on uniqueness of weak solutions to the Navier-Stokes equations. Analysis 16, 255–271 (1996)

[58] H. Kozono and Y. Taniuchi: Bilinear estimates in BMO and the Navier-Stokes equations. Math. Z. 235, 173–194 (2000)

[59] P. L. Lemarié-Rieusset: Recent Developments in the Navier-Stokes Problem. Research Notes in Mathematics 431. Chapman & Hall/CRC, Boca Raton, FL 2002

[60] P. L. Lemarié-Rieusset: The Navier-Stokes Problem in the 21st Century. Chapman and Hall/CRC, New York 2016

[61] J. Leray: Sur le mouvement d'un liquide visqueux emplissant l'espace. Acta Math. 63, 193–248 (1934)

[62] A. Lunardi: Interpolation Theory. Edizioni Della Normale, vol. 16, 3rd ed., Sc. Norm. Super. Pisa 2018

[63] P. Maremonti and V. A. Solonnikov: On nonstationary Stokes problem in exterior domains. Ann. Scuola Norm. Sup. Pisa Cl. Sci. (4) 24, 395–449 (1997)

[64] V. N. Maslennikova and M. E. Bogovskiĭ: Elliptic boundary value problems in unbounded domains with noncompact and nonsmooth boundaries. Rend. Sem. Mat. Fis. Milano LVI, 125–138 (1986)

[65] A. S. Mikhailov and T. N. Shilkin: $L_{3,\infty}$-solutions to the 3D-Navier-Stokes system in the domain with a curved boundary. Zap. Nauchn. Semin. POMI 336, 133–152 (2006) (Russian). J. Math. Sci. 143, 2924–2935 (2007) (English)

[66] T. Miyakawa: On the initial value problem for the Navier-Stokes equations in L^p spaces. Hiroshima Math. J. 11, 9–20 (1981)

[67] T. Miyakawa and H. Sohr: On energy inequality, smoothness and large time behavior in L^2 for weak solutions of the Navier-Stokes equations in exterior domains. Math. Z. 199, 455–478 (1988)

[68] J. Prüss and G. Simonett: Maximal regularity for evolution equations in weighted L_p-spaces. Arch. Math. 82, 415–431 (2004)

[69] G. Seregin: On smoothness of $L_{3,\infty}$-solutions to the Navier-Stokes equations up to boundary. Math. Ann. 332, 219–238 (2005)

[70] J. Serrin: The initial value problem for the Navier-Stokes equations. In: R. E. Langer (ed.), Nonlinear Problems, Univ. Wisconsin Press, 1963

[71] M. Shinbrot: The energy equation for the Navier-Stokes system. SIAM J. Math. Anal. 5, 948–954 (1974)

[72] C. G. Simader and H. Sohr: A new approach to the Helmholtz decomposition and the Neumann problem in L^q-spaces for bounded and exterior domains. Adv. Math. Appl. Sci. 11, World Scientific 1992, 1–35

[73] C. G. Simader, H. Sohr and W. Varnhorn: Necessary and sufficient conditions for the existence of the Helmholtz decomposition in general domains. Ann. Univ. Ferrara Sez. VII Sci. Mat. 60, 245–262 (2014)

[74] H. Sohr: The Navier-Stokes Equations. An Elementary Functional Analytic Approach. Birkhäuser Advanced Texts, Birkhäuser Verlag, Basel 2001

[75] H. Sohr: A regularity class for the Navier-Stokes equations in Lorentz spaces. J. Evol. Equ. 1, 441–467 (2001)

[76] V. A. Solonnikov: Estimates for solutions of nonstationary Navier-Stokes equations. J. Soviet Math. 8, 467–529 (1977)

[77] E. M. Stein and G. Weiss: Fractional integrals on n-dimensional Euclidean space. J. Math. Mech. 7, 503–514 (1958)

[78] R. S. Strichartz: L^p estimates for integral transforms. Trans. Amer. Math. Soc. 136, 33–50 (1969)

[79] H. Triebel: Interpolation Theory, Function Spaces, Differential Operators. North-Holland, Amsterdam 1978

[80] B. Wang: Ill-posedness for the Navier-Stokes equations in critical Besov spaces $\dot{B}_{\infty,q}^{-1}$. Adv. Math. 268, 350–372 (2015)

[81] T. Yoneda: Ill-posedness of the 3D-Navier-Stokes equations in a generalized Besov space near BMO^{-1}. J. Funct. Anal. 258, 3376–3387 (2010)

Lectures on the Analysis of Nonlinear
Partial Differential Equations Vol. 6
MLM 6, pp. 227–278

Ten Lectures on Elliptic Partial Differential Equations

Fanghua Lin*

Abstract

This note is a collection of my teaching materials in the past decades and a
deepgoing extension of the preceding Courant Lecture notes by Q. Han and I. It
exposes the systematic development of the general theory of second order elliptic
equations, with emphasis on the Krylov-Safonov theorem and De Giorgi (Nash-
Moser) theorem.

1 Introduction

This write-up covers my ten-hours-lectures given at the 2016 winter school of
the South University of Science and Technology of China. I wish to take this
opportunity to thank Professor Tao Tang, Xue-Feng Wang and Lin-lin Su for
the invitation and the warm hospitality. The present notes have included also
discussions on the topics of the existence of the Green functions and the Wiener
criterion for the regularity of boundary points in the Dirichlet problem which
were not presented at the winter school. This expanded version, in fact, has been
used by the author for roughly the 7-weeks material of the elliptic PDE course
at the Courant Institute over the past decades. I should point out that though
there are some overlaps between topics discussed in the current notes and that it
has been covered in the Courant Lecture notes by Q. Han and I, the latter was
published in the Courant Lecture Notes series in 1997 first, and then by the AMS
in 2001, I have purposely presented these overlapped material from different views
and often with new proofs. Indeed, one of the purposes of this set of lectures is
to provide a somewhat unified approach to both the Krylov-Safonov theorem for
non-divergence form uniformly elliptic equations and the De Giorgi (Nash-Moser)
theorem for the divergence case. Another goal is to provide discussions on some
closely related topics which have been omitted in our earlier book [5], in particular

---

*Courant Institute of Mathematical Sciences, New York University.

the existence and properties of the Green functions, the Wiener criterion as well as the classical theory of the Martin boundary.

1.1 A few important examples of elliptic PDEs

Most of studies of the elliptic partial differential equations are centered on the following four important examples which represent four categories of elliptic PDEs according to the order of the nonlinearity. In some sense, if one understands these four examples thoroughly, then one has more or less the whole view of the theory of elliptic PDEs.

(1) Linear elliptic PDEs, a best example is the Laplace equation

$$\Delta u = 0.$$

(2) Semilinear elliptic PDEs, a simple example is the following

$$\Delta u + f(u) = 0.$$

(3) Quasilinear elliptic PDEs, for example the minimal-surface equation

$$\mathcal{M}[u] \equiv \operatorname{div}\left(\frac{\nabla u}{\sqrt{1 + |\nabla u|^2}}\right) = 0.$$

This equation can also be written into an equation in the non-divergence form

$$\left(\delta_{ij} - \frac{u_{x_i} u_{x_j}}{1 + |\nabla u|^2}\right) u_{x_i x_j} = 0.$$

(4) Fully nonlinear elliptic PDEs, for example the Monge-Ampère equation

$$\det(u_{x_i x_j}) = 1.$$

The LHS of it is the multiple of all the eigenvalues of the Hessian $(u_{x_i x_j})_{1 \leq i,j \leq n}$.

The above equation is elliptic if all the eigenvalues of the Hessian $(u_{x_i x_j})_{1 \leq i,j \leq n}$ is positive. One can use the Crammer role to write the equation as a non-divergence form elliptic equation. By a classical calculation for Jacobian, one can also write this equation in a divergence form. However, one has not found yet that the latter fact would provide more than what have been already learned and understood as far as the regularity theory is concerned. On the other hand, the connections between the Monge-Ampère type equations and the optimal transport maps have been a fascinating topic of researches.

From another perspective, according to the form of the equation, one can divide the second order elliptic PDEs into two subclasses (according to the leading terms), the terminology has already been applied in above descriptions:

- Non-divergence form equations, i.e.

$$a_{ij}(x) u_{x_i x_j}(x) = 0.$$

For this type of equations, we shall see the notion of the viscosity solutions and, the Hopf and Alexandroff maximum principles would play a crucial role.

- Divergence form equations, i.e.

$$\partial_{x_i}(a_{ij}(x)u_{x_j}) = 0.$$

Here the notion of weak solutions defined by integration by parts would be natural, and we shall use the variational method and energy estimates.

Note that, in both cases the coefficients $(a_{ij}(x))$ should satisfy an uniform elliptic assumption, i.e. there exists a $\lambda \in (0,1)$ such that

$$\lambda I \leq (a_{ij}(x)) \leq \frac{1}{\lambda} I$$

for almost all x in the domain of consideration.

It is interesting that in a way all these four examples could be viewed in both the divergence and non-divergence forms. However much of the theory for the minimal surface equation is based on the divergence structure while that for the Monge-Ampère equation, it uses mainly the non-divergence form. Among these four basic elliptic equations, the Laplace equation is particularly important and fundamental. It has the intrinsic geometry nature in the sense that there is a natural elliptic operator (called the Laplace-Beltrami operator) depending only on the intrinsic metric. Hence there is also a close connection to the Brownian motion on Riemannian manifolds. The Laplace equation has a natural algebraic structure which can be seen clearly from the finite difference method to calculate approximations (or numerical) solutions. Indeed, one can derive formally the Laplace equations on graphs and groups. On the other hand, the Laplace equation can be derived from the basic variational principle: the Dirichlet principle through the so-called Euler-Lagrange equations. Hence it can be viewed as an operator between Hilbert spaces. Thus the energy method and also the spectral theory could be applied. Of course, when the Laplace equation is viewed as an elliptic equation of non-divergence type, the classical Hopf and Alexandroff maximum principles would be useful and powerful for deriving many desired a priori estimates.

The study of the minimal surface equation has a great impact not only for the developments of a general theory for quasi-linear (not necessarily uniformly elliptic) elliptic equations, but also it provides many fundamental ideas for the whole subject of geometric analysis. One of the items that would be discussed in this set of notes is the De Giorgi theory (which leads to a solution of Hilbert 19th problem by itself). It is intimately connected to problems from the geometric measure theory and its application to the regularity of geometric variational problems.

The semi-linear elliptic equations arise naturally in many classical problems in physics and geometry, for example, the famous Yamabe problem leads to equation of this type with critical Sobolev exponents. It has also been used widely in modelings of phase transition phenomena that result in the study of new sets of well-known problems such as free boundary problems and De Giorgi conjecture. These equations have both natural variational structure, hence its connection to the critical point theory and Morse theory, and also non-divergence structure where one can use maximum principle to derive estimates that are useful for geometric controls of solutions, for examples, monotone, symmetry properties as well as the geometry of its level sets.

Our discussions on Krylov-Safonov theorem can be applied to the study of fully nonlinear elliptic equations, see [2]. It will be useful for the understanding of Monge-Ampère type equations.

1.2 Function theory and elliptic PDEs

Much of the theory of elliptic partial differential equations before 1960 has its roots in the function theory of complex variables as well as the classical potential theory. For examples, from the Cauchy integral formula, one can derive the maximum principle, interior gradient estimate, the analyticity (power series expansions) and unique continuation. To give an illustration, we start with a holomorphic function $f(z)$ on a domain that is enclosed by a simply closed curve Γ in the complex plane. The Cauchy integral formula gives

$$f(z) = \frac{1}{2\pi i} \oint_{\Gamma} \frac{f(\xi)}{\xi - z} d\xi$$

where Γ is a Jordan curve of finite length, z is in its interior domain and here one takes the contour integration in counterclockwise direction. It is clear then that $|f(z)|$ is bounded by $\frac{|\Gamma| \max_{\Gamma} |f|}{2\pi dist(z,\Gamma)}$. One applies this estimate for the holomorphic functions $f^n(z)$, for $n = 1, 2, 3, \ldots$, and then takes n-th roots on both sides, and let n go to infinite. One obtains the maximum principle, that is, $|f(z)|$ achieves its maximum on the boundary of the domain, Γ. This proof is due to Landau. The gradient estimate in interior and power series expansions can be simply derived from this explicit formula. Thus to find such a representation for solutions of a giving elliptic PDE would be useful, and the latter could be done for an elliptic equation with sufficiently smooth coefficients (say Dini continuous) on a sufficiently smooth domain (say Lipschitz domains). One can use Green functions to obtain a representation for solutions for general elliptic equations (even when the coefficients of the equations are only measurable). However, one would not be able (and it is impossible in general) to obtain information for gradients of solutions when the coefficients are merely bounded and measurable. These are very long stories and often they are included in the classical potential theory, see for examples, [6], [1], [4].

The classical function theory also suggests topics like smooth solutions with isolated singularities, Liouville type theorems and Harnack's inequality. How to get the latter for general elliptic equations with uniformly elliptic and otherwise measurable coefficients is one of the main goals of this set of lectures.

1.2.1 Fundamental works

The founding work traces back to E. De Giorgi (1957), J. Nash (1958) and J. Moser (1961) for elliptic equations of divergence type. The utmost important tool in the development of regularity theory is the following Harnack's inequality:

$$\max_{D_{1/2}} u \leq C_0 \min_{D_{1/2}} u, \quad C_0 = C_0(n, \lambda) > 0 \tag{1.1}$$

for positive solutions to elliptic PDEs in the unit ball D_1.

We will present the proof by De Giorgi (with some simplifications). In both [5] and [4] the proof of Moser was carefully presented. It uses the beautiful estimate of F. John and L. Nirenberg for BMO (Bounded Mean Oscillation) functions at one critical step. The space of BMO was introduced in John-Nirenberg's work motivated from earlier work of John on the nonlinear elasticity. In 1972, C. Fefferman demonstrated the important duality between the Hardy space H^1 and the BMO space, it has since then became one of the most fundamental results in real and harmonic analysis.

The theory for elliptic equations of non-divergence type has to wait till early 1980 after Krylov-Safonov proved the fundamental Harnack estimate for positive solutions. Harnack's inequality leads to the following Hölder estimate:

$$||u||_{C^\alpha(B_{1/2})} \leq C(n, \lambda)||u||_{L^\infty(B_1)} \tag{1.2}$$

for the solution u to non-divergence uniformly elliptic PDEs and for some $\alpha > 0$ depending only on n and λ. We will present proofs for these results due to Krylov-Safonov before that for De Giorgi theorem. And we shall see some quite interesting connections between these two proofs. De Giorgi's work have led to a much further development for the theory of quasilinear elliptic equations, and equations of minimal surfaces type in particular, see [4]. And the work of Krylov-Safonov is a milestone in the theory of fully nonlinear elliptic PDEs, see [2] for further readings.

1.2.2 Perturbation theory

The classical regularity for the linear elliptic equations with sufficiently smooth coefficients is mostly based on some perturbation arguments (Schauder theory), such as that in [1], and it uses the classical potential theory. An example for the Schauder estimates can be described roughly as follows. It is to bound the $C^{2,\alpha}$ norm of a solution in terms of the source term. A typical statement is as follows:

Theorem 1.1. *Suppose that*

$$a_{ij}(x)u_{x_i x_j} = f(x) \quad in \ B_1 \subset \mathbb{R}^n$$

where

$$||a_{ij}(x)||_{C^\alpha(B_1)} \leq M$$

for some $M > 0$ and $\alpha \in (0,1)$. Then we have the estimate

$$||u||_{C^{2,\alpha}(B_{1/2})} \leq C(n, \lambda, \alpha, M)\{||f||_{C^\alpha(B_1)} + ||u||_{C^\alpha(B_1)}\}. \tag{1.3}$$

Idea of proof. Fix an $x_0 \in B_{1/2}$. For any $r_0 \in (0, \frac{1}{4})$ and $x \in B_{r_0}(x_0)$ we have

$$a_{ij}(x_0)u_{x_i x_j}(x) = (a_{ij}(x_0) - a_{ij}(x))u_{x_i x_j}(x) + f(x).$$

Taking C^α norm on both sides and using estimates which are available when the coefficients are constants, we get

$$\|u\|_{C^{2,\alpha}(B_{r_0}(x_0))} \leq C\|a_{ij}(x) - a_{ij}(x_0)\|_{L^\infty}\|u_{x_ix_j}\|_{C^\alpha}$$
$$+ C\|a_{ij}(x) - a_{ij}(x_0)\|_{C^\alpha}\|u_{x_ix_j}\|_{L^\infty}$$
$$+ C\|f\|_{C^\alpha}.$$

The right side norms are taking on a double sized ball $B_{2r_0}(x_0)$. There are some extra arguments needed in order to close the above estimate. If it is on a closed torus T^n, then one simply chooses a cover of T^n by r_0-size balls so that the overlaps of these balls (with the same centers) of size $2r_0$ is bounded by a dimensional constant (independent on the size of r_0). Then the result follows quickly from the assumption on (a_{ij}). For an interior estimate as stated above, one does a similar argument for solution u multiply by a cut-off function.   □

Remark 1.2. *We have seen that, if $a_{ij} \in C^\alpha$ for some $\alpha \in (0,1)$, then $u \in C^{2,\alpha}$. However it is not true for the case $\alpha = 0$: if $a_{ij} \in C^0$ one cannot guarantee that $u_{x_ix_j} \in C^0$; instead, the conclusion would be $u \in W^{2,p}$ for any $1 < p < +\infty$.*

In fact, one can further weaken the condition on a_{ij} and to find almost "the best possible" assumption for $a_{ij}(x)$ so that the $W^{2,p}$ estimate would be true. The assumption is that coefficients are in the space of VMO (Vanishing Mean Oscillation) functions, see for example a paper by Caffarelli in Annals of Math. VMO consists of those functions in L^1 that

$$\fint_Q |f(x) - f_Q|dx \to 0 \qquad \text{where } f_Q := \fint_Q f(x)dx$$

uniformly as the sizes $l(Q) \to 0$, for cubes Q with diameter $l(Q)$.

So far we have illustrated the regularity theory for the case of non-divergence form elliptic equations with good coefficients.

For the elliptic PDEs in the divergence form:

$$\partial_{x_i}(a_{ij}(x)u_{x_j}) = 0$$

there are similar results, see [5]. (Note that there is only a bound for the first derivative of u.)

- If $a_{ij} \in C^\alpha$ for some $\alpha \in (0,1)$, then $u \in C^{1,\alpha}$.
- If $a_{ij} \in C^0$ (a weaker assumption would also be VMO), then $u \in W^{1,p}$, for any $p \in [1,+\infty)$.

Finally we note that in the 2D case, there are classical works of C. B. Morrey and L. Nirenberg for non-perturbative estimates for non-divergence form equations; and for the divergence case, there are works of R. Courant and Ahlfors-Bers, see [3] and [4].

2 Mean Value Property (MVP) and harmonic functions

Definition 2.1. *For a function $f \in L^1(\Omega)$ where $\Omega \subset \mathbb{R}^n$, it is said to satisfy MVP if:*
(a)

$$f(x) = \fint_{B_r(x)} f(y) dy \qquad (2.1)$$

or, (b)

$$f(x) = \fint_{\partial B_r(x)} f(y) d\mathcal{H}^{n-1}(y) \qquad (2.2)$$

for a.e. $(x, r) \in \Omega \times \mathbb{R}_+$, $B_r(x) \subset \Omega$. These two conditions (a) *and* (b) *are in fact equivalent.*

MVP leads to a series of good properties. We start with the following

Proposition 2.2. *If $u \in L^1(\Omega)$ with MVP, then $u \in C^\infty(\Omega)$.*

Proof. We do the standard smoothing on u: choose a radial non-negative mollifier

$$0 \le \rho(x) = \rho(|x|) \in C_0^\infty(B_1), \quad \text{with} \int_{B_1} \rho(x) dx = 1$$

and let

$$\rho_\epsilon(x) = \frac{1}{\epsilon^n} \rho(\frac{x}{\epsilon}), \quad u_\epsilon(x) = (\rho_\epsilon * u)(x)$$

for $0 < \epsilon < 1$. The definition of MVP immediately gives $(\rho_\epsilon * u)(x) = u(x)$, hence $u \in C^\infty(\Omega)$. $\qquad\square$

Corollary 2.3. (Weyl's lemma) *If u is a distribution and $\Delta u = 0$ (derivatives on distributions are defined via integration by parts), then u is a smooth **function** and $\Delta u = 0$ a.e. in Ω.*

Proof. For every $\epsilon > 0$ we have

$$0 = \rho_\epsilon * (\Delta u) = (\Delta \rho_\epsilon) * u = \Delta u_\epsilon.$$

Taking the limit $\epsilon \downarrow 0$ yields the result. $\qquad\square$

Now we shall show a well-known fact:

Theorem 2.4. *If u is harmonic (i.e. $\Delta u = 0$), then u satisfies MVP; conversely, if $u \in C^2(\Omega)$ satisfies MVP, then u is harmonic.*

Proof. If u is harmonic: by divergence theorem, for fixed $(x, r) \in \Omega \times \mathbb{R}_+$ that satisfies $B_r(x) \subset \Omega$,

$$\frac{d}{dr} \fint_{\partial B_r(x)} u(y) d\mathcal{H}^{n-1}(y) = \fint_{\partial B_r(x)} \frac{\partial u}{\partial \rho} d\mathcal{H}^{n-1} = \frac{\int_{B_r(x)} \Delta u(y) dy}{|\partial B_r(x)|} = 0.$$

So

$$\fint_{\partial B_r(x)} u(y)d\mathcal{H}^{n-1}(y) = [\fint_{\partial B_r(x)} u(y)d\mathcal{H}^{n-1}(y)]_{r\downarrow 0} = u(x).$$

Next, if $u \in C^2(\Omega)$ satisfies MVP, then for fixed $(x,r) \in \Omega \times \mathbb{R}_+$ s.t. $r << 1$ and $B_r(x) \subset \Omega$, from the Taylor expansion near x

$$u(y) = u(x) + u_{x_i}(x)(y_i - x_i) + \frac{1}{2}u_{x_i x_j}(x)(y_i - x_i)(y_j - x_j) + o(|y - x|^2)$$

one can deduce that

$$\fint_{\partial B_r(x)} u(y)d\mathcal{H}^{n-1}(y) = u(x) + \frac{r^2}{2n}\Delta u(x) + o(r^2).$$

Hence the MVP implies harmonicity by taking the limit as r goes to zero in the last equation above. $\qquad\square$

For harmonic functions (or L^1 functions with MVP—we have shown that they are equivalent), two most important things are the *Harnack's inequality* and the *maximum principle*.

Theorem 2.5. (Harnack's inequality for nonnegative harmonic functions) *If $\Delta u = 0$ and $u \geq 0$ in B_1, then there exists a dimensional constant $C(n)$, such that*

$$\max_{B_{1/2}(0)} u \leq C(n) \min_{B_{1/2}(0)} u. \tag{2.3}$$

This Harnack's inequality can be easily derived from the MVP. Here we use another method to prove Harnack's inequality, and it needs the following lemma.

Lemma 2.6. (Gradient estimate for harmonic functions) *If $\Delta u = 0$ in $B_R(0)$, then*

$$|\nabla u(0)| \leq \frac{C(n)}{R} \sup_{B_R(0)} |u|. \tag{2.4}$$

Proof. From $\Delta u = 0$ we know that, for any fixed $\mathbf{e} \in \mathbb{S}^{n-1}$, $\Delta u_e = 0$ where $u_e = \nabla u \cdot \mathbf{e}$. Now by the MVP of u_e, we have

$$\nabla u(0) \cdot \mathbf{e} = u_e(0) = \fint_{B_R(0)} u_e(x)dx = \fint_{B_R(0)} \nabla u \cdot \mathbf{e}\,dx$$

$$= \frac{1}{|B_R|}\int_{\partial B_R(0)} u(x)\mathbf{e} \cdot \nu d\mathcal{H}^{n-1}(x).$$

Therefore,

$$|\nabla u(0)| \leq \frac{1}{|B_R|}\int_{\partial B_R(0)} |u|d\mathcal{H}^{n-1}(x) \leq \frac{|\partial B_R|}{|B_R|} \sup_{\partial B_R} |u|.$$

The proof of the lemma is complete. $\qquad\square$

Proof of Theorem 2.5. For non-negative u, one may assume that $u > 0$ in B_1, otherwise $u \equiv 0$ due to maximum principle (which is to be stated below). The proof of Lemma 2.6 and MVP imply

$$|\nabla u(x)| \leq \frac{1}{|B_{1/2}|} \int_{\partial B_{1/2}(x)} |u| = \frac{|\partial B_{1/2}|}{|B_{1/2}|} \fint_{\partial B_{1/2}} u = C(n)u(x)$$

for any fixed $x \in B_{1/2}(0)$. As a result

$$|\nabla \log u|(x) \leq C(n),$$

and by mean value theorem in calculus, for any $x_1, x_2 \in B_{1/2}(0)$,

$$|\log u(x_2) - \log u(x_1)| \leq C \sup_{B_{1/2}(0)} |\nabla \log u| \leq C(n).$$

The latter yields Harnack's inequality.   □

Now we state the Maximum Principles (MP) that can be derived from the MVP:

- (Maximum principle) If $\Delta u = 0$ in Ω, then

$$\sup_{\Omega} |u| \leq \sup_{\partial \Omega} |u|.$$

- (Strong maximum principle) If Ω is connected, $\Delta u = 0$ in Ω and there exists $x_0 \in \Omega$ such that

$$u(x_0) = \max_{\Omega} u,$$

then u must be a constant.

We note that, as a consequence of the maximum principle, the unique solvability of the Dirichlet problem for the Laplace equation on a bounded domain. Other consequences include the Liouville-type theorems and the compactness for certain families of harmonic functions.

- (Hopf boundary point lemma) If $\Delta u = 0$ in a ball $B \subset \mathbb{R}^n$, and $p \in \partial B$ such that

$$u(p) > u(x), \quad \text{for all } x \in B,$$

then the normal derivative of u at p is positive, i.e.

$$\frac{\partial u}{\partial \nu}(p) > 0.$$

We shall prove a quantitative version of it later.

3 Martin boundary and representation of harmonic functions

3.1 Poisson's formula and minimal harmonic functions

We started with the following Dirichlet problem for the Laplace equation on the n-dimensional unit ball

$$\begin{cases} \Delta u = 0 & \text{in } B_1, \\ u = g & \text{on } \partial B_1. \end{cases}$$

The well-known Poisson's formula gives a representation of the solution

$$u(x) = C(n) \int_{\partial B_1} \frac{1 - |x|^2}{|x - y|^n} g(y) dy.$$

Let us denote the Poisson kernel

$$P(x, y) = C(n) \frac{1 - |x|^2}{|x - y|^n}$$

and its action on the boundary data. Here one may easily verify that for any fixed $y \in \partial B_1$, the function $V^y(x) = P(x, y)$ is harmonic in x. It is also "minimal" in the following sense.

Definition 3.1. *A non-negative harmonic function u defined on B_1 is called minimal, if for any non-negative harmonic v which satisfies $v(x) \leq u(x)$ we can always find a constant $c \in [0, 1]$ such that $v(x) = cu(x)$.*

Exercise 3.2. *Show that for fixed $y \in \partial B_1$, $P(x, y)$ is harmonic in x, non-negative and minimal.*

With the help of Poisson kernel, we can further establish the relationship between the solution and boundary data. Suppose $\Delta u = 0$ and $u \geq 0$ in B_1 with $u(0) = 1$, as points approach the boundary radially, then

$$\lim_{r \to 1^-} u(r, \omega) = d\mu(\omega)$$

where $(r, \omega) \in [0, 1] \times \mathbb{S}^{n-1}$, and the right hand side is a Radon measure. With suitable boundary values, similar to the latter, one has the Fatou type theorems. Let us describe for the case of the upper half-space $\mathbb{R}^n_+$ the Fatou type theorems, see [7]:

Theorem 3.3. *For a harmonic function $\Delta u = 0$ in $\mathbb{R}^n_+$, $u|_{y=0} = f$ (y is the last coordinate of $\mathbb{R}^n_+$), the representation*

$$u(x', y) = (P * f)(x', y), \ f \in L^p(\mathbb{R}^{n-1}) \tag{3.1}$$

holds if and only if

$$\sup_{y > 0} \int_{\mathbb{R}^{n-1}} |u(x', y)|^p dx' < +\infty \tag{3.2}$$

for any fixed $p \in (1, +\infty]$. For $p = 1$, the representation (3.1) becomes

$$u(x', y) = (P * \mu)(x', y) \tag{3.3}$$

where μ is a Radon measure.

3.2 Martin boundary I

Let us first recall some classical results from convex analysis.

Proposition 3.4. (Choquet and Krein-Milman theorem) *Define the set*

$$K = \{u : \Delta u = 0 \text{ and } u > 0 \text{ in } B_1, \ u(0) = 1\},$$

we have

- *K is convex.*
- *K is compact in $C^0_{\text{loc}}(B_1)$.*
- *Any non-negative harmonic function u can be represented as*

$$u = \int_{\Sigma \subseteq K} y d\mu(y) \tag{3.4}$$

  where μ is a non-negative Radon measure supported on Σ, the set of extremal points in K.

Remark 3.5. *The extremal points in K is nothing but all the minimal positive harmonic functions, i.e. the following are equivalent [exercise]:*

- *$v \in \Sigma$.*
- *v is a minimal positive harmonic function.*
- *$v = v(\cdot) = cP(\cdot, y)$.*

Remark 3.6. *The Choquet representation also holds for any bounded domain $\Omega \subset \mathbb{R}^n$ instead of B_1.*

3.3 Optimal constant in Harnack's inequality

By Poisson's formula, any harmonic function u in $B_1 \subset \mathbb{R}^n$ can be represented as

$$u(x) = \int_{\partial B_1} P(x, y) g(y) d\mathcal{H}^{n-1}(y)$$

up to a constant we may just consider $P(x, y) = \frac{1 - |x|^2}{|x - y|^n}$. Fix $y_0 \in \partial B_1$ and $r \in (0, 1)$, in B_r we have

$$\frac{1 - r^2}{(1 + r)^n} \leq \frac{1 - |x|^2}{|x - y_0|^n} \leq \frac{1 - r^2}{(1 - r)^n}.$$

As a result,

$$\sup_{B_r} P(x, y_0) \leq \left(\frac{1 + r}{1 - r}\right)^n \inf_{B_r} P(x, y_0)$$

and for any $x \in B_r$,

$$\frac{1 - r^2}{(1 + r)^n} \int_{\partial B_1} g(y) d\mathcal{H}^{n-1}(y) \leq u(x) \leq \frac{1 - r^2}{(1 - r)^n} \int_{\partial B_1} g(y) d\mathcal{H}^{n-1}(y).$$

One can verify that the constant

$$C(n, r) := (\frac{1 + r}{1 - r})^n \tag{3.5}$$

is indeed optimal for Harnack's inequality.

3.4   Choquet and Riesz representations

Consider the Dirichlet problem on a bounded domain $\Omega \subset \mathbb{R}^n$:

$$\begin{cases} \Delta u = 0 & \text{in } \Omega, \\ u|_{\partial\Omega} = g \in C(\partial\Omega). \end{cases}$$

We denote by u_g the unique solution. Now fix $x_0 \in \Omega$ and consider the map $\mathcal{L}$ from $C(\partial\Omega)$ to $\mathbb{R}$, mapping g to $u_g(x_0)$. Obviously $\mathcal{L}$ is linear; by maximum principle it is also bounded. Now Riesz representation gives

$$u(x_0) = \int_{\partial\Omega} g(y) d\omega_{x_0}(y)$$

where ω_{x_0} is a non-negative Radon measure. By Harnack's inequality, for any two points

$$x_0, x_1 \in K \subset\subset \Omega,$$

there exists a positive constant $C(K)$, such that

$$d\omega_{x_0} \leq C(K) d\omega_{x_1} \leq C^2(K) d\omega_{x_0}$$

which shows that $d\omega_{x_0}$ and $d\omega_{x_1}$ are mutually absolutely continuous.

Furthermore, for general Ω we only know that $\text{supp}(d\omega_{x_0}) \subseteq \partial\Omega$ while for the special case $\Omega = B_1$, a comparison with Poisson's formula implies

$$d\omega_{x_0}(y) = P(x_0, y) d\mathcal{H}^{n-1}(y)$$

so we have $\text{supp}(d\omega_{x_0}) = \partial\Omega$. There is a large literature devoted to the study of these harmonic-measures. In particular it is an interesting problem to study the geometry of the support of harmonic measures and the precise Hausdorff dimension of the support. One could derive the Laplace equation and above representation formula from the theory of Brown motion, then these harmonic measures are natural probability measures for exits of random paths through the boundary of the domain Ω.

3.5   Martin boundary II

We have seen from discussions in the previous section that

- (Choquet) Suppose K is the set of positive harmonic functions on $\Omega \subseteq \mathbb{R}^n$. Then any $u \in K$ can be represented as

$$u(x) = \int_\Sigma u d\mu_x(y) \tag{3.6}$$

  where Σ is a subset of the extremal points in K and μ_x is a probability measure on Σ.

- (Riesz) The solution to the problem

$$\begin{cases} \Delta u = 0, \\ u|_{\partial\Omega} = f \in C(\partial\Omega) \end{cases}$$

  can be represented as

$$u(x) = \int_{\partial\Omega} f(y) d\omega_x(y) \tag{3.7}$$

  where $\{\omega_x(y)\}$ is a probability measure with $\mathrm{spt}(\omega_x) \subset \partial\Omega$.

Now we show a few simple and concrete examples to illustrate results. (One may also refer to Perron's method for very general domains.)

- $\Omega = \mathbb{R}^n$, then $K = \{\text{const}\}$.
- $\Omega = \mathbb{R}^n_+ = \mathbb{R}^n \cap \{x_n > 0\}$, then $K = \{cx_n : c \in \mathbb{R}\}$.
- $\Omega = \mathbb{R}^n$ with the equation

$$(-\Delta + 1)u = 0. \tag{3.8}$$

Note that for any $\xi \in \mathbb{S}^{n-1}$, $u_\xi = e^{x \cdot \xi}$ is always a solution with $u_\xi(0) = 1$. Naturally one would guess that any solution to (3.8) can be represented with $\{u_\xi : \xi \in \mathbb{S}^{n-1}\}$. In fact we have the following theorem:

Theorem 3.7. (Caffarelli - Littman theorem) *Let*

$$K = \{\text{positive solutions of } (-\Delta + 1)u = 0 \text{ on } \mathbb{R}^n, \ u(0) = 1\},$$

then for any $u \in K$,

$$u(x) = \int_{\mathbb{S}^{n-1}} e^{x \cdot \xi} d\mu(\xi) \tag{3.9}$$

where μ is a probability measure supported on $\mathbb{S}^{n-1}$.

This statement can be shown directly by Fourier series expansions on large balls $B_R(0)$, then taking the limit as R goes to infinite. In a recent work of the author and Q. Zhang, there is a general approach that works on much more general cases. It is based on a similar idea as the construction of the Martin boundary. Following the above example, now we shall present the rigorous definition of Martin boundary.

Theorem 3.8. (Constantinescu - Cornea theorem) *Let Ω be a topological space and*

$$\mathcal{F} = \{f \in C(\Omega, \bar{\mathbb{R}})\},$$

then there is a compactification of Ω, which is denoted by Ω^*, with respect to $\mathcal{F}$, such that

- Ω *is open and dense in* $\Omega^* = \Omega \cup \Sigma$ *(here Σ is the Martin boundary).*
- *Each $f \in \mathcal{F}$ can be extended to $f^* : \Omega^* \to [-\infty, \infty]$ and f^* is continuous on Ω^*.*
- Ω^* *is unique up to homomorphism.*

To get a better understanding of the definition, we study the Green function in Ω. First recall the fundamental solution for Laplace operator in $\mathbb{R}^n$:

$$\Gamma(x, a) = \begin{cases} \dfrac{1}{2\pi} \log \dfrac{1}{|x-a|}, & n = 2, \\[2ex] \dfrac{1}{(n-2)|\mathbb{S}^{n-1}|} \cdot \dfrac{1}{|x-a|^{n-2}}, & n \geq 3 \end{cases}$$

which has the property $-\Delta\Gamma = \delta_a$. Now if $\Delta u = 0$ in Ω,

$$\begin{aligned} u(a) &= \int_\Omega (-\Delta\Gamma_a(x))u(x)dx \\ &= \int_\Omega [(\Delta u)\Gamma_a - (\Delta\Gamma_a)u]dx \qquad\qquad (3.10) \\ &= \int_{\partial\Omega} \left(\frac{\partial u}{\partial\nu}\Gamma_a - \frac{\partial\Gamma_a}{\partial\nu}u\right)d\mathcal{H}^{n-1}, \end{aligned}$$

therefore u can be represented with the boundary data if Γ_a vanishes on $\partial\Omega$. This leads to the definition of Green function: $G_a(x)$ is the solution to

$$\begin{cases} -\Delta G_a(x) = \delta_a & \text{in } \Omega, \\ G_a(x)|_{\partial\Omega} = 0. \end{cases}$$

Replacing Γ_a with G_a in (3.10), the first term of RHS disappears and we have

$$u(a) = -\int_{\partial\Omega} \frac{\partial G_a}{\partial\nu} u \, d\mathcal{H}^{n-1}. \qquad\qquad (3.11)$$

Note that $G_a = \Gamma_a + h_a$ where

$$\begin{cases} -\Delta h_a = 0 & \text{in } \Omega, \\ h_a|_{\partial\Omega} = -\Gamma_a|_{\partial\Omega}. \end{cases}$$

Next, let's consider the class of functions

$$\mathcal{F} = \left\{ \frac{G(x, y)}{G(x_0, y)} : y \in \Omega \right\}$$

for a fixed $x_0 \in \Omega$, with each element

$$P(x, y) = \frac{G(x, y)}{G(x_0, y)} \in C(\Omega, [0, \infty]).$$

Definition 3.9. (Concrete definition of Martin boundary) *Let $\{y_j\}$ be a sequence in Ω which has no accumulation points in Ω (there are subsequences which converge to some point on $\partial\Omega$). It is called a fundamental sequence, if $\{P(x, y_j)\}$ converges in $C^0_{loc}(\Omega)$ to a positive harmonic function.*

For two fundamental sequences $\{y_j\}$ and $\{y'_j\}$, say they are equivalent if $\{P(x, y_j)\}$ and $\{P(x, y'_j)\}$ converge to the same limit. Now the Martin boundary can be defined as

$$\Sigma = \{\text{all equivalent classes of fundamental sequences}\}.$$

Remark 3.10. *From the Harnack's inequality, one knows that $\{P(x, y_j)\}$ is locally uniformly bounded; therefore the derivative is also locally bounded, and the existence of a convergent subsequence in C^0_{loc} is ensured by Ascoli-Arzelà theorem.*

On $\Omega^* = \Omega \coprod \Sigma$, we can also define the distance

$$D(z_1, z_2) = \int_\Omega \frac{|P(x, z_1) - P(x, z_2)|}{1 + |P(x, z_1) - P(x, z_2)|} dx.$$

Exercise 3.11. *Check that D is a distance on Ω^*.*

Now we have the general theorem of representation:

Theorem 3.12. *With the above definitions, the following properties hold:*

(1) *D is a distance on $\Omega \coprod \Sigma$, $(\Omega \coprod \Sigma, D)$ is a compact metric space with Σ being the boundary.*
(2) *$P(x, z)$ is continuous on $\Omega \times (\Omega \coprod \Sigma)$ except when $x = z$ where it blows up.*
(3) *Any minimal positive harmonic function on Ω is given by $P(x, w)$ for some $w \in \Sigma$.*
(4) *Let*

$$\Sigma_0 = \{w \in \Sigma : P(x, w) \text{ is a minimal positive harmonic function}\},$$

then for any positive harmonic function h on Ω, there exists a unique positive Borel measure μ, such that $\mu(\Sigma \setminus \Sigma_0) = 0$ and

$$h(x) = \int_{\Sigma_0} P(x, w) d\mu(w) \tag{3.12}$$

i.e. h is a linear combination of all the minimal positive harmonic functions.

The proof of this statement can be left as an exercise.

3.6 Further remarks on Martin boundary

Proposition 3.13. *For hyperbolic space $(\mathbf{B}^n_1(0), g_H)$, where*

$$g_H(x) = \frac{4(dx)^2}{(1 - |x|^2)^2}, \quad |ds|(x) = \frac{2|d_0S|(x)}{1 - |x|^2}$$

and

$$\Delta_g u = \frac{1}{\sqrt{g(x)}} \frac{\partial}{\partial x_i} (\sqrt{g(x)} g^{ij}(x) u_{x_j}),$$

the Martin boundary

$$\mathbb{S}^{n-1}_\infty = \partial_H \mathbf{B}^n_1$$

is diffeomorphic to $\mathbb{S}^{n-1}$.

This gives rise to an important open problem posed by Anderson and Schoen:

Open Problem 3.14. *Let* (M^n, g_H) *be a smooth compact Riemannian manifold, with positive constants* $A > B > 0$ *such that the curvature* $-A \le K_{g_H} \le -B$. *Anderson and Schoen have already shown that the Martin boundary is bi-Hölder to* $\mathbb{S}^{n-1}$, *is it also bi-Lipschitz to* $\mathbb{S}^{n-1}$?

Another problem is geometrically identifying the Martin boundary for the operator

$$\mathcal{L}_s := -\Delta_g + s \quad (s > 0)$$

on complete Riemannian manifold (M, g) with $\mathrm{Ric}(g) \ge 0$. (For Δ_g: one can show that if $\Delta_g u = 0$ and $u > 0$ then u must be a constant.) See my recent joint work with Q. Zhang for details.

4 Harmonic functions with isolated singularity

We start with a proposition in complex variables (Riemann's removable singularity theorem). Some general facts can be derived from the Laurent expansion:

Proposition 4.1. *Let* $f(z)$ *be holomorphic in* $D_1 \backslash \{0\}$. *If* $|f(z)|$ *is bounded in a neighborhood of* 0, *then the singularity at* 0 *is removable; if* $f(z) = o(\frac{1}{|z|})$ *near* 0, *then* 0 *is also removable.*

The most important harmonic function in $\mathbb{R}^n \backslash \{0\}$ is the fundamental solution of Laplace equation taking the form

$$u(x) = \frac{C(n)}{|x|^{n-2}},$$

with a singularity at 0. Similar to Proposition 4.1, we have the following theorem.

Theorem 4.2. *Let* $\Delta u = 0$ *in* $B_1 \backslash \{0\} \subset \mathbb{R}^n$ $(n \ge 3)$. *If* $u(x) = o(\frac{1}{|x|^{n-2}})$, *then* u *is harmonic (no singularity at* 0*).*

Proof. (This is a very important application of maximum principle!) Let h solve the boundary value problem

$$\begin{cases} \Delta h = 0 & \text{in } B_1, \\ h|_{\partial B_1} = u, \end{cases}$$

then h is a smooth harmonic function. Let $v = u - h$. It suffices to show that $v \equiv 0$ in B_1.

Fix any $x_0 \in B_1 \backslash \{0\}$. As $|v(x)| = o(\frac{1}{|x|^{n-2}})$ near 0, for any $\epsilon > 0$ there is a $\delta \in (0, |x_0|)$ such that

$$v(x) \le \epsilon(\frac{1}{|x|^{n-2}} - 1), \quad \text{for all } |x| = \delta.$$

Consider the harmonic function

$$w_\epsilon(x) = v(x) - \epsilon(\frac{1}{|x|^{n-2}} - 1) \tag{4.1}$$

in the region $B_{\delta,1} := \{x : \delta \le |x| \le 1\}$. We have shown that it is non-positive on the inner boundary; obviously it takes zero value on the outer boundary. Therefore by maximum principle, w_ϵ is always non-positive in $B_{\delta,1}$. As a result

$$v(x_0) \le \epsilon(\frac{1}{|x_0|^{n-2}} - 1).$$

Letting $\epsilon \downarrow 0$ we know that $v(x_0) \le 0$. Applying the same procedure to $-v$ we have $v(x_0) \ge 0$, so $v \equiv 0$ everywhere in $B_1 \backslash \{0\}$ and we are done. $\qquad \square$

Now we are able to understand the blowing-up behavior of all the non-negative harmonic functions in $B_1 \backslash \{0\}$ with possible singularity at 0.

Lemma 4.3. *Let $u \in C(B_1 \backslash \{0\})$, $u \ge 0$ and $\Delta u = 0$. Assume that*

$$\limsup_{x \to 0} u(x) = +\infty,$$

then we have the decomposition

$$u(x) = h(x) + \frac{c_0}{|x|^{n-2}} \tag{4.2}$$

where $h(x)$ is harmonic in B_1 with no singularity and c_0 is a constant.

A special case of the lemma is

Corollary 4.4. *For $u \in C(B_1 \backslash \{0\})$, $u \ge 0$ and $\Delta u = 0$, if*

$$u(x) = O(\frac{1}{|x|^{n-2}}) \tag{4.3}$$

then we also have (4.2).

Remark 4.5. *One can also find a complex version of the above lemma in the analysis of meromorphic functions: suppose $f(z)$ is holomorphic in $D_1 \backslash \{0\}$ and there exists a non-negative integer m such that $z^m f(z)$ is bounded near $z = 0$. Then the Laurent series of f at 0 can be written as*

$$f(z) = \frac{b_m}{z^m} + \frac{b_{m-1}}{z^{m-1}} + \cdots + \frac{b_1}{z} + a_0 + a_1 z + \cdots$$

with the first m terms depicting the singular image at the pole, and the rest are holomorphic.

For general uniformly elliptic PDEs

$$a_{ij}(x)u_{x_i x_j} = 0 \quad \text{in } B_1 \backslash \{0\}$$

where a_{ij} are Hölder continuous, if we have

$$u(x) = O(\frac{1}{|x|^{n-2+k}})$$

for some positive integer k, then the solution takes the form

$$u(x) = \frac{h(\frac{x}{|x|})}{|x|^{n-2+k}} + l.o.t.,$$

i.e. the solution is asymptotically spherical harmonic.

Proof of Lemma 4.3. Let

$$M = \limsup_{x \to 0} \frac{u(x)}{\frac{1}{|x|^{n-2}}}, \quad m = \liminf_{x \to 0} \frac{u(x)}{\frac{1}{|x|^{n-2}}} \tag{4.4}$$

then $0 \le m \le M \le +\infty$. Note that for each $r \in (0, \frac{1}{2})$, we can cover ∂B_r by a finite union of balls (the number is only dependent on n), each of them is centered on ∂B_r and has radius $\frac{r}{2}$. By Harnack's inequality applied on each of the balls, we know that the minimum and maximum values of u on ∂B_r are comparable. Hence there exists a dimensional constant $C(n)$, such that

$$m \ge C(n)M. \tag{4.5}$$

We assert that $M \ne +\infty$. In fact, if $M = +\infty$ then by (4.5) $m = +\infty$, so for a fixed $x_0 \in B_1$, for any $\epsilon > 0$ there exists $\delta_\epsilon \in (0, |x_0|)$ such that

$$\frac{u(x)}{\frac{1}{|x|^{n-2}}} \ge \frac{1}{\epsilon}, \quad \text{for any } x \in B_{\delta_\epsilon} \backslash \{0\},$$

hence on $\partial B_{\delta_\epsilon, 1}$ there is

$$\frac{1}{|x|^{n-2}} - 1 - \epsilon u(x) \le 0.$$

Note that LHS is harmonic, so by maximum principle LHS is non-positive in $B_{\delta_\epsilon, 1}$. Specifically,

$$\frac{1}{|x_0|^{n-2}} - 1 \le \epsilon u(x_0),$$

as $\epsilon \downarrow 0$ we have a contradiction. So $M \ne +\infty$; similarly we can prove that $M \ne 0$.

Now based on $0 < m \le M < +\infty$, we consider $v := u - h$, where h is the harmonic function in B_1 which coincides with u on ∂B_1. We shall show that

$$v(x) \equiv M(\frac{1}{|x|^{n-2}} - 1) \tag{4.6}$$

which would complete the proof.

To prove (4.6), we first note that for any $\epsilon > 0$,

$$v(x) \leq (M + \epsilon)(\frac{1}{|x|^{n-2}} - 1)$$

from the maximum principle and the definition of $\limsup$. Letting $\epsilon \downarrow 0$ we have

$$v(x) \leq M(\frac{1}{|x|^{n-2}} - 1).$$

Now consider the non-negative harmonic function in $B_1 \backslash \{0\}$

$$w(x) = M(\frac{1}{|x|^{n-2}} - 1) - v(x), \tag{4.7}$$

we show that w satisfies the condition of Theorem 4.2. In fact, suppose that

$$\limsup_{x \to 0} \frac{w(x)}{\frac{1}{|x|^{n-2}} - 1} = \epsilon_0 > 0 \tag{4.8}$$

then from (4.5) we know

$$\liminf_{x \to 0} \frac{w(x)}{\frac{1}{|x|^{n-2}} - 1} = \epsilon_1 > 0$$

and (again as a result of maximum principle)

$$w(x) \geq \epsilon_1(\frac{1}{|x|^{n-2}} - 1) \Rightarrow v(x) \leq (M - \epsilon_1)(\frac{1}{|x|^{n-2}} - 1)$$

which contradicts the choice of M. Hence (4.8) cannot hold and $w(x) = o(|x|^{2-n})$ as $x \to 0$. By Theorem 4.2 w is harmonic, since it takes zero value on ∂B_1, it must be identically zero. The proof is complete. $\qquad \square$

Remark 4.6. *The above theorem essentially tells us that, if $\Delta u = 0$, $u(x) \geq 0$ in $B_1 \backslash \{0\}$ and $u(x) \to +\infty$ as $|x| \to 0$, then u is **comparable** to $u_0(x) = |x|^{2-n}$, i.e.*

$$\lim_{x \to 0} \frac{u(x)}{u_0(x)} = C_0 > 0.$$

The isolated singularities of solutions to general elliptic PDEs can also be studied in the similar manners. Consider $u \in C^2_{\text{loc}}(B_1 \backslash \{0\})$ satisfying

$$\mathcal{L}u := a_{ij}(x)u_{x_i x_j} + b_i(x)u_{x_i} = 0 \text{ in } B_1 \backslash \{0\}. \tag{4.9}$$

The results are:

Proposition 4.7. *Whenever one has $\mathcal{L}g = 0$ in $B_1\backslash\{0\}$ and $g(x) \to +\infty$ as $|x| \to 0$, then for any solution u with $u(x \to 0) = +\infty$, there holds*

$$0 < A \leq u/g \leq B < +\infty$$

for x in a neighborhood of 0. Here $A < B$ are positive constants.

Remark 4.8. *The above proposition essentially says that the blow-up rate of solutions at an isolated singularity is unique. The proof only involves Hopf maximum principle.*

Proposition 4.9. *Following the assumptions of Proposition 4.7, if the Harnack's inequality is valid for $\mathcal{L}$, then there exists a positive constant C_0, such that*

$$\frac{u(x)}{g(x)} \to C_0 \text{ as } x \to 0.$$

At last, we introduce a result related to the Yamabe conjecture. Readers can view it as an exercise:

Proposition 4.10. (this is due to W. X. Chen and C. M. Li) *Suppose*

$$\Delta u + u^{\frac{n+2}{n-2}} = 0 \text{ in } B_1\backslash\{0\}, \ 0 < u \in C^2_{\text{loc}}(B_1\backslash\{0\})$$

or

$$\Delta u + f = 0 \text{ in } B_1\backslash\{0\}, \ 0 \leq f \in L^1_{\text{loc}}(B_1\backslash\{0\}), \ u > 0.$$

Then

$$f \in L^1(B_1); \ 0 < u \in L^p(B_1), \text{ for any } 1 \leq p < \frac{n}{n-2}$$

and u satisfies

$$u(x) = \frac{m}{|x|^{n-2}} + h(x) + \int_{B_1} G_\Delta(x,y) u^{\frac{n+2}{n-2}}(y)dy \tag{4.10}$$

where h is a harmonic function in B_1.

5   Maximum principles and Bernstein-Bochner theory

First we introduce two maximum principles for elliptic equations and systems.

Theorem 5.1. (Hopf maximum principle) *Suppose that $u(x) \in C^2(\Omega)\cap C(\bar{\Omega})$ (Ω is a domain in $\mathbb{R}^n$) satisfies the elliptic equation*

$$0 = \mathcal{L}u := a_{ij}(x)u_{x_ix_j} + b_i(x)u_{x_i} + c(x)u$$

where $c(x) \leq 0$ in Ω, then

$$\sup_\Omega u(x) \leq \sup_{\partial\Omega} u^+(x). \tag{5.1}$$

The maximum principle would fail for most elliptic systems. For these systems we have the following

Theorem 5.2. (Miranda maximum principle) *For elliptic systems with Hölder-continuous* (C^α) *coefficients, i.e.*

$$(\mathcal{L}\vec{u})_i = a_{ij}^{\alpha\beta}(x)u_{x_\alpha x_\beta}^j$$

where the ellipticity is given by

$$a_{ij}^{\alpha\beta}(x)\xi_i^\alpha\xi_j^\beta \sim |\xi|^2,$$

suppose

$$\begin{cases} \mathcal{L}u = 0 \text{ in } B_1, \\ u = \phi \text{ on } \partial B_1, \end{cases}$$

then

$$\|u\|_{L^\infty(B_1)} \leq M\|u\|_{L^\infty(\partial B_1)}. \tag{5.2}$$

A very important result of interior gradient estimate was carried out by S. Bernstein (1930's) and S. Bochner (1950's), followed by the work of H. Weinberger (around 1965) and S. T. Yau (around 1975). The original statement is

Theorem 5.3. (Interior gradient estimate) *Suppose* $\Delta u = 0$ *in* B_1, *then*

$$\|\nabla u\|_{L^\infty(B_{1/2})} \leq C(n)\|u\|_{L^\infty(B_1)}. \tag{5.3}$$

Proof. Note that

$$\Delta(|\nabla u|^2) = 2(\nabla u) \cdot (\nabla \Delta u) + 2\sum_{i,j} u_{x_i x_j}^2$$

$$= 2|\text{Hess}(u)|^2 \geq 0.$$

Let $\eta(x) \in C_0^2(B_1)$ be a cutoff function, we calculate

$$\Delta(\eta^2|\nabla u|^2) = 2\eta^2|\nabla^2 u|^2 + (\Delta(\eta^2))|\nabla u|^2 + 2(\nabla(\eta^2)) \cdot \nabla(|\nabla u|^2)$$

$$\geq 2\eta^2|\nabla^2 u|^2 - C_1|\nabla u|^2 + (-\eta^2|\nabla^2 u|^2 - C_2|\nabla\eta|^2|\nabla u|^2)$$

$$\geq \eta^2|\nabla^2 u|^2 - C_3|\nabla u|^2.$$

Meanwhile $\Delta(u^2) = 2|\nabla u|^2$, so for some (dimensional constant) M,

$$w := \eta^2|\nabla u|^2 + Mu^2$$

satisfies $\Delta w \geq 0$. By maximum principle,

$$\max_{B_1} w \leq \max_{\partial B_1} w$$

which implies (5.3). The proof is complete.   □

Remark 5.4. *The gradient estimate can be generalized to Riemannian manifold* (M, g): *for* $\Delta = \Delta_g$, *Bochner's identity tells*

$$\Delta_g |\nabla_g u|^2 = 2|\nabla^2 u|^2 + 2\mathrm{Ric}(\nabla u, \nabla u).$$

If $\mathrm{Ric}_g \geq -C_0 I$, *we have*

$$\Delta |\nabla u|^2 \geq 2|\nabla^2 u|^2 - C_0 |\nabla u|^2.$$

We can also choose η *and* $M(C_0)$, *such that*

$$\Delta(\eta^2 |\nabla u|^2 + M(C_0)u^2) \geq 0$$

and the similar interior gradient estimate still holds. (*Caution: we may use MVP to quickly prove Theorem 2.5; but MVP cannot be used on general manifolds.*)

In the special case $\mathrm{Ric}_g \geq 0$ *and* $\Delta_g u = 0$, *Bochner's identity gives*

$$\Delta(|\nabla u|^2) \geq 0,$$

one can prove that, if u *is bounded on* (M, g) *which is complete, then* u *is a constant.*

The Harnack's inequality for positive solutions can also be proved with Bernstein-Bochner method.

Proposition 5.5. *If* $\Delta u = 0$ *and* $u > 0$ *in* B_1, *then* $|\nabla \log u| \leq C(n)$ *in* $B_{1/2}$.

Proof. Denote $v = \log u$, then $\Delta v = -|\nabla v|^2$. Let $w = |\nabla v|^2$, direct calculation shows

$$\Delta w + 2 \sum_i D_i v D_i w = 2 \sum_{i,j} |D_{ij} v|^2$$

$$\geq \frac{1}{n} |\Delta v|^2 = \frac{|\nabla v|^4}{n} = \frac{w^2}{n}.$$

Take a cutoff function $\phi = \eta^4, \eta \in C_0^1(B_1)$. We have

$$\Delta(\phi w) + 2 \sum_i D_i v D_i(\phi w) \geq \phi \sum_{i,j} |D_{ij} v|^2 - 2|\nabla \phi||\nabla v|^3 - (\Delta \phi + \frac{C|\nabla \phi|^2}{\phi})|\nabla v|^2$$

which leads to

$$\Delta(\eta^4 w) + 2 D_i v D_i(\eta^4 w) \geq \frac{\eta^4 w^2}{2n} - C(n, \eta).$$

Hence, at the maximum of $\eta^4 w$,

$$\eta^4(x_0) w^2(x_0) \leq C(n, \eta)$$

and the proof is complete. $\square$

6   Moving plane method

6.1   Introduction

Following are the most important landmarks in the development of moving plane method:

- Alexandroff: Let M be a compact embedded hypersurface of constant mean-curvature in $\mathbb{R}^n (n \geq 2)$, then M is a standard sphere.
- H. Wente (1972—1974): The existence of immersed tori in $\mathbb{R}^3$ with constant mean curvature.
- H. Hopf: If $M^2 \hookrightarrow \mathbb{R}^3$ is an immersed sphere of constant mean-curvature, then M is a standard sphere.
- J. Serrin/ H. Weinberger (1972): If $\Omega \subset \mathbb{R}^n$ is a $C^{1,\alpha}$ bounded domain with

$$\begin{cases} -\Delta u = 1 \text{ in } \Omega, \\ \quad u = 0 \text{ on } \partial\Omega, \\ \dfrac{\partial u}{\partial \nu}|_{\partial\Omega} = \text{const}, \end{cases}$$

  then Ω is a ball. Their proof was simple: only using integration by parts!
- Gidas-Ni-Nirenberg (1978—1979): If

$$\begin{cases} -\Delta u = f(u) \text{ in } B, \\ \quad u = 0 \quad \text{ on } \partial B \end{cases}$$

  where B is a ball in $\mathbb{R}^n$, $u > 0$ inside of B and f is Lipschitz, then u is radial, i.e.

$$u(x) = \tilde{u}(r(x)).$$

6.2   Hopf maximum principle and Hölder continuity

Proposition 6.1. (E. Hopf's strong maximum principle—also known as boundary point lemma) *If* $a_{ij}(x)u_{x_i x_j}(x) \geq 0$ *in* B_1, *and for some* $x_0 \in \partial B_1$ *there is* $u(x) < u(x_0)$ *for all* $x \in B_1$, *then*

$$\frac{\partial u}{\partial \nu}(x_0) > 0. \tag{6.1}$$

Its proof involves the following lemma:

Lemma 6.2. (Quantitative Hopf lemma) *Assume* $\Delta u = 0$ *in* B_1 *and* $u \in C(\bar{B}_1)$; *for some* $x_0 \in \partial B_1$ *there is* $u(x) \leq u(x_0)$ *for all* $x \in B_1$. *Then*

$$\frac{\partial u}{\partial \nu}(x_0) \geq C(n)(u(x_0) - u(0)). \tag{6.2}$$

Proof. Let

$$v(x) = e^{-\alpha|x|^2} - e^{-\alpha} \ (\alpha > 0),$$

then $v = 0$ on ∂B_1 and $v > 0$ in B_1. Choose α large (for example $\alpha = 2n + 1$) such that

$$\Delta v(x) = e^{-\alpha|x|^2}(-2n\alpha + 4\alpha^2|x|^2) > 0, \text{ for any } |x| \geq \frac{1}{2}.$$

For $\epsilon > 0$, consider

$$h_\epsilon(x) = u(x) - u(x_0) + \epsilon v(x)$$

on $B_1 \backslash B_{1/2}$. We have:

- $h_\epsilon(x)|_{\partial B_1} \leq 0$.
- $\Delta h_\epsilon > 0$.
- $h_\epsilon(x) < 0$ on $\partial B_{1/2}$ (choose ϵ sufficiently small).

By maximum principle we know that $h_\epsilon(x) < 0$ in $B_1 \backslash B_{1/2}$. Now note that $h_\epsilon(x_0) = 0$, therefore

$$\begin{aligned}
\frac{\partial h_\epsilon}{\partial \nu}(x_0) \geq 0 &\Rightarrow \frac{\partial u}{\partial \nu}(x_0) + \epsilon \frac{\partial v}{\partial \nu}(x_0) \geq 0 \\
&\Rightarrow \frac{\partial u}{\partial \nu}(x_0) \geq \epsilon \cdot C(n).
\end{aligned} \tag{6.3}$$

Meanwhile, applying Harnack's inequality to $u(x_0) - u(x)$ (which is harmonic and non-negative in B_1) yields

$$\inf_{B_{1/2}} (u(x_0) - u(x)) > \tilde{C}(n)(u(x_0) - u(0))$$

so we can take

$$\epsilon = \tilde{C}(n)(u(x_0) - u(0))$$

and the result follows from (6.3).   □

Besides the boundary-point lemma, the Harnack's inequality also implies Hölder continuity.

Proposition 6.3. (Hölder continuity by Harnack's inequality) *If*

$$\mathcal{L}u := a_{ij}(x)u_{x_ix_j}(x) - 0 \text{ in } B_1, \ 0 < \lambda I \leq (a_{ij}) \leq \frac{1}{\lambda}I,$$

then

$$\|u\|_{C^\alpha(B_{1/2})} \leq C(n, \lambda)\|u\|_{L^\infty(B_1)}. \tag{6.4}$$

Proof. Let

$$M(r) = \max_{|x| \leq r} u(x), \ m(r) = \min_{|x| \leq r} u(x).$$

We need to prove that, for some $0 < \beta < 1$ and $\gamma \in (0, \frac{2}{3}]$,

$$M(\frac{\gamma}{2}) - m(\frac{\gamma}{2}) \leq \beta(M(\gamma) - m(\gamma)) \tag{6.5}$$

which would lead to

$$M(\rho) - m(\rho) \leq C\rho^\alpha(M(1) - m(1))$$

yielding the result. Now, to prove (6.5), we consider the non-negative function $M(r) - u(x)$ in B_r satisfying

$$\mathcal{L}(M(r) - u) = 0.$$

By Harnack's inequality,

$$\max_{|x| \leq r/2}(M(r) - u(x)) \leq C(n, \lambda) \min_{|x| \leq r/2}(M(r) - u(x)).$$

Specifically,

$$M(r) - m(\frac{r}{2}) \leq C(n, \lambda)(M(r) - M(\frac{r}{2})).$$

Applying the same process on $u(x) - m(r)$ in B_r yields

$$M(\frac{r}{2}) - m(r) \leq C(n, \lambda)(m(\frac{r}{2}) - m(r)).$$

Adding up the above two inequalities we get (6.5), and the proof is complete.   □

Remark 6.4. *The Quantitative Hopf lemma also implies Hölder continuity: from*

$$\frac{\partial u}{\partial \nu}(x_0) \geq C(n, \lambda)(u(x_0) - u(0))$$

we know that

$$\frac{\partial M(r)}{\partial r} \geq C(n, \lambda)r^{-1}(M(r) - M(0))$$

by a simple ODE approach, and the Hölder continuity follows.

6.3 Alexandroff maximum principle

Theorem 6.5. (Alexandroff maximum principle) *Suppose that*

$$\begin{cases} a_{ij}(x)u_{x_i x_j}(x) \geq f(x) & \text{in } \Omega \subseteq \mathbb{R}^n, \\ u|_{\partial \Omega} \leq 0 \end{cases}$$

where $f \in L^n(\Omega)$ and $u \in W^{2,n}(\Omega)$. Then

$$\sup_{\Omega} u(x) \leq C(n)\text{diam}(\Omega)(\int_{\Omega} |\frac{f^-(x)}{\mathcal{D}^*(x)}|^n dx)^{\frac{1}{n}} \tag{6.6}$$

where

$$f^- = \min\{0, f\}, \quad \mathcal{D}^*(x) = [\det(a_{ij}(x))]^{\frac{1}{n}}.$$

Example. Consider a simple case

$$\begin{cases} \Delta u = -1 & \text{in } B_R, \\ u = 0 & \text{on } \partial B_R. \end{cases}$$

The solution is

$$u(x) = \frac{1}{2n}(R^2 - |x|^2)$$

with $\sup u = \frac{1}{2n}R^2$, $\text{diam}(\Omega) = 2R$ and $(\int_{B_R} dx)^{1/n} = C(n)R$, satisfying (6.6).

Proof. Let Γ_u be the concave upper-envelope of the graph u, suppose $M = \max \Gamma_u$ and $d = \text{diam}\Omega$.

Let $\vec{p} \in B^n_{M/d}(0)$. Consider the hyper-plane

$$P_\lambda : \ y - \vec{p} \cdot \vec{x} = \lambda, \tag{6.7}$$

note that its slope $|\vec{p}| \leq \frac{M}{d}$. We move λ from $+\infty$ down, until P_λ touches Γ_u (this always happens before P_λ first touches $\partial\Omega$). A key observation is that, by the definition of concave upper-envelope, whenever P_λ first touches Γ_u, the touching point is always in the set

$$\Lambda_u = \{(x, u(x)) : u(x) = \Gamma_u(x)\}.$$

Therefore, the normal of the graph u (or Γ_u) is given by

$$(-\nabla u(x), 1) = (-\vec{p}, 1) \ \Rightarrow \ \vec{p} = \nabla u(x).$$

Now we know that: for any $\vec{p} \in B^n_{M/d}(0)$, there exists $x_0 \in \Lambda_u$ such that $Du(x_0) = \vec{p}$. As a result,

$$C(n)(\frac{M}{d})^n = |B^n_{M/d}(0)| \leq |Du(\Lambda_u)|$$
$$\leq \int_{\Lambda_u} |\text{Jac}(Du(x))|dx = \int_{\Lambda_u} |\det(u_{x_i x_j}(x))|dx, \tag{6.8}$$

here we used the fact

$$\text{for any } f : A \subseteq \mathbb{R}^n \to \mathbb{R}^n, \ \ |f(A)| \leq \int_A |\text{Jac}(f(x))|dx.$$

On the other hand, for any $x \in \Lambda_u$, $(u_{x_i x_j}(x)) \leq 0$ and $(a_{ij}(x)) \geq 0$. By AM-GM,

$$-f(x) \geq -a_{ij}(x)u_{x_i x_j}(x) \geq n[(\det(u_{x_i x_j}))(\det(a_{ij}))]^{\frac{1}{n}}. \tag{6.9}$$

The desired (6.6) follows from (6.8) and (6.9). $\square$

An important application of Alexandroff maximum principle is the following

Proposition 6.6. (Berestycki/ Nirenberg-Varadhan: maximum principle for narrow domains) *Let*

$$a_{ij}u_{x_i x_j} + cu = 0 \ \text{ in } \Omega, \quad u(x) \leq 0 \ \text{ on } \partial\Omega.$$

Assume that
$$|c(x)| \leq \Lambda, \ \det(a_{ij}) \geq \lambda^n, \ \text{diam}(\Omega) \leq d.$$

Then $u \leq 0$ in Ω whenever

$$|\Omega| \leq \delta = \delta(n, \lambda, \Lambda, d).$$

Proof. By Alexandroff maximum principle,

$$\sup_{\Omega} u^+ \leq C(n) \cdot \frac{d}{\lambda} \|cu\|_{L^n(\Omega)}$$

$$\leq C(n) \cdot \frac{d \cdot \Lambda}{\lambda} \|u\|_{L^n(\Omega)}$$

$$\leq C(n) \cdot \frac{d \cdot \Lambda}{\lambda} \cdot |\Omega|^{\frac{1}{n}} \sup_{\Omega} u^+.$$

Therefore, if $|\Omega|$ is too small then $\sup_{\Omega} u^+ = 0$. $\qquad\square$

Previously we have noted that the quantitative Hopf lemma implies Hölder continuity. Indeed, by the quantitative Hopf lemma:

$$\frac{\partial u}{\partial \nu}(x_0) \geq C(n, \lambda)(u(x_0) - u(0))$$

for

$$\mathcal{L}u = a_{ij}(x)u_{x_i x_j} \geq 0, \quad u(x_0) = \max_{\partial B_1} u,$$

and if we denote

$$M(R) = \max_{\partial B_R} u \geq 0,$$

then we have

$$R\frac{\partial M(R)}{\partial R} \geq C(n)M(R).$$

Solving the ODE we get

$$M(r) \leq M(R)(\frac{r}{R})^{C(n)}, \quad \text{for any } r \in (0, R).$$

6.4 The Gidas-Ni-Nirenberg theorem

As another application of moving plane method, we introduce the following important work:

Theorem 6.7. (Gidas-Ni-Nirenberg) *Suppose*

$$\begin{cases} \Delta u + f(u) = 0 & \text{in } B_1, \\ \qquad\quad u > 0 & \text{in } B_1, \\ \qquad\quad u = 0 & \text{on } \partial B_1 \end{cases}$$

where f is Lipschitz. Then $u(x)$ must be radial, i.e.

$$u(x) = u(|x|).$$

Proof. for any $0 \leq \lambda < 1$, let

- $\Pi_\lambda = \{x_1 = \lambda\} \cap B_1$ be a hyperplane in $\mathbb{R}^n$, here x_1 may be replaced with any direction ($x \cdot \vec{e}$ for $\vec{e} \in \mathbb{S}^{n-1}$);
- $\Sigma_\lambda = \{x_1 > \lambda\} \cap B_1$;

- Σ_λ^R be the reflection of Σ_λ (denote by R_λ the reflection operator) with respect to Π_λ.

For $x \in \Sigma_\lambda^R$, consider $v(x) = u(R_\lambda x)$ and $w(x) = u(x) - v(x)$. Clearly

$$w|_{\Pi_\lambda} = 0, \quad w|_{\partial\Sigma_\lambda^R - \Pi_\lambda} > 0.$$

Meanwhile,

$$0 = \Delta w + (f(u) - f(v)) = \Delta w + f'(\tilde{u})w$$

where $\tilde{u}(x)$ is some value between $u(x)$ and $v(x)$. As f is Lipschitz, we can write $c(x) = f'(\tilde{u}(x))$, and

$$\Delta w + c(x)w = 0, \quad |c(x)| \leq L.$$

By the maximum principle for narrow domain Σ_λ^R, we conclude that $w > 0$ in Σ_λ^R when $1 - \delta_0 \leq \lambda < 1$ for some small δ_0. (Note that the diameter of Σ_λ^R is bounded from above while its area goes to zero.) We claim that

$$w_\lambda \geq 0 \text{ in } \Sigma_\lambda^R, \text{ for any } 0 \leq \lambda < 1 \tag{6.10}$$

which would sufficiently finish the proof. To prove (6.10), let us suppose that

$$0 < \lambda_0 = \inf\{\lambda \in (0,1), \ s.t. \text{ for any } \mu \in [\lambda,1), \ w_\mu > 0 \text{ in } \Sigma_\mu^R\}. \tag{6.11}$$

Our aim is to show that $\lambda_0 = 0$. Suppose $\lambda_0 > 0$, from $w_{\lambda_0} \geq 0$ inside $\Sigma_{\lambda_0}^R$ and $w_{\lambda_0} > 0$ on $\partial\Sigma_{\lambda_0}^R \backslash \Pi_{\lambda_0}$, we know that $w_{\lambda_0} > 0$ inside $\Sigma_{\lambda_0}^R$. Let

$$K_\delta = \Sigma_{\lambda_0}^R \backslash (\delta - \text{neighborhood of } \Pi_{\lambda_0})$$

then on K_δ we have $w_{\lambda_0} \geq c(\delta) > 0$. Now take $\epsilon > 0$ sufficiently small and consider w_λ for $\lambda \in (\lambda_0 - \epsilon, \lambda_0)$:

- For fixed $x \in K_\delta$, $w_\lambda(x)$ is continuous in λ, so $w_\lambda \geq c(\delta)/2 > 0$ on K_δ.
- We assert that

$$w_\lambda|_{\Sigma_\lambda^R \backslash K_\delta} > 0$$

which is also a consequence of the maximum principle for small volume domain with bounded diameter, since

$$\text{diam}(\Sigma_\lambda^R \backslash K_\delta) \leq 2$$

and

$$\text{vol}(\Sigma_\lambda^R \backslash K_\delta) \leq c(\delta, \epsilon) << 1.$$

Therefore $w_\lambda > 0$ in Σ_λ^R, contradicting the choice of λ_0. The proof is hence complete. $\square$

7 Krylov-Safonov theory

7.1 Motivation

Consider a smooth uniformly elliptic function $F: Symm^{n \times n} \to \mathbb{R}$. Here uniform ellipticity means

$$\lambda|P| \leq F(M+P) - F(M) \leq \Lambda|P|, \text{ or } \lambda I \leq F_{P_{ij}}(D^2 u) \leq \Lambda I.$$

Let u satisfy $F(D^2 u) = 0$, then for every direction $e \in \mathbb{S}^{2^2}$, the directional derivative $u_e = \frac{\partial u}{\partial e}$ is a solution of the uniformly elliptic equation

$$F_{P_{ij}}(D^2 u)(u_e)_{ij} = 0. \tag{7.1}$$

Now we give the Krylov-Safonov theorem in the simplest setting:

Theorem 7.1. (Krylov-Safonov theorem) *If (a_{ij}) is uniformly elliptic, v solves*

$$a_{ij}\partial_{ij}v = 0 \text{ in } B_1,$$

then

$$\|v\|_{C^\alpha(B_{\frac{1}{2}})} \leq C(n, \alpha, \lambda, \Lambda)\|v\|_{L^\infty(B_1)}.$$

By Krylov-Safonov theorem, we can get $C^{1,\alpha}$ estimates for u satisfying $F(D^2 u) = 0$. Moreover, if we further assume F is concave, and differentiate it w.r.t direction e once again, we get

$$\begin{aligned}
0 &= \frac{\partial^2}{\partial e^2}F(D^2 u) \\
&= F_{ij}(D^2 u)(u_{ee})_{ij} + F_{ij,kl}(D^2 u)(u_e)_{ij}(u_e)_{kl} \\
&\leq F_{ij}(D^2 u)(u_{ee})_{ij}.
\end{aligned}$$

Thus for every $e \in \mathbb{S}^{2^2}$, u_{ee} is a subsolution of a uniformly elliptic equation. Using this fact, Evans and Krylov independently obtained $C^{2,\alpha}$ estimates for u.

7.2 Proof of Krylov-Safonov theorem

To prove the Hölder continuity, it suffices to show oscillation decays:

$$Osc_{B_{\frac{1}{2}}}u \leq (1-\theta)Osc_{B_1}u \tag{7.2}$$

for some $\theta > 0$. The reason is that once we get such estimates, we can apply it repeatedly, in balls of radius $\frac{1}{2^n}$ centered at a given point $x \in B_{1/2}$, to obtain that the oscillation of u around x decreases as a power of the radius. And this is just the definition of the Hölder continuity. Such oscillation decay can be deduced from the following key lemma:

Lemma 7.2. (Key lemma) *Let* $Lu = a_{ij}\partial_{ij}u \geq 0$, $u \leq 1$ *on* B_1. *There exist constants* r_0, θ_0 *depending only on* n, λ, Λ *such that if*

$$|\{x \in B_{r_0} : u(x) \leq 0\}| \geq \frac{1}{2}|B_{r_0}|,$$

then

$$\sup_{B_{\frac{1}{2}}} u \leq 1 - \theta_0.$$

Now we use an easy argument to show how to get oscillation decay (7.2) by the key lemma. Assume $Lu = 0$ in B_1, and $\inf_{B_1} u = -1$, $\sup_{B_1} u = 1$. Then one of the following two cases will happen:

1. $|\{x \in B_{r_0} : u(x) \leq 0\}| \geq \frac{1}{2}|B_{r_0}|$.
2. $|\{x \in B_{r_0} : u(x) \leq 0\}| \leq \frac{1}{2}|B_{r_0}|$.

By Lemma 7.2, case 1 implies $\sup_{B_{\frac{1}{2}}} u \leq 1 - \theta_0$, while case 2 implies $\inf_{B_{\frac{1}{2}}} u \geq -1 + \theta_0$. In either case we will obtain the desired decay of oscillation.

From the above deduction, we only need to prove the key lemma. Firstly we have the following lemma:

Lemma 7.3. *If* $Lu \leq 0$, $u \geq 0$ *in* B_1 *and* $\inf_{B_{\frac{1}{2}}} u \leq 1$, *then there exist* M_0, μ, r_0 *such that*

$$|\{x : u(x) \geq M_0\} \cap B_{r_0}| \leq (1 - \mu)|B_{r_0}|, \tag{7.3}$$

where constants M_0, μ, r_0 *only depend on* n, λ, Λ.

Proof. First take $r_0 < \frac{1}{2}$ to be a suitably small number (actually there is lots of flexibility in choosing r_0, for example we can set $r_0 = \frac{1}{16}$). Then we can find a radial function $\phi(x)$ such that $0 \leq \phi(x) \leq M_0$ in B_1, $\phi(x) = 0$ on ∂B_1, $L\phi(x) \geq 0$ in $\{r_0 \leq |x| \leq 1\}$. (For instance, we can take $\phi(x) = |x|^{-m} - 1$ in $\{r_0 \leq |x| \leq 1\}$.) M_0 is chosen to be suitably large to guarantee the existence of such $\phi(x)$.

Let $v = \phi - u \leq 0$ on ∂B_1, then we apply Alexandroff maximum principle to v,

$$1 \leq \sup_{B_1} v(x) \leq C_1(n, \lambda, \Lambda)\left(\int_{\Gamma(v)} (Lv)_-^n\right)^{1/n} \leq C_2(n, \lambda, \Lambda)|\Gamma(v)|^{1/n}. \tag{7.4}$$

where $\Gamma(v)$ is the contact set. Note that $Lv = L\phi - Lu$, therefore $Lv(x) \leq 0$ implies $x \in B_{r_0}$. At touch point x, $v(x) > 0$, which implies $u < \phi \leq M_0$. Then by (7.4) we have

$$1 \leq C(n, \lambda, \Lambda)|\{x \in B_{r_0} : u(x) < M_0\}|^{1/n}.$$

Then we obtain the desired estimate (7.3). $\qquad\qquad\qquad\qquad\qquad\qquad\square$

Remark 7.4. *Consider constant μ in Lemma 7.3, if $\mu > \frac{1}{2}$, then Lemma 7.3 implies Lemma 7.2 directly. To see this, we consider*

$$v(x) = M_0(1 - u) \geq 0, \quad Lv \leq 0.$$

Then by condition of Lemma 7.2,

$$|\{x \in B_{r_0} : v(x) \geq M_0\}| = |\{x \in B_{r_0} : u(x) \leq 0\}| \geq \frac{1}{2}|B_{r_0}|.$$

This contradicts with the estimate (7.3) in Lemma 7.3, therefore the condition of Lemma 7.3 must not hold. Thus we have $\inf_{B_{\frac{1}{2}}} v > 1$, which further implies $\sup_{B_{\frac{1}{2}}} u < 1 - \frac{1}{M_0} = 1 - \theta_0$.

Remark 7.5. *If for some ball $B_\rho \subset B_{r_0}$ satisfies*

$$|\{x : u(x) \geq K\} \cap B_\rho| > (1 - \mu)|B_\rho|,$$

then $u \geq \frac{K}{M_0}$ on $B_{\rho/8r_0} \supset B_{8\rho}$. This can be deduced directly from Lemma 7.3 by a scaling argument.

Recall Calderón-Zygmund decomposition:

Lemma 7.6. *Let $f : \mathbb{R}^d \to \mathbb{R}$ be integrable and α be a positive constant, then there exists an open set Ω such that*

(1) Ω is a disjoint union of open dyadic cubes, $\Omega = \bigcup_k Q_k$ such that for each Q_k,

$$\alpha \leq \fint_{Q_k} |f(x)| dx \leq 2^n \alpha.$$

(2) $|f(x)| \leq \alpha$ almost everywhere in Ω^c.

We use Calderón-Zygmund decomposition to prove following lemma:

Lemma 7.7. *Given Q_0 a cube, let $A \subset B \subset Q_0$ such that*

(1) $|A| \leq (1 - \mu)|Q_0|$.
(2) Whenever $|Q_\rho \cap A| \geq (1 - \mu)|Q_\rho|$ for some small cube $Q_\rho \subset Q_0$, then $Q_{4\rho} \subset B$.

Then $|A| \leq (1 - \mu)|B|$.

Proof. Let $f(x) = \chi_A(x)$ for $x \in Q_0$, then $\fint_{Q_0} f(x) dx \leq \alpha = 1 - \mu$. By Calderón-Zygmund decomposition, $Q_0 = \bigcup_{i=1}^{\infty} Q_i \bigcup F$, where $f(x) \leq \alpha < 1$ for a.e. $x \in F$. By Lebesgue differentiation theorem, $|F \cap A| = 0$. Therefore,

$$|A| = \sum_{i=1}^{\infty} |Q_i \cap A|.$$

Given each Q_i, let $\tilde{Q}_i$ be its predecessor. Then

$$\fint_{Q_i} f(x) dx > \alpha \Rightarrow |A \cap Q_i| > \alpha|Q_i| \Rightarrow \tilde{Q}_i \subset B.$$

Then one can choose a disjoint collection of $\{\tilde{Q}_j\}$ such that

$$\bigcup_{i=1}^{\infty} Q_i \subset \bigcup_j \tilde{Q}_j \subset B.$$

Then we have

$$|A| = \sum_{i=1}^{\infty} |Q_i \cap A| \le \sum_j |\tilde{Q}_j \cap A| \le \alpha \sum_j |\tilde{Q}_j| \le \alpha |B|.$$

$\square$

Remark 7.8. *In the above lemma we can substitute cubes with balls and the result still holds (with some minor change of constants $\sqrt{n}$).*

Combining Lemma 7.7 and Remark 7.5, we can iterate Lemma 7.3 to get

$$|\{x \in B_{r_0} : u(x) \ge M_0^k\}| \le (1-\mu)^k |B_{r_0}|.$$

Once $(1-\mu)^k < \frac{1}{2}$, by previous deduction in Remark 7.4 we prove Lemma 7.2 and then finish the proof of the Krylov-Safonov theorem.

8 De Giorgi theory

8.1 Motivation and history

Hilbert's nineteenth problem: Solutions of regular variational problems are analytic.

Some history:

(1) Bernstein (1904): C^3 solutions are analytic (for 2 variables);
(2) Petrowsky (1939): $C^{1,\alpha}$ solutions are analytic;
(3) C. B. Morrey and Nirenberg (1940s): 2-D scalar equation;
(4) De Giorgi (1956, 1957), Nash (1957, 1958): The solutions have first derivatives which are Hölder continuous;
(5) V. Maz'ya (1968): Counterexamples for higher order elliptic equation;
(6) De Giorgi (1968), Giusti-Miranda (1970), Nečas (1977): Counterexamples for system.

Hilbert's problem asks whether the solution to minimization problem

$$\min_{v|_{\partial\Omega}=g} \int_{\Omega} F(\nabla v)dx, \quad F \text{ is strictly convex and smooth}$$

is analytic. Let $u \in H_g^1(\Omega)$ be a minimizer, the Euler-Lagrange equation is

$$\nabla \cdot F_p(\nabla u) = 0 \text{ in } \Omega, \quad u = g \text{ on } \partial\Omega.$$

Let $v = u_e = \partial_e u$ for some direction $e \in \mathbb{S}^{n-1}$, then u_e solves

$$\frac{\partial}{\partial x_i}(a_{ij}(x)(u_e)_{x_j}) = 0$$

where $a_{ij}(x) = F_{p_i p_j}(\nabla u(x))$. Note that here we already know $u_e \in W^{1,2}_{loc}(\Omega)$ by differential quotient method. Once we know $u_e \in C^\alpha$, then coefficient $a_{ij}(x) \in C^\alpha$. Applying Schauder estimates and bootstrap argument will show that u is smooth.

Theorem 8.1. (De Giorgi theorem) *Suppose that $a_{ij} \in L^\infty(B_1)$ satisfies*

$$\lambda|\xi|^2 \le a_{ij}(x)\xi_i\xi_j \le \Lambda|\xi|^2, \quad |a_{ij}(x)| \le \Lambda, \quad \textit{for any } x \in B_1, \, \xi \in \mathbb{R}^n.$$

And suppose that $u \in H^1(B_1)$ solves

$$\partial_i(a_{ij}(x)\partial_j u) = 0 \textit{ in } B_1, \tag{8.1}$$

then $u \in C^\alpha(B_{1/2})$ for some $\alpha = \alpha(n, \lambda, \Lambda) > 0$ and

$$\|u\|_{C^\alpha(B_{1/2})} \le c(n, \lambda, \Lambda)\|u\|_{L^2(B_1)}.$$

Remark 8.2. *If $u \in W^{1,1}(B_1)$, $\partial_i(a_{ij}(x)\partial_j u) = 0$ makes sense in distributional sense. De Giorgi theorem is not true for $W^{1,1}$ solution (by J. Serrin).*

8.2   Proof of De Giorgi theorem

Suppose $Lu = \partial_i(a_{ij}(x)\partial_j u) = 0$.

  Part I: L^∞ estimate.

Theorem 8.3. *Let $u \in H^1(B_1)$ be a subsolution of (8.1), i.e. $Lu \ge 0$. Then*

$$\|u\|_{L^\infty(B_{\frac{1}{2}})} \le C(n, \lambda, \Lambda)\|u\|_{L^2(B_1)}.$$

Proof. Let $k \ge 0$, consider $v := (u - k)_+$. It holds that

$$Lv \ge 0, \quad v \ge 0.$$

Take a smooth cut-off function $\xi \in C_0^1(B_R)$ such that $\xi \equiv 1$ on B_r for $0 < r < R \le 1$, we calculate

$$0 \le \int_{B_R} (v\xi^2)Lv \, dx$$

$$= -\int_{B_R} \xi^2 a_{ij}(x)v_{x_i}v_{x_j} \, dx - \int_{B_R} 2\xi a_{ij}(x)\xi_{x_i}v_{x_j}v \, dx$$

$$\le -\lambda \int_{B_R} \xi^2 |\nabla v|^2 \, dx + \Lambda \int_{B_R} 2\xi|\nabla\xi||\nabla v|v \, dx$$

$$\le -\lambda \int_{B_R} \xi^2 |\nabla v|^2 \, dx + \frac{\lambda}{2} \int_{B_R} \xi^2 |\nabla v|^2 \, dx + C(\lambda, \Lambda) \int_{B_R} |\nabla\xi|^2 v^2 \, dx.$$

From this one can get

$$\int_{B_R} \xi^2 |\nabla v|^2 \, dx \leq C(\lambda, \Lambda) \int_{B_R} |\nabla \xi|^2 v^2 \, dx. \tag{8.2}$$

By Sobolev embedding, it holds that

$$\left(\int_{B_R} (\xi v)^{2^*} \, dx\right)^{2/2^*} \leq C(n) \int_{B_R} |\nabla(\xi v)|^2 \, dx, \tag{8.3}$$

where $2^* = \frac{2n}{n-2}$. Combining (8.2) and (8.3) gives that

$$\left(\int_{B_R} (\xi v)^{2^*} \, dx\right)^{2/2^*} \leq C(n, \lambda, \Lambda) \int_{B_R} |\nabla \xi|^2 v^2 \, dx.$$

Also note that using Hölder's inequality one can get

$$\int_{B_r} v^2 \, dx \leq \int_{B_R} \xi^2 v^2 \, dx \leq \left(\int_{B_R} (\xi v)^{2^*} \, dx\right)^{2/2^*} |\{x : v(x) \neq 0\}|^{2/n}. \tag{8.4}$$

Define

$$A(k, r) := \{x \in B_r : u(x) \geq k\}.$$

Then (8.4) can be written as

$$\int_{A(k,r)} (u-k)^2 \, dx \leq C |A(k, R)|^{2/n} \frac{1}{(R-r)^2} \int_{A(k,R)} (u-k)^2 \, dx. \tag{8.5}$$

Now we point out three observations about $A(k, r)$:
(1) $|A(k, R)| \leq \frac{1}{k^2} \int_{B_R} |u|^2 \, dx \to 0$ as $k \to \infty$.
(2) If $k < h$, then $\int_{A(h,r)} (u-h)^2 \, dx \leq \int_{A(k,r)} (u-k)^2 \, dx$.
(3) $|A(h, r)| = |B_r \cap \{u - k \geq h - k\}| \leq \frac{\int_{A(k,r)} (u-k)^2 \, dx}{(h-k)^2}$.

Therefore by (8.5) and above observations we obtain

$$\|(u-h)_+\|_{L^2(B_r)} \leq \frac{C}{R-r} \frac{1}{(h-k)^{2/n}} \|(u-k)_+\|_{L^2(B_R)}^{1+\frac{2}{n}}, \tag{8.6}$$

where $h > k$ and $R > r$. Define $\phi(k, r) = \|(u-k)_+\|_{L^2(B_r)}$, we set two sequences $\{k_l\}_0^\infty$ and $\{r_l\}_0^\infty$ as following:

$$k_l = k_0 + k(1 - 2^{-l}), \quad r_l = \frac{1}{2} + 2^{-(l+1)}, \text{ for } l = 0, 1, 2, \ldots.$$

Note that $k_l - k_{l-1} = k \cdot 2^{-l}$ and $r_{l-1} - r_l = 2^{-(l+1)}$. Then (8.6) implies that

$$\phi(k_l, r_l) \leq C \cdot 2^{l+1} \frac{2^{\varepsilon_0 l}}{k^{\varepsilon_0}} \phi(k_{l-1}, r_{l-1})^{1+\varepsilon_0}$$

$$\leq C \frac{2^{l(1+\varepsilon_0)}}{k^{\varepsilon_0}} \phi(k_{l-1}, r_{l-1})^{1+\varepsilon_0}.$$

We claim that we can choose k suitably large, and there exists an $A > 1$ such that

$$\phi(k_l, r_l) \leq \phi(k_0, r_0)A^{-l}. \tag{8.7}$$

Suppose (8.7) is true for l, we require that

$$C\frac{2^{l(1+\varepsilon_0)}}{k^{\varepsilon_0}}\left(\frac{\phi(k_0,r_0)}{A^l}\right)^{1+\varepsilon_0} \leq \frac{\phi(k_0,r_0)}{A^{l+1}}$$

$$\Leftrightarrow C\frac{2^{l(1+\varepsilon_0)}}{k^{\varepsilon_0}} \leq \frac{A^{l\varepsilon_0-1}}{\phi(k_0,r_0)^{\varepsilon_0}}.$$

We can take $A > 2^{\frac{1+\varepsilon_0}{\varepsilon_0}}$ and take k suitably large to make the second line holds for any $l \geq 0$. As $l \to \infty$, $k_l \to k_0 + k$ and $r \to \frac{1}{2}$, while $\phi(k_l, r_l) \to 0$. Thus we conclude that $\|u\|_{L^\infty(B_{1/2})} \leq k_0 + k$, which finishes the proof. □

Once we proved that L^∞ norm of the solution in $B_{1/2}$ can be controlled by its L^2 norm in B_1, we can further prove that actually the $L^\infty(B_{1/2})$ norm can be bounded by $L^p(B_1)$ norm for any $p \in (0, \infty)$ (note that when $p < 1$, it is not a norm but we can still define it formally). We have the following theorem.

Theorem 8.4. Let $u \in H^1(B_1)$ be a subsolution of (8.1), i.e. $Lu \geq 0$. Then

$$\|u\|_{L^\infty(B_{\frac{1}{2}})} \leq C(n,\lambda,\Lambda,p)\|u\|_{L^p(B_1)}, \quad \text{for any } p \in (0,\infty).$$

Proof. It suffices to prove for the case $p \in (0,2)$. By Theorem 8.3, we have for any ball $B_r \subset B_1$, it holds that

$$\|u\|_{L^\infty(B_{\frac{r}{2}})} \leq C(\fint_{B_r} u^2\, dx)^{1/2}.$$

We can assume $u \in L^\infty(B_1)$ for simplicity. Take $r_k = 1 - \frac{1}{2^{k+1}}$, define

$$M_k = \sup_{|x|\leq r_k} |u(x)|.$$

Let $\delta_k = r_{k+1} - r_k = 2^{-(k+2)}$. Suppose for every k, $M_k = u(x_k)$ for some $|x_k| \leq r_k$. Applying De Giorgi theorem to $B_{\delta_k}(x_k)$ gives

$$M_k \leq C(\fint_{B_{\delta_k}(x_k)} u^2\, dx)^{\frac{1}{2}}.$$

Since for $x \in B_{\delta_k}(x_k)$, $|u(x)|^2 \leq M_{k+1}^{2-p}|u|^p$, we have

$$M_k \leq C_1^k M_{k+1}^{1-\frac{p}{2}}\|u\|_{L^p(B_{\delta_k}(x_k))}^{p/2} \leq C_1^k M_{k+1}^{1-\theta},$$

where we assume $\|u\|_{L^p(B_1)} = 1$ and denote $\theta = \frac{p}{2}$.

Claim. There exist A, k_0 such that

$$M_k \leq A^{k+k_0} \text{ for } k = 0, 1, 2, \ldots. \tag{8.8}$$

Proof of the Claim. Since $u \in L^\infty(B_1)$, then for some $k \gg 1$, (8.8) is true. We want to show if (8.8) holds for k, then it must hold for $k-1$. Suppose that (8.8) is true for k, we have

$$M_{k-1} \le C_1^{k-1} M_k^{1-\theta} \le C_1^{k-1} A^{k(1-\theta)} A^{k_0(1-\theta)}.$$

We expect the following is true:

$$C_1^{k-1} A^{k(1-\theta)} A^{k_0(1-\theta)} \le A^{k-1+k_0}$$
$$\Leftrightarrow A^{1-(k+k_0)\theta} C_1^{k-1} \le 1.$$

We may simply take $k_0 > \frac{1}{\theta}$ and $A^\theta > C_1$. Note that these constants only depend on p, n, λ, Λ, and the proof is finished. $\square$

Remark 8.5. *For non-divergence equation $a_{ij}(x)u_{x_ix_j} = 0$ in B_1 with $a_{ij} \in L^\infty$ which satisfies the uniform ellipticity condition, we still have similar estimates as Theorem 8.4.*

Proof. Define $v = \xi u$ where $\xi = (1 - |x|^2)^\beta$ is a C^2 cutoff function for some $\beta \ge 1$, then we have

$$a_{ij}(x)v_{x_ix_j} = a_{ij}\xi u_{x_ix_j} + a_{ij}\xi_{x_ix_j}u + a_{ij}\xi_{x_i}u_{x_j} + a_{ij}\xi_{x_j}u_{x_i}$$
$$\ge -2\Lambda|D\xi||Du| + uL\xi.$$

By differentiation we have

$$\partial_i\xi = -2\beta x_i(1 - |x|^2)^{\beta-1},$$
$$\partial_{ij}\xi = -2\beta\delta_{ij}(1 - |x|^2)^{\beta-1} + 4\beta(\beta - 1)x_ix_j(1 - |x|^2)^{\beta-2}.$$

Now on the contact set Γ_v of v in B_1, we have $u > 0$ and

$$|Du| = \frac{1}{\xi}|Dv - uD\xi|$$
$$\le \frac{1}{\xi}(|Dv| + u|D\xi|)$$
$$< \frac{1}{\xi}(\frac{v}{1 - |x|} + u|D\xi|)$$
$$= (\frac{1}{1 - |x|} + \frac{|D\xi|}{\xi})u$$
$$\le C\xi^{-1/\beta}u.$$

So we have

$$Lv \ge -C\xi^{1-2/\beta}u - C\xi^{1-1/\beta}u \ge -C(n, \beta, \Lambda)\xi^{-2/\beta}v.$$

By Alexandroff maximum principle, we obtain for $\beta \ge 2$,

$$\sup_{B_1} v \le C(n, \beta, \Lambda)\|\xi^{-2/\beta}v^+\|_{L^n(B_1)}$$
$$\le C(n, \beta, \Lambda)(\sup_{B_1} v)^{1-2/\beta}\|(u^+)^{2/\beta}\|_{L^n(B_1)}.$$

Take $\beta = \frac{2n}{p}$ for some $p \leq n$, we get

$$\sup_{B_{1/2}} u \leq C(n,p)\sup_{B_1} v \leq C(n,p,\Lambda)\|u^+\|_{L^p(B_1)}.$$

This finishes the proof of the remark. $\qquad\square$

Part II: C^α estimate and Harnack's inequality.

Theorem 8.6. (Harnack's inequality) *Suppose that $u \in H^1(B_2)$ solves following equation*

$$Lu = \frac{\partial}{\partial x_i}(a_{ij}(x)u_{x_j}) = 0 \ in \ B_2,$$

$$u > 0 \ in \ B_2,$$

where a_{ij} satisfies

$$\lambda I \leq (a_{ij}) \leq \frac{I}{\lambda}, \ for \ some \ \lambda > 0.$$

Then there exists a constant $c(n,\lambda)$ such that

$$\sup_{B_{1/8}} u(x) \leq c(n,\lambda)\inf_{B_{1/8}} u(x).$$

Note that in the theorem the constant is set to be $\frac{1}{8}$ for convenience, actually it can be any constant less than 2. We next show that Harnack's inequality can be proved by a weak Harnack estimate, which is the following theorem:

Theorem 8.7. *Let $Lu \leq 0$ and $u > 0$ in B_2, and let $\varepsilon > 0$ such that*

$$|\{x \in B_1 : u(x) \geq 1\}| \geq \varepsilon.$$

Then there exists constant $c(n,\lambda,\varepsilon)$ such that

$$\inf_{B_{1/2}} u \geq c(n,\lambda,\varepsilon) > 0.$$

Proof. Let $v = \log\frac{1}{u}$. Then we multiply positive test function $\phi \in C_0^\infty(B_2)$,

$$\int \phi Lv = \int \partial_i\phi a_{ij}(x)\frac{u_{x_j}}{u}$$

$$= \int a_{ij}\partial_i(\frac{\phi}{u})u_{x_j} + \int a_{ij}\frac{u_{x_i}u_{x_j}}{u^2}\phi \geq 0.$$

Therefore $Lv^+ \geq 0$. Note that $\{x \in B_1 : u(x) \geq 1\} = \{x \in B_1 : v^+(x) = 0\}$, we get

$$|\{u(x) \geq 1\}| > \varepsilon \Rightarrow |\{v^+(x) = 0\}| \geq \varepsilon.$$

We have the following lemma by Poincaré:

Lemma 8.8. (Poincaré) *For any $\varepsilon > 0$ there exists a constant $C(\varepsilon, n)$ such that, for any $v \in H^1(B_1)$ satisfying $|\{x \in B_1 : v(x) = 0\}| \geq \varepsilon$, there holds*

$$\int_{B_1} v^2(x)\, dx \leq C(\varepsilon, n) \int_{B_1} |\nabla v|^2\, dx.$$

Proof. Suppose the statement in the lemma is not true. Then there exists a sequence $\{v_i\} \subset H^1(B_1)$ such that

$$|\{v_i = 0\}| \geq \varepsilon, \quad \int_{B_1} |\nabla v_i|^2\, dx = 1, \quad \lim_{i \to \infty} \int_{B_1} |v_i|^2\, dx = \infty.$$

Let $h_i = v_i - \bar{v}_i$. By Poincaré's inequality h_i is uniformly bounded in H^1 and up to a subsequence we have

$$h_i \rightharpoonup h \text{ weakly in } H^1, \text{ strongly in } L^2, \quad \bar{v}_i \to \infty.$$

On $E_i = \{v_i(x) = 0\}$, $h_i(x) = -\bar{v}_i$, and $|E_i| \geq 0$ for all i. This contradicts with the fact h_i is bounded in L^2. □

Then by Theorem 8.3, we have

$$\sup_{B_{1/2}} v^+ \leq C(n, \lambda) \left(\int_{B_1} (v^+)^2 \right)^{1/2} \leq C(n, \lambda, \varepsilon) \left(\int_{B_1} |\nabla v^+|^2 \right)^{1/2}.$$

So once $\int_{B_1} |\nabla v^+|^2$ is bounded, weak Harnack estimate follows. Let $\xi \in C_0^2(B_2)$ be a cut-off function such that $\xi \equiv 1$ in B_1. We calculate

$$0 \geq \int_{B_2} \frac{\xi^2}{u} Lu\, dx$$

$$= -\int_{B_2} \frac{2\xi \xi_{x_i}}{u} a_{ij} u_{x_j} + \int_{B_2} \frac{\xi^2}{u^2} a_{ij} u_{x_i} u_{x_j}.$$

This implies

$$\lambda \int_{B_2} \xi^2 \frac{|\nabla u|^2}{u^2} < \int_{B_2} \frac{\xi^2}{u^2} a_{ij} u_{x_i} u_{x_j} \leq \int_{B_2} \frac{2\xi \xi_{x_i}}{u} a_{ij} u_{x_j}.$$

Using Schwartz's inequality we conclude that

$$\int_{B_1} \sum_i \left| \frac{u_{x_i}}{u} \right|^2 \leq C(n, \lambda).$$

This finishes the proof. □

Remark 8.9. *We just proved that $u > 0$ solves $Lu = 0$ in B_2, $v = -\log u$ satisfies $\int_{B_1} |\nabla v|^2 \leq C(n, \lambda)$. By scaling, we have if $Lu = 0$ in B_{2r} then $\frac{1}{r^{n-2}} \int_{B_r} |\nabla v|^2\, dx \leq C(n, \lambda)$. Poincaré's inequality implies*

$$\int_{B_r} |v - \bar{v}_r|^2\, dx \leq Cr^n \Rightarrow v \in BMO.$$

The next lemma is a direct consequence of the weak Harnack estimates.

Lemma 8.10. *Let $u > 0$ and $Lu \leq 0$ in B_2. There exists $M_0 > 0$ such that if $\int_{B_{1/2}} u \leq 1$, then $\left|\{x \in B_{1/4} : u(x) \geq M_0\}\right| \leq \frac{1}{4}|B_{1/4}|$.*

Proof. Let $\varepsilon_0 = \frac{1}{8}|B_{1/4}|$, $v = \frac{u}{M_0}$ for some M_0 to be determined later. Suppose the conclusion in the lemma is not true, then

$$\left|\{x \in B_{1/4} : u \geq M_0\}\right| > \frac{1}{4}|B_{1/4}| \Rightarrow \left|\{x \in B_1 : v \geq 1\}\right| \geq \varepsilon_0.$$

By Theorem 8.7 we obtain

$$\inf_{B_{1/2}} v \geq C(n, \lambda) \Rightarrow \inf_{B_{1/2}} u \geq C(n, \lambda)M_0.$$

We choose $M_0 > \frac{1}{C(n,\lambda)}$ and arrive at a contradiction with the assumption $\int_{B_{1/2}} u \leq 1$. $\qquad\square$

Lemma 8.11. *Let $u > 0$ and $Lu \leq 0$ in B_2. $\inf_{B_{1/2}} u \leq 1$. M_0 is the constant in Lemma 8.10. Let $A_{k+1} = \{x \in B_{1/4} : u(x) \geq M_0^{k+1}\}$, then there holds that*

$$|A_{k+1}| \leq \frac{1}{4}|A_k| \leq \cdots \leq \frac{1}{4^{k+1}}|B_{1/4}|.$$

Proof. The proof of this lemma is the same as the proof of corresponding lemma in Krylov-Safonov theory. We only utilize a key observation and Calderón-Zygmund decomposition. For details see Remark 7.5 and Lemma 7.7. $\qquad\square$

Once we have Lemma 8.11, it is straightforward to obtain for some $p_0 > 0$, we have

$$\left(\int_{B_{1/4}} u^{p_0}\, dx\right)^{1/p_0} \leq C(n, \lambda) \inf_{B_{1/8}} u.$$

On the other hand, by Theorem 8.4, we have

$$\sup_{B_{1/8}} u \leq C(n, \lambda)\left(\int_{B_{1/4}} u^{p_0}\, dx\right)^{1/p_0}.$$

This gives Harnack's inequality for divergence form uniformly elliptic equations with bounded measurable coefficients. With Harnack's inequality, we are ready to prove C^α continuity, and therefore it finishes the proof of De Giorgi theorem. Suppose $Lu = 0$ in B_2 and we have Harnack's inequality:

$$\sup_{B_1} u \leq C \inf_{B_1} u.$$

Then $\sup_{B_1} u - u$ and $u - \inf_{B_1} u$ also solves $Lu = 0$ in B_1 and they are strictly positive in $B_{1/2}$ by strong maximum principle. By Harnack's inequality, we have

$$\sup_{B_1} u - \inf_{B_{1/2}} u \leq C(\sup_{B_1} u - \sup_{B_{1/2}} u),$$

$$\sup_{B_{1/2}} u - \inf_{B_1} u \leq C(\inf_{B_{1/2}} u - \inf_{B_1} u).$$

Adding up above two inequalities gives

$$Osc_{B_{1/2}} u \le \frac{C-1}{C+1} Osc_{B_1} u.$$

Finally Hölder continuity follows easily from the oscillation decay. The proof of Theorem 8.1 is complete.

8.3 More on isolated singularities of solutions

Denote $Lu = \frac{\partial}{\partial x_i}(a_{ij} u_{x_j})$. The Green function $G^y(x)$ is defined as

$$LG^y(x) = -\delta_y \text{ in } \Omega \subset \mathbb{R}^n, \quad G^y(x)\big|_{\partial \Omega} = 0.$$

The existence of fundamental solution of L-operator (symmetric case) was established by Littman-Stampacchia-Weinberger in 1963. They used Moser-Harnack's inequality and classical potential theory. On a neighborhood of y, $G^y(x)$ satisfies

$$G^y(x) \ge 0,$$
$$C_1(n, \lambda)|x - y|^{2-n} \le G^y(x) \le C_2(n, \lambda)|x - y|^{2-n}.$$

Nonsymmetric case is proved by Royden.

For removable singularities of operator L, we start with the following lemma by Serrin:

Lemma 8.12. (Serrin) *If $Lu = 0$ in $B_{r_0}\backslash\{0\}$ and $u(x) = O(|x|^{2-n})$ then there exists a constant a such that*

$$u(x) = w(x) + aG^0(x),$$

where w is a regular solution of

$$Lw = 0 \text{ in } B_{r_0}, \quad w = u \text{ on } \partial B_{r_0}.$$

Proof. Let $v = u - w$, then v satisfies

$$Lv = 0 \text{ in } B_{r_0}\backslash\{0\},$$
$$v\big|_{\partial B_{r_0}} = 0,$$
$$v = O(|x|^{2-n}),$$
$$G^0(x) \simeq |x|^{2-n}.$$

By maximum principle, there holds that

$$-C_* G^0(x) \le v \le C_* G^0(x).$$

Denote $M = \limsup_{|x| \to 0} \frac{v(x)}{G^0(x)}$. Consider function $w = MG^0(x) - v(x)$, then by definition of M, $w > 0$ in $B_{r_0}\backslash\{0\}$, and there exist sequences $r_i \to 0$, $\varepsilon_i \to 0$ and $x_i \in B_{r_0}\backslash\{0\}$ such that $|x_i| = r_i$ and $w(x_i) \le \varepsilon_i G^0(x_i)$. By Harnack's inequality, we have $w(x) > c(n)\varepsilon_i G^0(x)$ for any $|x| = r_i$ and every i. Note that we also have $w(x) = 0$ on ∂B_{r_0}. Let $i \to \infty$ and using maximum principle we conclude that $w(x) \equiv 0$ on $B_{r_0}\backslash\{0\}$. This finishes the proof. $\square$

Lemma 8.13. (Moser) *Suppose $Lu = 0$ in $B_{R_0}^c$ and $u(x)$ is bounded, then*

$$u(x) = u_\infty + O(|x|^{2-n}) \text{ for some constant } u_\infty.$$

Proof. Denote $M = \limsup_{|x| \to \infty} u(x) < \infty$. Then consider $v = M + \varepsilon - u$ for some constant $\varepsilon << 1$. Then v satisfies

$$Lv = 0 \text{ in } B_{R_0}^c, \quad v(x) > 0 \text{ for } |x| \geq R^\varepsilon$$

and there exists $x_i \to \infty$ such that $v(x_i) \to \varepsilon$. Let $R_i = |x_i| > R^\varepsilon$. According to Harnack's inequality, we have

$$\sup_{|x|=R_i} v(x) \leq C(n,\lambda) \inf_{|x|=R_i} v(x) < 2C(n,\lambda)\varepsilon.$$

By maximum principle we have

$$\varepsilon \leq \limsup_{|x| \to \infty} v(x) \leq 2C(n,\lambda)\varepsilon.$$

Let $\varepsilon \to 0$ we obtain that

$$0 \leq \liminf_{|x| \to \infty} M - u \leq \limsup_{|x| \to \infty} M - u \leq 0,$$

which leads to $\lim_{|x| \to \infty} u(x) = M$. Suppose $u(x) \to u_\infty$ as x goes to infinity, there exist C and some suitably large R_1 such that for any $\varepsilon > 0$ it holds that

$$u - u_\infty + \varepsilon + \frac{C}{|x|^{n-2}} > 0 \text{ on } \partial B_{R_1} \text{ and when } x \to \infty,$$

$$u - u_\infty - \varepsilon - \frac{C}{|x|^{n-2}} < 0 \text{ on } \partial B_{R_1} \text{ and when } x \to \infty.$$

Using maximum principle and notice that ε is arbitrary, we conclude that $|u(x) - u_\infty| \leq \frac{C}{|x|^{n-2}}$ for $|x| \geq R_1$. $\qquad\square$

Some Remarks:

(1) Consider transformation $x \to y = \frac{x}{|x|^2}$ and $u(x) \to v(y) = \frac{u(y/|y|^2)}{G(y/|y|^2)}$. If u satisfies the following equation

$$L_G\left(\frac{u}{G}\right) = \frac{\partial}{\partial x_i}\left(G^2(x)a_{ij}(x)\left(\frac{u}{G}\right)_{x_j}\right) = 0,$$

then by direct calculation, v will solve

$$\frac{\partial}{\partial y_i}(\tilde{a}_{ij}(y)v_{y_j}) = 0,$$

for some uniformly elliptic $\tilde{a}$ defined by $\tilde{a}_{ij} = G^2 a_{kl}(x)\frac{\partial y_i}{\partial x_k}\frac{\partial y_j}{\partial x_l}J(y)$. This transformation links Lemma 8.12 and Lemma 8.13.

(2) The statement of Lemma 8.12 is also true under the assumption $u(x) = O(|x|^{2-n-\delta})$.

(3) For $n \geq 3$, either

$$u(x) = u_\infty + O(|x|^{2-n}) \text{ one sided bounded,}$$

or

$$M(\sigma) = \sup_{|x|=\sigma} u(x) \geq A\sigma^\alpha \text{ and } m(\sigma) = \inf_{|x|=\sigma} u(x) \leq -A\sigma^\alpha$$

for some positive α, A.

For **non-divergence case**, there are some examples of isolated singularities whose blow up rate is not the same as r^{2-n}. Consider the equation

$$a_{ij}(x)u_{x_i x_j}(x) = 0, \quad a_{ij} = \delta_{ij} + g(r)\frac{x_i x_j}{r^2}.$$

It is obvious uniformly elliptic. Consider the radial functions $u(r)$, then the equation becomes

$$u_{rr} + \frac{n-1}{r}\frac{u_r}{1+g(r)} = 0.$$

The solution can be written explicitly as

$$u(r) = \int^r \exp\left(\int^s \frac{1-n}{1+g(\rho)}\frac{d\rho}{\rho}\right)ds.$$

Then we examine following examples:

(1) $n = 2$, $1 + g(r) = \frac{\ln r}{2+\ln r}$ for $0 < r < e^{-3}$, $u(r) = a + \frac{b}{\ln r}$, there is no positive solution blowing up at origin.

(2) $n = 2$, $g(r) = \frac{2}{\ln r - 2}$, $u(r) = (\ln\frac{1}{r})^3$.

(3) $n \geq 3$, $g(r) = \frac{-1}{1+(n-1)\ln r}$, $u(r) = \frac{r^{2-n}}{\ln\frac{1}{r}}(1+\varepsilon r)$, in this case, for any $\varepsilon > 0$, near the origin it is always true that

$$r^{2-n+\varepsilon} \leq u(r) \leq r^{2-n-\varepsilon}.$$

(4) $n \geq 2$, $g(r) = \frac{(n-2)\ln r - 2}{\ln r + 2}$, $u(r) = a + \frac{b}{\ln r}$, 2D case for the first example.

(5) If we take g to be Dini continuous, then the solution always behaves like $O(r^{2-n})$ near origin.

9 Wiener criterion

9.1 Motivation and history

Let $\Omega \subset \mathbb{R}^n$ be an open bounded set and g be a continuous function on $\partial\Omega$, we consider the Dirichlet problem with boundary value g:

$$\begin{cases} \Delta u &= 0 \text{ in } \Omega, \\ u &= g \text{ on } \partial\Omega. \end{cases} \tag{9.1}$$

Perron's method provides a way to show the existence of solution for equation (9.1). The solution is defined as

$$u(x) = \sup\{v(x) : v \in C(\bar{\Omega}),\ v \text{ is subharmonic in } \Omega,\ \text{and } v|_{\partial\Omega}(x) \le g(x)\}.$$

Then one has

    (a) $u(x)$ is harmonic.

    (b) $\displaystyle\lim_{x\in\Omega,x\to x_0\in\partial\Omega} u(x) = g(x_0)$ if x_0 is a regular point.

The problem is when is (b) achieved? There is a long story.

(1) In $2-D$ case, suppose Ω is simply connected, if $x_0 \in \partial\Omega$ can be connected to ∞ by a continuous path in Ω^c, then x_0 is regular. This is a necessary and sufficient condition. The proof uses conformal mapping.

(2) If $\partial\Omega$ is locally Lipschitz, then $\partial\Omega$ is regular.

(3) **(Wiener Criteria)** For $x_0 \in \partial\Omega$, take $0 < \rho < r_0$ small. Denote

$$\Omega_\rho^c(x_0) = B_\rho(x_0) \cap \Omega^c,$$

$$E_\rho(x_0) = \Omega_\rho^c(x_0) \cap \{|x - x_0| \ge \tfrac{\rho}{2}\},$$

$$C_\rho(x_0) = Cap(E_\rho(x_0)).$$

Theorem 9.1. (Wiener) x_0 *is a regular point* (*for* Δ) $\Leftrightarrow \int_0^{r_0} \frac{C_\rho(x_0)}{\rho^{n-2}} \frac{d\rho}{\rho} = +\infty.$

(4) Puschol (Tantz, Oleĭnik) proved similar criterion for second order elliptic equations with $a_{ij} \in C^2(+lower\ orders)$.

(5) Herve proved that it suffices to assume a_{ij} is Hölder.

(6) Krylov weakened the smoothness assumption on a_{ij} to Dini continuity.

(7) Littman-Stampacchia-Weinberger proved the same criterion for uniformly elliptic operator $L = \partial_{x_i}(a_{ij}(x)u_{x_j})$ with bounded measurable coefficients a_{ij}.

9.2 Preliminaries

Now we state some well-known results for equation

$$Lu = \frac{\partial}{\partial x_i}(a_{ij}(x)u_{x_j}) = 0.$$

(1) By Lax-Milgram theorem, for bounded domain Ω, the equation

$$\begin{cases} Lu = \operatorname{div}\vec{f}, \quad \vec{f}\in L^2(\Omega,\mathbb{R}^n), \\ u = h \text{ on } \partial\Omega \end{cases}$$

    admits a unique solution u such that $u - h \in H_0^1(\Omega)$.

(2) Caccioppoli's inequality: for any ball $B_R(y) \subset \Omega$,

$$\int_{B_\rho(y)} |\nabla u|^2\,dx \le \frac{C\lambda^{-2}}{(R-\rho)^2} \int_{B_R(y)} u^2\,dx.$$

(3) De Giorgi theorem:

$$\|u\|_{C^\alpha(B_{1/2})} \leq C(\alpha, n, \lambda)\|u\|_{L^2(B_1)}.$$

(4) Harnack's inequality: if $Lu = 0$ in B_1 and $u > 0$, then

$$\sup_{B_{1/2}} u \leq C(n, \lambda) \inf_{B_{1/2}} u.$$

(5) Morrey-Stampacchia: Assume $\partial\Omega$ is Lipschitz. If $\vec{f} \in L^p$, $h \in W^{1,p}(\Omega)$ with $p > n$, then $u \in C^\alpha(\bar{\Omega})$ for $\alpha = \min\{1 - \frac{n}{p}, \alpha_0\}$ where α_0 is the De Giorgi power.

(6) Weinberger, Talenti: If $Lu = \operatorname{div} \vec{f}$ and $\vec{f} \in L^p$ for some $p > n$, then

$$\max_\Omega |u| \leq C_0(n, p)\lambda^{-1}|\Omega|^{\frac{1}{n} - \frac{1}{p}}\|\vec{f}\|_{L^p}.$$

If $Lu = f$ and $f \in L^p$ for some $p > \frac{n}{2}$, then

$$\max_\Omega |u| \leq C_1(n, p)\lambda^{-1}|\Omega|^{\frac{2}{n} - \frac{1}{p}}\|\vec{f}\|_{L^p}.$$

Consider equation

$$\begin{cases} Lu & = 0 \text{ in } \Omega, \\ u & = h \text{ on } \partial\Omega, \text{ for } h \in C(\partial\Omega). \end{cases} \tag{9.2}$$

There are two observations:

- If $h \in H^{1/2}(\partial\Omega)$, then there exists a unique solution $u \in H^1(\Omega)$ and $u - h \in H_0^1(\Omega)$.
- The maximum principle still holds, i.e.

$$\min_{\partial\Omega} h \leq u \leq \max_{\partial\Omega} h.$$

A quick proof: consider $v = (u - M)_+$ where $M = \max_{\partial\Omega} h$. Then $(u - M)_+ \in H_0^1(\Omega)$ and $Lv \geq 0$. Then we have

$$0 \leq \int_\Omega vLv \, dx = -\int_\Omega a_{ij}(x)v_{x_i}v_{x_j} \, dx \leq 0,$$

which implies $v \equiv 0$ in Ω.

Define a linear operator $B : C(\partial\Omega) \cap H^{1/2}(\partial\Omega) \to H^1(\Omega) \cap L^\infty(\Omega)$ which maps h to u. By above observation we have that B is a bounded operator. For $v \in H_{loc}^1(\Omega) \cap L^\infty(\Omega)$, define a norm

$$\||v\|| = \|v\|_{L^\infty} + \sup_{\delta>0}(\delta^2 \int_{\Omega_\delta} |\nabla v|^2 \, dx)^{1/2},$$

where $\Omega_\delta = \{x \in \Omega : d(x, \partial\Omega) > \delta\}$.

Given $h \in C(\partial\Omega)$, consider mollification of h: $\rho_\varepsilon * h = h_\varepsilon \in C(\partial\Omega) \cap Lip(\partial\Omega)$. As $\varepsilon \to 0$, $h_\varepsilon \to h$ uniformly. Let $u_\varepsilon = Bh_\varepsilon$, then there holds that

$$\min_{\partial\Omega} h \leq \min_{\partial\Omega} h_\varepsilon \leq u_\varepsilon \leq \max_{\partial\Omega} h_\varepsilon \leq \max_{\partial\Omega} h.$$

By Caccioppoli's inequality, we have

$$\int_{\Omega_\delta} |\nabla u_\varepsilon|^2 \, dx \leq \frac{C}{\delta^2} \int_\Omega u_\varepsilon^2 \, dx,$$

which leads to

$$\limsup_{\varepsilon \to 0} \||Bh_\varepsilon|\| \leq C\|h\|_{L^\infty}.$$

As $\varepsilon \to 0$, $Bh_\varepsilon \to u = Bh$, which is well defined for $h \in C(\partial\Omega)$, and we also have

$$\||Bh|\| \leq C(n, \lambda)\|h\|_{L^\infty}.$$

Now we give the definitions of regular points and barrier functions.

Definition 9.2. $x_0 \in \partial\Omega$ is called a regular point of L if for every $h \in C(\partial\Omega)$ that $u = Bh$ satisfies

$$\lim_{x \in \Omega, \, x \to x_0} u(x) = h(x_0).$$

Definition 9.3. A barrier function at x_0 is a function $V^{x_0}(x) \in H^1(\Omega)$, such that
(1) $LV^{x_0}(x) = 0$ in Ω, $V^{x_0} \geq 0$.
(2) For any $\rho > 0$, there exists $m > 0$ such that $V^{x_0}(x) \geq m$ for $x \in \partial\Omega \cap \{|x - x_0| > \rho\}$.
(3) $\lim_{x \in \Omega, \, x \to x_0} V^{x_0}(x) = 0$.

One important observation is the following lemma:

Lemma 9.4. $x_0 \in \partial\Omega$ is regular $\Leftrightarrow$ There is a barrier function $V^{x_0}(x)$ at x_0.

Proof. If $x_0 \in \partial\Omega$ is regular, then $V^{x_0}(x) = B(|x - x_0|)$ is a barrier function. On the other hand, suppose at $x_0 \in \partial\Omega$, there exists a barrier function $V^{x_0}(x)$. Consider $h(x) \in C(\partial\Omega)$ such that $|h(x)| \leq M$, then for some small $\varepsilon > 0$, by continuity of $h(x)$ there exists $\rho > 0$ such that for $x \in \partial\Omega \cap \{|x - x_0| \leq \rho\}$, we have $|h(x) - h(x_0)| \leq \varepsilon$. Outside the ball $B_\rho(x_0)$, by definition of barrier function, there exists $m > 0$ such that $V^{x_0}(x) > m$. Combine the two estimates we obtain

$$h(x_0) - \varepsilon - \frac{2M}{m}V^{x_0}(x) \leq h(x) \leq h(x_0) + \varepsilon + \frac{2M}{m}V^{x_0}(x), \quad \text{for any } x \in \partial\Omega.$$

Then by maximum principle, we conclude that Bh satisfies

$$h(x_0) - \varepsilon - \frac{2M}{m}V^{x_0}(x) \leq Bh(x) \leq h(x_0) + \varepsilon + \frac{2M}{m}V^{x_0}(x), \quad \text{for any } x \in \Omega.$$

As $x \to x_0$, we have

$$\liminf_{x \to x_0} Bh(x) \geq h(x_0) - \varepsilon, \quad \limsup_{x \to x_0} Bh(x) \leq h(x_0) + \varepsilon.$$

Since ε can be arbitrarily small, we have $\lim_{x \to x_0} Bh(x) = h(x_0)$ and x_0 is a regular point. $\qquad\square$

Another important observation is

Lemma 9.5. *Let $\Omega' \subset \Omega$, $x_0 \in \partial\Omega' \cap \partial\Omega$. Assume for some $\rho_0 > 0$, $\partial\Omega \cap B_{\rho_0}(x_0) = \partial\Omega' \cap B_{\rho_0}(x_0)$, then x_0 is a regular point of L on Ω if and only if x_0 is a regular point of L on Ω' .*

Proof. Suppose x_0 is a regular point of Ω', we consider $B_\Omega(|x - x_0|)$ on Ω. It suffices to show it is a barrier function for Ω. Note that on $\partial\Omega'$, we have

$$B_\Omega(|x - x_0|) \leq C(\operatorname{diam}\Omega, \rho_0)|x - x_0| \text{ on } \partial\Omega'\backslash B_{\rho_0}(x_0).$$

By maximum principle, we obtain that

$$0 \leq B_\Omega(|x - x_0|) \leq C(\rho, \operatorname{diam}\Omega)B_{\Omega'}.$$

Consequently, $B_\Omega(|x - x_0|)$ is a barrier function and x_0 is a regular point for Ω.

Now suppose that x_0 is a regular point of Ω, let D be a large ball that contains Ω, for $\rho \leq \rho_0$, we define

$$E_\rho = D\backslash\Omega \cap B_\rho(x_0), \quad \Omega_\rho = D\backslash E_\rho.$$

Note that $\partial\Omega_\rho = \partial\Omega$ inside $B_\rho(x_0)$, thus $x_0 \in \partial\Omega_\rho$ is a regular point for Ω_ρ. Let u_ρ be the solution of the following problem

$$Lu_\rho = 0 \text{ in } \Omega_\rho,$$
$$u_\rho = 0 \text{ on } \partial D, \quad u_\rho = 1 \text{ on } \partial E_\rho.$$

Define function $V^{x_0}(x) = \sum\limits_{k=2}^{\infty} 2^{-k}(1 - u_{\rho_0/k}(x))$. Such sequence converges uniformly and each term $1 - u_{\rho_0/k}(x)$ is continuous at $x = x_0$. We claim that it is a barrier function at x_0 for Ω'. Firstly it holds that $V^{x_0} \geq 0$ and solves $LV^{x_0}(x) = 0$ in Ω. Also as $x \in \Omega \to x_0$, we have $V^{x_0}(x) \to 0$. Finally, by Harnack's inequality, for any $k \geq 2$, there exists $m_k > 0$ such that

$$1 - u_{\rho_0/k}(x) \geq 2^k m_k \text{ on } \partial\Omega'\backslash B_{2\rho_0/k}(x_0).$$

This implies that $V^{x_0}(x)$ is uniformly away from 0 on $\partial\Omega'\backslash B_\varepsilon$. The proof is complete. □

9.3 Capacity

Definition 9.6. *Let Ω be an open bounded domain, $E \subset\subset \Omega$, the capacity of E with respect to L, Ω is defined as*

$$Cap_{L,\Omega}(E) = \inf\{\int_\Omega a_{ij}(x)v_{x_i}v_{x_j}\, dx : v \in H_0^1(\Omega),\ v \geq 1 \text{ on } E\}.$$

Denote $K = \{v \in H_0^1(\Omega),\ v \geq 1 \text{ on } E\}$ is a convex, closed subset of $H_0^1(\Omega)$, $I(v) = \int_\Omega a_{ij}(x)v_{x_i}v_{x_j}\, dx \cong \|v\|_{H^1}^2$ for any $v \in H_0^1(\Omega)$. It is obvious that $I(v)$ is a strictly convex functional, so there exists a unique $u \in H_0^1(\Omega)$ such that $u \geq 1$ on E, and

$$Cap_{L,\Omega}(E) = \int_\Omega a_{ij}(x)u_{x_i}u_{x_j}\, dx.$$

For such minimizer u, let $u_t(x) = u(x) + t\xi(x)$ where $\xi \geq 0$ and $\xi \in H_0^1(\Omega)$, by minimality of u we have

$$\frac{d}{dt}\Big|_{t=0+} I(u_t) \geq 0$$

$$\Rightarrow \int_\Omega a_{ij}(x)u_{x_i}\xi_{x_j}\, dx \geq 0$$

$$\Rightarrow -Lu = \mu \geq 0 \text{ in } \mathcal{D}(\Omega).$$

Therefore we obtain that u is a supersolution.

Remark 9.7.

(1) *In fact u is the minimal supersolution in $H_0^1(\Omega)$ above $\mathcal{X}_E$. (If v is another supersolution, take $\min(u,v)$, energy will drop.)*
(2) *u is a solution of $Lu = 0$ on $\Omega \setminus \bar{E}$.*

Question: How to solve $-Lu = \mu \geq 0$ where μ is only a Radon measure?

Definition 9.8. *u is called a weak solution of*

$$\begin{cases} -Lu = \mu \text{ in } \Omega, \\ u = 0 \text{ on } \partial\Omega \end{cases}$$

if and only if $u \in L^1(\Omega)$ and for any $\xi \in C_0^\infty(\Omega)$, it holds that

$$\int_\Omega u(-L\xi)dx = \int_\Omega \xi d\mu(x).$$

Lemma 9.9. *There exists a continuous linear operator G from $H^{-1}(\Omega)$ into $H_0^1(\Omega)$ such that for any $f \in H^{-1}(\Omega)$, $u = G(f) \in H_0^1(\Omega)$ and $-Lu = f$.*

G is called Green operator. Note that $G : W^{-1,p} \to C_0(\Omega)(C^\beta)$ for $p > n$ and there holds that

$$\|G(f)\|_{L^\infty(\Omega)} \leq C\Lambda|\Omega|^{\frac{1}{n}-\frac{1}{p}}\|f\|_{W^{-1,p}}.$$

Now recall that u solves $-Lu = \mu$ and $u|_{\partial\Omega} = 0$, test the equation with $G(\phi)$ for $\phi \in W^{-1,p}$, we get

$$\int_\Omega u\phi\, dx = \int_\Omega G(\phi)d\mu.$$

It follows that

$$\left| \int_\Omega u\phi\, dx \right| \leq C\Lambda|\Omega|^{\frac{1}{n}-\frac{1}{p}}\|\phi\|_{W^{-1,p}}\|\mu\|$$

$$\Rightarrow \|u\|_{W^{1,p'}} \leq C\Lambda|\Omega|^{\frac{1}{n}-\frac{1}{p}}\|\mu\|.$$

Note that $\mathcal{M}(\Omega) \subset G(W^{-1,p})^*$, we can conclude that $u = G^*(\mu) \in W^{1,p'}$ is a well defined solution in distributional sense. Now we state some properties of solution u and Green function.

(1) $\mu \in \mathcal{M}(\Omega) \cap H^{-1}(\Omega) \Rightarrow u \in H_0^1(\Omega)$.
(2) If $\mu \in \mathcal{M}(\Omega)$ then $\nabla u \in L^{\frac{n}{n-1}-}$, $u \in L^{\frac{n}{n-2}-}$.
(3) If $0 \le \mu^* \le \mu \in H^{-1}(\Omega)$, then $\mu^* \in H^{-1}(\Omega)$, since

$$|\mu^*(\phi)| \le \mu^*(|\phi|) \le \mu(|\phi|) \le \|\mu\|_{H^{-1}(\Omega)} \|\phi\|_{H_0^1(\Omega)}.$$

(4) Consider Green function $-Lg^y(x) = \delta_y (y \in \Omega)$ in Ω, such that $g^y(x)|_{\partial\Omega} = 0$. Denote $g(x,y) = g^y(x)$, $x \in \Omega, y \in \Omega$, $g(x,y) \ge 0$ and $g(x,y)$ is continuous in x for $x \ne y$.
(5) $g(x,y)$ is symmetric.
   Let $-Lu = \xi$, $-Lv = \eta$ in Ω and $u|_{\partial\Omega} = v|_{\partial\Omega} = 0$.

$$\begin{aligned} v(x) &= \langle v(y), \delta_x(y) \rangle \\ &= \langle v(y), -L_y g^x(y) \rangle \\ &= \langle -Lv, g^x(y) \rangle \\ &= \langle \eta, g^x(y) \rangle. \end{aligned}$$

Note that $\langle -Lu, v \rangle = \langle u, -Lv \rangle$, consequently we obtain

$$\iint_{\Omega \times \Omega} g(x,y)\xi(x)\eta(y)\,dxdy = \iint_{\Omega \times \Omega} g(y,x)\xi(x)\eta(y)\,dxdy \text{ for any } \xi, \eta.$$

This implies $g(x,y) = g(y,x)$.

(6) $\lim\limits_{x \to y \in \Omega} g(x,y) = +\infty$. Assume $y = 0$, and $\lim\limits_{x \to 0} g(x,0) < \infty$, then by Harnack's inequality $g(x,0)$ is bounded near 0. Let $Lu = 0$ in $\Omega \backslash \{0\}$ and u be bounded, then one can show that $Lu = 0$ in Ω. It is equivalent to prove that $Cap_L(\{0\}) = 0$ for any L.

For some fixed $y \in \Omega$, define $D_a = \{x \in \Omega : g(x,y) \ge a\}$ for some $0 < a < \infty$. Then we define

$$u_a = \begin{cases} 1 \text{ in } D_a, \\ \frac{g(x,y)}{a} \text{ in } \Omega \backslash D_a. \end{cases}$$

Then u_a is the capacity potential of D_a. Assume that u_a satisfies

$$-Lu_a = \mu_a, \quad u_a = 0 \text{ on } \partial\Omega,$$

we have

$$\int_\Omega |\nabla u_a|^2 \, dx \cong \langle -Lu_a, u_a \rangle = \langle \mu_a, u_a \rangle = Cap_L(D_a) < \infty.$$

Note that $D_a \backslash \partial D_a = \{x \in \Omega, g(x,y) > a\}$ is an open set, and $u_a \equiv 1$ on D_a, it holds that $Lu_a = 0$ in $D_a \backslash \partial D_a$ and therefore μ_a is supported on ∂D_a. We calculate

$$1 = u_a(y) = \int_\Omega g(x,y)d\mu_a(x) = \int_{\partial D_a} g(x,y)d\mu_a(x) = a \cdot Cap_L(D_a)$$

$$\Rightarrow Cap_L(D_a) = \frac{1}{a}.$$

Let $B_\rho(y) \subset \Omega$, $a = \min\limits_{x \in \partial B_\rho(y)} g(x,y)$. Then $B_\rho(y) \subset D_a$. It follows that

$$Cap_L(B_\rho(y)) \le Cap_L(D_a) = \frac{1}{a}.$$

Therefore we get

$$Cap_L(B_\rho(y)) \le \frac{1}{\min_{\partial B_\rho(y)} g(x,y)}.$$

Similarly, we also have

$$\frac{1}{\max_{\partial B_\rho(y)} g(x,y)} \le Cap_L(B_\rho(y)).$$

By Harnack's inequality, there exists $C(n,\lambda)$ such that

$$\max_{\partial B_\rho(y)} g(x,y) \le C(n,\lambda) \min_{\partial B_\rho(y)} g(x,y),$$

we then conclude that

$$\frac{1}{C(n,\lambda)} \frac{1}{Cap_L(B_\rho(y))} \le g(x,y) \le \frac{C(n,\lambda)}{Cap_L(B_\rho(y))}.$$

Also note that by definition it is easy to check that

$$Cap_L(B_\rho) \approx Cap_\Delta(B_\rho),$$

Therefore, we have $g_L(x,y) \approx g_\Delta(x,y)$, which is the following theorem:

Theorem 9.10. *Let $g(x,y)$ be the Green function of L on Ω, and $B_\rho(y) \subset \Omega$, then there exists constant $C(n,\lambda)$ such that*

$$C(n,\lambda)^{-1} g_\Delta(x,y) \le g_L(x,y) \le C(n,\lambda) g_\Delta(x,y).$$

Since $g_\Delta(x,y) = c(n)|x-y|^{2-n} + o(|x-y|^{2-n})$, it holds that

$$g_L(x,y) \approx |x-y|^{2-n}.$$

Now we are ready to prove Wiener criteria for L.

Theorem 9.11. *Consider that Ω is an open bounded domain in $\mathbb{R}^n$ and $\Omega \subset B_{R_0}$ for some large R_0. Suppose $x_0 \in \partial\Omega$. Let $E_\rho = (B_{R_0}\backslash\Omega) \cap B_\rho(x_0)$, and $u_\rho(x)$ be capacity potential of E_ρ. Then the following are equivalent:*

(1) x_0 *is a regular point;*

(2) $u_\rho(x_0) = 1$, for any ρ satisfying $0 < \rho < r_0$ for some small r_0;
(3) $\int_0^{r_0} c(r)r^{1-n}\,dr = +\infty$, where $c(r) = Cap_{L,\Omega}(E_r) \approx Cap_\Delta(E_r)$.

On the other hand, for irregular points, the following are equivalent:

(1) x_0 is an irregular point;
(2) $u_\rho(x_0) = 0$, for any ρ satisfying $0 < \rho < r_0$ for some small r_0;
(3) $\int_0^{r_0} c(r)r^{1-n}\,dr < +\infty$.

Note that here $u_\rho(x) = \lim_{\sigma \to 0} \int_{\Omega \cap \{|y-x|>\sigma\}} g(y,x)d\gamma_\rho(y)$, where γ_ρ is the capacity distribution for E_ρ, i.e. $-Lu_\rho = \gamma_\rho$.

Proof. x_0 is regular $\Leftrightarrow u^\rho(x_0) = 1$ is easy to prove, one just use the method in the proof of Lemma 9.5 to construct the barrier function for x_0, the details will be omitted here.

$$u_\rho(x) = \int_{E_\rho} g(y,x)d\gamma_\rho(y)$$

$$u_\rho(x_0) = \int_{E_\rho} g(y,x_0)d\gamma_\rho(y) = C_n \int_0^\rho r^{2-n}d\gamma_\rho(E_r) \tag{9.3}$$

$$= C_n\rho^{2-n}\gamma_\rho(E_\rho) + (n-2)C_n \int_0^\rho r^{1-n}\gamma_\rho(E_r)dr.$$

For $0 < r \le \rho$, we have

$$c(r) = \gamma_r(E_r) = \int_\Omega u_\rho d\gamma_r$$

$$= \int_\Omega u_r d\gamma_\rho$$

$$= \gamma_\rho(E_r) + \int_{E_\rho \setminus E_r} u_r d\gamma_\rho.$$

This implies that

$$\gamma_\rho(E_r) \le \gamma_r(E_r) = c(r).$$

If $\int_0^{r_0} c(r)r^{1-n}\,dr < +\infty$, then there exists a sequence $\rho_i \to 0$ such that

$$\frac{c(\rho_i)}{\rho_i^{n-2}} << 1.$$

Then by (9.3) we have $u^{\rho_i} << 1$ as $\rho_i \to 0$, which implies that x_0 is irregular.

Assume that x_0 is an irregular point, $u_{\rho_i}(x_0) < 1$ for a sequence $\rho_i \to 0$. We observe that

$$c(\frac{r}{2}) = \gamma_{r/2}(E_{r/2}) \tag{9.4}$$

$$\le \gamma_\rho(E_r) + \int_{E_\rho \setminus E_r} u_{r/2}(y)d\gamma_\rho.$$

Note that $u_{r/2}(x) = C(n) \int_{E_{r/2}} |x - y|^{2-n} d\gamma_{r/2}(y)$, and for any $x \in E_\rho \backslash E_r$, $y \in E_{r/2}$, it holds that

$$|x - y| \geq |x - x_0| - \frac{r}{2} \geq \frac{|x - x_0|}{2}.$$

Thus we have

$$u_{r/2}(x) \leq C(n)2^{n-2}|x - x_0|^{2-n}\gamma_{r/2}(E_{r/2}).$$

Substituting this into (9.4), we get

$$c(r/2) \leq \gamma_\rho(E_r) + C_1(n)u_\rho(x_0)c(r/2), \quad C_1(n) = C(n)2^{n-2}.$$

Suppose we have $u_\rho(x_0) << 1$ for all ρ near 0, then we have $c(r/2) \leq 2\gamma_\rho(E_r)$, and then

$$\int_0^{r_0} c(r)r^{1-n}dr \leq C \int_0^\rho \gamma_\rho(E_r)r^{1-n}dr$$

$$\leq C \int_0^\rho r^{2-n}d\gamma_\rho(E_r) < \infty.$$

This is the expected estimates for Wiener criteria. $\qquad\square$

Now we are left to prove the final lemma:

Lemma 9.12. *If $u_\rho(x_0) < 1$, then there exists constant c_n such that for any $\varepsilon > 0$, we have*

$$u_\tau(x_0) \leq c_n\varepsilon \text{ whenever } \tau << \rho.$$

Proof. We have that

$$u_\rho(x) = c_n \int_{|y-x_0|\leq\rho} |y - x|^{2-n} d\gamma_\rho(y)$$

$$= c_n \int_{|y-x|\leq\sigma} |y - x|^{2-n}d\gamma_\rho(y) + c_n \int_{|y-x|>\sigma} |y - x|^{2-n} d\gamma_\rho(y)$$

$$= v(x) + w(x).$$

Since $w(x_0) \leq u_\rho(x_0) < 1$, there exists $0 < \tau < \sigma$ such that

$$w(x) \leq \frac{1}{2}(1 + u_\rho(x_0)), \quad \text{for any } x \in E_\tau.$$

We also have $v(x) = 1 - w(x)$ in H^1 sense for $x \in E_\tau$, it follows that

$$v(x) \geq \frac{1}{2}(1 - u_\rho(x_0)) \text{ on } E_\tau$$

$$\geq \frac{1}{2}(1 - u_\rho(x_0))u_\tau(x) \text{ on } E_\tau.$$

Therefore we can choose σ suitably small to make sure $u_\tau(x_0) \leq cv(x_0) \leq c\varepsilon$. This finishes the proof. $\qquad\square$

10 Acknowledgements

Author wishes to thank Dr. Zhiyuan Geng and Yunzi Ding for taking the lecture notes. And he thanks Professor Ping Zhang and Jun Geng for careful readings of the first draft of these notes. The author is partially supported by an NSF grant, DMS-1955249.

References

[1] S. Agmon, A. Douglis and L. Nirenberg, Estimates near the boundary for solutions of elliptic partial differential equations satisfying general boundary conditions. I, *Comm. Pure Appl. Math.*, **12** (1959), 623–727.

[2] L. A. Caffarelli and X. Cabré, *Fully nonlinear elliptic equations*, American Mathematical Society Colloquium Publications, **43**, American Mathematical Society, Providence, RI, 1995.

[3] R. Courant and D. Hilbert, *Methods of mathematical physics*, Vol. II, Partial differential equations, Reprint of the 1962 original. Wiley Classics Library, John Wiley & Sons, Inc., New York, 1989.

[4] D. Gilbarg and N. S. Trudinger, *Elliptic partial differential equations of second order*, Second edition, Grundlehren der Mathematischen Wissenschaften, **224**, Springer-Verlag, Berlin, 1983.

[5] Q. Han and F. Lin, *Elliptic partial differential equations*, Second edition, Courant Lecture Notes in Mathematics, **1**, Courant Institute of Mathematical Sciences, New York, American Mathematical Society, Providence, RI, 2011.

[6] C. E. Kenig, *Harmonic analysis techniques for second order elliptic boundary value problems*, CBMS Regional Conference Series in Mathematics, **83**, Published for the Conference Board of the Mathematical Sciences, Washington, DC, American Mathematical Society, Providence, RI, 1994.

[7] E. M. Stein, *Singular integrals and differentiability properties of functions*, Princeton Mathematical Series, **30**, Princeton University Press, Princeton, N.J., 1970.

Lectures on the Analysis of Nonlinear Partial Differential Equations Vol. 6
MLM 6, pp. 279–324

Linear Inviscid Damping for Shear Flows

Dongyi Wei[*] and Zhifei Zhang[†]

Abstract

In these notes, we will review some recent progress on the linear inviscid damping for shear flows. For monotone shear flows, we will present a complete proof of linear inviscid damping, which is based on the limiting absorption principle and the vector field method. For non-monotone shear flows, we will present a proof of linear inviscid damping in the sense of L^2 space-time estimate.

1 Introduction

We consider the 2D incompressible Euler equations in a finite channel $\Omega = \{(x, y) : x \in \mathbb{T}, y \in [-1, 1]\}$:

$$\begin{cases} \partial_t V + V \cdot \nabla V + \nabla P = 0, \\ \nabla \cdot V = 0, \\ V^2(t, x, \pm 1) = 0, \\ V|_{t=0} = V_0(x, y). \end{cases} \tag{1.1}$$

Here $V = (V^1, V^2)$ and P denote the velocity and the pressure of the fluid respectively. We introduce the vorticity $\omega = \partial_x V^2 - \partial_y V^1$, which satisfies

$$\omega_t + V \cdot \nabla \omega = 0, \quad \omega(0, x, y) = \omega_0(x, y). \tag{1.2}$$

A special solution of (1.1) is $(u(y), 0)$ called shear flow. To study the stability of this solution, we linearize the equation (1.2) around it:

$$\begin{cases} \partial_t \omega + \mathcal{L}\omega = 0, \\ \omega|_{t=0} = \omega_0(x, y), \end{cases} \tag{1.3}$$

where

$$\mathcal{L} = u(y)\partial_x + u''(y)\partial_x(-\Delta)^{-1}. \tag{1.4}$$

*School of Mathematical Sciences, Peking University, 100871, Beijing, P. R. China E-mail jnwdyi@pku.edu.cn

†School of Mathematical Sciences, Peking University, 100871, Beijing, P. R. China E-mail zfzhang@math.pku.edu.cn

See [11] and references therein about the linear stability/instability results for shear flows.

For the Couette flow (i.e., $u(y) = y$), (1.3) is reduced to a passive transport equation

$$\partial_t \omega + y \partial_x \omega = 0, \quad \omega|_{t=0} = \omega_0(x, y).$$

In [16], Orr observed an important phenomena that the velocity will tend to 0 as $t \to \infty$, although the Euler equations are a conserved system. This phenomena is so-called inviscid damping, which is the analogue in hydrodynamics of Landau damping found by Landau [10]. In this case, the mechanism leading to the damping is the vorticity mixing driven by shear flow. See [20] for similar phenomena in various system.

Recently, there are a lot of works devoting to linear/nonlinear inviscid damping for the Couette flow. Lin and Zeng [13] proved that if $\int_{\mathbb{T}} \omega_0(x, y) dx = 0$, then the velocity of the linearized Euler equations decays as

$$\|V(t)\|_{L^2_{x,y}} \leq \|\omega(t)\|_{L^2_x H^{-1}_y} \leq \frac{C}{\langle t \rangle} \|\omega_0\|_{H^{-1}_x H^1_y}.$$

Due to possible nonlinear transient growth, it is a challenging task from linear damping to nonlinear damping. Moreover, nonlinear damping is sensitive to the topology of the perturbation. Indeed, Lin and Zeng [13] proved that nonlinear inviscid damping is not true for the perturbation in H^s for $s < \frac{3}{2}$. Bedrossian and Masmoudi [2] proved nonlinear inviscid damping around the Couette flow in Gevrey class $2 - \varepsilon$ in a special domain $\Omega = \mathbb{T} \times \mathbb{R}$, which is an analogue of nonlinear Landau damping proved by Mouhot and Villani [15]. Deng and Masmoudi [5] proved that the Gevrey regularity in [2] seems optimal for nonlinear inviscid damping in some sense. Ionescu and Jia [7] proved nonlinear inviscid damping in a finite channel when the data is in Gevrey class 2 and the support of the initial vorticity is away from the boundary.

For general shear flows, the linear inviscid damping is also a challenging problem due to the presence of nonlocal operator $u''(y)\partial_x(-\Delta)^{-1}$. In this case, the linear dynamics is associated with the singularities at the critical layer $u - c$ of the solution of the Rayleigh equation

$$(u - c)(\phi'' - \alpha^2 \phi) - u'' \phi = f.$$

Based on the Laplace transform and analyzing the singularity of the solution ϕ at the critical layer, Case [4] gave a first prediction of linear damping for monotone shear flow. However, there are few rigorous mathematical results. Rosencrans and Sattinger [19] gave t^{-1} decay of the stream function for analytic monotone shear flow. Stepin [21] proved $t^{-\nu}(\nu < \mu_0)$ decay of the stream function for monotone shear flow $u(y) \in C^{2+\mu_0}(\mu_0 > \frac{1}{2})$ without inflection point. Zillinger [27] proved t^{-1} decay of $\|V(t)\|_{L^2}$ for a class of monotone shear flows, which require that $L\|u''\|_{W^{3,\infty}} \ll 1$.

Case's formal argument does not work for non-monotone flows such as Poiseuille flow $u(y) = y^2$ and Kolmogorov flow $u(y) = \cos y$. Bouchet and Morita [3] may be the first to study the linear damping for non-monotone shear flows. Based on Laplace tools and numerical computations, they found a new dynamic phenomena: **vorticity depletion phenomena**: for large time,

$$\widehat{\omega}(t, \alpha, y) \sim \omega_\infty(y) \exp(-i\alpha u(y)t) + O(t^{-\gamma}).$$

The vorticity depletion means that $\omega_\infty(y)$ vanishes at stationary points of $u(y)$. This is another important mechanism leading to the damping for non-monotone shear flows. Based on this observation and using stationary phase expansion, they predicted similar decay rates of the velocity as in the monotonic case.

In a series of works [22, 23, 24], we mathematically confirm Case's prediction on linear damping for monotone shear flows and Bouchet-Morita's prediction for non-monotone shear flows including Poiseuille flow and Kolmogorov flow.

Let us first review our main results. The first result is the linear inviscid damping for monotone flows from [22, 25]. Here $\Omega = \big\{(x, y) : x \in \mathbb{T}, y \in [0, 1]\big\}$.

Theorem 1.1. *Let $u(y) \in C^4([0, 1])$ be a monotone function. Suppose that the linearized operator $\mathcal{L}$ has no embedding eigenvalues. Assume that $\int_\mathbb{T} \omega_0(x, y)dx = 0$ and $P_\mathcal{L}\omega_0 = 0$, where $P_\mathcal{L}$ is the spectral projection to $\sigma_d(\mathcal{L})$. Then it holds that*

1. *if $\omega_0(x, y) \in H_x^{-1}H_y^1$, then*

$$\|V(t)\|_{L^2} \leq \frac{C}{\langle t \rangle}\|\omega_0\|_{H_x^{-1}H_y^1};$$

2. *if $\omega_0(x, y) \in H_x^{-1}H_y^2$, then*

$$\|V^2(t)\|_{L^2} \leq \frac{C}{\langle t \rangle^2}\|\omega_0\|_{H_x^{-1}H_y^2};$$

3. *if $\omega_0(x, y) \in H_x^{-1}H_y^k$ for $k = 0, 1$, there exists $\omega_\infty(x, y) \in H_x^{-1}H_y^k$ such that*

$$\|\omega(t, x + tu(y), y) - \omega_\infty\|_{H_x^{-1}L_y^2} \longrightarrow 0 \quad as \quad t \to +\infty.$$

Recently, for monotone stable shear flows, Masmoudi-Zhao [14] and Ionescu-Jia [8] independently proved nonlinear inviscid damping in the Gevrey class.

In [23], we introduce a class of non-monotone flows denoted by $\mathcal{K}$, which consists of the function $u(y)$ satisfying $u(y) \in H^3(-1, 1)$, and $u''(y) \neq 0$ for critical points (i.e., $u'(y) = 0$) and $u'(\pm 1) \neq 0$. For the flows in $\mathcal{K}$, we have the following linear damping result.

Theorem 1.2. *Assume that $u(y) \in \mathcal{K}$ and the linearized operator $\mathcal{R}_\alpha$ defined by (3.2) has no embedding eigenvalues. Assume that $\widehat{\omega}_0(\alpha, y) \in H_y^1(-1, 1)$ and $P_{\mathcal{R}_\alpha}\widehat{\psi}_0(\alpha, y) = 0$, where ψ_0 is the stream function and $P_{\mathcal{R}_\alpha}$ is the spectral projection to $\sigma_d(\mathcal{R}_\alpha)$. Then it holds that*

$$\|\widehat{V}(\cdot, \alpha, \cdot)\|_{L_t^2 L_y^2} + \|\partial_t\widehat{V}(\cdot, \alpha, \cdot)\|_{L_t^2 L_y^2} \leq C_\alpha\|\widehat{\omega}_0(\alpha, \cdot)\|_{H_y^1}.$$

In particular, $\lim\limits_{t \to +\infty} \|\widehat{V}(t, \alpha, \cdot)\|_{L_y^2} = 0.$

The following theorem confirms the vorticity depletion phenomena for the flows in $\mathcal{K}$.

Theorem 1.3. *Under the same assumptions as in Theorem 1.2, if $u'(y_0) = 0$, then*

$$\lim_{t \to +\infty} \widehat{\omega}(t, \alpha, y_0) = 0.$$

Formally, Theorem 1.2 implies that the velocity should have at least $t^{-\frac{1}{2}}$ decay. To obtain the explicit decay rate of the velocity, we consider a class of symmetric shear flows:

(S) $\quad u(y) = u(-y), \quad u'(y) > 0$ for $y > 0, \quad u'(0) = 0$ and $u''(0) > 0$.

An important example is the Poiseuille flow $u(y) = y^2$.

Theorem 1.4. *Assume that $u(y) \in C^4([-1,1])$ satisfies (S) and the linearized operator $\mathcal{L}$ has no embedding eigenvalues. Assume that $\int_{\mathbb{T}} \omega_0(x, y) dx = 0$ and $P_{\mathcal{L}} \omega_0 = 0$, where $P_{\mathcal{L}}$ is the spectral projection to $\sigma_d(\mathcal{L})$. Then it holds that*

1. *if $\omega_0(x, y) \in H_x^{-\frac{1}{2}} H_y^1$, then*

$$\|V(t)\|_{L^2} \leq \frac{C}{\langle t \rangle} \|\omega_0\|_{H_x^{-\frac{1}{2}} H_y^1};$$

2. *if $\omega_0(x, y) \in H_x^{\frac{1}{2}} H_y^2$, then*

$$\|V^2(t)\|_{L^2} \leq \frac{C}{\langle t \rangle^2} \|\omega_0\|_{H_x^{\frac{1}{2}} H_y^2};$$

3. *if $\omega_0(x, y) \in H_x^{-\frac{1}{2}+k} H_y^k$ for $k = 0, 1$, there exists $\omega_\infty(x, y) \in H_x^{-\frac{1}{2}+k} H_y^k$ such that*

$$\|\omega(t, x + tu(y), y) - \omega_\infty\|_{H_x^{-\frac{1}{2}+k} L_y^2} \longrightarrow 0 \quad as \quad t \to +\infty.$$

In [24], we proved the linear inviscid damping and vorticity depletion phenomena for the Kolmogorov flow. Here $\Omega = \mathbb{T}_{2\pi\delta} \times \mathbb{T}_{2\pi}$.

Theorem 1.5. *Let $\omega(t, x, y)$ be the solution of (1.3) with $u(y) = -\cos y$. If $\delta \in (0, 1)$ and $\int_{\mathbb{T}_{2\pi\delta}} \omega_0(x, y) dx = 0$, then it holds that*

1. *if $\omega_0(x, y) \in H_x^{-\frac{1}{2}} H_y^1$, then*

$$\|V(t)\|_{L^2} \leq \frac{C}{\langle t \rangle} \|\omega_0\|_{H_x^{-\frac{1}{2}} H_y^1};$$

2. *if $\omega_0(x, y) \in H_x^{\frac{1}{2}} H_y^2$, then*

$$\|V^2(t)\|_{L^2} \leq \frac{C}{\langle t \rangle^2} \|\omega_0\|_{H_x^{\frac{1}{2}} H_y^2};$$

3. *if $\omega_0(x,y) \in H_x^{\frac{1}{2}} H_y^k$ for $k = 0,1$, there exists $\omega_\infty(x,y) \in H_x^{\frac{1}{2}} H_y^k$ such that*

$$\|\omega(t, x + tu(y), y) - \omega_\infty\|_{L^2} \longrightarrow 0 \quad as \quad t \to +\infty.$$

For $\delta = 1$, it holds that

1. *if $\omega_0(x,y) \in H_x^{-\frac{1}{2}} H_y^1$, then*

$$\|P_{\geq 2} V(t)\|_{L^2} \leq \frac{C}{\langle t \rangle} \|\omega_0\|_{H_x^{-\frac{1}{2}} H_y^1};$$

2. *if $\omega_0(x,y) \in H_x^{\frac{1}{2}} H_y^2$, then*

$$\|P_{\geq 2} V^2(t)\|_{L^2} \leq \frac{C}{\langle t \rangle^2} \|\omega_0\|_{H_x^{\frac{1}{2}} H_y^2}.$$

Here $P_{\geq 2}$ denotes the orthogonal projection of $L^2(\mathbb{T}^2)$ to the subspace $\{f \in L^2(\mathbb{T}^2) : f(x,y) = \sum\limits_{|\alpha| \geq 2} \hat{f}(\alpha, y) e^{i\alpha x}\}$.

The following theorem gives the explicit decay rate of the vorticity at critical points.

Theorem 1.6. *Let $\delta \in (0,1)$ and $\omega(t,x,y)$ be the solution of (1.3) with $u(y) = -\cos y$. If $\int_{\mathbb{T}_{2\pi\delta}} \omega_0(x,y) dx = 0$, then for $s > \frac{1}{2}$, it holds that for all $k \in \mathbb{Z}$,*

1. *if $\omega_e(x,y) \in H_x^{1+s} H_y^3$, then for $x \in \mathbb{T}_{2\pi\delta}$, $|t| > 1$,*

$$|\omega(t, x, k\pi)| \leq \frac{C}{|t|} \|\omega_e\|_{H_x^{1+s} H_y^3};$$

2. *if $\omega_e(x,y) \in H_x^{1+s} H_y^3$, then for $x \in \mathbb{T}_{2\pi\delta}$, $|t| > 1$,*

$$|V^2(t, x, k\pi)| \leq \frac{C}{|t|^2} \|\omega_e\|_{H_x^{1+s} H_y^3};$$

3. *if $\omega_o(x,y) \in H_x^{s-\frac{1}{2}} H_y^3$, then for $x \in \mathbb{T}_{2\pi\delta}$, $|t| > 1$,*

$$|V^1(t, x, k\pi)| \leq \frac{C}{|t|^{\frac{3}{2}}} \|\omega_0\|_{H_x^{s-\frac{1}{2}} H_y^3}.$$

Here $\omega_e(x,y) = \frac{1}{2}(\omega_0(x,y) + \omega_0(x,-y))$ and $\omega_o(x,y) = \frac{1}{2}(\omega_0(x,y) - \omega_0(x,-y))$.

Finally, let us mention many important results about the inviscid damping, see [1, 6, 9, 12, 18, 25, 26, 28] and references therein.

2 Mixing driven by shear flow

In this section, we consider a passive transport equation on $\mathbb{T} \times [-1, 1]$:

$$\partial_t \omega + u(y) \partial_x \omega = 0, \quad \omega(0) = \omega_{in}. \tag{2.1}$$

We have the following mixing result driven by the shear flow $u(y)$.

Proposition 2.1. *Let* $u \in C^{m+1}([-1,1])$ *and* $k \geq 0$ *be an integer. Let* $y_1, ..., y_k \in$ $[-1,1]$ *be* k *different points and* $1 \leq m_1, ..., m_k \leq m-1$ *be integer. Suppose that* u *satisfies for any* $l = 1, 2, ..., k$,

$$\sum_{i=1}^{m_l} |u^{(i)}(y_l)| = 0, \ u^{(m_l+1)}(y_l) \neq 0,$$

and $u'(y) \neq 0$ *for any* $y \neq y_l$. *Here* $u^{(n)}(y) = \frac{du^{(n-1)}(y)}{dy}$ *and* $u^{(0)}(y) = u(y)$. *Then for any* ω_{in} *with* $\int_{\mathbb{T}} \omega_{in}(x, y)dx = 0$, *it holds that*

$$\|\omega(t)\|_{L_x^2 H_y^{-1}} \leq C \max_{l=1,...,k} \left\{ (1 + |t|)^{-\frac{1}{m_l+1}} \right\} \|\omega_{in}\|_{L_x^2 H_y^1}.$$

Especially, if u *is strictly monotone, then*

$$\|\omega(t)\|_{L_x^2 H_y^{-1}} \leq C(1 + |t|)^{-1} \|\omega_{in}\|_{L_x^2 H_y^1}.$$

Proof. Taking Fourier transform in x variable, we obtain

$$\partial_t \widehat{\omega}(t, \alpha, y) + i\alpha u(y)\widehat{\omega}(t, \alpha, y) = 0.$$

Thus, $\widehat{\omega}(t, \alpha, y) = \widehat{\omega}_{in}(\alpha, y)e^{-i\alpha u(y)t}$. Let $\chi_\delta(y)$ be the smooth cut-off function such that $\chi_\delta(y) = 0$ for $|y| \geq 2\delta$ and $\chi_\delta(y) = 1$ for $|y| \leq \delta$. We can choose δ small enough such that

$$\min_{l=1,...,k} \min_{|y-y_l| \leq 2\delta} |u^{(m_l+1)}(y)| \geq c_0.$$

For any smooth function $\eta(y)$, we have

$$\left| \int_{-1}^1 \widehat{\omega}(t, \alpha, y)\eta(y)dy \right|$$

$$\leq \sum_{l=1}^k \left| \int_{-1}^1 \widehat{\omega}_{in}(\alpha, y)\chi_\delta(y - y_l)\eta(y)e^{-i\alpha tu(y)}dy \right|$$

$$+ \left| \int_{-1}^1 \widehat{\omega}_{in}(\alpha, y)\eta(y)e^{-i\alpha tu(y)} \left(1 - \sum_{l=1}^k \chi_\delta(y - y_l) \right) dy \right|.$$

For the second term, we get by integration by parts that for $t \geq 1$,

$$\left| \int_{-1}^1 \widehat{\omega}_{in}(\alpha, y)\eta(y)e^{-i\alpha tu(y)} \left(1 - \sum_{l=1}^k \chi_\delta(y - y_l) \right) dy \right|$$

$$\leq \left| \int_{-1}^1 \frac{d}{dy} \left(\frac{1}{i\alpha u'(y)t} \widehat{\omega}(\alpha, y)\eta(y) \left(1 - \sum_{l=1}^k \chi_\delta(y - y_l) \right) \right) e^{-i\alpha tu(y)}dy \right|$$

$$+ \frac{C}{\alpha t} \|\widehat{\omega}_{in}(\alpha, \cdot)\|_{L^\infty} \|\eta\|_{L^\infty} \leq \frac{C}{\alpha t} \|\widehat{\omega}_{in}(\alpha, \cdot)\|_{H^1} \|\eta\|_{H^1}.$$

For the first term, we consider each neighborhood of y_l. Let t be large enough so that $(\alpha t)^{-\frac{1}{m_l+1}} \leq \frac{1}{2}\delta$. Then for each $l = 1, ..., k$,

$$\left| \int_{-1}^{1} \widehat{\omega}_{in}(\alpha, y)\chi_\delta(y - y_l)\eta(y)e^{-i\alpha tu(y)} dy \right|$$

$$\leq C \left| \int_{|y-y_l|\leq(\alpha t)^{-\frac{1}{m_l+1}}} \widehat{\omega}_{in}(\alpha, y)\chi_\delta(y - y_l)\eta(y)e^{-i\alpha tu(y)} dy \right|$$

$$+ \frac{C}{\alpha t} \left| \int_{2\delta\geq|y-y_l|\geq(\alpha t)^{-\frac{1}{m_l+1}}} \frac{d}{dy}\left(\frac{\widehat{\omega}_{in}(\alpha, y)\chi_\delta(y - y_l)\eta(y)}{u'(y)} \right)e^{-i\alpha tu(y)} dy \right|.$$

Notice that for δ small, $k = 0, 1$,

$$|u^{(k+1)}(y)| = |(y - y_l)^{m_l-k}T(u^{(m_l+1)}, m_l - k - 1)(y, y_l)| \sim |y - y_l|^{m_l-k},$$

where $T(u^{(m_l+1)}, s)(y, y_l) = \int_0^1 \frac{(1-t)^s}{s!}u^{(m_l+1)}(y_l + t(y - y_l))dt$ for $s \geq 0$ and $T(u^{(m_l+1)}, s)(y, y_l) = u^{(m_l+1)}(y)$ for $s = -1$. Then we infer that

$$\left| \int_{-1}^{1} \widehat{\omega}_{in}(\alpha, y)\chi_\delta(y - y_l)\eta(y)e^{-i\alpha tu(y)} dy \right| \leq C(\alpha t)^{-\frac{1}{m_l+1}} \|\widehat{\omega}_{in}(\alpha, \cdot)\|_{H^1}\|\eta\|_{H^1}.$$

This proves the proposition.   □

When $u(y) \in$ (S), Proposition 2.1 shows that

$$\|\omega(t)\|_{L_x^2 H_y^{-1}} \leq C(1 + |t|)^{-\frac{1}{2}}.$$

However, for the linearized Euler equations, Theorem 1.4 implies that

$$\|\omega(t)\|_{L_x^2 H_y^{-1}} \leq C(1 + |t|)^{-1}.$$

This unexpected decay rate is due to a combined effect of the vorticity mixing and vorticity depletion. In this case, the nonlocal part $u''(y)\partial_x(-\Delta)^{-1}$ of $\mathcal{L}$ plays an important role near the critical point.

3 Linear inviscid damping via the Dunford integral

In this section, we present a sketch of the proof for the linear inviscid damping via the Dunford integral.

3.1 Solution representation formula

In terms of the stream function ψ, the linearized Euler equations (1.3) take

$$\partial_t\Delta\psi + u(y)\partial_x\Delta\psi - u''(y)\partial_x\psi = 0.$$

Taking the Fourier transform in x, we get

$$(\partial_y^2 - \alpha^2)\partial_t\widehat{\psi} = i\alpha\big(u''(y) - u(y)(\partial_y^2 - \alpha^2)\big)\widehat{\psi}.$$

Inverting the operator $(\partial_y^2 - \alpha^2)$, we find

$$-\frac{1}{i\alpha}\partial_t\widehat{\psi} = \mathcal{R}_\alpha\widehat{\psi}, \tag{3.1}$$

where

$$\mathcal{R}_\alpha\widehat{\psi} = -(\partial_y^2 - \alpha^2)^{-1}\big(u''(y) - u(y)(\partial_y^2 - \alpha^2)\big)\widehat{\psi}. \tag{3.2}$$

Let us recall some facts about the spectrum $\sigma(\mathcal{R}_\alpha)$ of the operator $\mathcal{R}_\alpha$(see [19, 21, 22] for more details).

1. The spectrum $\sigma(\mathcal{R}_\alpha)$ is compact.
2. The continuous spectrum $\sigma_c(\mathcal{R}_\alpha)$ is contained in the range Ran u of $u(y)$.
3. The eigenvalues of $\mathcal{R}_\alpha$ can not cluster except possibly along on Ran u.
4. If $u(y)$ has no infection points, then $\mathcal{R}_\alpha$ has no embedding eigenvalues.
5. If $u(y)$ has inflection points, then $\mathcal{R}_\alpha$ has no embedding eigenvalues for $\alpha^2 > \alpha_{max}^2$, where

$$\alpha_{max}^2 \overset{\text{def}}{=} -\inf_{y_c:u''(y_c)=0}\inf_{\phi\in H_0^1(0,1)}\frac{\int_0^1 |\phi'(y)|^2 - \frac{u''(y)}{u(y)-u(y_c)}|\phi(y)|^2 dy}{\int_0^1 |\phi(y)|^2 dy}.$$

Let Ω_ϵ be a simply connected domain including the spectrum $\sigma(\mathcal{R}_\alpha)$ of $\mathcal{R}_\alpha$. Then the solution $\widehat{\psi}(t,\alpha,y)$ is given by the following Dunford integral:

$$\widehat{\psi}(t,\alpha,y) = \frac{1}{2\pi i}\int_{\partial\Omega_\epsilon} e^{-i\alpha tc}(c - \mathcal{R}_\alpha)^{-1}\widehat{\psi}(0,\alpha,y)dc. \tag{3.3}$$

Thus, the large time behaviour of $\widehat{\psi}(t,\alpha,y)$ is related to the properties of the resolvent $(c - \mathcal{R}_\alpha)^{-1}$.

3.2 Rayleigh equation

Let $\Phi(\alpha,y,c)$ be the solution of the inhomogeneous Rayleigh equation with $f(\alpha,y,c) = \frac{\widehat{\omega}_0(\alpha,y)}{i\alpha(u-c)}$ and $c \in \Omega_\epsilon$:

$$\begin{cases} \Phi'' - \alpha^2\Phi - \dfrac{u''}{u - c}\Phi = f, \\ \Phi(-1) = \Phi(1) = 0. \end{cases} \tag{3.4}$$

Then we find that

$$(c - \mathcal{R}_\alpha)^{-1}\widehat{\psi}(0,\alpha,y) = i\alpha\Phi(\alpha,y,c).$$

This shows that

$$\widehat{\psi}(t,\alpha,y) = \frac{1}{2\pi}\int_{\partial\Omega_\epsilon} \alpha\Phi(\alpha,y,c)e^{-i\alpha ct}dc. \tag{3.5}$$

To solve the inhomogeneous Rayleigh equation (3.4), we first construct two independent solutions of the homogeneous Rayleigh equation

$$\phi'' - \alpha^2 \phi - \frac{u''}{u-c}\phi = 0. \tag{3.6}$$

Here $c \in \Omega_\epsilon$. In fact, it suffices to consider $c = c_r + c_\delta$, where c_r is chosen so that $|c_\delta| = dis(c, \mathrm{Ran}\, u)$.

First of all, $u(y) - c$ and $(u(y) - c) \int^y \frac{1}{(u(y')-c)^2} dy'$ are two solutions to (3.6) for $\alpha = 0$. For y away form $u(y) = c_r$, the solution behaves as one of $\phi'' - \alpha^2\phi = 0$. Thus, it is natural to expect that the solution behaves as $u(y) - c$ and $(u(y) - c) \int^y \frac{1}{(u(y')-c)^2} dy'$ near $u(y) = c_r$. This leads us to choose $\phi = (u(y) - c)\phi_1$ as the form of the solution. Then ϕ_1 satisfies

$$((u(y) - c)^2 \phi_1')' = \alpha^2 \phi_1 (u(y) - c)^2.$$

If $\phi_1(y_c, c) = 1$ and $\phi_1'(y_c, c) = 0$ at y_c, then we have

$$\phi_1(y, c) = 1 + \int_{y_c}^y \frac{\alpha^2}{(u(y')-c)^2} \int_{y_c}^{y'} \phi_1(z, c)(u(z) - c)^2 dz dy'$$

$$= 1 + \alpha^2 T\phi_1(y, c). \tag{3.7}$$

Assume that u is monotone in the integral range of T and let $y_c = u^{-1}(c_r)$ so that for $y_c \leq z \leq y' \leq y$ or $y \leq y' \leq z \leq y_c$,

$$\left| \frac{(u(z) - c)^2}{(u(y') - c)^2} \right| \leq 1,$$

which implies the following lemma.

Lemma 3.1. *There exists a constant C independent of A so that*

$$\left\| \frac{Tf(y,c)}{\cosh A(y - y_c)} \right\|_{L_{y,c}^\infty} \leq \frac{C}{A^2} \left\| \frac{f(y,c)}{\cosh A(y - y_c)} \right\|_{L_{y,c}^\infty}.$$

By taking A large enough, we have $\phi_1(y, c) = \sum_{k=0}^\infty (\alpha^2 T(1))^k$, and thus the existence of ϕ_1. More precisely, assume that u is monotone in $[0, 1]$, and let Ω_{ϵ_0} be an ϵ_0 neighborhood of the imagine of $u(y)$ for $y \in [0, 1]$.

Proposition 3.2. *There exists $\phi_1(y, c) \in C([0, 1] \times \Omega_{\epsilon_0})$ such that $\phi(y, c) = (u(y) - c)\phi_1(y, c)$ is a solution of the Rayleigh equation:*

$$\begin{cases} \phi'' - \alpha^2\phi - \frac{u''}{u-c}\phi = 0, \\ \phi(y_c, c) = u(y_c) - c, \quad \phi'(y, c)|_{y=y_c} = u'(y_c). \end{cases} \tag{3.8}$$

Moreover, $\partial_y \phi_1(y, c) \in C([0, 1] \times \Omega_{\epsilon_0})$, and there exists $\epsilon_1 > 0, C > 0$ such that for any $\epsilon_0 \in [0, \epsilon_1)$ and $(y, c) \in [0, 1] \times \Omega_{\epsilon_0}$,

$$|\phi_1(y, c)| \geq \frac{1}{2}, \quad |\phi_1(y, c) - 1| \leq C|y - y_c|^2,$$

where the constants ϵ_1, C may depend on α.

In order to obtain the decay estimates, we need to establish some uniform estimates of $\phi_1(y, c)$ for $c \in \operatorname{Ran} u$. See [22] for more details.

The solution of inhomogeneous Rayleigh equation (3.4) must take the form:

$$\Phi(y, c) = \phi(y, c) \int^y \frac{\int^{y'} f_0(y'')\phi(y'', c)dy''}{\phi(y', c)^2} dy' \quad \text{(inhomogeneous part)}$$

$$+ \beta(c) \int^y \frac{\phi(y, c)}{\phi(y', c)^2} dy' + \gamma(c)\phi(y, c). \quad \text{(special solution)}$$

The main task is to determine the coefficients $\beta(c)$ and $\gamma(c)$ so that $\Phi(y, c)$ satisfies the boundary conditions (or the periodic condition) and is C^2 continuous at each critical point $u'(y) = 0$.

Precisely, we consider the channel case (the periodic case is similar) and let $-1 < y_1 < y_2 < \cdots < y_m < 1$ be m zero points of $u'(y)$. Let $\{\phi^{(k)}\}_{k=0}^m$ be $m+1$ solutions of homogeneous Rayleigh equation in $\{[y_k, y_{k+1}]\}_{k=0}^m$ with $y_0 = -1$, $y_{m+1} = 1$. Define $y_c^{(k)} \in [y_k, y_{k+1}]$ in the following way: if $c \in u([y_k, y_{k+1}])$, let $u(y_c^{(k)}) = c_r$; otherwise, $y_c^{(k)} = y_k$. Then let $\Phi^{(k)}$ be in the form for $y \in [y_k, y_{k+1}]$:

$$\Phi^{(k)}(y, c) = \phi^{(k)}(y, c) \int_{y_k}^y \frac{\int_{y_c^{(k)}}^{y'} f_0(y'')\phi^{(k)}(y'', c)dy''}{\phi^{(k)}(y', c)^2} dy'$$

$$+ \beta_k(c) \int_{y_k}^y \frac{\phi^{(k)}(y, c)}{\phi^{(k)}(y', c)^2} dy' + \gamma_k(c)\phi^{(k)}(y, c).$$

Then $\Phi^{(k)}(y, c)$ is required to satisfy

$$\begin{cases} \Phi^{(0)}(-1, c) = 0, \\ \Phi^{(k)}(y_{k+1}, c) = \Phi^{(k+1)}(y_{k+1}, c), \\ \partial_y \Phi^{(k)}(y_{k+1}, c) = \partial_y \Phi^{(k+1)}(y_{k+1}, c), \\ \Phi^{(m)}(1, c) = 0. \end{cases} \tag{3.9}$$

This leads to solving the following linear system:

$$M \begin{bmatrix} \beta_0 \\ \gamma_0 \\ \vdots \\ \beta_k \\ \gamma_k \\ \vdots \\ \beta_m \\ \gamma_m \end{bmatrix} = F(\phi^{(k)}, f_0),$$

where M is $(2m+2) \times (2m+2)$ matrix determined by the coefficients of $\{\beta_k, \gamma_k\}_{k=0}^m$. Moreover, the determinant $\det(M) \neq 0$ for $c \in \Omega$ is equivalent to that there is no eigenvalue of $\mathcal{R}_\alpha$ in Ω.

Our spectral assumption on $\mathcal{L}$ ensures that

$$\left.\begin{array}{l} \text{No embedding eigenvalue} \Rightarrow \lim_{c \to c_r} \det(M)(c) \neq 0 \\ \det(M)(c) \text{ is continuous for } c \notin \text{Ran } u \end{array}\right\} \Rightarrow \det(M) \neq 0 \text{ near Ran } u.$$

3.3 Linear inviscid damping for monotone shear flows

Here we present a sketch of the proof in [22] for monotone shear flows with the spectral assumption in Theorem 1.1. When $\mathcal{L}$(thus, $\mathcal{R}_\alpha$) has no embedding eigenvalues, we have

$$\Phi(y,c) = \phi(y,c) \int_0^y \frac{1}{\phi(z,c)^2} \int_{y_c}^z \phi(y',c)f(y',c)dy'dz + \mu(c)\phi(y,c) \int_0^y \frac{1}{\phi(y',c)^2}dy',$$

where $y_c = u^{-1}(c_r)$ and

$$\mu(c) = -\frac{\int_0^1 \frac{1}{\phi(z,c)^2} \int_{y_c}^z \phi f(y',c)dy'dz}{\int_0^1 \frac{1}{\phi(y,c)^2}dy}.$$

Moreover, it holds that as $\varepsilon \to 0$,

$$\Phi(\alpha, y, c \pm i\varepsilon) \to \Phi_\pm(\alpha, y, c) \quad \text{for } c \in \text{Ran } u.$$

Thus, we obtain

$$\widehat{\psi}(t, \alpha, y) = \frac{1}{2\pi} \int_{\partial \Omega_\epsilon} \alpha \Phi(\alpha, y, c)e^{-i\alpha ct}dc$$

$$= \lim_{\epsilon \to 0^+} \frac{1}{2\pi} \int_{\partial \Omega_\epsilon} \alpha \Phi(\alpha, y, c)e^{-i\alpha ct}dc$$

$$= \frac{1}{2\pi} \int_{u(0)}^{u(1)} \alpha \widetilde{\Phi}(y, c)e^{-i\alpha ct}dc, \qquad (3.10)$$

where

$$\widetilde{\Phi}(y,c) = \begin{cases} (\mu_-(c) - \mu_+(c))\phi(y,c) \displaystyle\int_0^y \frac{1}{\phi(z,c)^2}dz, & 0 \leq y < y_c, \\ (\mu_-(c) - \mu_+(c))\phi(y,c) \displaystyle\int_1^y \frac{1}{\phi(z,c)^2}dz, & y_c < y \leq 1, \end{cases}$$

where we denote

$$\mu_-(c) - \mu_+(c) = \frac{2}{\alpha}\rho(c)\overline{\mu}(c), \quad \rho(c) = (c - u(0))(u(1) - c). \qquad (3.11)$$

For $\mu(c)$, we have the following regularity estimates.

Proposition 3.3. *There exists a constant C independent of α such that*

1. L^2 *estimates*

$$\|\rho\overline{\mu}\|_{L^2} \leq \frac{C}{|\alpha|}\|\widehat{\omega}_0(\alpha,\cdot)\|_{L^2},$$

$$\|\overline{\mu}\|_{L^2} \leq C\|\widehat{\omega}_0(\alpha,\cdot)\|_{L^2}.$$

2. $W^{1,2}$ *estimates*

$$\|\partial_c(\rho\overline{\mu})\|_{L^2} \leq C\|\widehat{\omega}_0(\alpha,\cdot)\|_{H^1},$$

$$\|\partial_c\overline{\mu}\|_{L^2} \leq C(1+|\alpha|)\|\widehat{\omega}_0(\alpha,\cdot)\|_{H^1}.$$

3. $W^{2,2}$ *estimates*

$$\|\partial_c^2(\rho\overline{\mu})\|_{L^2} \leq C(1+|\alpha|)\|\widehat{\omega}_0(\alpha,\cdot)\|_{H^2},$$

$$\|\rho\partial_c^2(\rho\overline{\mu})\|_{L^2} \leq C\|\widehat{\omega}_0(\alpha,\cdot)\|_{H^2}.$$

Recall that $\widehat{\omega}(t,\alpha,y)$ satisfies

$$\begin{cases} \partial_t\widehat{\omega} + i\alpha u\widehat{\omega} + i\alpha u''\widehat{\psi} = 0, \\ \widehat{\omega}|_{t=0} = \mathcal{F}\omega_0(\alpha,y). \end{cases}$$

This is equivalent to

$$\left(e^{i\alpha t u(y)}\widehat{\omega}(t,\alpha,y)\right)_t = -i\alpha e^{i\alpha t u(y)}u''(y)\widehat{\psi}(t,\alpha,y).$$

Integration in t gives

$$e^{i\alpha t u(y)}\widehat{\omega}(t,\alpha,y) = \widehat{\omega}_0(\alpha,y) - i\alpha u''(y)\int_0^t e^{i\alpha\tau u(y)}\widehat{\psi}(\tau,\alpha,y)d\tau.$$

We denote $W(t,x,y) \triangleq \omega(t,x+u(y)t,y)$. We find that

$$\widehat{W}(t,\alpha,y) = e^{i\alpha t u(y)}\widehat{\omega}(t,\alpha,y).$$

Then we get by (3.10) that

$$\widehat{W}(t,\alpha,y) = \widehat{\omega}_0(\alpha,y) - \frac{u''(y)}{\pi}\int_{u(0)}^{u(1)}\frac{(e^{it(u(y)-c)\alpha}-1)\rho(c)\overline{\mu}(c)\Gamma(y,c)}{u(y)-c}dc,$$

where

$$\Gamma(y,c) = \begin{cases} \phi(y,c)\displaystyle\int_0^y \frac{1}{\phi(z,c)^2}dz, & 0 \leq y < y_c, \\ \phi(y,c)\displaystyle\int_1^y \frac{1}{\phi(z,c)^2}dz, & y_c < y \leq 1. \end{cases}$$

We have the following uniform estimates in Sobolev space for $W(t,x,y)$.

Proposition 3.4. *It holds that*

$$\|W(t)\|_{H_x^{-1}L_y^2} \le C\|\omega_0\|_{H_x^{-1}L_y^2},$$

$$\|W(t)\|_{H_x^{-1}H_y^1} \le C\|\omega_0\|_{H_x^{-1}H_y^1},$$

$$\|\rho(u(y))\partial_y^2 W(t)\|_{H_x^{-1}L_y^2} \le C\|\omega_0\|_{H_x^{-1}H_y^2}.$$

With the uniform Sobolev estimates of the vorticity, the decay estimates of the velocity can be deduced by a dual method. By the elliptic estimate, we get

$$\|V\|_{L^2} \le C\|\omega\|_{H^{-1}}.$$

We get by duality that

$$\|V(t)\|_{L^2} \le \sup_{\varphi \in C_0^\infty, \, \|\varphi\|_{H^1} \le 1} \left| \int_0^1 \int_{-\pi}^{\pi} \varphi \omega(t) dx dy \right|$$

$$\le C \sup_{\varphi \in C_0^\infty, \, \|\varphi\|_{H^1} \le 1} \left| \sum_{|\alpha| \ne 0} \int_0^1 \overline{\widehat{\varphi}(\alpha, y)} \widehat{W}(t, \alpha, y) e^{-i\alpha u(y)t} dy \right|. \qquad (3.12)$$

Integration by parts gives

$$\int_0^1 \overline{\widehat{\varphi}(\alpha, y)} \widehat{W}(t, \alpha, y) e^{-i\alpha u(y)t} dy = \frac{e^{-i\alpha u(y)t}}{-i\alpha t u'(y)} \overline{\widehat{\varphi}(\alpha, y)} \widehat{W}(t, \alpha, y) \Big|_{y=0}^1$$

$$- \int_0^1 \frac{e^{-i\alpha u(y)t}}{-i\alpha t} \partial_y \left(\frac{\overline{\widehat{\varphi}(\alpha, y)} \widehat{W}(t, \alpha, y)}{u'(y)} \right) dy$$

$$= - \int_0^1 \frac{e^{-i\alpha u(y)t}}{-i\alpha t} \partial_y \left(\frac{\overline{\widehat{\varphi}(\alpha, y)} \widehat{W}(t, \alpha, y)}{u'(y)} \right) dy,$$

from which and Proposition 3.4, we infer that

$$\|V(t)\|_{L^2} \le C|t|^{-1} \sup_{\varphi \in C_0^\infty, \, \|\varphi\|_{H^1} \le 1} \|W(t)\|_{H_x^{-1}H_y^1} \|\varphi\|_{L_x^2 H_y^1}$$

$$\le C|t|^{-1} \|\omega_0\|_{H_x^{-1}H_y^1}.$$

Also we get from (3.12) that

$$\|V(t)\|_{L^2} \le C \sup_{\varphi \in C_0^\infty, \, \|\varphi\|_{H^1} \le 1} \|W(t)\|_{H_x^{-1}L_y^2} \|\varphi\|_{H_x^1 L_y^2}$$

$$\le C \|\omega_0\|_{H_x^{-1}L_y^2}.$$

Recall that $-\Delta V^2 = \partial_x \omega$ and $V^2(x, 0) = V^2(x, 1) = 0$. We define

$$-\Delta \vartheta = V^2, \quad \vartheta(t, x, 0) = \vartheta(t, x, 1) = 0.$$

We get by the elliptic estimate that

$$\|\vartheta\|_{H^2} \le C\|V^2\|_{L^2}.$$

Integration by parts gives

$$\|V^2\|_{L^2}^2 = \iint \partial_x \omega \vartheta \, dx \, dy = \sum_{|\alpha| \neq 0} \int_0^1 i\alpha e^{-i\alpha u(y)t} \widehat{W}(t, \alpha, y) \overline{\widehat{\vartheta}}(t, \alpha, y) dy$$

$$= \sum_{|\alpha| \neq 0} \int_0^1 \frac{e^{-i\alpha u(y)t}}{t} \partial_y \left(\frac{\widehat{W}(t, \alpha, y) \overline{\widehat{\vartheta}}(t, \alpha, y)}{u'(y)} \right) dy$$

$$= \sum_{|\alpha| \neq 0} \frac{e^{-i\alpha u(y)t}}{-i\alpha t^2 u'} \partial_y \left(\frac{\widehat{W}(t, \alpha, y) \overline{\widehat{\vartheta}}(t, \alpha, y)}{u'(y)} \right) \Big|_{y=0}^1$$

$$+ \sum_{|\alpha| \neq 0} \int_0^1 \frac{e^{-i\alpha u(y)t}}{i\alpha t^2} \partial_y \left(\frac{1}{u'} \partial_y \left(\frac{\widehat{W}(t, \alpha, y) \overline{\widehat{\vartheta}}(t, \alpha, y)}{u'(y)} \right) \right) dy.$$

By Sobolev embedding and Proposition 3.4, the first term on the right hand side is bounded by

$$Ct^{-2} \sum_{|\alpha| \neq 0} \frac{1}{|\alpha|} \|\widehat{W}(t, \alpha, \cdot) \vartheta(t, \alpha, \cdot)/(u'\rho)\|_{L_y^\infty}$$

$$\leq Ct^{-2} \sum_{|\alpha| \neq 0} \frac{1}{|\alpha|} \|\widehat{W}(t, \alpha, \cdot)\|_{L_y^\infty} \|\vartheta(t, \alpha, \cdot)/\rho\|_{L_y^\infty}$$

$$\leq Ct^{-2} \sum_{|\alpha| \neq 0} \frac{1}{|\alpha|} \|\widehat{W}(t, \alpha, \cdot)\|_{H_y^1} \|\vartheta(t, \alpha, \cdot)\|_{H_y^2}$$

$$\leq Ct^{-2} \|\omega_0\|_{H_x^{-1} H_y^2} \|V^2\|_{L^2}.$$

Using Proposition 3.4 again and Hardy's inequality, we get

$$\left\| \partial_y \left(\frac{1}{u'} \partial_y \left(\frac{\widehat{W}(t, \alpha, y) \overline{\widehat{\vartheta}}(t, \alpha, y)}{u'(y)} \right) \right) \right\|_{L_y^1} \leq C \|\rho(u(y)) \partial_y^2 \widehat{W}(t, \alpha, y)\|_{L_y^2} \left\| \frac{\vartheta(t, \alpha, y)}{\rho(u(y))} \right\|_{L_y^2}$$

$$+ C \|\partial_y \widehat{W}(t, \alpha, y)\|_{L_y^2} \|\vartheta(t, \alpha, y)\|_{H_y^1}$$

$$+ C \|\widehat{W}(t, \alpha, y)\|_{L_y^2} \|\vartheta(t, \alpha, y)\|_{H_y^2}$$

$$\leq C \|\omega_0(\alpha, \cdot)\|_{H_y^2} \|\vartheta(t, \alpha, y)\|_{H_y^2}.$$

This implies that the second term is bounded by

$$Ct^{-2} \|\omega_0\|_{H_x^{-1} H_y^2} \|V^2\|_{L^2}.$$

Thus, we show that

$$\|V^2\|_{L^2} \leq Ct^{-2} \|\omega_0\|_{H_x^{-1} H_y^2}.$$

3.4   Linear inviscid damping for symmetric flows

For symmetric flows, the solution of (3.4) can be decomposed into the odd part and even part due to the symmetry of $u(y)$ (so that the matrix M is almost diagonal):

$$\begin{cases} \Phi_o'' - \alpha^2 \Phi_o - \dfrac{u''}{u-c} \Phi_o = f_o, \\ \Phi_o(0) = \Phi_o(1) = 0, \end{cases}$$

and

$$\begin{cases} \Phi_e'' - \alpha^2 \Phi_e - \dfrac{u''}{u-c} \Phi_e = f_e, \\ \Phi_e(0) = \Phi_e(1) = 0, \end{cases}$$

where f_o and f_e are the odd part and even part of f respectively. These two equations can be dealt as in the monotone case. Then we can give the precise formula of $\Phi_{\pm}(\alpha, y, c)$ and obtain the representation formula of the stream function

$$\begin{aligned} \widehat{\psi}(t, \alpha, y) =& \frac{1}{2\pi} \int_{u(0)}^{u(1)} \alpha \widetilde{\Phi}_o(y, c) e^{-i\alpha ct} dc + \frac{1}{2\pi} \int_{u(0)}^{u(1)} \alpha \widetilde{\Phi}_e(y, c) e^{-i\alpha ct} dc \\ =& \widehat{\psi}_o(t, \alpha, y) + \widehat{\psi}_e(t, \alpha, y). \end{aligned}$$

In the monotone case, we first derive the formula of the vorticity from the stream function. Then we prove that the vorticity is bounded in Sobolev space. Finally, the decay estimate of the velocity is proved by using a dual argument. For symmetric flows and the Kolmogorov flow, we find that it is difficult to follow the same procedure. New idea is that we directly derive the decay estimates of the velocity based on the formulation of stream function and the dual argument. More precisely, we will derive the following two important formulas: for $f = g'' - \alpha^2 g$ with $g \in H^2(0, 1) \cap H_0^1(0, 1)$

$$\int_0^1 \widehat{\psi}_o(t, \alpha, y) f(y) dy = - \int_{u(0)}^{u(1)} K_o(c, \alpha) e^{-i\alpha ct} dc,$$

and for $f = g'' - \alpha^2 g$ with $g \in H^2(0, 1)$ and $g'(0) = g(1) = 0$

$$\int_0^1 \widehat{\psi}_e(t, \alpha, y) f(y) dy = - \int_{u(0)}^{u(1)} K_e(c, \alpha) e^{-i\alpha ct} dc.$$

Then the decay estimates are based on the following $W^{2,1}$ estimates of the kernel.

Proposition 3.5. *Assume that* $f = g'' - \alpha^2 g$ *with* $g \in H^2(0,1) \cap H_0^1(0,1)$ *and* $\widehat{\omega}_o(\alpha, y) = \frac{1}{2}(\widehat{\omega}_0(\alpha, y) - \widehat{\omega}_0(\alpha, -y)) \in H^2(0,1)$. *Then it holds that*

$$K_o(u(0), \alpha) = K_o(u(1), \alpha) = 0,$$

and there exists a constant C independent of α so that

$$\|K_o(\cdot, \alpha)\|_{L_c^1} \le C\|\widehat{\omega}_o(\alpha, \cdot)\|_{L_y^2}\|g\|_{L^2},$$

$$\|(\partial_c K_o)(\cdot, \alpha)\|_{L_c^1} \le C\|\widehat{\omega}_o(\alpha, \cdot)\|_{H_y^1}\|g\|_{H^1},$$

$$\|(\partial_c^2 K_o)(\cdot, \alpha)\|_{L_c^1} \le C|\alpha|^{\frac{1}{2}}\|\widehat{\omega}_o(\alpha, \cdot)\|_{H_y^2}\|f\|_{L^2}.$$

Proposition 3.6. *Assume that $f = g'' - \alpha^2 g$ with $g \in H^2(0,1)$ and $g'(0) = g(1) = 0$, and $\widehat{\omega}_e(\alpha, y) = \frac{1}{2}(\widehat{\omega}_0(\alpha, y) + \widehat{\omega}_0(\alpha, -y)) \in H^2(0,1)$. Then we have*

$$K_e(u(0), \alpha) = K_e(u(1), \alpha) = 0,$$

and there exists a constant C independent of α such that

$$\|K_e(\cdot, \alpha)\|_{L_c^1} \le C\|\widehat{\omega}_e(\alpha, \cdot)\|_{L_y^2}\|g\|_{L^2},$$

$$\|(\partial_c K_e)(\cdot, \alpha)\|_{L_c^1} \le C|\alpha|^{\frac{1}{2}}\|\widehat{\omega}_e(\alpha, \cdot)\|_{H_y^1}(\|g'\|_{L^2} + |\alpha|\|g\|_{L^2}),$$

$$\|(\partial_c^2 K_e)(\cdot, \alpha)\|_{L_c^1} \le C|\alpha|^{\frac{3}{2}}\|\widehat{\omega}_e(\alpha, \cdot)\|_{H_y^2}\|f\|_{L^2}.$$

Now it is easy to derive the decay estimate of the velocity. Using Proposition 3.5 and Proposition 3.6, we get by integration by parts that

$$\|\widehat{\psi}_o(t, \alpha, \cdot)\|_{L_y^2} = 2 \sup_{\|f\|_{L^2}=1} \left| \int_0^1 \widehat{\psi}_o(t, \alpha, y) f(y) dy \right|$$

$$= 2 \sup_{\|f\|_{L^2}=1} \left| \int_{u(0)}^{u(1)} K_o(c, \alpha) e^{-i\alpha ct} dc \right|$$

$$\le C \frac{1}{|\alpha|^{\frac{3}{2}} t^2} \|\widehat{\omega}_o(\alpha, \cdot)\|_{H_y^2},$$

and

$$\|\widehat{\psi}_e(t, \alpha, \cdot)\|_{L_y^2} = 2 \sup_{\|f\|_{L^2}=1} \left| \int_0^1 \widehat{\psi}_e(t, \alpha, y) f(y) dy \right|$$

$$= 2 \sup_{\|f\|_{L^2}=1} \left| \int_{u(0)}^{u(1)} K_e(c, \alpha) e^{-i\alpha ct} dc \right|$$

$$\le C \frac{1}{|\alpha|^{\frac{1}{2}} t^2} \|\widehat{\omega}_o(\alpha, \cdot)\|_{H_y^2}.$$

Similarly, we have

$$\alpha^2 \|\widehat{\psi}_o(t, \alpha, \cdot)\|_{L_y^2}^2 + \|\partial_y \widehat{\psi}_o(t, \alpha, \cdot)\|_{L_y^2}^2$$

$$= -2 \int_0^1 \widehat{\psi}_o(t, \alpha, y) (\overline{\widehat{\psi}_o}'' - \alpha^2 \overline{\widehat{\psi}_o})(t, \alpha, y) dy \le \begin{cases} C\|\widehat{\omega}_o(\alpha, \cdot)\|_{L_y^2}\|\widehat{\psi}_o(t, \alpha, \cdot)\|_{L_y^2}, \\ C\dfrac{1}{|\alpha t|}\|\widehat{\omega}_o(\alpha, \cdot)\|_{H_y^1}\|\widehat{\psi}_o(t, \alpha, \cdot)\|_{H_y^1}, \end{cases}$$

and

$$\alpha^2 \|\widehat{\psi_e}(t,\alpha,\cdot)\|_{L^2}^2 + \|\partial_y \widehat{\psi_e}(t,\alpha,\cdot)\|_{L^2}^2$$

$$= -2 \int_0^\pi \widehat{\psi_e}(t,\alpha,y) \overline{(\widehat{\psi_e}'' - \alpha^2 \widehat{\psi_e})}(t,\alpha,y) dy$$

$$\leq \begin{cases} C \|\widehat{\omega_e}(\alpha,\cdot)\|_{L_y^2} \|\widehat{\psi_e}(t,\alpha,\cdot)\|_{L_y^2}, \\ C \dfrac{1}{|\alpha|^{\frac{1}{2}}|t|} \|\widehat{\omega_e}(\alpha,\cdot)\|_{H_y^1} (|\alpha| \|\widehat{\psi_e}(t,\alpha,\cdot)\|_{L_y^2} + \|\partial_y \widehat{\psi_e}(t,\alpha,\cdot)\|_{L_y^2}). \end{cases}$$

Let

$$V_o = \nabla^\perp \psi_o, \quad V_e = \nabla^\perp \psi_e.$$

Thus, we deduce that for $t \geq 1$,

$$\|\widehat{V_o}(t,\alpha,\cdot)\|_{L_y^2} \leq |\alpha| \|\widehat{\psi_o}(t,\alpha,\cdot)\|_{L_y^2} + \|\partial_y \widehat{\psi_o}(t,\alpha,\cdot)\|_{L_y^2}$$

$$\leq C \frac{1}{|\alpha t|} \|\widehat{\omega_o}(\alpha,\cdot)\|_{H_y^1},$$

$$\|\widehat{V_e}(t,\alpha,\cdot)\|_{L_y^2} \leq |\alpha| \|\widehat{\psi_e}(t,\alpha,\cdot)\|_{L_y^2} + \|\partial_y \widehat{\psi_e}(t,\alpha,\cdot)\|_{L_y^2}$$

$$\leq C \frac{1}{|\alpha|^{\frac{1}{2}}|t|} \|\widehat{\omega_e}(\alpha,\cdot)\|_{H_y^1}.$$

Thanks to $V = V_o + V_e$, we get

$$\|V(t)\|_{L_{x,y}^2} \leq \frac{C}{\langle t \rangle} \|\omega_0\|_{H_x^{-\frac{1}{2}} H_y^1}.$$

Using $\|\widehat{V^2}\|_{L^2(I_0)} \leq C|\alpha| \|\widehat{\psi}\|_{L^2(I_0)} \leq C \dfrac{|\alpha|^{\frac{1}{2}}}{t^2} \|\widehat{\omega_0}\|_{H^2(I_0)}$, we get

$$\|V^2(t)\|_{L_{x,y}^2} \leq \frac{C}{\langle t \rangle^2} \|\omega_0\|_{H_x^{\frac{1}{2}} H_y^2}.$$

To obtain the decay of the vorticity at critical points, we give a formula of the vorticity at critical points via the formula of the stream function ψ. Let $W(t,x,y) = \omega(t, x + u(y)t, y)$. Then $\widehat{W}(t,\alpha,y) = e^{i\alpha t u(y)} \widehat{\omega}(t,\alpha,y)$ and

$$\partial_t \widehat{W} = -i\alpha e^{i\alpha u(y)t} u'' \widehat{\psi}.$$

Thanks to $\widehat{\omega}(t,\alpha,0) \to 0$ as $t \to \infty$, so does $\widehat{W}(t,\alpha,0)$. Thus,

$$\widehat{W}(t,\alpha,0) = \int_t^\infty i\alpha e^{i\alpha u(0)s} u''(0) \widehat{\psi}(s,\alpha,0) ds.$$

Roughly, we have

$$\|\psi\|_{L^\infty} \leq C \|\psi\|_{L^2}^{1/2} \|\psi\|_{H^1}^{1/2},$$

which implies

$$|\widehat{\psi}(s,\alpha,0)| \le Cs^{-\frac{3}{2}}, \quad |\widehat{\omega}(t,\alpha,0)| \le Ct^{-\frac{1}{2}}.$$

In fact, we can prove that $|\widehat{\psi}(s,\alpha,0)| \le Cs^{-2}$ and $|\widehat{\omega}(t,\alpha,0)| \le Ct^{-1}$.

Similar arguments could be used to prove the linear damping for the Kolmogorov flow [24].

4 Limiting absorption principle

In this section, we first prove the limiting absorption principle for the flows in $\mathcal{K}$. Then we apply the limiting absorption principle to prove the linear inviscid damping in the sense of the L^2 space-time estimate.

Consider the inhomogeneous Rayleigh equation:

$$(u-c)(\Phi'' - \alpha^2\Phi) - u''\Phi = \omega, \quad \Phi(-1) = \Phi(1) = 0, \tag{4.1}$$

where $c \in \Omega \setminus D_0$, $D_0 = \mathrm{Ran}\, u$. For linear inviscid damping, one of key ingredients is to establish the limiting absorption principle: as $\varepsilon \to 0$,

$$\Phi(\alpha, y, c \pm i\varepsilon) \to \Phi_\pm(\alpha, y, c) \quad \text{for } c \in \mathrm{Ran}\, u.$$

For monotone shear flows, $\Phi(y,c) \sim a(y,c) + b(y,c)|y - y_c|\ln|y - y_c|$. Thus, $\Phi(y,c)$ is uniformly bounded. In this case, the limiting absorption principle can be proved directly. If u has critical points, then $\frac{1}{u(y)-c}$ is more singular so that the problem becomes highly nontrivial.

Using blow-up analysis and compactness argument, we prove the limiting absorption principle for shear flows $u \in \mathcal{K}$.

Proposition 4.1. *If $\mathcal{R}_\alpha$ has no embedding eigenvalues, then there exists ϵ_0 such that for $c \in \Omega_{\epsilon_0} \setminus D_0$, Φ has the following uniform bound*

$$\|\Phi\|_{H^1(-1,1)} \le C\|\omega\|_{H^1(-1,1)}.$$

Here C is a constant independent of ϵ_0. Moreover, there exists $\Phi_\pm(\alpha, y, c) \in H_0^1(-1,1)$ for $c \in \mathrm{Ran}\, u$, such that $\Phi(\alpha, \cdot, c \pm i\epsilon) \to \Phi_\pm(\alpha, \cdot, c)$ in $C([-1,1])$ as $\epsilon \to 0+$ and

$$\|\Phi_\pm(\alpha, \cdot, c)\|_{H^1(-1,1)} \le C\|\omega\|_{H^1(-1,1)}.$$

Let us recall the definition of embedding eigenvalue.

Definition 4.2. *We say that $c \in \mathrm{Ran}\, u$ is an embedding eigenvalue of $\mathcal{R}_\alpha$ if there exists $0 \ne \psi \in H_0^1(-1,1)$ such that for all $\varphi \in H_0^1(-1,1)$,*

$$\int_{-1}^1 (\psi'\varphi' + \alpha^2\psi\varphi)dy + p.v. \int_{-1}^1 \frac{u''\psi\varphi}{u-c}dy + i\pi \sum_{y \in u^{-1}\{c\}, u'(y) \ne 0} \frac{(u''\psi\varphi)(y)}{|u'(y)|} = 0.$$

$$\tag{4.2}$$

It is easy to see from (4.2) that

$$-\psi'' + \alpha^2\psi + \frac{u''\psi}{u-c} = 0 \quad \text{in} \quad [-1,1] \setminus u^{-1}\{c\}. \tag{4.3}$$

We use the contradiction argument. Assume that there exists $\psi_n \in H_0^1(-1,1)$, $\omega_n \in H^1(-1,1)$, and c_n with $\text{Im}\, c_n > 0$, such that $\|\psi_n\|_{H^1(-1,1)} = 1$, $\|\omega_n\|_{H^1(-1,1)} \to 0$, $c_n \to c \in \text{Ran}\, u$, and

$$(u - c_n)(\psi_n'' - \alpha^2\psi_n) - u''\psi_n = \omega_n.$$

The goal is to show that $\psi_n \to 0$, which contradicts with the fact $\|\psi_n\|_{H^1(-1,1)} = 1$.

Since ψ_n in bounded in H^1, then there exists a subsequence of $\{\psi_n\}$ (still denoted by $\{\psi_n\}$) and $\psi \in H_0^1(-1,1)$ such that $\psi_n \rightharpoonup \psi$ weakly in $H^1(-1,1)$. Since ψ_n'' is bounded in $L^\infty_{loc}((-1,1) \setminus u^{-1}\{c\})$, we have $\psi_n \to \psi$ in $H^1_{loc}((-1,1) \setminus u^{-1}\{c\})$ and ψ satisfies (4.3) in $[-1,1] \setminus u^{-1}\{c\}$ and (4.2) holds for $\varphi \in H_0^1(-1,1)$, $\text{supp}\, \varphi \subset [-1,1] \setminus u^{-1}\{c\}$.

The proof of Proposition 4.1 is based on the following two key lemmas. The first lemma deals with the case of $u'(y_0) \neq 0$ with $c = u(y_0)$.

Lemma 4.3. *Let ψ_n, $\omega_n \in H^1(a,b)$, $u_n \in H^3(a,b)$ be sequences, which satisfy*

$$\psi_n \rightharpoonup \psi, \ \omega_n \to \omega \quad \text{in} \quad H^1(a,b),$$
$$u_n \to u \quad \text{in} \quad H^3(a,b),$$
$$u_n(\psi_n'' - \alpha_n^2\psi_n) - u_n''\psi_n = \omega_n,$$

and $\alpha_n \to \alpha$, $\text{Im}\, u_n < 0$. Moreover, $\text{Im}\, u = 0$ in $[a,b]$, $u(y_0) = 0$, $y_0 \in [a,b]$, $u'(y)u'(y_0) > 0$ in $[a,b]$. Then we have $\psi_n \to \psi$ in $H^1(a,b)$, and for all $\varphi \in H_0^1(a,b)$,

$$\int_a^b (\psi'\varphi' + \alpha^2\psi\varphi)dy + p.v. \int_a^b \frac{(u_0''\psi + \omega)\varphi}{u_0}dy + i\pi\frac{((u_0''\psi + \omega)\varphi)(y_0)}{|u'(y_0)|} = 0. \tag{4.4}$$

Proof. Without loss of generality, we may assume that $u_0'(y) > 0$ for $y \in [a,b]$. Otherwise, we can consider $\bar{\psi}_n, -\bar{\omega}_n, -\bar{u}_n$.

As $\psi_n \rightharpoonup \psi$, $\omega_n \to \omega$ in $H^1(a,b)$ and $u_n \to u_0$ in $H^3(a,b)$, ψ_n, ω_n, u_n'' are uniformly bounded in $H^1(a,b)$. Let $g_n = u_n''\psi_n + \omega_n$. Then g_n is uniformly bounded in $H^1(a,b)$ and $\psi_n'' - \alpha_n^2\psi_n = g_n/u_n$. Choose N large enough so that $\text{Re}\, u_n'(y) \geq c_0 > 0$ for $n \geq N$ and $y \in [a,b]$.

It is easy to see that

$$\left(\psi_n' - \frac{g_n}{u_n'}\ln u_n\right)' = \alpha_n^2\psi_n - \left(\frac{g_n}{u_n'}\right)'\ln u_n.$$

Here $\ln(re^{i\theta}) = \ln r + i\theta$ for $r > 0, \theta \in [-\pi,\pi]$. We know that $\left(\frac{g_n}{u_n'}\right)'$ is uniformly bounded in $L^2(a,b)$, and $\frac{g_n}{u_n'}$ is uniformly bounded in $L^\infty(a,b)$. Due to $\text{Re}\, u_n'(y) \geq c_0 > 0$ for $y \in [a,b]$, $\ln u_n$ is uniformly bounded in $L^p(a,b), p > 2$. Therefore,

$\psi'_n - \frac{g_n}{u'_n}\ln u_n$ is uniformly bounded in $L^2(a,b)$ and $\dot{W}^{1,1}(a,b)$, thus bounded in $L^\infty(a,b)$. Thus, we conclude

$$\limsup_{\delta \to 0} \|\psi'_n\|_{L^2((y_0-\delta,y_0+\delta)\cap(a,b))} = 0. \qquad (4.5)$$

On the other hand, ψ''_n is bounded in $L^\infty((a,b)\setminus(y_0-\delta,y_0+\delta))$. So, $\psi_n \to \psi$ in $H^1((a,b)\setminus(y_0-\delta,y_0+\delta))$, which together with (4.5) shows that $\psi_n \to \psi$ in $H^1(a,b)$ and $g_n \to g_0 = u''_0\psi + \omega$ in $H^1(a,b)$.

Now we prove (4.4). As $\psi''_n - \alpha_n^2\psi_n = g_n/u_n$, we have

$$\int_a^b (\psi'_n\varphi' + \alpha_n^2\psi_n\varphi)dy + \int_a^b \frac{g_n\varphi}{u_n}dy = 0$$

for any $\varphi \in H_0^1(a,b)$. Obviously,

$$\lim_{n\to\infty} \int_a^b (\psi'_n\varphi' + \alpha_n^2\psi_n\varphi)dy = \int_a^b (\psi'\varphi' + \alpha^2\psi\varphi)dy.$$

It remains to show that

$$\lim_{n\to\infty} \int_a^b \frac{g_n\varphi}{u_n}dy = p.v. \int_a^b \frac{g_0\varphi}{u_0}dy + i\pi\frac{(g_0\varphi)(y_0)}{|u'_0(y_0)|}.$$

Using the facts that

$$\int_a^b \frac{g_n\varphi}{u_n}dy = -\int_a^b \left(\frac{g_n\varphi}{u'_n}\right)'\ln u_n dy, \quad \left(\frac{g_n\varphi}{u'_n}\right)' \to \left(\frac{g_0\varphi}{u'_0}\right)',$$
$$\ln u_n \to \ln|u_0| - i\pi\chi_{[a,y_0)} \quad \text{in} \quad L^2(a,b),$$

we infer that

$$\int_a^b \frac{g_n\varphi}{u_n}dy \to -\int_a^b \left(\frac{g_0\varphi}{u'_0}\right)'\ln|u_0|dy + i\pi\int_a^{y_0} \left(\frac{g_0\varphi}{u'_0}\right)'dy$$
$$= p.v. \int_a^b \frac{g_0\varphi}{u_0}dy + i\pi\frac{(g_0\varphi)(y_0)}{|u'_0(y_0)|}.$$

This proves (4.4). $\qquad\square$

The second lemma deals with the case of $u'(y_0) = 0$.

Lemma 4.4. *Let $\psi_n, \omega_n \in H^1(a,b)$ be sequences, $u \in H^3(a,b)$, which satisfy*

$$\psi_n \rightharpoonup 0, \ \omega_n \to 0 \quad in \ H^1(a,b),$$
$$(u-c_n)(\psi''_n - \alpha^2\psi_n) - u''\psi_n = \omega_n,$$

and $\operatorname{Im} c_n > 0$, $c_n \to u(y_0)$, $u'(y_0) = 0$, $y_0 = \frac{a+b}{2} \in (a,b)$, $\delta = y_0 - a \in (0,1)$, $u''(y)u''(y_0) > 0$ in $[a,b]$. Then we have $\psi_n \to 0$ in $H^1(a,b)$.

Proof. Without loss of generality, we may assume that $y_0 = 0, u''(y_0) = 2, u(y_0) = 0$. Then $[a, b] = [-\delta, \delta]$. Let $c_n = r_n^2 e^{2i\theta_n}$, $r_n > 0$, $0 < \theta_n < \pi/2$. So, $r_n \to 0$.

First of all, we consider the case of $w_n(0) = 0$. It is not difficult to prove that (see Lemma 6.3 in [23])

$$|(\psi_n'' - \alpha^2 \psi_n)(0)| \leq C r_n^{-\frac{3}{2}}, \quad |\psi_n(0)| \leq C r_n^{\frac{1}{2}}, \quad |(u''\psi_n + w_n)(0)| \leq C r_n^{\frac{1}{2}}.$$

We introduce

$$\widetilde{\psi}_n(y) = r_n^{-\frac{1}{2}} \psi_n(r_n y), \quad \widetilde{w}_n(y) = r_n^{-\frac{1}{2}} w_n(r_n y), \quad u_n(y) = r_n^{-2}(u(r_n y) - u(0)).$$

It holds that

$$(u_n - e^{2i\theta_n})(\widetilde{\psi}_n'' - (\alpha r_n)^2 \widetilde{\psi}_n) - u_n'' \widetilde{\psi}_n = \widetilde{w}_n,$$

and

$$|\widetilde{\psi}_n(0)| = |r_n^{-\frac{1}{2}} \psi_n(0)| \leq C, \quad \|\widetilde{\psi}_n'\|_{L^2(a/r_n, b/r_n)} = \|\psi_n'\|_{L^2(a,b)} \leq C,$$
$$|\widetilde{w}_n(0)| = 0, \quad \|\widetilde{w}_n'\|_{L^2(a/r_n, b/r_n)} = \|w_n'\|_{L^2(a,b)} \to 0.$$

Thus, $\widetilde{\psi}_n$ is bounded in $H^1_{loc}(\mathbb{R})$, and $\widetilde{w}_n \to 0$ in $H^1_{loc}(\mathbb{R})$. Up to a subsequence, we may assume that $\widetilde{\psi}_n \rightharpoonup \widetilde{\psi}_0$ in $H^1_{loc}(\mathbb{R})$, $\theta_n \to \theta_0$ with $\widetilde{\psi}_0' \in L^2(\mathbb{R})$, $\theta_0 \in [0, \pi/2]$.

Using the facts that

$$u_n(y) = y^2 \int_0^1 \int_0^1 t u''(r_n y t s) ds dt, \quad u_n'(y) = y \int_0^1 u''(r_n y t) dt,$$

and $u_n''(y) = u''(r_n y)$, $u_n'''(y) = r_n u'''(r_n y)$, we can deduce that $u_n \to y^2$ in $H^3_{loc}(\mathbb{R})$.

Now, if $\theta_0 \neq 0$, then $\widetilde{\psi}_n''$ is bounded in $L^\infty_{loc}(\mathbb{R})$ and $\widetilde{\psi}_n \to \widetilde{\psi}_0$ in $C^1_{loc}(\mathbb{R})$, and

$$(y^2 - e^{2i\theta_0})\widetilde{\psi}_0'' = 2\widetilde{\psi}_0.$$

If $\theta_0 = 0$, then $\widetilde{\psi}_n''$ is bounded in $L^\infty_{loc}(\mathbb{R} \setminus \{\pm 1\})$ and $\widetilde{\psi}_n \to \widetilde{\psi}_0$ in $C^1_{loc}(\mathbb{R} \setminus \{\pm 1\})$, and

$$(y^2 - 1)\widetilde{\psi}_0'' = 2\widetilde{\psi}_0 \quad \text{in } \mathbb{R} \setminus \{\pm 1\}.$$

Lemma 4.3 also ensures that $\widetilde{\psi}_n \to \widetilde{\psi}_0$ in $H^1(1-\delta, 1+\delta) \cap H^1(-1-\delta, -1+\delta)$, thus,

$$\widetilde{\psi}_n \to \widetilde{\psi}_0 \quad \text{in } H^1_{loc}(\mathbb{R}) \cap C^1_{loc}(\mathbb{R} \setminus \{\pm 1\}).$$

Thanks to $(y^2 - e^{2i\theta_0})\widetilde{\psi}_0'' = 2\widetilde{\psi}_0$, we deduce that for $y > 1$,

$$\widetilde{\psi}_0(y) = (y^2 - e^{2i\theta_0})\left(C_1 + C_2 \int_y^{+\infty} \frac{dz}{(z^2 - e^{2i\theta_0})^2}\right),$$

where C_1, C_2 are constants. As $\widetilde{\psi}'_0 \in L^2(\mathbb{R})$, we have $C_1 = 0$, thus $\widetilde{\psi}_0 \in L^2(2, +\infty)$. Similarly, $\widetilde{\psi}_0 \in L^2(-\infty, -2)$. Hence, $\widetilde{\psi}_0 \in L^2(\mathbb{R})$, $\widetilde{\psi}_0 \in H^1(\mathbb{R})$.

If $\theta_0 \neq 0$, then

$$\int_{\mathbb{R}} |\widetilde{\psi}'_0|^2 dy = -\int_{\mathbb{R}} \widetilde{\psi}''_0 \overline{\widetilde{\psi}_0} dy = -\int_{\mathbb{R}} \frac{2|\widetilde{\psi}_0|^2}{y^2 - e^{2i\theta_0}} dy.$$

Multiplying $e^{i\theta_0}$ on both sides, and then taking the imaginary part, we obtain

$$\sin\theta_0 \int_{\mathbb{R}} |\widetilde{\psi}'_0|^2 dy = -\sin\theta_0 \int_{\mathbb{R}} \frac{2(y^2 + 1)|\widetilde{\psi}_0|^2}{|y^2 - e^{2i\theta_0}|^2} dy,$$

which implies $\widetilde{\psi}_0 = 0$.

If $\theta_0 = 0$, we first claim that for any $\varphi \in H^1(\mathbb{R})$,

$$\int_{\mathbb{R}} \widetilde{\psi}'_0 \varphi' dy + \text{p.v.} \int_{\mathbb{R}} \frac{2\widetilde{\psi}_0 \varphi}{y^2 - 1} dy + i\pi \sum_{y=\pm 1} \widetilde{\psi}_0 \varphi = 0. \tag{4.6}$$

Indeed, it holds that for $y \in \mathbb{R} \setminus \{\pm 1\}$, $(y^2 - 1)\widetilde{\psi}''_0 = 2\widetilde{\psi}_0$, thus (4.6) holds for $\varphi \in H^1(\mathbb{R})$, supp $\varphi \subset \mathbb{R} \setminus \{\pm 1\}$. Lemma 4.3 ensures that (4.6) holds for $\varphi \in H^1(\mathbb{R})$, supp $\varphi \subset [1 - \delta, 1 + \delta]$ or $[-1 - \delta, -1 + \delta]$. Therefore, (4.6) holds for any $\varphi \in H^1(\mathbb{R})$.

Taking $\varphi = \overline{\widetilde{\psi}_0}$ in (4.6) and taking the imaginary part, we deduce that $\widetilde{\psi}_0(\pm 1) = 0$, which implies $\widetilde{\psi}_0 \in H^2(\mathbb{R})$. As $((y^2 - 1)\widetilde{\psi}'_0 - 2y\widetilde{\psi}_0)' = 0$ and $((y^2 - 1)\widetilde{\psi}'_0 - 2y\widetilde{\psi}_0)|_{\pm 1} = 0$, we infer that

$$(y^2 - 1)\widetilde{\psi}'_0 - 2y\widetilde{\psi}_0 = 0, \quad ((y^2 - 1)^{-1}\widetilde{\psi}_0)' = 0 \quad \text{in } \mathbb{R} \setminus \{\pm 1\},$$

which mean that there exist constants C_3, C_4, C_5 so that $\widetilde{\psi}_0 = C_3(y^2 - 1)$ in $(-\infty, -1)$, $\widetilde{\psi}_0 = C_4(y^2 - 1)$ in $(-1, 1)$, $\widetilde{\psi}_0 = C_5(y^2 - 1)$ in $(1, +\infty)$. However, $\widetilde{\psi}_0 \in H^2(\mathbb{R})$, thus $C_3 = C_4 = C_5 = 0$. Then $\widetilde{\psi}_0 = 0$.

Up to now, we proved that $\widetilde{\psi}_n \to 0$ in $H^1_{loc}(\mathbb{R}) \cap C^1_{loc}(\mathbb{R} \setminus \{\pm 1\})$, thus $\|\psi'_n\|_{L^2(-2r_n, 2r_n)} = \|\widetilde{\psi}'_n\|_{L^2(-2,2)} \to 0$. Since ψ''_n is bounded in $L^\infty_{loc}((a, b) \setminus \{0\})$ and $\psi_n \rightharpoonup 0$ in $H^1(a, b)$, we have $\psi_n \to 0$ in $C^1_{loc}((a, b) \setminus \{0\})$. Integration by parts gives

$$\int_{2r_n}^b \left(|\psi'_n|^2 + \alpha^2 |\psi_n|^2 + \frac{u''|\psi_n|^2}{u - c_n} \right) dy = -\int_{2r_n}^b \frac{u''\omega_n \overline{\psi_n}}{u - c_n} dy + \psi'_n \overline{\psi_n}\big|_{2r_n}^b. \tag{4.7}$$

Notice that for n sufficiently large and $y \in [2r_n, b]$, we have $u(y) \geq u(2r_n) > 2r_n^2 = 2|c_n|$. Then we have

$$\text{Re}\frac{1}{u(y) - c_n} \geq \frac{1}{2(u(y) + |c_n|)} \geq \frac{1}{4u(y)} \geq \frac{1}{Cy^2},$$

$$u''(y) \geq C^{-1}, \quad \left|\frac{1}{u(y) - c_n}\right| \leq \frac{1}{u(y) - |c_n|} \leq \frac{2}{u(y)} \leq \frac{C}{y^2}.$$

Taking the real part of (4.7), we get

$$\int_{2r_n}^{b} (|\psi_n'|^2 + \alpha^2 |\psi_n|^2) dy + C^{-1} \left\| \frac{\psi_n}{y} \right\|_{L^2(2r_n, b)}^2 \leq C \left\| \frac{\psi_n}{y} \right\|_{L^2(2r_n, b)} \left\| \frac{\omega_n}{y} \right\|_{L^2(2r_n, b)}$$
$$+ \left| \psi_n' \overline{\psi_n} \right|_{2r_n}^{b} \Big|,$$

which implies that

$$\|\psi_n'\|_{L^2(2r_n, b)}^2 \leq C \|\omega_n/y\|_{L^2(2r_n, b)}^2 + \left| \psi_n' \overline{\psi_n} \right|_{2r_n}^{b} \Big|.$$

Notice that $\psi_n' \overline{\psi_n} \big|_{2r_n}^{b} = \psi_n' \overline{\psi_n}(b) - \widetilde{\psi_n'} \widetilde{\psi_n}(2) \to 0$, and by Hardy's inequality, $\|\omega_n/y\|_{L^2(2r_n, b)} \leq C \|\omega_n\|_{H^1(a,b)} \to 0$. Thus, $\|\psi_n'\|_{L^2(2r_n, b)} \to 0$. Similarly, we have $\|\psi_n'\|_{L^2(a, -2r_n)} \to 0$. This shows that $\|\psi_n'\|_{L^2(a,b)} \to 0$. As $\psi_n \to 0$ in $H^1(a, b)$, we have $\|\psi_n\|_{L^2(a,b)} \to 0$ and then $\|\psi_n\|_{H^1(a,b)} \to 0$.

For general case, we consider

$$\psi_{n*}(y) = \psi_n(y) + \frac{\omega_n(y_0)}{u''(y_0)} \cosh \alpha(y - y_0),$$

$$\omega_{n*}(y) = \omega_n(y) - u''(y) \frac{\omega_n(y_0)}{u''(y_0)} \cosh \alpha(y - y_0).$$

Then $\psi_{n*}, \omega_{n*} \in H^1(a, b)$, $\omega_{n*}(y_0) = 0$ and

$$(u - c_n)(\psi_{n*}'' - \alpha^2 \psi_{n*}) - u'' \psi_{n*} = \omega_{n*}.$$

Notice that

$$\|\omega_{n*}\|_{H^1(a,b)} \leq \|\omega_n\|_{H^1(a,b)} + C|\omega_n(y_0)| \leq C\|\omega_n\|_{H^1(a,b)} \to 0,$$
$$\|\psi_n - \psi_{n*}\|_{H^1(a,b)} \leq C|\omega_n(y_0)| \leq C\|\omega_n\|_{H^1(a,b)} \to 0,$$

thus, $\psi_{n*} \to 0$ in $H^1(a, b)$. This is reduced to the case of $\omega(y_0) = 0$. Thus, $\|\psi_{n*}\|_{H^1(a,b)} \to 0$, hence $\|\psi_n\|_{H^1(a,b)} \to 0$. $\square$

Now we prove Proposition 4.1.

Proof. We only consider $\mathrm{Im}\, c > 0$, and the case of $\mathrm{Im}\, c < 0$ can be proved by taking conjugation. We use the contradiction argument.

Assume that there exists $\psi_n \in H_0^1(-1, 1)$, $\omega_n \in H^1(-1, 1)$, and c_n with $\mathrm{Im}\, c_n > 0$, such that $\|\psi_n\|_{H^1(-1,1)} = 1$, $\|\omega_n\|_{H^1(-1,1)} \to 0$, $c_n \to c \in \mathrm{Ran}\, u$, and

$$(u - c_n)(\psi_n'' - \alpha^2 \psi_n) - u'' \psi_n = \omega_n.$$

Then there exists a subsequence of $\{\psi_n\}$ (still denoted by $\{\psi_n\}$) and $\psi \in H_0^1(-1, 1)$ such that $\psi_n \rightharpoonup \psi$ weakly in $H^1(-1, 1)$. Since ψ_n'' is bounded in $L_{loc}^\infty((-1, 1) \setminus u^{-1}\{c\})$, we have $\psi_n \to \psi$ in $H_{loc}^1((-1, 1) \setminus u^{-1}\{c\})$ and ψ satisfies (4.3) in $[-1, 1] \setminus u^{-1}\{c\}$ and (4.2) holds for $\varphi \in H_0^1(-1, 1)$, $\mathrm{supp}\, \varphi \subset [-1, 1] \setminus u^{-1}\{c\}$.

For $y_0 \in u^{-1}\{c\}$, if $u'(y_0) = 0$, then $y_0 \neq \pm 1$, $u''(y_0) \neq 0$, and there exists $0 < \delta < 1 - |y_0|$ so that $u''(y)u''(y_0) > 0$ for $|y - y_0| \leq \delta$. It is easy to see that $\psi/(u-c)$ is bounded in $[y_0 - \delta, y_0 + \delta]$, and then (4.2) holds for $\varphi \in H_0^1(-1, 1)$, supp $\varphi \subset (y_0 - \delta, y_0 + \delta)$.

If $u'(y_0) \neq 0$, there exists $0 < \delta < 1$ so that $u'(y)u'(y_0) > 0$ for $|y - y_0(1 - \delta)| \leq \delta$. Then Lemma 4.3 ensures that (4.2) holds for $\varphi \in H_0^1(-1, 1)$, supp $\varphi \subseteq [y_0(1 - \delta) - \delta, y_0(1 - \delta) + \delta]$. Thus, (4.2) holds for all $\varphi \in H_0^1(-1, 1)$. Since $\mathcal{R}_\alpha$ has no embedding eigenvalues, $\psi = 0$. Thus, $\psi_n \to 0$ in $H_{loc}^1((-1, 1) \setminus u^{-1}\{c\})$. Furthermore, if $u'(y_0) = 0$, Lemma 4.4 gives $\psi_n \to 0$ in $H^1(y_0 - \delta, y_0 + \delta)$; if $u'(y_0) \neq 0$, Lemma 4.3 gives $\psi_n \to 0$ in $H^1(y_0(1 - \delta) - \delta, y_0(1 - \delta) + \delta)$. Thus, $\psi_n \to 0$ in $H^1(-1, 1)$, which leads to a contradiction.

See Proposition 6.5 in [23] for the proof of the second statement. □

Now we are in a position to prove Theorem 1.2 and Theorem 1.3.

Proof. First of all, we have

$$\widehat{\psi}(t, \alpha, y) = \lim_{\epsilon \to 0+} \frac{1}{2\pi i} \int_{\partial\Omega_\epsilon} e^{-i\alpha tc} i\alpha \Phi(\alpha, y, c) dc$$

$$= \frac{1}{2\pi i} \int_{\text{Ran } u} e^{-i\alpha tc} i\alpha \big(\Phi_-(\alpha, y, c) - \Phi_+(\alpha, y, c)\big) dc$$

$$= \frac{1}{2\pi} \int_{\text{Ran } u} e^{-i\alpha tc} \widetilde{\Phi}(\alpha, y, c) dc,$$

where $\widetilde{\Phi}(\alpha, y, c) = \alpha\big(\Phi_-(\alpha, y, c) - \Phi_+(\alpha, y, c)\big)$.

For $k, l = 0, 1$, we have

$$\partial_t^k \partial_y^l \widehat{\psi}(t, \alpha, y) = \frac{1}{2\pi} \int_{\text{Ran } u} (-i\alpha c)^k e^{-i\alpha tc} \partial_y^l \widetilde{\Phi}(\alpha, y, c) dc,$$

from which and Plancherel's formula, we infer that

$$\|\widehat{V}(t, \alpha, y)\|_{H_t^1 L_y^2}^2 = \int_{\mathbb{R}} \big(\|\widehat{V}(t, \alpha, \cdot)\|_{L_y^2}^2 + \|\partial_t \widehat{V}(t, \alpha, \cdot)\|_{L_y^2}^2\big) dt$$

$$= \int_{-1}^1 \int_{\mathbb{R}} \big(\alpha^2|\widehat{\psi}(t, \alpha, y)|^2 + |\partial_y\widehat{\psi}(t, \alpha, \cdot)|^2 + \alpha^2|\partial_t\widehat{\psi}(t, \alpha, y)|^2 + |\partial_t\partial_y\widehat{\psi}(t, \alpha, y)|^2\big) dt dy$$

$$= \frac{1}{2\pi|\alpha|} \int_{-1}^1 \int_{\text{Ran } u} (1 + |\alpha c|^2)\big(\alpha^2|\widetilde{\Phi}(\alpha, y, c)|^2 + |\partial_y\widetilde{\Phi}(\alpha, y, c)|^2\big) dc dy$$

$$\leq C \int_{\text{Ran } u} \|\widetilde{\Phi}(\alpha, \cdot, c)\|_{H_y^1}^2 dc \leq C \int_{\text{Ran } u} \|\widehat{\omega}_0(\alpha, \cdot)\|_{H_y^1}^2 dc = C\|\widehat{\omega}_0(\alpha, \cdot)\|_{H_y^1}^2.$$

Now we turn to the vorticity depletion. Let $W(\alpha, y, c) = i\alpha(\alpha^2 - \partial_y^2)\Phi(\alpha, y, c)$. It follows from (3.5) that

$$\widehat{\omega}(t, \alpha, y) = \frac{1}{2\pi i} \int_{\partial\Omega_\epsilon} e^{-i\alpha tc} W(\alpha, y, c) dc.$$

It can be proved that $W(\alpha, y_0, \cdot \pm i\epsilon)$ is bounded in $L_c^p(\mathrm{Ran}\, u)(1 < p < 4/3)$ for $0 < \epsilon < \epsilon_0$. Thus, there exists a subsequence $\epsilon_n \to 0+$ and $W_\pm(c) \in L_c^p(\mathrm{Ran}\, u)(1 < p < 4/3)$ so that $W(\alpha, y_0, \cdot \pm i\epsilon_n) \rightharpoonup W_\pm$ weakly in $L_c^p(\mathrm{Ran}\, u)$. Then we deduce from Riemann-Lebesgue lemma that

$$\hat{\omega}(t, \alpha, y_0) = \lim_{\epsilon \to 0+} \frac{1}{2\pi i} \int_{\partial \Omega_\epsilon} e^{-i\alpha tc} W(\alpha, y_0, c) dc$$

$$= \lim_{n \to \infty} \frac{1}{2\pi i} \int_{\mathrm{Ran}\, u} \left(e^{-i\alpha t(c - i\epsilon_n)} W(\alpha, y_0, c - i\epsilon_n) \right.$$
$$\left. - e^{-i\alpha t(c + i\epsilon_n)} W(\alpha, y_0, c + i\epsilon_n) \right) dc$$

$$= \frac{1}{2\pi i} \int_{\mathrm{Ran}\, u} e^{-i\alpha tc} \left(W_-(c) - W_+(c) \right) dc \to 0 \quad \text{as} \quad t \to \infty.$$

This could be seen from the following formal argument. If $u'(y_0) = 0$ and $y_c \sim y_0$, then $\Phi(y, c) \sim \frac{\beta(c)}{y - y_0} \notin L^2$ for $y \sim y_0$ and $c \sim u(y_0)$. Thus, $\beta(c) = 0$ when $y_c \sim y_0$. Then as $y \to y_0$ and $c \to u(y_0)$, we have $(u(y) - c)\Phi''(y, c) \to 0$, which implies that

$$i\alpha u'' \Phi(y, c) + \hat{\omega}_0(y) \to 0 \quad \text{as} \quad y \to y_0 \text{ and } c \to u(y_0).$$

On the other hand, we formally have

$$\hat{\omega}(t, \alpha, y) = \frac{1}{2\pi i} \int_{\partial \Omega} e^{-i\alpha tc} i\alpha (\Phi'' - \alpha^2 \Phi)(y, c) dc$$

$$= \frac{1}{2\pi i} \int_{\partial \Omega} \frac{e^{-i\alpha tc} (i\alpha u'' \Phi(y, c) + \hat{\omega}_0(\alpha, y))}{u - c} dc$$
$$\sim e^{-i\alpha tu(y)} (i\alpha u'' \Phi(y, u(y)) + \hat{\omega}_0(\alpha, y)) \to 0 \quad \text{as} \quad y \to y_0.$$

$$\square$$

5 Linear inviscid damping via vector field method

In this section, we assume that $\Omega = \mathbb{T}_L \times [0, 1]$, $\Lambda := \{2\pi k/L | k \in \mathbb{Z}_+\}$, $\sigma_d(\mathcal{R}_\alpha) = \emptyset$, $u'(y) \geq c_0 > 0$, $2\pi/L \geq c_0$. We present a new proof of linear inviscid damping for monotone shear flows, which is based on the vector field method in the spirit of the wave equation.

5.1 Limiting absorption principle

In this subsection, we establish the limiting absorption principle for monotone flows. The main difference is that we present a uniform H^1 bound of Φ in the wave number α. We denote by $c^i = \mathrm{Im}(c)$ and $c^r = \mathrm{Re}(c)$ for $c \in \mathbb{C}$ in the sequel.

Proposition 5.1. *Assume that $\mathcal{R}_\alpha$ has no embedding eigenvalues for any α. Then there exists $\varepsilon_0 > 0$ such that for any $c \in \mathbb{C}$, $0 < \operatorname{Im}(c) < \varepsilon_0$ and $\alpha \in \Lambda$, the unique solution Φ to the boundary value problem*

$$(u - c)(\partial_y^2 \Phi - \alpha^2 \Phi) - u'' \Phi = \omega, \quad \Phi(0) = \Phi(1) = 0$$

has the following uniform H^1 bound

$$\|\partial_y \Phi\|_{L^2} + \alpha \|\Phi\|_{L^2} \leq C\alpha^{-1}(\|\partial_y \omega\|_{L^2} + \alpha \|\omega\|_{L^2}).$$

Moreover, if $\omega(0) = \omega(1) = 0$, then we have

$$|\partial_y \Phi(0)| + |\partial_y \Phi(1)| \leq C\alpha^{-\frac{1}{2}}(\|\partial_y \omega\|_{L^2} + \alpha \|\omega\|_{L^2}).$$

We need the following lemmas.

Lemma 5.2. *If $f \in H_0^1(0, 1)$, then for $c^i > 0$,*

$$\left| \int_0^1 \frac{f(y)}{u(y) - c} dy \right| \leq C\alpha^{-\frac{1}{2}}(\|f'\|_{L^2} + \alpha \|f\|_{L^2}),$$

where the constant C only depends on c_0.

Proof. Due to $c^i > 0$, we have

$$\int_0^1 \frac{f(y)}{u(y) - c} dy = i \int_0^1 f(y) \int_0^{+\infty} e^{-it(u(y) - c)} dt dy$$

$$= i \int_0^{+\infty} e^{itc} \int_0^1 f(y) e^{-itu(y)} dy dt.$$

Let $g(t) = \int_0^1 f(y) e^{-itu(y)} dy$. Then we get

$$\left| \int_0^1 \frac{f(y)}{u(y) - c} dy \right| \leq \int_0^{+\infty} |e^{itc}| \left| \int_0^1 f(y) e^{-itu(y)} dy \right| dt$$

$$= \int_0^{+\infty} e^{-tc^i} |g(t)| dt \leq \|g\|_{L^1(\mathbb{R})}.$$

Due to $f \in H_0^1(0, 1)$, we have

$$g(t) = \int_0^1 f(y) e^{-itu(y)} dy = \int_{u(0)}^{u(1)} e^{-itz}(f/u') \circ u^{-1}(z) dz,$$

$$itg(t) = \int_{u(0)}^{u(1)} e^{-itz}((f/u')'/u') \circ u^{-1}(z) dz,$$

from which and Plancherel's formula, we infer that

$$\|g\|_{L^2(\mathbb{R})}^2 = 2\pi \|(f/u') \circ u^{-1}\|_{L^2(u(0), u(1))}^2 = 2\pi \||f|^2/u'\|_{L^1(0,1)} \leq (2\pi/c_0)\|f\|_{L^2}^2,$$

$$\|tg(t)\|_{L^2(\mathbb{R})}^2 = 2\pi \||(f/u')'|^2/u'\|_{L^1(0,1)} \leq C(\|f'\|_{L^2} + \|f\|_{L^2})^2.$$

Thus, we obtain

$$\|(\alpha^2 + t^2)^{\frac{1}{2}} g(t)\|^2_{L^2(\mathbb{R})} = \alpha^2 \|g\|^2_{L^2(\mathbb{R})} + \|tg(t)\|^2_{L^2(\mathbb{R})}$$
$$\leq C\alpha^2 \|f\|^2_{L^2} + C(\|f'\|_{L^2} + \|f\|_{L^2})^2 \leq C(\|f'\|_{L^2} + \alpha\|f\|_{L^2})^2,$$

and

$$\left| \int_0^1 \frac{f(y)}{u(y) - c} dy \right| \leq \|g\|_{L^1(\mathbb{R})} \leq \|(\alpha^2 + t^2)^{\frac{1}{2}} g(t)\|_{L^2(\mathbb{R})} \|(\alpha^2 + t^2)^{-\frac{1}{2}}\|_{L^2(\mathbb{R})}$$
$$\leq C(\|f'\|_{L^2} + \alpha\|f\|_{L^2})\alpha^{-\frac{1}{2}}.$$

This completes the proof.   □

Lemma 5.3. *Let* $\alpha \geq c_0 > 0$, $c^i > 0$. *Then the unique solution* Φ *to the boundary value problem*

$$(u - c)(\partial_y^2 \Phi - \alpha^2 \Phi) = \omega, \quad \Phi(0) = \Phi(1) = 0$$

has the uniform H^1 *bound*

$$\|\partial_y \Phi\|_{L^2} + \alpha\|\Phi\|_{L^2} \leq C\alpha^{-1}(\|\partial_y \omega\|_{L^2} + \alpha\|\omega\|_{L^2}).$$

Moreover, if $\omega(0) = \omega(1) = 0$, *then we have*

$$|\partial_y \Phi(0)| + |\partial_y \Phi(1)| \leq C\alpha^{-\frac{1}{2}}(\|\partial_y \omega\|_{L^2} + \alpha\|\omega\|_{L^2}).$$

Proof. By Gagliardo-Nirenberg inequality, we get

$$\|\omega\|_{L^\infty} \leq C\|\omega\|^{\frac{1}{2}}_{L^2} \|\omega\|^{\frac{1}{2}}_{H^1} \leq C\alpha^{-\frac{1}{2}}(\|\omega\|_{H^1} + \alpha\|\omega\|_{L^2}) \leq C\alpha^{-\frac{1}{2}}(\|\partial_y \omega\|_{L^2} + \alpha\|\omega\|_{L^2}),$$

and similarly $\|\Phi\|_{L^\infty} \leq C\alpha^{-\frac{1}{2}}(\|\partial_y \Phi\|_{L^2} + \alpha\|\Phi\|_{L^2})$. Since

$$\|\partial_y \Phi\|^2_{L^2} + \alpha^2 \|\Phi\|^2_{L^2} = -\langle \partial_y^2 \Phi - \alpha^2 \Phi, \Phi \rangle = -\langle \omega/(u - c), \Phi \rangle = - \int_0^1 \frac{\omega(y)\overline{\Phi(y)}}{u(y) - c} dy,$$

and $\omega\overline{\Phi}(0) = \omega\overline{\Phi}(1) = 0$, we get by Lemma 5.2 that

$$\|\partial_y \Phi\|^2_{L^2} + \alpha^2 \|\Phi\|^2_{L^2}$$
$$\leq C\alpha^{-\frac{1}{2}}(\|\partial_y(\omega\overline{\Phi})\|_{L^2} + \alpha\|\omega\overline{\Phi}\|_{L^2})$$
$$\leq C\alpha^{-\frac{1}{2}}(\|\partial_y \omega\|_{L^2}\|\Phi\|_{L^\infty} + \|\omega\|_{L^\infty}\|\partial_y \Phi\|_{L^2} + \alpha\|\omega\|_{L^\infty}\|\Phi\|_{L^2})$$
$$\leq C\alpha^{-\frac{1}{2}}(\alpha^{-\frac{1}{2}}\|\partial_y \omega\|_{L^2} + \|\omega\|_{L^\infty})(\alpha^{\frac{1}{2}}\|\Phi\|_{L^\infty} + \|\partial_y \Phi\|_{L^2} + \alpha\|\Phi\|_{L^2})$$
$$\leq C\alpha^{-\frac{1}{2}}(\alpha^{-\frac{1}{2}}\|\partial_y \omega\|_{L^2} + \alpha^{\frac{1}{2}}\|\omega\|_{L^2})(\|\partial_y \Phi\|_{L^2} + \alpha\|\Phi\|_{L^2}),$$

which implies the first inequality.

Let

$$\gamma_1(y) = \frac{\sinh(\alpha y)}{\sinh \alpha}, \quad \gamma_0(y) = \frac{\sinh(\alpha(1 - y))}{\sinh \alpha}. \tag{5.1}$$

Then we have $|\gamma_j| \leq 1$, $|\gamma_j'| \leq C\alpha$ and

$$|\partial_y\Phi(j)| = |\langle\partial_y^2\Phi - \alpha^2\Phi, \gamma_j\rangle| = |\langle\omega/(u-c), \gamma_j\rangle| = \left|\int_0^1 \frac{\omega(y)\gamma_j(y)}{u(y)-c}dy\right|, \quad j = 0, 1.$$

If $\omega(0) = \omega(1) = 0$, then $\omega\gamma_j \in H_0^1(0,1)$, and by Lemma 5.2, we have

$$|\partial_y\Phi(j)| \leq C\alpha^{-\frac{1}{2}}\left(\|\partial_y(\omega\gamma_j)\|_{L^2} + \alpha\|\omega\gamma_j\|_{L^2}\right) \leq C\alpha^{-\frac{1}{2}}\left(\|\partial_y\omega\|_{L^2} + \alpha\|\omega\|_{L^2}\right),$$
$$j = 0, 1,$$

which gives the second inequality. $\qquad\square$

Now we are in a position to prove Proposition 5.1.

Proof. Suppose that the first inequality is not true. Then there exist $\Phi_n \in H_0^1(0,1)$, $\omega_n \in H^1(0,1)$ and $c_n \in \mathbb{C}$, $\alpha_n \in \Lambda$ with $c_n^i > 0$ such that $\|\partial_y\Phi_n\|_{L^2} + \alpha_n\|\Phi_n\|_{L^2} = \alpha_n^{-1}$, $\|\partial_y\omega_n\|_{L^2} + \alpha_n\|\omega_n\|_{L^2} = \delta_n \to 0$, $c_n^i \to 0$, $c_n \to c_\infty \in \mathbb{R} \cup \{\pm\infty\}$ and

$$(u - c_n)(\partial_y^2\Phi_n - \alpha_n^2\Phi_n) - u''\Phi_n = \omega_n.$$

By Lemma 5.3, we have

$$1 = \alpha_n(\|\partial_y\Phi_n\|_{L^2} + \alpha_n\|\Phi_n\|_{L^2}) \leq C\left(\|\partial_y(\omega_n + u''\Phi_n)\|_{L^2} + \alpha_n\|\omega_n + u''\Phi_n\|_{L^2}\right)$$
$$\leq C\left(\|\partial_y\omega_n\|_{L^2} + \|\partial_y\Phi_n\|_{L^2} + \alpha_n\|\omega_n\|_{L^2} + \alpha_n\|\Phi_n\|_{L^2}\right) \leq C(\alpha_n^{-1} + \delta_n).$$

Since $\delta_n \to 0$, this implies that α_n is uniformly bounded. Up to a subsequence, we may assume that α_n is constant ($\alpha_n = \alpha > 0$) and that there exists $\Phi_0 \in H_0^1(0,1)$ so that $\Phi_n \rightharpoonup \Phi_0$ in $H^1(0,1)$.

If $c_n \to \pm\infty$, then $\|(u - c_n)^{-1}\|_{L^\infty} \to 0$ and

$$\|\partial_y\Phi_n\|_{L^2}^2 + \alpha_n^2\|\Phi_n\|_{L^2}^2 = -\langle\partial_y^2\Phi_n - \alpha_n^2\Phi_n, \Phi_n\rangle = -\langle(u-c_n)^{-1}(\omega_n + u''\Phi_n), \Phi_n\rangle$$
$$\leq \|(u-c_n)^{-1}\|_{L^\infty}(\|\omega_n\|_{L^2}\|\Phi_n\|_{L^2} + C\|\Phi_n\|_{L^2}^2) \to 0,$$

which contradicts with $\|\partial_y\Phi_n\|_{L^2} + \alpha_n\|\Phi_n\|_{L^2} = \alpha_n^{-1}$, $\alpha_n = \alpha$, $n \geq 1$.

If $c_n \to c_\infty \in \mathbb{R} \setminus [u(0), u(1)]$, then $\Phi_n \to \Phi_0$ in $H^1(0,1)$ and Φ_0 satisfies

$$(u - c_0)(\Phi_0'' - \alpha^2\Phi_0) - u''\Phi_0 = 0, \quad \Phi_0(y_1) = \Phi_0(y_2) = 0.$$

Thus, $\|\partial_y\Phi_0\|_{L^2} + \alpha\|\Phi_0\|_{L^2} = \alpha^{-1}$ and c_0 is an eigenvalue of $\mathcal{R}_\alpha$, which is a contradiction.

If $c_n \to c_\infty \in [u(0), u(1)]$, as in the proof of Proposition 4.1, we know that Φ_0 satisfies (4.2) for any $\varphi \in H_0^1(0,1)$ with supp $\varphi \in (0,1)$, that $\Phi_0 \equiv 0$ on $[0,1]$ (since $\mathcal{R}_\alpha$ has no embedding eigenvalues), and that $\Phi_n \to 0$ in $H^1(0,1)$, which contradicts with $\|\partial_y\Phi_n\|_{L^2} + \alpha\|\Phi_n\|_{L^2} = \alpha^{-1}$. In summary, this shows the first inequality.

If $\omega(0) = \omega(1) = 0$, then $\omega + u''\Phi = 0$ at $y = 0, 1$. Then from Lemma 5.3 and the first inequality, we deduce that

$$|\partial_y\Phi(j)| \leq C\alpha^{-\frac{1}{2}}\left(\|\partial_y(\omega + u''\Phi)\|_{L^2} + \alpha\|\omega + u''\Phi\|_{L^2}\right)$$
$$\leq C\alpha^{-\frac{1}{2}}\left(\|\partial_y\omega\|_{L^2} + \alpha\|\omega\|_{L^2} + \|\partial_y\Phi\|_{L^2} + \alpha\|\Phi\|_{L^2}\right)$$
$$\leq C\alpha^{-\frac{1}{2}}\left(\|\partial_y\omega\|_{L^2} + \alpha\|\omega\|_{L^2}\right), \quad j = 0, 1,$$

which gives the second inequality. $\qquad\square$

5.2 Space-time estimate

For any fixed $\alpha \neq 0$, we define

$$\mathcal{R}'_\alpha \widehat{\omega} = -(u''(\partial_y^2 - \alpha^2)^{-1} - u)\widehat{\omega}.$$

Then we have $\mathcal{R}'_\alpha (\partial_y^2 - \alpha^2) = (\partial_y^2 - \alpha^2)\mathcal{R}_\alpha$ in $H_0^1(0,1)$ and

$$\partial_t \widehat{\omega} = -i\alpha \mathcal{R}'_\alpha \widehat{\omega}.$$

Without loss of generality, we may assume $\alpha > 0$ in the sequel.

Proposition 5.4. *Assume that $\mathcal{R}_\alpha$ has no embedding eigenvalues or eigenvalues. Let $\psi = -(\partial_y^2 - \alpha^2)^{-1}\omega$ and $\omega(t,y)$ solve*

$$\partial_t \omega + i\alpha \mathcal{R}'_\alpha \omega + f = 0$$

for $t \in [0,T]$ and $y \in [0,1]$, where $f \in L^2((0,T), H^1(0,1))$. Then we have

$$\|\omega(T)\|_{L^2}^2 + \alpha^2 \int_0^T \left(\|\partial_y \psi(t)\|_{L^2}^2 + \alpha^2 \|\psi(t)\|_{L^2}^2 \right) dt$$

$$\leq C\|\omega(0)\|_{L^2}^2 + C\alpha^{-2} \int_0^T \left(\|\partial_y f(t)\|_{L^2}^2 + \alpha^2 \|f(t)\|_{L^2}^2 \right) dt.$$

Moreover, if $f(t,0) = f(t,1) = 0$, then

$$\alpha \int_0^T \left(|\partial_y \psi(t,0)|^2 + |\partial_y \psi(t,1)|^2 \right) dt \leq C\|\omega(0)\|_{L^2}^2$$

$$+ C \int_0^T \left(\alpha^{-2} \|\partial_y f(t)\|_{L^2}^2 + \|f(t)\|_{L^2}^2 \right) dt.$$

Here the constant C only depends on u.

We need the following lemmas.

Lemma 5.5. *Let $\psi_1 = -(\partial_y^2 - \alpha^2)^{-1}\omega_1$ and $\omega_1(t,y)$ solve $\partial_t \omega_1 + i\alpha u \omega_1 = 0$ for $t \in \mathbb{R}$ and $y \in [0,1]$. Then we have*

$$\alpha^2 \int_\mathbb{R} \left(\|\partial_y \psi_1(t)\|_{L^2}^2 + \alpha^2 \|\psi_1(t)\|_{L^2}^2 \right) dt \leq C\|\omega_1(0)\|_{L^2}^2,$$

$$\alpha \int_\mathbb{R} \left(|\partial_y \psi_1(t,0)|^2 + |\partial_y \psi_1(t,1)|^2 \right) dt \leq C\|\omega_1(0)\|_{L^2}^2,$$

where the constant C only depends on c_0.

Proof. We use the basis in $L^2(0,1)$: $\varphi_k(y) = \sin(\pi k y)$, $k \in \mathbb{Z}_+$. Then we have

$$\omega_1 = \sum_{k=1}^{+\infty} 2\langle \omega_1, \varphi_k \rangle \varphi_k, \quad \|\omega_1\|_{L^2}^2 = \sum_{k=1}^{+\infty} 2|\langle \omega_1, \varphi_k \rangle|^2, \quad \psi_1 = \sum_{k=1}^{+\infty} \frac{2\langle \omega_1, \varphi_k \rangle}{(\pi k)^2 + \alpha^2} \varphi_k,$$

$$\|\partial_y \psi_1(t)\|_{L^2}^2 + \alpha^2 \|\psi_1(t)\|_{L^2}^2 = \langle \psi_1(t), \omega_1(t) \rangle = \sum_{k=1}^{+\infty} \frac{2|\langle \omega_1(t), \varphi_k \rangle|^2}{(\pi k)^2 + \alpha^2}.$$

Since $\partial_t \omega_1 + i\alpha u \omega_1 = 0$, the solution is given by $\omega_1(t,y) = e^{-i\alpha t u(y)}\omega_1(0,y)$. So,

$$
\langle \omega_1(t), \varphi_k \rangle = \int_0^1 e^{-i\alpha t u(y)} \omega_1(0,y) \varphi_k(y) dy
$$

$$
= \int_{u(0)}^{u(1)} e^{-i\alpha t z} \frac{\omega_1(0, u^{-1}(z)) \varphi_k(u^{-1}(z))}{u'(u^{-1}(z))} dz,
$$

from which and Plancherel's formula, we infer that

$$
\int_{\mathbb{R}} |\langle \omega_1(t), \varphi_k \rangle|^2 dt = \frac{2\pi}{\alpha} \int_{u(0)}^{u(1)} \left| \frac{\omega_1(0, u^{-1}(z)) \varphi_k(u^{-1}(z))}{u'(u^{-1}(z))} \right|^2 dz
$$

$$
= \frac{2\pi}{\alpha} \int_0^1 \frac{|\omega_1(0,y) \varphi_k(y)|^2}{u'(y)} dy
$$

$$
\leq \frac{2\pi}{\alpha} \int_0^1 \frac{|\omega_1(0,y)|^2}{u'(y)} dy \leq \frac{2\pi}{\alpha c_0} \|\omega_1(0)\|_{L^2}^2.
$$

Therefore,

$$
\int_{\mathbb{R}} (\|\partial_y \psi_1(t)\|_{L^2}^2 + \alpha^2 \|\psi_1(t)\|_{L^2}^2) dt = \sum_{k=1}^{+\infty} \int_{\mathbb{R}} \frac{2|\langle \omega_1(t), \varphi_k \rangle|^2}{(\pi k)^2 + \alpha^2} dt
$$

$$
\leq \sum_{k=1}^{+\infty} \frac{2\pi}{\alpha c_0} \frac{2\|\omega_1(0)\|_{L^2}^2}{(\pi k)^2 + \alpha^2}
$$

$$
\leq \int_0^{+\infty} \frac{2\pi}{\alpha c_0} \frac{2\|\omega_1(0)\|_{L^2}^2}{(\pi z)^2 + \alpha^2} dz = \frac{2\pi}{\alpha^2 c_0} \|\omega_1(0)\|_{L^2}^2,
$$

which gives the first inequality.

Recall that γ_1 and γ_2 are defined in (5.1). Then we have

$$
\langle \omega_1, \gamma_1 \rangle = -\langle (\partial_y^2 - \alpha^2)\psi_1, \gamma_1 \rangle = -\langle \psi_1, (\partial_y^2 - \alpha^2)\gamma_1 \rangle - (\partial_y \psi_1 \gamma_1 - \psi_1 \gamma_1')|_0^1
$$

$$
= -\partial_y \psi_1(t,1),
$$

and $\langle \omega_1, \gamma_0 \rangle = \partial_y \psi_1(t,0)$. As in the proof of the first inequality, we have

$$
\int_{\mathbb{R}} |\partial_y \psi_1(t,j)|^2 dt = \int_{\mathbb{R}} |\langle \omega_1(t), \gamma_j \rangle|^2 dt = \frac{2\pi}{\alpha} \int_0^1 \frac{|\omega_1(0,y)\gamma_j(y)|^2}{u'(y)} dy
$$

$$
\leq \frac{2\pi}{\alpha c_0} \|\omega_1(0)\|_{L^2}^2, \quad j = 0, 1,
$$

which gives the second inequality.   □

Lemma 5.6. Let $\psi = -(\partial_y^2 - \alpha^2)^{-1}\omega$ and $\omega(t,y)$ solve $\partial_t \omega + i\alpha u \omega + f = 0$ for $t \in [0,T]$ and $y \in [0,1]$ and $\omega(0) = 0$, where $f \in L^2((0,T), H^1(0,1))$. Then we have

$$
\|\omega(T)\|_{L^2}^2 \leq C \int_0^T (\alpha^{-2} \|\partial_y f(t)\|_{L^2}^2 + \|f(t)\|_{L^2}^2) dt,
$$

where the constant C only depends on c_0.

Proof. Let $\omega_1(t,y) = e^{i\alpha(T-t)u(y)}\omega(T,y)$ and $\psi_1 = -(\partial_y^2 - \alpha^2)^{-1}\omega_1$. Then we have

$$\partial_t\omega_1 + i\alpha u\omega_1 = 0, \quad \omega_1(T) = \omega(T), \quad \|\omega_1(0)\|_{L^2} = \|\omega(T)\|_{L^2}.$$

Then it follows from Lemma 5.5 that

$$\alpha^2\int_0^T\left(\|\partial_y\psi_1(t)\|_{L^2}^2 + \alpha^2\|\psi_1(t)\|_{L^2}^2\right)dt \leq C\|\omega_1(0)\|_{L^2}^2 = C\|\omega(T)\|_{L^2}^2,$$

$$\alpha\int_0^T\left(|\partial_y\psi_1(t,0)|^2 + |\partial_y\psi_1(t,1)|^2\right)dt \leq C\|\omega_1(0)\|_{L^2}^2 = C\|\omega(T)\|_{L^2}^2.$$

Noticing that

$$\partial_t\langle\omega_1,\omega\rangle = \langle\partial_t\omega_1,\omega\rangle + \langle\omega_1,\partial_t\omega\rangle = -\langle i\alpha u\omega_1,\omega\rangle - \langle\omega_1, f + i\alpha u\omega\rangle = -\langle\omega_1, f\rangle$$
$$= \langle(\partial_y^2 - \alpha^2)\psi_1, f\rangle = -\langle\partial_y\psi_1, \partial_y f\rangle - \alpha^2\langle\psi_1, f\rangle + (\partial_y\psi_1\overline{f})|_{y=0}^1,$$

we infer that

$$|\partial_t\langle\omega_1,\omega\rangle| \leq \|\partial_y f\|_{L^2}\|\partial_y\psi_1\|_{L^2} + \alpha^2\|f\|_{L^2}\|\psi_1\|_{L^2}$$
$$+ \|f\|_{L^\infty}\left(|\partial_y\psi_1(t,0)| + |\partial_y\psi_1(t,1)|\right).$$

By Gagliardo-Nirenberg inequality, we get

$$\|f\|_{L^\infty}^2 \leq C\|f\|_{L^2}\|f\|_{H^1} \leq C\alpha^{-1}(\|f\|_{H^1}^2 + \alpha^2\|f\|_{L^2}^2) \leq C\alpha^{-1}(\|\partial_y f\|_{L^2}^2 + \alpha^2\|f\|_{L^2}^2).$$

In summary, we obtain

$$\|\omega(T)\|_{L^2}^2 = \langle\omega_1,\omega\rangle|_0^T \leq \int_0^T|\partial_t\langle\omega_1,\omega\rangle|dt$$

$$\leq \int_0^T\left(\|\partial_y f\|_{L^2}\|\partial_y\psi_1\|_{L^2} + \alpha^2\|f\|_{L^2}\|\psi_1\|_{L^2} + \|f\|_{L^\infty}(|\partial_y\psi_1(t,0)| + |\partial_y\psi_1(t,1)|)\right)dt$$

$$\leq \left(\int_0^T(\|\partial_y f(t)\|_{L^2}^2 + \alpha^2\|f(t)\|_{L^2}^2 + 2\alpha\|f(t)\|_{L^\infty}^2)dt\right)^{\frac{1}{2}}$$

$$\times \left(\int_0^T(\|\partial_y\psi_1(t)\|_{L^2}^2 + \alpha^2\|\psi_1(t)\|_{L^2}^2 + \alpha^{-1}|\partial_y\psi_1(t,0)|^2 + \alpha^{-1}|\partial_y\psi_1(t,1)|^2)dt\right)^{\frac{1}{2}}$$

$$\leq \left(C\int_0^T(\|\partial_y f(t)\|_{L^2}^2 + \alpha^2\|f(t)\|_{L^2}^2)dt\right)^{\frac{1}{2}}\left(C\alpha^{-2}\|\omega(T)\|_{L^2}^2\right)^{\frac{1}{2}},$$

which gives our result. $\qquad\square$

Now we are in a position to prove Proposition 5.4.

Proof. **Step 1.** We introduce

$$\omega_1(t,y) = e^{-i\alpha t u(y)}\omega(0,y), \quad \omega_2 = \omega - \omega_1, \quad \psi_j = -(\partial_y^2 - \alpha^2)^{-1}\omega_j, \ j = 1, 2,$$

for $t \in [0, T]$. Then we have $\omega = \omega_1 + \omega_2$, $\psi = \psi_1 + \psi_2$ and

$$\partial_t \omega_1 + i\alpha u \omega_1 = 0.$$

By Lemma 5.5, we have

$$\alpha^2 \int_0^T (\|\partial_y \psi_1(t)\|_{L^2}^2 + \alpha^2 \|\psi_1(t)\|_{L^2}^2) dt \leq C\|\omega_1(0)\|_{L^2}^2 = C\|\omega(0)\|_{L^2}^2, \qquad (5.2)$$

$$\alpha \int_0^T (|\partial_y \psi_1(t,0)|^2 + |\partial_y \psi_1(t,1)|^2) dt \leq C\|\omega_1(0)\|_{L^2}^2 = C\|\omega(0)\|_{L^2}^2. \qquad (5.3)$$

Also we have $\|\omega_1(T)\|_{L^2}^2 = \|\omega(0)\|_{L^2}^2$.

Thanks to the definition of $\mathcal{R}'_\alpha$ and ψ_1, we have $\mathcal{R}'_\alpha \omega_1 = u''\psi_1 + u\omega_1$. Thus, $\partial_t \omega_1 + i\alpha \mathcal{R}'_\alpha \omega_1 = i\alpha u''\psi_1$, $\omega_2(0) = \omega(0) - \omega_1(0) = 0$, and

$$\partial_t \omega_2 + i\alpha \mathcal{R}'_\alpha \omega_2 = -f - i\alpha u''\psi_1 := f_1.$$

Moreover, we have

$$\|\partial_y f_1\|_{L^2} + \alpha \|f_1\|_{L^2} \leq \|\partial_y f\|_{L^2} + \alpha \|f\|_{L^2} + \alpha \|\partial_y (u''\psi_1)\|_{L^2} + \alpha^2 \|u''\psi_1\|_{L^2}$$
$$\leq \|\partial_y f\|_{L^2} + \alpha \|f\|_{L^2} + C\alpha(\|\partial_y \psi_1\|_{L^2} + \alpha\|\psi_1\|_{L^2}),$$

which gives

$$\int_0^T (\|\partial_y f_1(t)\|_{L^2}^2 + \alpha^2 \|f_1(t)\|_{L^2}^2) dt \qquad (5.4)$$

$$\leq C \int_0^T (\|\partial_y f(t)\|_{L^2}^2 + \alpha^2 \|f(t)\|_{L^2}^2) dt + C\alpha^2 \int_0^T (\|\partial_y \psi_1(t)\|_{L^2}^2 + \alpha^2 \|\psi_1(t)\|_{L^2}^2) dt$$

$$\leq C \int_0^T (\|\partial_y f(t)\|_{L^2}^2 + \alpha^2 \|f(t)\|_{L^2}^2) dt + C\|\omega(0)\|_{L^2}^2.$$

This means that $f_1 \in L^2((0,T); H^1(0,1))$.

Step 2. Now we extend ω_2, ψ_2, f_1 from $t \in [0, T]$ to $t \in [0, +\infty)$ in the following way:

$$\omega_2(t) = e^{-i(t-T)\alpha \mathcal{R}'_\alpha} \omega_2(T), \quad \psi_2(t) = -(\partial_y^2 - \alpha^2)^{-1}\omega_2(t), \quad f_1(t) = 0 \text{ for } t > T.$$

Then $\partial_t \omega_2 + i\alpha \mathcal{R}'_\alpha \omega_2 = f_1$ for $t \in [0, +\infty)$. Since $\mathcal{R}'_\alpha$ is a bounded operator on $H^1(0,1)$ and $\omega_2(0) = 0$, we have $\omega_2 \in C([0,T]; H^1(0,1))$. Since $\omega_2(t) = e^{-i(t-T)\alpha \mathcal{R}'_\alpha} \omega_2(T)$, $\psi_2(t) = -(\partial_y^2 - \alpha^2)^{-1}\omega_2(t)$ for $t > T$, by Theorem 1.2, we have $\psi_2 \in L^2((T, +\infty); H^1(0,1))$. Thanks to $\mathcal{R}'_\alpha(\partial_y^2 - \alpha^2) = (\partial_y^2 - \alpha^2)\mathcal{R}_\alpha$, we find

$$\partial_t \psi_2 + i\alpha \mathcal{R}_\alpha \psi_2 = f_2, \quad \psi_2(0) = 0,$$

with $f_2 = -(\partial_y^2 - \alpha^2)^{-1} f_1$. Let $f_3 = i\alpha u''\psi_2 - f_1$. Then

$$\partial_t \omega_2 + i\alpha u \omega_2 + f_3 = 0,$$

where f_3 satisfies

$$\|\partial_y f_3\|_{L^2} + \alpha\|f_3\|_{L^2} \leq \|\partial_y f_1\|_{L^2} + \alpha\|f_1\|_{L^2} + \alpha\|\partial_y(u''\psi_2)\|_{L^2} + \alpha^2\|u''\psi_2\|_{L^2}$$
$$\leq \|\partial_y f_1\|_{L^2} + \alpha\|f_1\|_{L^2} + C\alpha(\|\partial_y\psi_2\|_{L^2} + \alpha\|\psi_2\|_{L^2}).$$

Then it follows from Lemma 5.6 that for any $s > 0$,

$$\|\omega_2(s)\|_{L^2}^2 \leq C\int_0^s \left(\alpha^{-2}\|\partial_y f_3(t)\|_{L^2}^2 + \|f_3(t)\|_{L^2}^2\right)dt \qquad (5.5)$$

$$\leq C\int_0^{+\infty} \left(\alpha^{-2}\|\partial_y f_1(t)\|_{L^2}^2 + \|f_1(t)\|_{L^2}^2 + \|\partial_y\psi_2(t)\|_{L^2}^2 + \alpha^2\|\psi_2(t)\|_{L^2}^2\right)dt < +\infty.$$

Thus, $\omega_2 \in L^\infty((0,+\infty); L^2(0,1))$ and $\psi_2 = -(\partial_y^2 - \alpha^2)^{-1}\omega_2 \in L^\infty((0,+\infty); H^2(0,1))$.

Now we can take Laplace transform in t. For $\mathrm{Re}(\lambda) > 0$, let

$$\Phi(\lambda,y) = \int_0^{+\infty} \psi_2(t,y)e^{-\lambda t}dt, \quad F_j(\lambda,y) = \int_0^T f_j(t,y)e^{-\lambda t}dt, \ j = 1,2.$$

Then $\Phi(\lambda,\cdot) \in H^2(0,1)$, $F_1(\lambda,\cdot) \in H^1(0,1)$ for $\mathrm{Re}(\lambda) > 0$. Using Plancherel's formula, we know that for $\varepsilon > 0$, $j = 0,1$,

$$\int_{\mathbb{R}} \|\partial_y^j\Phi(\varepsilon + is)\|_{L^2}^2 ds = 2\pi\int_0^{+\infty} e^{-2\varepsilon t}\|\partial_y^j\psi_2(t)\|_{L^2}^2 dt, \qquad (5.6)$$

$$\int_{\mathbb{R}} |\partial_y\Phi(\varepsilon + is, j)|^2 ds = 2\pi\int_0^{+\infty} e^{-2\varepsilon t}|\partial_y\psi_2(t,j)|^2 dt, \qquad (5.7)$$

$$\int_{\mathbb{R}} \|\partial_y^j F_1(\varepsilon + is)\|_{L^2}^2 ds = 2\pi\int_0^T e^{-2\varepsilon t}\|\partial_y^j f_1(t)\|_{L^2}^2 dt. \qquad (5.8)$$

Furthermore, Φ satisfies

$$(u - i\lambda/\alpha)(\partial_y^2\Phi - \alpha^2\Phi) - u''\Phi = W, \quad \Phi(\lambda,0) = \Phi(\lambda,1) = 0 \qquad (5.9)$$

with $W = -(i/\alpha)(\partial_y^2 - \alpha^2)F_2 = (i/\alpha)F_1$.

If $\mathrm{Re}(\lambda) \in (0,\alpha\varepsilon_0)$, then $\mathrm{Im}(i\lambda/\alpha) \in (0,\varepsilon_0)$, and by Proposition 5.1,

$$\|\partial_y\Phi(\lambda)\|_{L^2}^2 + \alpha^2\|\Phi(\lambda)\|_{L^2}^2 \leq C\alpha^{-4}(\|\partial_y F_1(\lambda)\|_{L^2}^2 + \alpha^2\|F_1(\lambda)\|_{L^2}^2).$$

Integrating this over $\mathrm{Re}(\lambda) = \varepsilon \in (0,\alpha\varepsilon_0)$ and using (5.6), (5.8), we deduce that

$$\int_0^{+\infty} e^{-2\varepsilon t}\left(\|\partial_y\psi_2(t)\|_{L^2}^2 + \alpha^2\|\psi_2(t)\|_{L^2}^2\right)dt$$

$$\leq C\alpha^{-4}\int_0^T e^{-2\varepsilon t}\left(\|\partial_y f_1(t)\|_{L^2}^2 + \alpha^2\|f_1(t)\|_{L^2}^2\right)dt.$$

Letting $\varepsilon \to 0+$, we obtain

$$\int_0^{+\infty} \left(\|\partial_y\psi_2(t)\|_{L^2}^2 + \alpha^2\|\psi_2(t)\|_{L^2}^2\right)dt \leq C\alpha^{-4}\int_0^T \left(\|\partial_y f_1(t)\|_{L^2}^2 + \alpha^2\|f_1(t)\|_{L^2}^2\right)dt.$$

$$(5.10)$$

Step 3. Recall that $\omega = \omega_1 + \omega_2$, $\psi = \psi_1 + \psi_2$ for $t \in [0, T]$. It follows from (5.2), (5.5), (5.10) and (5.4) that

$$\|\omega(T)\|_{L^2}^2 + \alpha^2 \int_0^T \left(\|\partial_y \psi(t)\|_{L^2}^2 + \alpha^2 \|\psi(t)\|_{L^2}^2 \right) dt$$

$$\leq 2 \sum_{j=1}^2 \|\omega_j(T)\|_{L^2}^2 + 2\alpha^2 \sum_{j=1}^2 \int_0^T \left(\|\partial_y \psi_j(t)\|_{L^2}^2 + \alpha^2 \|\psi_j(t)\|_{L^2}^2 \right) dt$$

$$\leq C\|\omega(0)\|_{L^2}^2 + C \int_0^{+\infty} \left(\alpha^{-2} \|\partial_y f_1(t)\|_{L^2}^2 + \|f_1(t)\|_{L^2}^2 \right.$$

$$\left. + \alpha^2 \|\partial_y \psi_2(t)\|_{L^2}^2 + \alpha^4 \|\psi_2(t)\|_{L^2}^2 \right) dt$$

$$\leq C\|\omega(0)\|_{L^2}^2 + C\alpha^{-2} \int_0^T \left(\|\partial_y f_1(t)\|_{L^2}^2 + \alpha^2 \|f_1(t)\|_{L^2}^2 \right) dt$$

$$\leq C\|\omega(0)\|_{L^2}^2 + C\alpha^{-2} \int_0^T \left(\|\partial_y f(t)\|_{L^2}^2 + \alpha^2 \|f(t)\|_{L^2}^2 \right) dt,$$

which gives the first inequality.

If $f(t, 0) = f(t, 1) = 0$, then $f_1 = 0$, $F_1 = 0$ and $W = 0$ at $y = 0, 1$. Thus, by Proposition 5.1 and (5.9), we deduce that for $\mathrm{Re}(\lambda) \in (0, \alpha\varepsilon_0)$, $j = 0, 1$,

$$|\partial_y \Phi(\lambda, j)| \leq C\alpha^{-\frac{1}{2}} \left(\|\partial_y W\|_{L^2} + \alpha\|W\|_{L^2} \right) = C\alpha^{-\frac{3}{2}} \left(\|\partial_y F_1(\lambda)\|_{L^2} + \alpha\|F_1(\lambda)\|_{L^2} \right).$$

Hence,

$$\alpha|\partial_y \Phi(\lambda, j)|^2 \leq C\alpha^{-2} (\|\partial_y F_1(\lambda)\|_{L^2}^2 + \alpha^2 \|F_1(\lambda)\|_{L^2}^2).$$

Integrating this over $\mathrm{Re}(\lambda) = \varepsilon \in (0, \alpha\varepsilon_0)$ and using (5.7), (5.8), we obtain

$$\alpha \int_0^{+\infty} e^{-2\varepsilon t} |\partial_y \psi_2(t, j)|^2 dt \leq C\alpha^{-2} \int_0^T e^{-2\varepsilon t} \left(\|\partial_y f_1(t)\|_{L^2}^2 + \alpha^2 \|f_1(t)\|_{L^2}^2 \right) dt,$$

$$j = 0, 1.$$

Letting $\varepsilon \to 0+$, we get

$$\alpha \int_0^{+\infty} |\partial_y \psi_2(t, j)|^2 dt \leq C\alpha^{-2} \int_0^T \left(\|\partial_y f_1(t)\|_{L^2}^2 + \alpha^2 \|f_1(t)\|_{L^2}^2 \right) dt, \quad j = 0, 1.$$

$$(5.11)$$

Now the second inequality follows from (5.3), (5.11) and (5.4). $\qquad \square$

5.3 Decay estimates

Let $\psi = -(\partial_y^2 - \alpha^2)^{-1}\omega$ and $\omega(t, y)$ solve $\partial_t \omega + i\alpha \mathcal{R}'_\alpha \omega = 0$ for $t \in [0, +\infty)$, $y \in [0, 1]$.

First of all, it follows from Proposition 5.4 that

$$\sup_{t>0} \|\omega(t)\|_{L^2}^2 + \alpha^2 \int_0^{+\infty} (\|\partial_y\psi(t)\|_{L^2}^2 + \alpha^2\|\psi(t)\|_{L^2}^2)dt \le C\|\omega(0)\|_{L^2}^2, \qquad (5.12)$$

$$\alpha \int_0^{+\infty} (|\partial_y\psi(t,0)|^2 + |\partial_y\psi(t,1)|^2)dt \le C\|\omega(0)\|_{L^2}^2. \qquad (5.13)$$

We introduce the vector field $X = (1/u')\partial_y + i\alpha t$, which commutes with $\partial_t + i\alpha u$. Then we have

$$(\partial_t + i\alpha u)X\omega = X(\partial_t\omega + i\alpha u\omega) = -i\alpha((1/u')\partial_y + i\alpha t)(u''\psi)$$
$$= -i\alpha(u'''/u')\psi - i\alpha u''X\psi.$$

We denote

$$\omega_1 = X\omega, \quad \psi_1 = -(\partial_y^2 - \alpha^2)^{-1}\omega_1,$$
$$\psi_2 = -(\partial_y^2 - \alpha^2)^{-1}(\partial_y\omega/u'), \quad \psi_3 = \psi_2 - \partial_y\psi/u'.$$

Then we find

$$\psi_1 = \psi_2 + i\alpha t\psi, \quad X\psi = \psi_1 - \psi_3.$$

This shows that

$$(\partial_t + i\alpha u)\omega_1 = -i\alpha(u'''/u')\psi - i\alpha u''(\psi_1 - \psi_3),$$

which implies

$$\partial_t\omega_1 + i\alpha\mathcal{R}'_{\alpha,\beta}\omega_1 = (\partial_t + i\alpha u)\omega_1 + i\alpha u''\psi_1$$
$$= -i\alpha(u'''/u')\psi + i\alpha u''\psi_3 := \psi_4.$$

Lemma 5.7. *It holds that for any $t > 0$,*

$$\alpha^2(\|\partial_y\psi(t)\|_{L^2} + \alpha\|\psi(t)\|_{L^2}) \le C(1+t)^{-1}(\|\partial_y\omega(0)\|_{L^2} + \alpha\|\omega(0)\|_{L^2}),$$

where the constant C only depends on u.

Proof. By Proposition 5.4 we have

$$\sup_{t>0} \|\omega_1(t)\|_{L^2}^2 \le C\|\omega_1(0)\|_{L^2}^2 + C\int_0^{+\infty} (\alpha^{-2}\|\partial_y\psi_4(t)\|_{L^2}^2 + \|\psi_4(t)\|_{L^2}^2)dt. \quad (5.14)$$

To proceed, let us first claim that

$$\alpha^{-1}\|\partial_y\psi_4\|_{L^2} + \|\psi_4\|_{L^2} \le C(\|\partial_y\psi\|_{L^2} + \alpha\|\psi\|_{L^2}) + C\alpha^{\frac{1}{2}}(|\partial_y\psi(t,0)| + |\partial_y\psi(t,1)|). \qquad (5.15)$$

Using (5.14), (5.15), (5.12) and (5.13), we conclude that

$$\sup_{t>0} \|\omega_1(t)\|_{L^2}^2 \le C\|\omega_1(0)\|_{L^2}^2 + C\int_0^{+\infty} \left(\|\partial_y\psi(t)\|_{L^2}^2 + \alpha^2\|\psi(t)\|_{L^2}^2\right)dt \qquad (5.16)$$

$$+ C\alpha\int_0^{+\infty}\left(|\partial_y\psi(t,0)|^2 + |\partial_y\psi(t,1)|^2\right)dt$$

$$\le C\|\omega_1(0)\|_{L^2}^2 + C\alpha^{-2}\|\omega(0)\|_{L^2}^2 + C\|\omega(0)\|_{L^2}^2 \le C\|\omega(0)\|_{H^1}^2,$$

here we used the fact that

$$\|\omega_1(0)\|_{L^2} = \|(\partial_y\omega/u' + i\alpha t\omega)|_{t=0}\|_{L^2} = \|\partial_y\omega/u'|_{t=0}\|_{L^2} \le C\|\partial_y\omega|_{t=0}\|_{L^2}$$
$$\le C\|\omega(0)\|_{H^1}.$$

Since $\omega_1 = \partial_y\omega/u' + i\alpha t\omega$, $\psi = -(\partial_y^2 - \alpha^2)^{-1}\omega$, and $1/u' \in C^1([0,1])$, we have

$$\alpha t\left(\|\partial_y\psi\|_{L^2}^2 + \alpha^2\|\psi\|_{L^2}^2\right) = \alpha t\langle\psi,\omega\rangle = i\langle\psi,\omega_1 - \partial_y\omega/u'\rangle$$
$$= i\langle\psi,\omega_1\rangle + i\langle\partial_y(\psi/u'),\omega\rangle$$
$$\le \|\psi\|_{L^2}\|\omega_1\|_{L^2} + \|\partial_y(\psi/u')\|_{L^2}\|\omega\|_{L^2}$$
$$\le \|\psi\|_{L^2}\|\omega_1\|_{L^2} + C\left(\|\psi\|_{L^2} + \|\partial_y\psi\|_{L^2}\right)\|\omega\|_{L^2}$$
$$\le C\left(\alpha\|\psi\|_{L^2} + \|\partial_y\psi\|_{L^2}\right)\left(\alpha^{-1}\|\omega_1\|_{L^2} + \|\omega\|_{L^2}\right),$$

from which, (5.12) and (5.16), we infer that

$$\alpha t\left(\|\partial_y\psi(t)\|_{L^2} + \alpha\|\psi(t)\|_{L^2}\right) \le C\left(\alpha^{-1}\|\omega_1(t)\|_{L^2} + \|\omega(t)\|_{L^2}\right) \qquad (5.17)$$
$$\le C\left(\alpha^{-1}\|\omega(0)\|_{H^1} + \|\omega(0)\|_{L^2}\right)$$
$$\le C\alpha^{-1}\left(\|\partial_y\omega(0)\|_{L^2} + \alpha\|\omega(0)\|_{L^2}\right).$$

On the other hand, using $\|\partial_y\psi\|_{L^2}^2 + \alpha^2\|\psi\|_{L^2}^2 = \langle\psi,\omega\rangle \le \|\psi\|_{L^2}\|\omega\|_{L^2}$, we get

$$\|\partial_y\psi(t)\|_{L^2} + \alpha\|\psi(t)\|_{L^2} \le C\alpha^{-1}\|\omega(t)\|_{L^2} \le C\alpha^{-1}\|\omega(0)\|_{L^2}. \qquad (5.18)$$

Then the lemma is a consequence of (5.17) and (5.18).

It remains to prove (5.15). As $u(y) \in C^4([0,1])$ and $u'(y) \ge c_0$, we have u'''/u', $u'' \in C^1([0,1])$, and

$$\alpha^{-1}\|\partial_y\psi_4\|_{L^2} + \|\psi_4\|_{L^2} \le \|\partial_y((u'''/u')\psi)\|_{L^2} + \|\partial_y(u''\psi_3)\|_{L^2} \qquad (5.19)$$
$$+ \alpha\|(u'''/u')\psi\|_{L^2} + \alpha\|u''\psi_3\|_{L^2}$$
$$\le C\left(\|\partial_y\psi\|_{L^2} + \|\partial_y\psi_3\|_{L^2} + \alpha\|\psi\|_{L^2} + \alpha\|\psi_3\|_{L^2}\right).$$

To estimate ψ_3, we decompose $\psi_3 = \psi_{3,1} + \psi_{3,2}$, where

$$(\partial_y^2 - \alpha^2)\psi_{3,1} = (\partial_y^2 - \alpha^2)\psi_3, \quad (\partial_y^2 - \alpha^2)\psi_{3,2} = 0$$

with $\psi_{3,1} = 0$, $\psi_{3,2} = \psi_3$ at $y = 0,1$. Recall that

$$(\partial_y^2 - \alpha^2)\psi_2 = -\partial_y\omega/u', \quad (\partial_y^2 - \alpha^2)\psi = -\omega.$$

Then we have

$$
\begin{aligned}
(\partial_y^2 - \alpha^2)\psi_{3,1} &= (\partial_y^2 - \alpha^2)\psi_3 = (\partial_y^2 - \alpha^2)(\psi_2 - \partial_y\psi/u') && (5.20)\\
&= (\partial_y^2 - \alpha^2)\psi_2 - \partial_y(\partial_y^2 - \alpha^2)\psi/u' - 2\partial_y(\partial_y\psi(1/u')') + \partial_y\psi(1/u')''\\
&= -\partial_y\omega/u' - \partial_y(-\omega)/u' - 2\partial_y(\partial_y\psi(1/u')') + \partial_y\psi(1/u')''\\
&= -2\partial_y(\partial_y\psi(1/u')') + \partial_y\psi(1/u')'',
\end{aligned}
$$

which implies

$$
\begin{aligned}
\|\partial_y\psi_{3,1}\|_{L^2}^2 + \alpha^2\|\psi_{3,1}\|_{L^2}^2 &= -\langle\psi_{3,1}, (\partial_y^2 - \alpha^2)\psi_{3,1}\rangle = -\langle\psi_{3,1}, (\partial_y^2 - \alpha^2)\psi_3\rangle\\
&= -\langle\psi_{3,1}, -2\partial_y(\partial_y\psi(1/u')') + \partial_y\psi(1/u')''\rangle\\
&= -2\langle\partial_y\psi_{3,1}, \partial_y\psi(1/u')'\rangle - \langle\psi_{3,1}, \partial_y\psi(1/u')''\rangle\\
&\leq C\|\partial_y\psi_{3,1}\|_{L^2}\|\partial_y\psi\|_{L^2} + C\|\psi_{3,1}\|_{L^2}\|\partial_y\psi\|_{L^2}.
\end{aligned}
$$

This shows that

$$
\|\partial_y\psi_{3,1}\|_{L^2} + \alpha\|\psi_{3,1}\|_{L^2} \leq C\|\partial_y\psi\|_{L^2}. \qquad (5.21)
$$

To estimate $\psi_{3,2}$, we recall that

$$
(\partial_y^2 - \alpha^2)\gamma_j = 0, \quad \gamma_j(j) = 1, \quad \gamma_j(1-j) = 0 \text{ for } j \in \{0,1\},
$$

where γ_j is defined in (5.1). So, $\psi_{3,2} = \psi_{3,2}(t,0)\gamma_0 + \psi_{3,2}(t,1)\gamma_1$. Thanks to $|\gamma_j'(j)| = \alpha\coth\alpha \leq C\alpha$ for $j \in \{0,1\}$, we get

$$
\|\gamma_j'\|_{L^2}^2 + \alpha^2\|\gamma_j\|_{L^2}^2 = -\langle\gamma_j, (\partial_y^2 - \alpha^2)\gamma_j\rangle + \gamma_j'\gamma_j|_0^1 = |\gamma_j'\gamma_j(j)| = |\gamma_j'(j)| \leq C\alpha,
$$

which gives

$$
\begin{aligned}
\|\partial_y\psi_{3,2}\|_{L^2} + \alpha\|\psi_{3,2}\|_{L^2} &\leq |\psi_{3,2}(t,0)|(\|\gamma_0'\|_{L^2} + \alpha\|\gamma_0\|_{L^2})\\
&\quad + |\psi_{3,2}(t,1)|(\|\gamma_1'\|_{L^2} + \alpha\|\gamma_1\|_{L^2})\\
&\leq C\alpha^{\frac{1}{2}}\big(|\psi_{3,2}(t,0)| + |\psi_{3,2}(t,1)|\big).
\end{aligned}
$$

Thanks to $\psi_{3,2}(t,j) = \psi_3(t,j)$, $\psi_2(t,j) = 0$ for $j \in \{0,1\}$, and $\psi_3 = \psi_2 - \partial_y\psi/u'$, we get

$$
|\psi_{3,2}(t,j)| = |\psi_3(t,j)| = |\partial_y\psi(t,j)/u'(j)| \leq C|\partial_y\psi(t,j)|
$$

and hence,

$$
\begin{aligned}
\|\partial_y\psi_{3,2}\|_{L^2} + \alpha\|\psi_{3,2}\|_{L^2} &\leq C\alpha^{\frac{1}{2}}(|\psi_{3,2}(t,0)| + |\psi_{3,2}(t,1)|) && (5.22)\\
&\leq C\alpha^{\frac{1}{2}}(|\partial_y\psi(t,0)| + |\partial_y\psi(t,1)|).
\end{aligned}
$$

Now (5.15) follows from (5.19), (5.21) and (5.22). $\square$

Since $\psi(t, j) = 0$ for $j = 0, 1$, we have

$$\partial_t \omega(t, j) + i\alpha u(j)\omega(t, j) = 0, \quad \text{for } j \in \{0, 1\},$$

and $|\omega(t, j)| = |e^{-i\alpha t u(j)}\omega(0, j)| \leq \|\omega(0)\|_{L^\infty}$. With γ_j defined as above, using the fact that

$$\langle \omega, \gamma_1 \rangle = -\langle (\partial_y^2 - \alpha^2)\psi, \gamma_1 \rangle = -\langle \psi, (\partial_y^2 - \alpha^2)\gamma_1 \rangle - (\partial_y \psi \gamma_1 - \psi \gamma_1')|_0^1 = -\partial_y \psi(t, 1),$$

we infer that for any $t > 0$,

$$\begin{aligned}
\alpha t|\partial_y \psi(t, 1)| &= \alpha t|\langle \omega, \gamma_1 \rangle| = |\langle \omega_1 - \partial_y \omega/u', \gamma_1 \rangle| \\
&= \left|\langle \omega_1, \gamma_1 \rangle + \langle \omega, (\gamma_1/u')' \rangle - \omega \gamma_1/u'|_{y=0}^1\right| \\
&\leq \|\omega_1\|_{L^2}\|\gamma_1\|_{L^2} + \|\omega\|_{L^2}\|(\gamma_1/u')'\|_{L^2} + |\omega(t, 1)/u'(1)| \\
&\leq \|\omega_1\|_{L^2}\|\gamma_1\|_{L^2} + C\|\omega\|_{L^2}(\|\gamma_1\|_{L^2} + \|\gamma_1'\|_{L^2}) + C\|\omega(0)\|_{L^\infty} \\
&\leq C\|\omega_1\|_{L^2}\alpha^{-\frac{1}{2}} + C\|\omega\|_{L^2}(\alpha^{-\frac{1}{2}} + \alpha^{\frac{1}{2}}) + C\|\omega(0)\|_{L^\infty} \\
&\leq C\alpha^{\frac{1}{2}}(\alpha^{-1}\|\omega_1\|_{L^2} + \|\omega\|_{L^2}) + C\|\omega(0)\|_{H^1}^{\frac{1}{2}}\|\omega(0)\|_{L^2}^{\frac{1}{2}} \\
&\leq C\alpha^{\frac{1}{2}}(\alpha^{-1}\|\omega(0)\|_{H^1} + \|\omega(0)\|_{L^2}) \leq C\alpha^{-\frac{1}{2}}(\|\partial_y\omega(0)\|_{L^2} + \alpha\|\omega(0)\|_{L^2}).
\end{aligned}$$

On the other hand, we have

$$|\partial_y \psi(t, 1)| = |\langle \omega, \gamma_1 \rangle| \leq \|\omega\|_{L^2}\|\gamma_1\|_{L^2} \leq C\|\omega(t)\|_{L^2}\alpha^{-\frac{1}{2}} \leq C\|\omega(0)\|_{L^2}\alpha^{-\frac{1}{2}}.$$

This shows that

$$|\partial_y \psi(t, 1)| \leq C\alpha^{-\frac{3}{2}}(1 + t)^{-1}(\|\partial_y\omega(0)\|_{L^2} + \alpha\|\omega(0)\|_{L^2}). \tag{5.23}$$

Similarly, we have

$$|\partial_y \psi(t, 0)| = |\langle \omega, \gamma_0 \rangle| \leq C\alpha^{-\frac{3}{2}}(1 + t)^{-1}(\|\partial_y\omega(0)\|_{L^2} + \alpha\|\omega(0)\|_{L^2}). \tag{5.24}$$

The following lemma is devoted to the decay estimate for the second component of the velocity. For this, we introduce the following norms:

$$\|\omega\|_{-2}^2 = \|\psi\|_{L^2}^2, \quad \|\omega\|_{-1}^2 = \|\partial_y\psi\|_{L^2}^2 + \alpha^2\|\psi\|_{L^2}^2, \quad \|\omega\|_0^2 = \|\omega\|_{L^2}^2,$$

where $\psi = -(\partial_y^2 - \alpha^2)^{-1}\omega$ and

$$\|\omega\|_1^2 = \|\partial_y\omega\|_{L^2}^2 + \alpha^2\|\omega\|_{L^2}^2, \quad \|\omega\|_2^2 = \|\partial_y^2\omega\|_{L^2}^2 + 2\alpha^2\|\partial_y\omega\|_{L^2}^2 + \alpha^4\|\omega\|_{L^2}^2.$$

Since $\|\omega\|_0^2 = \|(\partial_y^2 - \alpha^2)\psi\|_{L^2}^2 = \|\partial_y^2\psi\|_{L^2}^2 + 2\alpha^2\|\partial_y\psi\|_{L^2}^2 + \alpha^4\|\psi\|_{L^2}^2$, we have

$$\alpha^{k-j}\|\omega\|_j \leq \|\omega\|_k \quad \text{for every} \ -2 \leq j \leq k \leq 2.$$

We denote by the semigroup $\omega(t) = e^{-it\alpha\mathcal{R}_\alpha'}\omega_0$ the solution to $\partial_t\omega + i\alpha\mathcal{R}_\alpha'\omega = 0$, $\omega(0) = \omega_0$. Then Lemma 5.7 and (5.12) imply that for every $f_0 \in H^1(0, 1)$ and $t > 0$,

$$\alpha^2\|e^{-it\alpha\mathcal{R}_\alpha'}f_0\|_{-1} \leq C(1 + t)^{-1}\|f_0\|_1, \quad \|e^{-it\alpha\mathcal{R}_\alpha'}f_0\|_0 \leq C\|f_0\|_0. \tag{5.25}$$

Lemma 5.8. *It holds that for any $t > 0$,*

$$\alpha^4 \|\psi(t)\|_{L^2} \leq C(1+t)^{-2}(\|\partial_y^2 \omega(0)\|_{L^2} + \alpha\|\partial_y\omega(0)\|_{L^2} + \alpha^2\|\omega(0)\|_{L^2}),$$

where the constant C only depends on u.

Proof. It suffices to show that for every $f \in H^2(0,1)$ and $t > 0$,

$$\alpha^4 \|e^{-it\alpha\mathcal{R}'_\alpha} f\|_{-2} \leq C(1+t)^{-2}\|f\|_2.$$

For $T > 0$, we define

$$M = M(T) := \sup\left\{\alpha^4(1+t)^2\|e^{-it\alpha\mathcal{R}'_\alpha} f\|_{-2} : 0 < t < T, \; f \in H^2(0,1), \; \|f\|_2 \leq 1\right\}.$$

First of all, we get by (5.25) that

$$
\begin{aligned}
\alpha^4(1+t)^2\|e^{-it\alpha\mathcal{R}'_\alpha} f\|_{-2} &\leq \alpha^2(1+t)^2\|e^{-it\alpha\mathcal{R}'_\alpha} f\|_0 \\
&\leq C\alpha^2(1+t)^2\|f\|_0 \leq C(1+T)^2\|f\|_2,
\end{aligned}
$$

which implies that $M(T) \leq C(1+T)^2$. Now we fix $T > 0$ and assume $M = M(T) > 1$. We will show that

$$M(T) \leq C\big(\ln(M(T)+1)+1\big)$$

with C independent of T and α.

Let us first claim that for $0 < t < T$ and $\omega_1 = X\omega$,

$$\|\omega_1(t)\|_{-2} \leq C\alpha^{-3}t^{-1}\|\omega(0)\|_2\big(1 + \ln(M+1)\big), \tag{5.26}$$

which will be proved in Lemma 5.9. Now we presume $f = \omega(0)$ from here on.

Recall that $\psi_1 = -(\partial_y^2 - \alpha^2)^{-1}\omega_1$, $\psi_3 = \psi_2 - \partial_y\psi/u'$, $\psi_1 = \psi_2 + i\alpha t\psi$, $\psi_3 = \psi_{3,1} + \psi_{3,2}$. Then by (5.21), (5.22), (5.17), (5.23) and (5.24), we get

$$
\begin{aligned}
\alpha t\|\psi\|_{L^2} &\leq \|\psi_1\|_{L^2} + \|\psi_2\|_{L^2} \leq \|\omega_1\|_{-2} + \|\psi_3\|_{L^2} + \|\partial_y\psi/u'\|_{L^2} \\
&\leq \|\omega_1\|_{-2} + \|\psi_{3,1}\|_{L^2} + \|\psi_{3,2}\|_{L^2} + C\|\partial_y\psi\|_{L^2} \\
&\leq \|\omega_1\|_{-2} + C\alpha^{-1}\|\partial_y\psi\|_{L^2} + C\alpha^{-\frac{1}{2}}(|\partial_y\psi(t,0)| + |\partial_y\psi(t,1)|) + C\|\partial_y\psi\|_{L^2} \\
&\leq \|\omega_1\|_{-2} + C\alpha^{-2}t^{-1}(\|\partial_y\omega(0)\|_{L^2} + \alpha\|\omega(0)\|_{L^2}).
\end{aligned}
$$

This means that

$$\alpha t\|\psi(t)\|_{L^2} \leq \|\omega_1(t)\|_{-2} + C\alpha^{-2}t^{-1}\|\omega(0)\|_1,$$

which along with (5.26) gives

$$
\begin{aligned}
\alpha t\|\psi(t)\|_{L^2} &\leq \|\omega_1(t)\|_{-2} + C\alpha^{-2}t^{-1}\|\omega(0)\|_1 \\
&\leq C\alpha^{-3}t^{-1}\|\omega(0)\|_2(1 + \ln(M+1)) + C\alpha^{-3}t^{-1}\|\omega(0)\|_2 \\
&\leq C\alpha^{-3}t^{-1}\|\omega(0)\|_2(1 + \ln(M+1)).
\end{aligned}
$$

And by (5.18), we have

$$\|\psi(t)\|_{L^2} \leq C\alpha^{-2}\|\omega(0)\|_{L^2} \leq C\alpha^{-4}\|\omega(0)\|_2.$$

Then we conclude that for $0 < t < T$,

$$\begin{aligned}
\|e^{-it\alpha\mathcal{R}'_\alpha}\omega(0)\|_{-2} = \|\omega(t)\|_{-2} &= \|\psi(t)\|_{L^2} \\
&\leq C\alpha^{-4}\|\omega(0)\|_2 \min(t^{-2}(1+\ln(M+1)),1) \\
&\leq C\alpha^{-4}(1+t)^{-2}\|\omega(0)\|_2 (1+\ln(M+1)).
\end{aligned}$$

Here C is a constant independent of T, α and $\omega(0)$. Thanks to the definition of $M(T)$, we have

$$M(T) \leq C(1+\ln(M(T)+1)).$$

Thus, there exists a constant $C_0 > 0$ independent of T and α so that if $M(T) > 1$, then $M(T) \leq C_0(1+\ln(M(T)+1))$. This implies the existence of a constant $C_1 > 1$ so that $M(T) < C_1$ for every $T > 0$. Now we have

$$\begin{aligned}
\alpha^4\|\psi(t)\|_{L^2} - \alpha^4\|e^{-it\alpha\mathcal{R}'_\alpha}\omega(0)\|_{-2} &\leq C_1(1+t)^{-2}\|\omega(0)\|_2 \\
&\leq C_1(1+t)^{-2}(\|\partial_y^2\omega(0)\|_{L^2} + 2\alpha\|\partial_y\omega(0)\|_{L^2} + \alpha^2\|\omega(0)\|_{L^2}),
\end{aligned}$$

which gives our result. □

Lemma 5.9. *Let $\omega_1 = X\omega$. It holds that for any $0 < t < T$,*

$$\|\omega_1(t)\|_{-2} \leq C\alpha^{-3}t^{-1}\|\omega(0)\|_2(1+\ln(M+1)),$$

where the constant C is independent of T and α.

Proof. Recall that $\partial_t\omega_1 + i\alpha\mathcal{R}'_\alpha\omega_1 = \psi_4$. By Duhamel's principle, we get

$$\omega_1(t) = e^{-it\alpha\mathcal{R}'_\alpha}\omega_1(0) + \int_0^t e^{-i(t-s)\alpha\mathcal{R}'_\alpha}\psi_4(s)ds,$$

from which, we infer that

$$\|\omega_1(t)\|_{-2} \leq \|e^{-it\alpha\mathcal{R}'_\alpha}\omega_1(0)\|_{-2} + \int_0^t \|e^{-i(t-s)\alpha\mathcal{R}'_\alpha}\psi_4(s)\|_{-2}ds. \tag{5.27}$$

Thanks to $\omega_1(0,y) = \partial_y\omega(0,y)/u'(y)$, we get

$$\|\omega_1(0)\|_1 = \|\partial_y\omega(0)/u'\|_1 \leq C\|\partial_y\omega(0)\|_1 \leq C\|\omega(0)\|_2,$$

which along with (5.25) gives

$$\begin{aligned}
\|e^{-it\alpha\mathcal{R}'_\alpha}\omega_1(0)\|_{-2} &\leq \alpha^{-1}\|e^{-it\alpha\mathcal{R}'_\alpha}\omega_1(0)\|_{-1} \leq C\alpha^{-3}t^{-1}\|\omega_1(0)\|_1 \tag{5.28} \\
&\leq C\alpha^{-3}t^{-1}\|\omega(0)\|_2.
\end{aligned}$$

By (5.15), (5.17), (5.23) and (5.24), we have

$$\begin{aligned}
\alpha^{-1}\|\psi_4(t)\|_1 &\leq C(\|\partial_y\psi(t)\|_{L^2} + \alpha\|\psi(t)\|_{L^2}) + C\alpha^{\frac12}(|\partial_y\psi(t,0)| + |\partial_y\psi(t,1)|) \\
&\leq C(\alpha^{-2}+\alpha^{-1})t^{-1}(\|\partial_y\omega(0)\|_{L^2} + \alpha\|\omega(0)\|_{L^2}) \leq C\alpha^{-1}t^{-1}\|\omega(0)\|_1.
\end{aligned}$$

from which and (5.25), we infer that for $t > s > 0$,

$$\|e^{-i(t-s)\alpha \mathcal{R}'_\alpha} \psi_4(s)\|_{-2} \le \alpha^{-1} \|e^{-i(t-s)\alpha \mathcal{R}'_\alpha} \psi_4(s)\|_{-1} \le C\alpha^{-3}(t-s)^{-1}\|\psi_4(s)\|_1 \tag{5.29}$$

$$\le C\alpha^{-3}(t-s)^{-1}s^{-1}\|\omega(0)\|_1.$$

As $(t-s)^{-1}s^{-1}$ is not integrable, we have to improve the estimate for s close to t or 0. To this end, we decompose $\psi_4 = -i\alpha(u'''/u')\psi + i\alpha u'' \psi_3 = \psi_{4,1} + \psi_{4,2} + \psi_{4,3} + \psi_{4,4}$, where

$$\psi_{4,1}(t, y) = -i\alpha \int_0^y (u'''/u')'(z)\psi(t, z)dz,$$

$$\psi_{4,2}(t, y) = -i\alpha \int_0^y (u'''/u')(z)\partial_y\psi(t, z)dz,$$

$$\psi_{4,3} = i\alpha u'' \psi_{3,1}, \quad \psi_{4,4} = i\alpha u'' \psi_{3,2}.$$

Then we have

$$\psi_{4,1} + \psi_{4,2} = -i\alpha(u'''/u')\psi, \quad \psi_{4,3} + \psi_{4,4} = i\alpha u'' \psi_3.$$

Thanks to the definition of $M = M(T)$, we deduce that for any $f \in H^2(0,1)$ and $0 < s < T$,

$$\|e^{-is\alpha \mathcal{R}'_\alpha} f\|_{-2} \le M\alpha^{-4}(1+s)^{-2}\|f\|_2,$$

and by (5.25), we have

$$\|e^{-is\alpha \mathcal{R}'_\alpha} f\|_{-2} \le \alpha^{-1}\|e^{-is\alpha \mathcal{R}'_\alpha} f\|_{-1} \le C\alpha^{-3}(1+s)^{-1}\|f\|_1$$

$$\le C\alpha^{-4}(1+s)^{-1}\|f\|_2.$$

Therefore,

$$\|e^{-is\alpha \mathcal{R}'_\alpha} f\|_{-2} \le \alpha^{-4}\min(M(1+s)^{-2}, C(1+s)^{-1})\|f\|_2 \tag{5.30}$$

$$\le CM\alpha^{-4}(1+s)^{-1}(1+M+s)^{-1}\|f\|_2,$$

which implies

$$\|\psi(s)\|_{L^2} = \|\omega(s)\|_{-2} = \|e^{-is\alpha \mathcal{R}'_\alpha} \omega(0)\|_{-2} \le \frac{CM\|\omega(0)\|_2}{\alpha^4(1+s)(1+M+s)}. \tag{5.31}$$

Let $\psi_{3,3} = (\partial_y^2 - \alpha^2)^{-1}\psi_{3,1}$. Using (5.20) and $\psi_{3,1} = 0$ at $y = 0, 1$, we get

$$\|\psi_{3,1}\|_{L^2}^2 = \langle \psi_{3,1}, (\partial_y^2 - \alpha^2)\psi_{3,3} \rangle = \langle (\partial_y^2 - \alpha^2)\psi_{3,1}, \psi_{3,3} \rangle$$

$$= \langle -2(\partial_y\psi(1/u')')' + \partial_y\psi(1/u')'', \psi_{3,3} \rangle$$

$$= -2\langle \psi, \partial_y(\partial_y\psi_{3,3}(1/u')') \rangle - \langle \psi, \partial_y(\psi_{3,3}(1/u')'') \rangle$$

$$= -\langle \psi, 2\partial_y^2\psi_{3,3}(1/u')' + 3\partial_y\psi_{3,3}(1/u')'' + \psi_{3,3}(1/u')''' \rangle$$

$$\le C\|\psi\|_{L^2}(\|\partial_y^2\psi_{3,3}\|_{L^2} + \|\partial_y\psi_{3,3}\|_{L^2} + \|\psi_{3,3}\|_{L^2}) \le C\|\psi\|_{L^2}\|\psi_{3,1}\|_{L^2},$$

here we used $\|\psi_{3,1}\|_{L^2}^2 = \|\partial_y^2\psi_{3,3}\|_{L^2}^2 + 2\alpha^2\|\partial_y\psi_{3,3}\|_{L^2}^2 + \alpha^4\|\psi_{3,3}\|_{L^2}^2$. This gives $\|\psi_{3,1}\|_{L^2} \le C\|\psi\|_{L^2}$, and then

$$\begin{aligned}
\|(\psi_{4,1} + \psi_{4,2} + \psi_{4,3})(s)\|_0 &= \| - i\alpha(u'''/u')\psi(s) + i\alpha u''\psi_{3,1}(s)\|_0 \qquad (5.32)\\
&\le C\alpha\|\psi(s)\|_{L^2} + C\alpha\|\psi_{3,1}(s)\|_{L^2}\\
&\le C\alpha\|\psi(s)\|_{L^2} \le CM\alpha^{-3}s^{-2}\|\omega(0)\|_2.
\end{aligned}$$

Using $\|\partial_y^2\psi_{3,2}\|_{L^2} = \alpha^2\|\psi_{3,2}\|_{L^2}$, (5.22), (5.23) and (5.24), we obtain

$$\begin{aligned}
\|\psi_{3,2}\|_2^2 &= \|\partial_y^2\psi_{3,2}\|_{L^2}^2 + 2\alpha^2\|\partial_y\psi_{3,2}\|_{L^2}^2 + \alpha^4\|\psi_{3,2}\|_{L^2}^2\\
&\le C\alpha^2(\|\partial_y\psi_{3,2}\|_{L^2} + \alpha\|\psi_{3,2}\|_{L^2})^2 \le C\alpha^3(|\partial_y\psi(t,0)| + |\partial_y\psi(t,1)|)^2\\
&\le C(1+t)^{-2}(\|\partial_y\omega(0)\|_{L^2} + \alpha\|\omega(0)\|_{L^2})^2 \le C(1+t)^{-2}\|\omega(0)\|_1^2,
\end{aligned}$$

which gives

$$\|\psi_{3,2}(t)\|_2 \le C(1+t)^{-1}\|\omega(0)\|_1,$$

and

$$\|\psi_{4,4}(s)\|_2 = \|i\alpha u''\psi_{3,2}(s)\|_2 \le C\alpha\|\psi_{3,2}(s)\|_2 \le C\alpha(1+s)^{-1}\|\omega(0)\|_1. \qquad (5.33)$$

Then by (5.25), (5.32), (5.33) and (5.30), we infer that for $0 < s < t < T$,

$$\begin{aligned}
\|e^{-i(t-s)\alpha\mathcal{R}'_\alpha}\psi_4(s)\|_{-2} &\le \|e^{-i(t-s)\alpha\mathcal{R}'_\alpha}(\psi_{4,1} + \psi_{4,2} + \psi_{4,3})(s)\|_{-2}\\
&\quad + \|e^{-i(t-s)\alpha\mathcal{R}'_\alpha}\psi_{4,4}(s)\|_{-2}\\
&\le \alpha^{-2}\|e^{-i(t-s)\alpha\mathcal{R}'_\alpha}(\psi_{4,1} + \psi_{4,2} + \psi_{4,3})(s)\|_0\\
&\quad + CM\alpha^{-4}(1+t-s)^{-1}(1+M+t-s)^{-1}\|\psi_{4,4}(s)\|_2\\
&\le C\alpha^{-2}\|(\psi_{4,1} + \psi_{4,2} + \psi_{4,3})(s)\|_0 \qquad (5.34)\\
&\quad + \frac{CM\|\omega(0)\|_1}{\alpha^3(1+t-s)(1+M+t-s)s}\\
&\le \frac{CM\|\omega(0)\|_2}{\alpha^5 s^2} + \frac{CM\|\omega(0)\|_1}{\alpha^3(1+t-s)(1+M+t-s)s}.
\end{aligned}$$

Thanks to $\psi_{3,1} = 0$, $\psi = 0$ at $y = 0, 1$, we get by (5.20) and (5.12) that

$$\begin{aligned}
\|\psi_{3,1}\|_2 &= \|(\partial_y^2 - \alpha^2)\psi_{3,1}\|_{L^2} = \| - 2\partial_y(\partial_y\psi(1/u')) + \partial_y\psi(1/u')''\|_{L^2}\\
&\le C(\|\partial_y^2\psi\|_{L^2} + \|\partial_y\psi\|_{L^2}) \le C\|(\partial_y^2 - \alpha^2)\psi\|_{L^2} = C\|\omega\|_0 \le C\|\omega(0)\|_0,
\end{aligned}$$

which gives

$$\|\psi_{4,3}(s)\|_2 = \|i\alpha u''\psi_{3,1}(s)\|_2 \le C\alpha\|\psi_{3,1}(s)\|_2 \le C\alpha\|\omega(0)\|_0. \qquad (5.35)$$

Since $\partial_y\psi_{4,1} = -i\alpha(u'''/u')'\psi$ and $\psi_{4,1}(t,0) = 0$, we get

$$\|\psi_{4,1}\|_{L^2} \le \|\partial_y\psi_{4,1}\|_{L^2} = \| - i\alpha(u'''/u')'\psi\|_{L^2} \le C\alpha\|\psi\|_{L^2}.$$

Thanks to $\partial_y \psi_{4,2} = -i\alpha(u'''/u')\partial_y\psi$ and $\psi_{4,1} + \psi_{4,2} = -i\alpha(u'''/u')\psi$, we get

$$\|\psi_{4,2}\|_{L^2} \leq \|\psi_{4,1}\|_{L^2} + \|\psi_{4,1} + \psi_{4,2}\|_{L^2}$$
$$\leq C\alpha\|\psi\|_{L^2} + \|\alpha(u'''/u')\psi\|_{L^2} \leq C\alpha\|\psi\|_{L^2},$$
$$\alpha\|\partial_y\psi_{4,2}\|_{L^2} + \|\partial_y^2\psi_{4,2}\|_{L^2} \leq C\|\partial_y\psi_{4,2}\|_1 = C\|\alpha(u'''/u')\partial_y\psi\|_1$$
$$\leq C\alpha\|\partial_y\psi\|_1 \leq C\alpha\|\psi\|_2.$$

Summing up, we conclude that

$$\|\psi_{4,1}\|_1 \leq C\alpha^2\|\psi\|_{L^2}, \ \|\psi_{4,2}\|_2 \leq C\alpha^3\|\psi\|_{L^2} + C\alpha\|\psi\|_2 \leq C\alpha\|\psi\|_2,$$

which together with (5.31) and (5.12) gives

$$\|\psi_{4,1}(s)\|_1 \leq C\alpha^2\|\psi(s)\|_{L^2} \leq CM\alpha^{-2}(1+s)^{-1}(1+M+s)^{-1}\|\omega(0)\|_2, \quad (5.36)$$
$$\|\psi_{4,2}(s)\|_2 \leq C\alpha\|\psi(s)\|_2 = C\alpha\|\omega(s)\|_0 \leq C\alpha\|\omega(0)\|_0. \quad (5.37)$$

It follows from (5.25), (5.33), (5.35), (5.36) and (5.37) that for $0 < s < t < T$,

$$\|e^{-i(t-s)\alpha\mathcal{R}'_\alpha}\psi_4(s)\|_{-2} \quad (5.38)$$

$$\leq \sum_{j=1}^{4} \|e^{-i(t-s)\alpha\mathcal{R}'_\alpha}\psi_{4,j}(s)\|_{-2}$$

$$\leq \alpha^{-1}\|e^{-i(t-s)\alpha\mathcal{R}'_\alpha}\psi_{4,1}(s)\|_{-1}$$
$$+ M\alpha^{-4}(t-s)^{-2}(\|\psi_{4,2}(s)\|_2 + \|\psi_{4,3}(s)\|_2 + \|\psi_{4,4}(s)\|_2)$$
$$\leq C\alpha^{-3}(t-s)^{-1}\|\psi_{4,1}(s)\|_1 + CM\alpha^{-4}(t-s)^{-2}\alpha\|\omega(0)\|_1$$
$$\leq \frac{CM\|\omega(0)\|_2}{\alpha^5(t-s)(1+s)(1+M+s)} + \frac{CM\|\omega(0)\|_1}{\alpha^3(t-s)^2}.$$

Then we infer from (5.27), (5.28), (5.29), (5.34) and (5.38) that

$$\|\omega_1(t)\|_{-2}$$

$$\leq \|e^{-it\alpha\mathcal{R}'_\alpha}\omega_1(0)\|_{-2} + \int_0^{\frac{t}{M+1}} \|e^{-i(t-s)\alpha\mathcal{R}'_\alpha}\psi_4(s)\|_{-2}ds$$

$$+ \int_{\frac{t}{M+1}}^{\frac{Mt}{M+1}} \|e^{-i(t-s)\alpha\mathcal{R}'_\alpha}\psi_4(s)\|_{-2}ds + \int_{\frac{Mt}{M+1}}^{t} \|e^{-i(t-s)\alpha\mathcal{R}'_\alpha}\psi_4(s)\|_{-2}ds$$

$$\leq C\alpha^{-3}t^{-1}\|\omega(0)\|_2 + \int_0^{\frac{t}{M+1}} \left(\frac{CM\|\omega(0)\|_2}{\alpha^5(t-s)(1+s)(1+M+s)} + \frac{CM\|\omega(0)\|_1}{\alpha^3(t-s)^2}\right) ds$$

$$+ \int_{\frac{t}{M+1}}^{\frac{Mt}{M+1}} C\alpha^{-3}(t-s)^{-1}s^{-1}\|\omega(0)\|_1 ds$$

$$+ \int_{\frac{Mt}{M+1}}^{t} \left(\frac{CM\|\omega(0)\|_2}{\alpha^5 s^2} + \frac{CM\|\omega(0)\|_1}{\alpha^3(1+t-s)(1+M+t-s)s}\right) ds$$

$$\leq C\alpha^{-3}t^{-1}\|\omega(0)\|_2 + \int_0^{\frac{t}{M+1}} \left(\frac{CM\|\omega(0)\|_2}{\alpha^5 t(1+s)(1+M+s)} + \frac{CM\|\omega(0)\|_1}{\alpha^3 t^2}\right) ds$$

$$+ \int_{\frac{t}{M+1}}^{\frac{t}{2}} C\alpha^{-3}t^{-1}s^{-1}\|\omega(0)\|_1 ds + \int_{\frac{t}{2}}^{\frac{Mt}{M+1}} C\alpha^{-3}(t-s)^{-1}t^{-1}\|\omega(0)\|_1 ds$$

$$+ \int_{\frac{Mt}{M+1}}^{t} \left(\frac{CM\|\omega(0)\|_2}{\alpha^5 t^2} + \frac{CM\|\omega(0)\|_1}{\alpha^3(1+t-s)(1+M+t-s)t} \right) ds$$

$$\leq C\alpha^{-3}t^{-1}\|\omega(0)\|_2 + C\alpha^{-5}t^{-1}\|\omega(0)\|_2 \ln(M+1) + \frac{CM\|\omega(0)\|_1}{\alpha^3 t^2} \frac{t}{M+1}$$

$$+ 2C\alpha^{-3}t^{-1}\|\omega(0)\|_1 \ln\frac{M+1}{2} + \frac{CM\|\omega(0)\|_2}{\alpha^5 t^2} \frac{t}{M+1} + \frac{C\|\omega(0)\|_1 \ln(M+1)}{\alpha^3 t}$$

$$\leq C\alpha^{-3}t^{-1}\|\omega(0)\|_2(1+\ln(M+1)).$$

Here we used the facts that $\|\omega(0)\|_1 \leq C\alpha^{-1}\|\omega(0)\|_2 \leq C\|\omega(0)\|_2$ and

$$\int_0^{+\infty} \frac{M}{(1+s)(1+M+s)} ds = \ln\frac{1+s}{1+M+s}\Big|_0^{+\infty} = \ln(M+1).$$

This completes the proof of the lemma.   □

Finally, we prove the linear inviscid damping for monotone shear flows.

Proof. Here we only need the following slightly weak results in Lemmas 5.7 and 5.8 (the case $\alpha < 0$ or $t < 0$ can be proved by taking conjugation):

$$|\alpha|(\|\partial_y\psi(t)\|_{L^2} + |\alpha|\|\psi(t)\|_{L^2}) \leq C\langle t\rangle^{-1}\|\omega(0)\|_{H^1},$$
$$|\alpha|^2\|\psi(t)\|_{L^2} \leq C\langle t\rangle^{-2}\|\omega(0)\|_{H^2}. \tag{5.39}$$

Thanks to $V = \nabla^\perp \psi = (\psi_y, -\psi_x)$, we get by (5.39) that

$$\|V(t)\|_{L^2_{x,y}}^2 = C\sum_{\alpha\neq 0}(\alpha^2\|\widehat{\psi}(t,\alpha,\cdot)\|_{L^2_y}^2 + \|\partial_y\widehat{\psi}(t,\alpha,\cdot)\|_{L^2_y}^2)$$

$$\leq C\sum_{\alpha\neq 0}|\alpha|^{-2}\langle t\rangle^{-2}\|\widehat{\omega}_0(\alpha,\cdot)\|_{H^1_y}^2 \leq C\langle t\rangle^{-2}\|\omega_0\|_{H^{-1}_x H^1_y}^2,$$

and

$$\|V^2(t)\|_{L^2_{x,y}}^2 = C\sum_{\alpha\neq 0}\alpha^2\|\widehat{\psi}(t,\alpha,\cdot)\|_{L^2_y}^2 \leq C\sum_{\alpha\neq 0} \frac{\|\widehat{\omega}_0(\alpha,\cdot)\|_{H^2_y}^2}{|\alpha|^2\langle t\rangle^4} \leq C\frac{\|\omega_0\|_{H^{-1}_x H^2_y}^2}{\langle t\rangle^4}.$$

This shows that

$$\|V(t)\|_{L^2_{x,y}} \leq C\langle t\rangle^{-1}\|\omega_0\|_{H^{-1}_x H^1_y}, \quad \|V^2(t)\|_{L^2_{x,y}} \leq C\langle t\rangle^{-2}\|\omega_0\|_{H^{-1}_x H^2_y}.$$

□

References

[1] J. Bedrossian, M. Coti Zelati and V. Vicol, *Vortex axisymmetrization, inviscid damping, and vorticity depletion in the linearized 2D Euler equations*, Ann. PDE, 5 (2019), Art. 4, 192 pp.

[2] J. Bedrossian and N. Masmoudi, *Inviscid damping and the asymptotic stability of planar shear flows in the 2D Euler equations*, Publ. Math. Inst. Hautes Études Sci., 122 (2015), 195–300.

[3] F. Bouchet and H. Morita, *Large time behavior and asymptotic stability of the 2D Euler and linearized Euler equations*, Physica D, 239 (2010), 948–966.

[4] K. Case, *Stability of inviscid plane Couette flow*, Phys. Fluids, 3 (1960), 143–148.

[5] Y. Deng and N. Masmoudi, *Long time instability of the Couette flow in low Gevrey spaces*, arXiv:1803.01246.

[6] E. Grenier, T. Nguyen, F. Rousset and A. Soffer, *Linear inviscid damping and enhanced viscous dissipation of shear flows by using the conjugate operator method*, J. Funct. Anal., 278 (2020), 108339, 27 pp.

[7] A. Ionescu and H. Jia, *Inviscid damping near the Couette flow in a channel*, Comm. Math. Phys., 374 (2020), 2015–2096.

[8] A. Ionescu and H. Jia, *Nonlinear inviscid damping near monotonic shear flows*, arXiv:2001.03087.

[9] A. Ionescu and H. Jia, *Axi-symmetrization near point vortex solutions for the 2D Euler equation*, arXiv:1904.09170.

[10] L. Landau, *On the vibration of the electronic plasma*, J. Phys. USSR, 10 (1946), 25.

[11] Z. Lin, *Instability of some ideal plane flows*, SIAM J. Math. Anal., 35 (2003), 318–356.

[12] Z. Lin and M. Xu, *Metastability of Kolmogorov flows and inviscid damping of shear flows*, Arch. Ration. Mech. Anal., 231 (2019), 1811–1852.

[13] Z. Lin and C. Zeng, *Inviscid dynamic structures near Couette flow*, Arch. Ration. Mech. Anal., 200 (2011), 1075–1097.

[14] N. Masmoudi and Z. Zhao, *Nonlinear inviscid damping for a class of monotone shear flows in finite channel*, arXiv:2001.08564.

[15] C. Mouhot and C. Villani, *On Landau damping*, Acta Math., 207 (2011), 29–201.

[16] W. Orr, *Stability and instability of steady motions of a perfect liquid*, Proc. Ir. Acad. Sect. A: Math Astron. Phys. Sci., 27(1907), 9–66.

[17] L. Rayleigh, *On the stability or instability of certain fluid motions*, Proc. London Math. Soc., 9 (1880), 57–70.

[18] S. Ren and W. Zhao, *Linear damping of Alfvén waves by phase mixing*, SIAM J. Math. Anal., 49 (2017), 2101–2137.

[19] S. I. Rosencrans and D. H. Sattinger, *On the spectrum of an operator occurring in the theory of hydrodynamics stability*, J. Math. Phys., 45 (1966), 289–300.

[20] D. Ryutov, *Landau damping: half a century with the great discovery*, Plasma Phys. Control Fusion, 41 (1999), A1–A12.

[21] S. Stepin, *Nonself-adjoint Friedrichs models in hydrodynamics stability*, Func-

tional Analysis and Its Applications, 29 (1995), 91–101.

[22] D. Wei, Z. Zhang and W. Zhao, *Linear inviscid damping for a class of monotone shear flow in Sobolev spaces*, Comm. Pure Appl. Math., 71 (2018), 617–687.

[23] D. Wei, Z. Zhang and W. Zhao, *Linear inviscid damping and vorticity depletion for shear flows*, Ann. PDE, 5 (2019), Paper No. 3, 101 pp.

[24] D. Wei, Z. Zhang and W. Zhao, *Linear inviscid damping and enhanced dissipation for the Kolmogorov flow*, Adv. Math., 362 (2020), 106963.

[25] D. Wei, Z. Zhang and H. Zhu, *Linear inviscid damping for the β-plane equation*, Comm. Math. Phys., 375 (2020), 127–174.

[26] C. Zillinger, *Linear inviscid damping for monotone shear flows in a finite periodic channel, boundary effects, blow-up and critical Sobolev regularity*, Arch. Ration. Mech. Anal., 221 (2016), 1449–1509.

[27] C. Zillinger, *Linear inviscid damping for monotone shear flows*, Trans. Amer. Math. Soc., 369 (2017), 8799–8855.

[28] C. Zillinger, *On circular flows: Linear stability and damping*, J. Differential Equations, 263 (2017), 7856–7899.